DIFFERENTIAL AND
INTEGRAL CALCULUS

DIFFERENTIAL AND INTEGRAL CALCULUS

BY

R. COURANT
Professor of Mathematics in New York University

TRANSLATED BY

E. J. McSHANE
Professor of Mathematics in the University of Virginia

VOLUME I

SECOND EDITION

Wiley Classics Library Edition Published 1988

WILEY

INTERSCIENCE PUBLISHERS
A DIVISION OF JOHN WILEY & SONS, INC.

First published 1934
Second edition 1937
Reprinted 1940, 1941, 1942 (twice)
1943, 1944, 1945, 1946, 1947, 1948
1949, 1950, 1951, 1952, 1953, 1954
1955, 1956 (twice), 1957 (twice)
1958, 1959, 1960, 1961, 1962, 1963
1964, 1965, 1966, 1967, 1968, 1970

10

ISBN 0 471 17820 9

ISBN 0-471-60842-4 (pbk.).

Printed in the United States of America

PREFACE
TO THE FIRST GERMAN EDITION

Although there is no lack of textbooks on the differential and integral calculus, the beginner will have difficulty in finding a book that leads him straight to the heart of the subject and gives him the power to apply it intelligently. He refuses to be bored by diffuseness and general statements which convey nothing to him, and will not tolerate a pedantry which makes no distinction between the essential and the non-essential, and which, for the sake of a systematic set of axioms, deliberately conceals the facts to which the growth of the subject is due.

True, it is easier to perceive defects than to remedy them. I make no claim to have presented the beginner with the ideal textbook. Yet I do not consider the publication of my lectures superfluous. In order and choice of material, in fundamental aim, and perhaps also in mode of presentation, they differ considerably from the current literature.

The reader will notice especially the complete break away from the out-of-date tradition of treating the differential calculus and the integral calculus separately. This separation, a mere result of historical accident, with no good foundation either in theory or in practical convenience in teaching, hinders the student from grasping the central point of the calculus, namely, the connexion between definite integral, indefinite integral, and derivative. With the backing of Felix Klein and others, the simultaneous treatment of differential calculus and integral calculus has steadily gained ground in lecture courses. I here attempt to give it a place in the literature. This first volume deals mainly with the integral and differential calculus for functions of one variable; a second volume will be devoted to functions of several variables and some other extensions of the calculus.

My aim is to exhibit the close connexion between analysis and its applications and, without loss of rigour and precision, to give due credit to intuition as the source of mathematical truth. The presentation of analysis as a closed system of truths without reference to their origin and purpose has, it is true, an æsthetic charm and satisfies a deep philosophical need. But the attitude of those who consider analysis solely as an abstractly logical, introverted science is not only highly unsuitable for beginners but endangers the future of the subject; for to pursue mathematical analysis while at the same time turning one's back on its applications and on intuition is to condemn it to hopeless atrophy. To me it seems extremely important that the student should be warned from the very beginning against a smug and presumptuous purism; this is not the least of my purposes in writing this book.

The book is intended for anyone who, having passed through an ordinary course of school mathematics, wishes to apply himself to the study of mathematics or its applications to science and engineering, no matter whether he is a student of a university or technical college, a teacher, or an engineer. I do not promise to save the reader the trouble of thinking, but I do seek to lead the way straight to useful knowledge, and aim at making the subject easier to grasp, not only by giving proofs step by step, but also by throwing light on the interconnexions and purposes of the whole.

The beginner should note that I have avoided blocking the entrance to the concrete facts of the differential and integral calculus by discussions of fundamental matters, for which he is not yet ready. Instead, these are collected in appendices to the chapters, and the student whose main purpose is to acquire the facts rapidly or to proceed to practical applications may postpone reading these until he feels the need for them. The appendices also contain some additions to the subject-matter; they have been made relatively concise. The reader will notice, too, that the general style of presentation, at first detailed, is more condensed towards the end of the book. He should not, however, let himself be disheartened by isolated difficulties which he may find in the concluding chapters. Such gaps in understanding, if not too frequent, usually fill up of their own accord.

PREFACE TO THE ENGLISH EDITION

When American colleagues urged me to publish an English edition of my lectures on the differential and integral calculus, I at first hesitated. I felt that owing to the difference between the methods of teaching the calculus in Germany and in Britain and America a simple translation was out of the question, and that fundamental changes would be required in order to meet the needs of English-speaking students.

My doubts were not laid to rest until I found a competent colleague in Professor E. J. McShane, of the University of Virginia, who was prepared not only to act as translator but also —after personal consultation with me—to make the improvements and alterations necessary for the English edition.

Apart from many matters of detail the principal changes are these: (1) the English edition contains a large number of classified examples; (2) the division of material between the two volumes differs somewhat from that in the German text. In addition to a detailed account of the theory of functions of one variable, the present volume contains (in Chapter X) a sketch of the differentiation and integration of functions of several variables. The second volume deals in full with functions of several independent variables, and includes the elements of vector analysis. There is also a more systematic discussion of differential equations, and an appendix on the foundations of the theory of real numbers.

Thus the first volume contains the material for a course in elementary calculus, while the subject-matter of the second volume is more advanced. In the first volume, however, there is much which should be omitted from a first course. These sections, intended for students wishing to penetrate more deeply into the theory, are collected in the appendices to the chapters, so that beginners can study the book without inconvenience, omitting or postponing the reading of these appendices.

The publication of this book in English has only been made possible by the generosity of my German publisher, Julius Springer, Berlin, to whom I wish to express my most cordial thanks. I have likewise to thank Blackie and Son, Ltd., who in spite of these difficult times have undertaken to publish this edition. My special thanks are due to the members of their technical staff for the excellent quality of their work, and to their mathematical editors, especially Miss W. M. Deans, who have relieved Prof. McShane and myself of much of the responsibility of preparing the manuscript for the press and reading the proofs. I am also indebted to many friends and colleagues, notably to Professor McClenon of Grinnell College, Iowa, to whose encouragement the English edition is due; and to Miss Margaret Kennedy, Newnham College, Cambridge, and Dr. Fritz John, who co-operated with the publisher's staff in the proof-reading.

R. COURANT.

Cambridge, England,
June, 1934.

PREFACE
TO THE SECOND ENGLISH EDITION

This second edition differs from the first chiefly in the improvement and rearrangement of the examples, the addition of many new examples at the end of the book, and the inclusion of some additional material on differential equations.

R. COURANT.

New Rochelle, N.Y.,
June, 1937.

CONTENTS

CHAPTER III

DIFFERENTIATION AND INTEGRATION OF THE
ELEMENTARY FUNCTIONS

APPENDIX

CHAPTER IV

FURTHER DEVELOPMENT OF THE INTEGRAL
CALCULUS

CHAPTER V

APPLICATIONS

CHAPTER VI

TAYLOR'S THEOREM AND THE APPROXIMATE
EXPRESSION OF FUNCTIONS BY POLYNOMIALS

APPENDIX

CHAPTER VII

NUMERICAL METHODS

APPENDIX

CHAPTER VIII

INFINITE SERIES AND OTHER LIMITING PROCESSES

APPENDIX

CONTENTS

DIFFERENTIAL AND INTEGRAL CALCULUS

Introductory Remarks

When the beginner comes in contact with the so-called higher mathematics for the first time, he is apt to be obsessed by the feeling that there is a certain discontinuity between school mathematics and university mathematics. This feeling ultimately rests on more than the historical circumstances which have caused university teaching to take a form differing so widely from that of the school. For the very *nature* of the higher mathematics, or rather, of the modern mathematics, developed during the last three centuries, distinguishes it from the elementary mathematics which wholly dominated the school curriculum until recently and whose subject-matter was often taken over almost directly from the mathematics of the ancient Greeks.

A leading characteristic of elementary mathematics is its intimate association with *geometry*. Even where the subject passes beyond the realm of geometry into that of arithmetic, the fundamental ideas still remain geometrical. Another feature of ancient mathematics is perhaps its tendency to concentrate on particular cases. Things which to-day we should regard as special cases of a general phenomenon are set down higgledy-piggledy without any visible relationship between them. Its intimate association with geometrical ideas and its stress on individual niceties give the older mathematics a charm of its own. Yet it was a definite advance when at the beginning of the modern age in mathematics quite different tendencies developed, acting as the stimulus for a great expansion of the subject, which in spite of many improvements in detail had in a sense stood still for centuries.

The fundamental tendency of all modern mathematics is towards the replacement of separate discussions of individual cases by more and more general systematic *methods*, which perhaps do not always do full justice to the individual features of a particular case, but which, owing to their generality and power, give promise of a wealth of new results. Again, the concept of number and the methods of analysis have come to occupy more and more independent positions and now dominate geometry entirely. These new tendencies towards the development of mathematics along a variety of lines are most clearly exhibited in the rise of analytical geometry, whose development is chiefly due to Fermat and Descartes, and of the differential and integral calculus, which is generally regarded as having originated with Newton and Leibnitz.

The three hundred years during which modern mathematics has existed have seen such important advances not only in pure mathematics, but in an immense variety of applications to science and engineering, that its fundamental ideas and above all the concept of a function have by degrees become very widely known and have eventually penetrated even into the school curriculum.

In this book my aim has been to develop the most important facts in the differential and integral calculus so far that at the close the reader, although he may have had no previous knowledge of higher mathematics, may be well equipped on the one hand for the study of the more advanced branches and of the foundations of the subject, or on the other hand, for the manipulation of the calculus in the varied realms in which it is applied.

I should like to warn the reader specially against a danger which arises from the discontinuity mentioned in the opening paragraph. The point of view of school mathematics tempts one to linger over details and to lose one's grasp of general relationships and systematic methods. On the other hand, in the "higher" point of view there lurks the opposite danger of getting out of touch with concrete details, so that one is left helpless when faced with the simplest cases of individual difficulty, because in the world of general ideas one has forgotten how to come to grips with the concrete. The reader must find his own way of meeting this dilemma. In this he can only succeed by repeatedly thinking out particular cases for himself and acquiring a firm grasp of the application of general principles in particular cases; herein lies the chief task of anyone who wishes to pursue the study of Science.

CHAPTER I

Introduction

The differential and integral calculus is based upon two concepts of outstanding importance, apart from the concept of number, namely, the concept of *function* and the concept of *limit*. These concepts can, it is true, be recognized here and there even in the mathematics of the ancients, but it is only in modern mathematics that their essential character and significance are fully brought out. In this introductory chapter we shall attempt to explain these concepts as simply and clearly as possible.

1. THE CONTINUUM OF NUMBERS

The question as to the real nature of numbers is one which concerns philosophers more than mathematicians, and philosophers have been much occupied with it. But mathematics must be carefully kept free from conflicting philosophical opinions; preliminary study of the essential nature of the concept of number from the point of view of the theory of knowledge is fortunately not required by the student of mathematics. We shall therefore take the numbers, and in the first place the natural numbers 1, 2, 3, . . ., as given, and we shall likewise take as given the rules * by which we calculate with these numbers; and we shall only briefly recall the way in which the concept of the positive integers (the natural numbers) has had to be extended.

---

* These rules are as follows: $(a + b) + c = a + (b + c)$. That is, if to the sum of two numbers a and b we add a third number c, we obtain the same result as when we add to a the sum of b and c. (This is called the associative law of addition.) Secondly, $a + b = b + a$ (the commutative law of addition). Thirdly, $(ab)c = a(bc)$ (the associative law of multiplication). Fourthly, $ab = ba$ (the commutative law of multiplication). Fifthly, $a(b + c) = ab + ac$ (the distributive law of multiplication).

1. The System of Rational Numbers and the Need for its Extension.

In the domain of the natural numbers the fundamental operations of addition and multiplication can always be performed without restriction; that is, the sum and the product of two natural numbers are themselves always natural numbers. But the inverses of these operations, subtraction and division, cannot invariably be performed within the domain of natural numbers; and because of this mathematicians were long ago obliged to invent the number 0, the negative integers, and positive and negative fractions. The totality of all these numbers is usually called the class of *rational numbers*, since they are all obtained from unity by using the " rational operations of calculation ", addition, multiplication, subtraction and division.

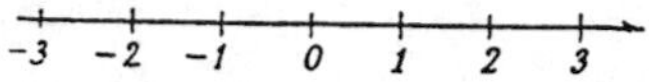

Fig. 1.—The number axis

Numbers are usually represented graphically by means of the points of a straight line, the " number axis ", by taking an arbitrary point of the line as the origin or zero point and another arbitrary point as the point 1; the distance between these two points (the length of the *unit interval*) then serves as a scale by which we can assign a point on the line to every rational number, positive or negative. It is customary to mark off the positive numbers to the right and the negative numbers to the left of the origin (cf. fig. 1). If, as usual, we define the absolute value (also called the numerical value or modulus) $|a|$ of a number a to be a itself when * $a \geqq 0$, and to be $-a$ when $a < 0$, then $|a|$ simply denotes the distance of the corresponding point on the number axis from the origin.

The geometrical representation of the rational numbers by points on the number axis suggests an important property which is usually stated as follows: *the set of rational numbers is everywhere dense.* This means that in every interval of the number axis, no matter how small, there are always rational numbers; geometrically, in the segment of the number axis between any two rational points, no matter how close together, there are points corresponding to rational numbers. This density of the rational

---

* By the sign $\geqq$ we mean that *either* the sign $>$ *or* the sign $=$ shall hold. A corresponding statement holds for the signs $\pm$ and $\mp$ which will be used later.

numbers at once becomes clear if we start from the fact that the numbers $\frac{1}{2}, \frac{1}{2^2}, \frac{1}{2^3}, \ldots, \frac{1}{2^n}, \ldots$ become steadily smaller and approach nearer and nearer to zero as n increases. If we now divide the number axis into equal parts of length $1/2^n$, beginning at the origin, the end-points $\frac{1}{2^n}, \frac{2}{2^n}, \frac{3}{2^n}, \ldots$ of these intervals represent rational numbers of the form $m/2^n$; here we still have the number n at our disposal. If now we are given a fixed interval of the number axis, no matter how small, we need only choose n so large that $1/2^n$ is less than the length of the interval; the intervals of the above subdivision are then small enough for us to be sure that at least one of the points of subdivision $m/2^n$ lies in the interval.

Yet in spite of this property of density the rational numbers are not sufficient to represent *every* point on the number axis. Even the Greek mathematicians recognized that when a given line segment of unit length is chosen there are intervals whose lengths cannot be represented by rational numbers; these are the so-called segments incommensurable with the unit. Thus, for example, the hypotenuse of a right-angled isosceles triangle with sides of unit length is not commensurable with the unit of length. For, by the theorem of Pythagoras, the square of this length l must be equal to 2. Therefore, if l were a rational number and consequently equal to p/q, where p and q are integers different from 0, we should have $p^2 = 2q^2$. We can assume that p and q have no common factors, for such common factors could be cancelled out to begin with. Since, according to the above equation, p^2 is an even number, p itself must be even, say $p = 2p'$. Substituting this expression for p gives us $4p'^2 = 2q^2$, or $q^2 = 2p'^2$; consequently q^2 is even, and so q is also even. Hence p and q both have the factor 2. But this contradicts our hypothesis that p and q have no common factor. Thus the assumption that the hypotenuse can be represented by a fraction p/q leads to contradiction and is therefore false.

The above reasoning, which is a characteristic example of an "indirect proof", shows that the symbol $\sqrt{2}$ cannot correspond to any rational number. Thus we see that if we insist that after choice of a unit interval every point of the number axis shall have a number corresponding to it, we are forced to extend the domain of rational numbers by the introduction of new "irrational" numbers. This system of rational and irrational numbers, such that each point on the axis corresponds to just one number and each number corresponds to just one point on the axis, is called the system of *real numbers*.*

2. Real Numbers and Infinite Decimals.

Our requirement that to each point of the axis there shall correspond one real number states nothing *a priori* about the possibility of calculating with these real numbers in the same way as with rational numbers. We establish our right to do this by showing that our requirement is equivalent to the following fact: the totality of all real numbers is represented by the totality of all finite and infinite decimals.

We first recall the fact, familiar from elementary mathematics, that every rational number can be represented by a terminating or by a recurring decimal; and conversely, that every such decimal represents a rational number. We shall now show that to *every* point of the number axis we can assign a uniquely determined decimal (usually infinite), so that we can represent the irrational points or irrational numbers by infinite decimals. (In accordance with the above remark the irrational numbers must be represented by infinite non-recurring decimals, for example, $0.101101110\ldots$).

Suppose that the points which correspond to the integers are marked on the number axis. By means of these points the axis is subdivided into intervals or segments of length 1. In what follows, we shall say that a point of the line belongs to an interval if it is an interior point or an end-point of the interval. Now let P be an arbitrary point of the number axis. Then the point belongs to one, or if it is a point of division to two, of the above intervals. If we agree that in the second case the right-hand one of the two intervals meeting at P is to be chosen, we have in all cases an interval with end-points g and $g+1$ to which P belongs, where g is an integer. This interval we subdivide into ten equal sub-intervals by means of the points corresponding to the numbers $g + \dfrac{1}{10}, g + \dfrac{2}{10}, \ldots, g + \dfrac{9}{10}$, and

---

* Thus named to distinguish it from the system of complex numbers, obtained by yet another extension.

we number these sub-intervals $0, 1, \ldots, 9$ in the natural order from left to right. The sub-interval with the number a then has the end-points $g + \dfrac{a}{10}$ and $g + \dfrac{a}{10} + \dfrac{1}{10}$. The point P must be contained in one of these sub-intervals. (If P is one of the new points of division it belongs to two consecutive intervals; as before, we choose the one on the right.) Suppose that the interval thus determined is associated with the number a_1. The end-points of this interval then correspond to the numbers $g + \dfrac{a_1}{10}$ and $g + \dfrac{a_1}{10} + \dfrac{1}{10}$. This sub-interval we again divide into ten equal parts and determine that one to which P belongs; as before, if P belongs to two sub-intervals we choose the one on the right. We thus obtain an interval with the end-points $g + \dfrac{a_1}{10} + \dfrac{a_2}{10^2}$ and $g + \dfrac{a_1}{10} + \dfrac{a_2}{10^2} + \dfrac{1}{10^2}$, where a_2 is one of the digits $0, 1, \ldots, 9$. This sub-interval we again subdivide, and continue to repeat the process. After n steps we arrive at a sub-interval containing P, having length $\dfrac{1}{10^n}$ and with end-points corresponding to the numbers

$$g + \frac{a_1}{10} + \frac{a_2}{10^2} + \ldots + \frac{a_n}{10^n} \text{ and } g + \frac{a_1}{10} + \frac{a_2}{10^2} + \ldots + \frac{a_n}{10^n} + \frac{1}{10^n}.$$

Here each a is one of the numbers $0, 1, \ldots, 9$. But

$$\frac{a_1}{10} + \frac{a_2}{10^2} + \ldots + \frac{a_n}{10^n}$$

is simply the decimal fraction $0 \cdot a_1 a_2 \ldots a_n$. The end-points of the interval, therefore, may also be written in the form

$$g + 0 \cdot a_1 a_2 \ldots a_n \text{ and } g + 0 \cdot a_1 a_2 \ldots a_n + \frac{1}{10^n}.$$

If we consider the above process repeated indefinitely, we obtain an *infinite decimal* $0 \cdot a_1 a_2 \ldots$, which has the following meaning. If we break off this decimal at any place, say the n-th, the point P will lie in the interval of length $\dfrac{1}{10^n}$ whose end-points (approximating points) are

$$g + 0 \cdot a_1 a_2 \ldots a_n \text{ and } g + 0 \cdot a_1 a_2 \ldots a_n + \frac{1}{10^n}.$$

In particular, the point corresponding to the rational number $g + 0·a_1a_2 \ldots a_n$ will lie arbitrarily near to the point P if only n is large enough; for this reason the points $g + 0·a_1a_2 \ldots a_n$ are called approximating points. *We say that the infinite decimal $g + 0·a_1a_2 \ldots$ is the real number corresponding to the point* P.

Here we would emphasize the fundamental *assumption* that we can calculate in the usual way with the real numbers, and hence with the decimals. It is possible to prove this using only the properties of the integers as a starting-point. But this is no light task; and rather than allow it to bar our progress at this early stage, we regard the fact that the ordinary rules of calculation apply to the real numbers as an axiom, on which we shall base the whole differential and integral calculus.

We here insert a remark concerning the possibility, in certain cases, of choosing the interval in *two* ways in the above scheme of expansion. From our construction it follows that the points of division arising in our repeated process of subdivision, and such points only, can be represented by finite decimals $g + 0·a_1a_2 \ldots a_n$. Let us suppose that such a point P first appears as a point of division at the n-th stage of the subdivision. Then according to the above process we have chosen at the n-th stage the interval to the right of P. In the following stages we must choose a sub-interval of this interval. But such an interval must have P as its left end-point. Therefore in all further stages of the subdivision we must choose the first sub-interval, which has the number 0. Thus the infinite decimal corresponding to P is $g + 0·a_1a_2 \ldots a_n000 \ldots$. If, on the other hand, we had at the n-th stage chosen the left-hand interval containing P, then in all later stages of subdivision we should have had to choose the sub-interval farthest to the right, which has P as its right end-point. Such a sub-interval has the number 9. Thus for P we should have obtained a decimal expansion in which all the digits from the $(n + 1)$-th onward are nines. The double possibility of choice in our construction therefore corresponds to the fact that for example the number $\frac{1}{4}$ has the two decimal expansions $0·25000 \ldots$ and $0·24999 \ldots$.

3. Expression of Numbers in Scales other than that of 10.

In our representation of the real numbers we made the number 10 play a special part, for each interval was subdivided into ten equal parts. The only reason for this is the widespread use of the decimal system. We could just as well have taken p equal sub-intervals, where p is an arbitrary integer greater than 1. We should then have obtained an expression of the form $g + \dfrac{b_1}{p} + \dfrac{b_2}{p^2} + \ldots$, where each b is one of the numbers

$0, 1, \ldots, p-1$. Here again we find that the rational numbers, and only the rational numbers, have recurring or terminating expansions of this kind. For theoretical purposes it is often convenient to choose $p = 2$. We then obtain the *binary expansion* of the real numbers,

$$g + \frac{b_1}{2} + \frac{b_2}{2^2} + \ldots,$$

where each b is either * 0 or 1.

For numerical calculations it is customary to express the whole number g, which for simplicity we here take to be positive, in the decimal system, that is, in the form

$$a_m 10^m + a_{m-1} 10^{m-1} + \ldots + a_1 10 + a_0,$$

where each a_ν is one of the digits $0, 1, \ldots, 9$. Then for $g + 0 \cdot a_1 a_2 \ldots$ we write simply

$$a_m a_{m-1} \cdots a_1 a_0 \cdot a_1 a_2 \ldots$$

Similarly, the positive whole number g can be written in one and only one way in the form

$$\beta_k p^k + \beta_{k-1} p^{k-1} + \ldots + \beta_1 p + \beta_0,$$

where each of the numbers β_ν is one of the numbers $0, 1, \ldots, p-1$. This, with our previous expression, gives the following result: every positive real number can be represented in the form

$$\beta_k p^k + \beta_{k-1} p^{k-1} + \ldots + \beta_1 p + \beta_0 + \frac{b_1}{p} + \frac{b_2}{p^2} + \ldots,$$

where β_ν and b_ν are whole numbers between 0 and $p-1$. Thus, for example, the binary expansion of the fraction $\frac{21}{4}$ is

$$\frac{21}{4} = 1 \times 2^2 + 0 \times 2 + 1 + \frac{0}{2} + \frac{1}{2^2}.$$

---

* Even for numerical calculations the decimal system is not the best. The sexagesimal system ($p = 60$), with which the Babylonians calculated, has the advantage that a comparatively large proportion of the rational numbers whose decimal expansions do not terminate possess terminating sexagesimal expansions.

4. Inequalities.

Calculation with inequalities plays a far larger part in higher mathematics than in elementary mathematics. We shall therefore briefly recall some of the simplest rules concerning them.

If $a > b$ and $c > d$ it follows that $a + c > b + d$, but *not* that $a - c > b - d$. Moreover, if $a > b$ it follows that $ac > bc$, provided c is positive. On multiplication by a negative number the sense of the inequality is reversed. If $a > b > 0$ and $c > d > 0$, it follows that $ac > bd$.

For the absolute values of numbers the following inequalities hold:

$$|a \pm b| \leq |a| + |b|, \quad |a \pm b| \geq |a| - |b|.$$

The square of any real number is greater than or equal to zero. Therefore, if x and y are arbitrary real numbers

$$(x - y)^2 = x^2 + y^2 - 2xy \geq 0,$$

or
$$2xy \leq x^2 + y^2.$$

5. Schwarz's Inequality.

Let $a_1, a_2, \ldots, a_n$ and $b_1, b_2, \ldots, b_n$ be any real numbers. In the preceding inequality we make the substitutions *

$$x = \frac{|a_i|}{\sqrt{(a_1^2 + a_2^2 + \ldots + a_n^2)}}, \quad y = \frac{|b_i|}{\sqrt{(b_1^2 + b_2^2 + \ldots + b_n^2)}}$$

for $i = 1$, $i = 2$, $\ldots$, $i = n$ successively and add the resulting inequalities. On the right we obtain the sum 2, for

$$\left(\frac{|a_1|}{\sqrt{(a_1^2 + \ldots + a_n^2)}}\right)^2 + \ldots + \left(\frac{|a_n|}{\sqrt{(a_1^2 + \ldots + a_n^2)}}\right)^2 = 1,$$

$$\left(\frac{|b_1|}{\sqrt{(b_1^2 + \ldots + b_n^2)}}\right)^2 + \ldots + \left(\frac{|b_n|}{\sqrt{(b_1^2 + \ldots + b_n^2)}}\right)^2 = 1.$$

If we divide both sides of the inequality by 2 we obtain

$$\frac{|a_1 b_1| + |a_2 b_2| + \ldots + |a_n b_n|}{\sqrt{(a_1^2 + \ldots + a_n^2)} \sqrt{(b_1^2 + \ldots + b_n^2)}} \leq 1,$$

---

* Here and hereafter the symbol $\sqrt{x}$, where $x > 0$, denotes that *positive* number whose square is x.

or finally

$$|a_1b_1| + |a_2b_2| + \ldots + |a_nb_n| \leq \sqrt{(a_1^2 + \ldots + a_n^2)}\sqrt{(b_1^2 + \ldots + b_n^2)}.$$

Since the expressions on both sides of this inequality are positive, we may square and then omit the modulus signs:

$$(a_1b_1 + a_2b_2 + \ldots + a_nb_n)^2 \leq (a_1^2 + \ldots + a_n^2)(b_1^2 + \ldots + b_n^2).$$

This is the Cauchy-Schwarz inequality.

EXAMPLES *

1. Prove that the following numbers are irrational: (a) $\sqrt{3}$. (b) $\sqrt{n}$, where n is not a perfect square. (c) $\sqrt[3]{3}$. (d)* $x = \sqrt{2} + \sqrt[3]{2}$. (e)* $x = \sqrt{3} + \sqrt[3]{2}$.

2.* In an ordinary system of rectangular co-ordinates, the points for which both co-ordinates are integers are called *lattice points*. Prove that a triangle whose vertices are lattice points cannot be equilateral.

3. Prove the inequalities:

(a) $x + \dfrac{1}{x} \geq 2$, $x > 0$.   (b) $x + \dfrac{1}{x} \leq -2$, $x < 0$.

(c) $\left| x + \dfrac{1}{x} \right| \geq 2$, $x \neq 0$.

4. Show that if $a > 0$, $ax^2 + 2bx + c \geq 0$ for all values of x if, and only if, $b^2 - ac \leq 0$.

5. Prove the following inequalities:

(a) $x^2 + xy + y^2 \geq 0$.
(b)* $x^{2n} + x^{2n-1}y + x^{2n-2}y^2 + \ldots + y^{2n} \geq 0$.
(c)* $x^4 - 3x^3 + 4x^2 - 3x + 1 \geq 0$.

6. Prove Schwarz's inequality by considering the expression

$$(a_1x + b_1)^2 + (a_2x + b_2)^2 + \ldots + (a_nx + b_n)^2,$$

collecting terms and applying Ex. 4.

7. Show that the equality sign in Schwarz's inequality holds if, and only if, the a's and b's are proportional; that is, $ca_\nu + db_\nu = 0$ for all ν's, where c, d are independent of ν and not both zero.

8. For $n = 2, 3$, state the geometrical interpretation of Schwarz's inequality.

9. The numbers γ_1, γ_2 are direction cosines of a line; that is, $\gamma_1^2 + \gamma_2^2 = 1$. Similarly, $\eta_1^2 + \eta_2^2 = 1$. Prove that the equation $\gamma_1\eta_1 + \gamma_2\eta_2 = 1$ implies the equations $\gamma_1 = \eta_1$, $\gamma_2 = \eta_2$.

10.* Prove the inequality

$$\sqrt{(a_1 - b_1)^2 + \ldots + (a_n - b_n)^2} \leq \sqrt{(a_1^2 + \ldots + a_n^2)} + \sqrt{(b_1^2 + \ldots + b_n^2)}$$

and state its geometrical interpretation.

* The more difficult examples are indicated by an asterisk.

2. The Concept of Function

1. Examples.

(*a*) If an ideal gas is compressed in a vessel by means of a piston, the temperature being kept constant, the pressure p and the volume v are connected by the relation

$$pv = C,$$

where C is a constant. This formula, called *Boyle's Law*, states nothing about the quantities v and p themselves, but has the following meaning: if p has a definite value, arbitrarily chosen in a certain range (the range being determined physically and not mathematically), then v can be determined, and conversely:

$$v = \frac{C}{p}, \quad p = \frac{C}{v}.$$

We then say that v is a function of p, or in the converse case that p is a function of v.

(*b*) If we heat a metal rod, which at temperature 0° has length l_0, to the temperature $\theta°$, then its length l will be given, on the simplest physical assumptions, by the law

$$l = l_0 (1 + \beta\theta),$$

where β, the " coefficient of expansion ", is a constant. Again we say that l is a function of θ.

(*c*) In a triangle let the lengths of two sides, say a and b, be given. If for the angle γ between these two sides we choose any arbitrary value less than 180° the triangle is completely determined; in particular, the third side c is determined. In this case we say that if a and b are given c is a function of the angle γ. As we know from trigonometry, this function is represented by the formula

$$c = \sqrt{(a^2 + b^2 - 2ab \cos \gamma)}.$$

2. Formulation of the Concept of Function.

In order to give a general definition of the mathematical concept of function, we fix upon a definite interval of our number scale, say the interval between the numbers a and b, and consider the totality of numbers x which belong to this interval, that is, which satisfy the relation

$$a \leqq x \leqq b.$$

If we consider the symbol x as denoting at will any of the numbers in this interval, we call it a *(continuous) variable* in the *interval*.

If now to each value of x in this interval there corresponds a single definite value y, where x and y are connected by any law whatsoever, we say that y *is a function of* x, and write symbolically

$$y = f(x), \quad y = F(x), \quad y = g(x),$$

or some similar expression. We then call x the *independent variable* and y the *dependent variable,* or we call x the *argument* of the function y.

It should be remarked that for certain purposes it makes a difference whether in the interval from a to b we include the end-points, as we have done above, or exclude them; in the latter case, the variable x is restricted by the inequalities

$$a < x < b.$$

To avoid misunderstanding we may call the first kind of interval (including end-points) a *closed* interval, the second kind an *open* interval. If only *one* end-point and not the other is included (as for example $a < x \leqq b$) we speak of an interval *open at one end* (in this case the end a). Finally, we may also consider open intervals which extend without bound in one direction or both. We then say that the variable x ranges over an *infinite* (open) interval, and write symbolically

$$a < x < \infty \quad \text{or} \quad -\infty < x < b \quad \text{or} \quad -\infty < x < \infty.$$

In the general definition of a function which is defined in an interval nothing is said about the nature of the relation by which the dependent variable is determined when the independent variable is given. This relation may be as complicated as we please, and in theoretical investigations this wide generality is an advantage. But in applications, and in particular in the differential and integral calculus, the functions with which we have to deal are not of the widest generality; on the contrary, the laws of correspondence by which a value of y is assigned to each x are subject to certain simplifying restrictions.

3. Graphical Representation. Continuity. Monotonic Functions.

Natural restrictions of the general function-concept are suggested if we consider the connexion with geometry. The fundamental idea of analytical geometry is in fact that of giving a curve defined by some geometrical property a characteristic analytical representation by regarding one of the rectangular co-ordinates, say y, as a function $y = f(x)$ of the other co-ordinate x; for example, a parabola is represented by the function $y = x^2$, the circle with radius 1 about the origin by the two functions $y = \sqrt{(1 - x^2)}$ and $y = -\sqrt{(1 - x^2)}$. In the first example we may think of the function as defined in the infinite interval $-\infty < x < \infty$; in the second we must restrict ourselves to the interval $-1 \leqq x \leqq 1$, since outside this interval the function has no meaning (when x and y are real).

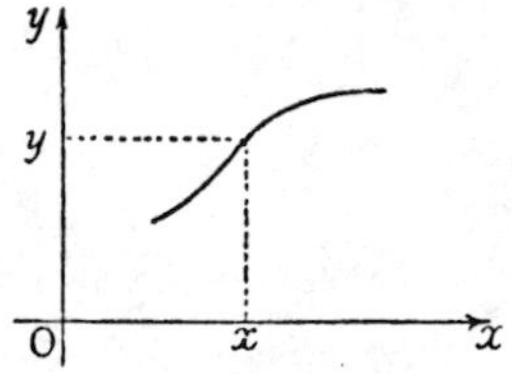

Fig. 2.—Rectangular axes

Conversely, if instead of starting with a curve determined geometrically we consider a given function $y = f(x)$, we can represent the functional dependence of y on x graphically by making use of a rectangular co-ordinate system in the usual way (cf. fig. 2). If for each abscissa x we lay off the corresponding ordinate $y = f(x)$, we obtain the geometrical representation of the function. The restriction which we now wish to impose on the function-concept is this: the geometrical representation of the function shall take the form of a " reasonable " geometrical curve. This, it is true, implies a vague general idea rather than a strict mathematical condition. But we shall soon formulate conditions, such as continuity, differentiability, &c., which will ensure that the graph of a function has the character of a curve capable of being visualized geometrically. At any rate, we shall exclude a function such as the following: for every rational value of x, the function y has the value 1; for every irrational value of x, the value of y is 0. This assigns a definite value of y to each x; but in every interval of x, no matter how small, the value of y jumps from 0 to 1 and back an infinite number of times.

Unless the contrary is expressly stated, it will always be assumed that the law which assigns a value of the function to each value of x assigns just one value of y to each value of x, as for example $y = x^2$ or $y = \sin x$. If we begin with a curve given geometrically it may happen, as in the case of the circle $x^2 + y^2 = 1$, that the whole course of the curve is not given by one single (one-valued) function, but requires several functions—in the case of the circle, the two functions $y = \sqrt{(1 - x^2)}$ and $y = -\sqrt{(1 - x^2)}$. The same is true for the hyperbola $y^2 - x^2 = 1$, which is represented by the two functions $y = \sqrt{(1 + x^2)}$ and $y = -\sqrt{(1 + x^2)}$. Such curves therefore do not determine the corresponding functions uniquely. Consequently it is sometimes said that the function corresponding to

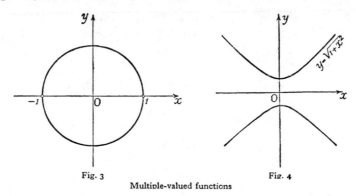

Fig. 3   Fig. 4

Multiple-valued functions

the curve is *multiple-valued*. The separate functions representing the curve are then called the single-valued *branches* belonging to the curve. For the sake of clearness we shall henceforth use the word "function" to mean a single-valued function. In conformity with this, the symbol $\sqrt{x}$ (for $x \geqq 0$) will always denote the *non-negative* number whose square is x.

If a curve is the geometrical representation of *one* function it will be cut by any parallel to the y-axis in at most one point, since to each point x in the interval of definition there corresponds just one value of y. Otherwise, as for example in the case of the circle which is represented by the two functions

$$y = \sqrt{(1 - x^2)} \text{ and } y = -\sqrt{(1 - x^2)},$$

such parallels to the y-axis may intersect the curve in more than one point. The portions of a curve corresponding to different single-valued branches are sometimes so connected with each

2

other that the complete curve is a single figure which can be drawn with one stroke of the pen, e.g. the circle (cf. fig. 3), or, on the other hand, the branches may be completely separated, e.g. the hyperbola (cf. fig. 4).

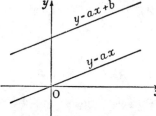

Fig. 5.—Linear functions

Here follow some further examples of the graphical representation of functions.

(a) $y = ax$.

y is proportional to x. The graph (cf. fig. 5) is a straight line through the origin of the co-ordinate system.

(b) $y = ax + b$.

y is a "linear function" of x. The graph is a straight line through the point $x = 0$, $y = b$, which, if $a \neq 0$, also passes through the point $x = -b/a$, $y = 0$, and if $a = 0$ runs horizontally.

(c) $$y = \frac{a}{x}.$$

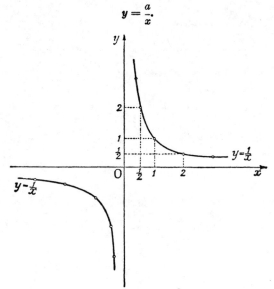

Fig. 6.—Infinite discontinuities

y is inversely proportional to x. If in particular $a = 1$, so that

$$y = \frac{1}{x},$$

we find, for example, that

$y = 1$ for $x = 1$, $y = 2$ for $x = \frac{1}{2}$, $y = \frac{1}{2}$ for $x = 2$.

The graph (cf. fig. 6) is a curve, a rectangular hyperbola, symmetrical with respect to the bisectors of the angles between the co-ordinate axes.

This last function is obviously not defined for the value $x = 0$, since division by zero has no meaning. The exceptional point $x = 0$, in whose neighbourhood there occur arbitrarily large values of the function, both positive and negative, is the simplest example of an *infinite discontinuity*, a subject to which we shall return later (cf. p. 51).

(d) $y = x^2$.

As is well known, this function is represented by a parabola (cf. fig. 7).

Similarly, the function $y = x^3$ is represented by the so-called cubical parabola (cf. fig. 8).

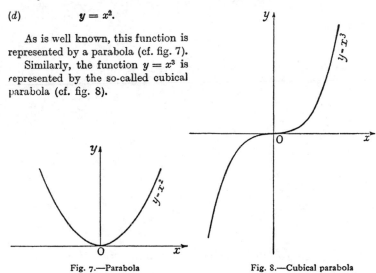

Fig. 7.—Parabola   Fig. 8.—Cubical parabola

The curves just considered and their graphs exhibit a property which is of the greatest importance in the discussion of functions, namely, the property of *continuity*. We shall later (§ 8, p. 49) analyse this concept in more detail; intuitively it comes to this, that a small change in x causes only a small change in y and not a sudden jump in its value; that is, the graph is not broken off. More exactly, the change in y remains less than any arbitrarily chosen positive bound, provided that the change in x is correspondingly small.

A function which for all values of x in an interval has the same value $y = a$ is called a *constant*; it is graphically represented by a horizontal straight line. A function $y = f(x)$ such that throughout the interval in which it is defined an increase in the value of x always causes an increase in the value of y is called a *monotonic increasing* function; if, on the other hand, an increase in the value of x always causes a decrease in the value of y, the function is called a *monotonic decreasing* function. Such functions are represented graphically by curves which in the corresponding interval always rise (from left to right) or always fall (cf. fig. 9).

If the curve represented by $y = f(x)$ is symmetrical with respect to the y-axis, that is, if $x = -a$ and $x = a$ give the same value for the function, or

$$f(-x) = f(x),$$

we say that the function is an *even* function. For example, the function $y = x^2$ is even (cf. fig. 7). If, on the other hand, the

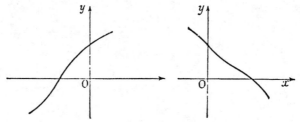

Fig. 9.—Monotonic functions

curve is symmetrical with respect to the origin, that is, if

$$f(-x) = -f(x),$$

we call the function an *odd* function; for example, the functions $y = x$ and $y = x^3$ (cf. fig. 8) and $y = 1/x$ are odd.

4. Inverse Functions.

Even in our first example on p. 14 it was made evident that a formal relationship between two quantities may be regarded in two different ways, since it is possible either to consider the first variable as a function of the second or to consider the second as a function of the first. If, for example, $y = ax + b$, where we assume that $a \neq 0$, x is represented as a function of y by the equation $x = (y - b)/a$. Again, the functional relationship represented by the equation $y = x^2$ can also be represented by the equation $x = \pm\sqrt{y}$, so that the function $y = x^2$ amounts to the same thing as the two functions $x = \sqrt{y}$ and $x = -\sqrt{y}$.

Thus, when an arbitrary function $y = f(x)$ is given we can attempt to determine x as a function of y, or, as we shall say, to replace the function $y = f(x)$ by the *inverse function* $x = \phi(y)$.

Geometrically this has the following meaning: we consider the curve obtained by reflecting the graph of $y = f(x)$ in the line bisecting the angle between the positive x-axis and the positive y-axis * (cf. fig. 10). This at once gives us a graphical representation of x as a function of y and thus represents the inverse function $x = \phi(y)$.

These geometrical ideas, however, show us at once that a

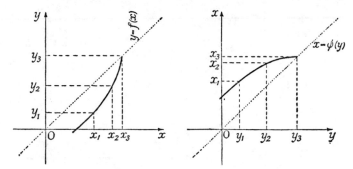

Fig. 10.—Inversion of a function

function $y = f(x)$ defined in an interval has not a single-valued inverse function unless certain conditions hold. If the graph of the function is cut by a line $y = c$ parallel to the x-axis in more than one point, the value $y = c$ will correspond to more than one value of x, so that the function cannot have a single-valued inverse function. This case cannot occur if $y = f(x)$ is continuous and monotonic. For then fig. 10 shows us that to each value of y in the interval y_1yy_3 there corresponds just *one* value of x in the interval x_1xx_3, and from the figure we infer that *a function which is continuous and monotonic in an interval always has a single-valued inverse function, and this inverse function is also continuous and monotonic.* (For a rigorous proof, see p. 67.)

* Instead of reflecting the graph in this way, we could first rotate the co-ordinate axes and the curve $y = f(x)$ through a right angle and then reflect the graph in the x-axis.

3. MORE DETAILED STUDY OF THE ELEMENTARY FUNCTIONS

1. The Rational Functions.

We now pass on to a brief review of the elementary functions which the reader has already met with in his previous studies. The simplest types of function are obtained by repeated application of the elementary operations: addition, multiplication, subtraction. If we apply these operations to an independent

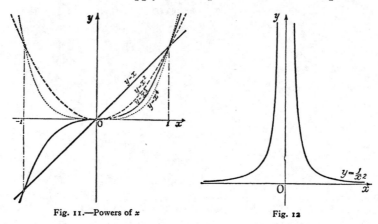

Fig. 11.—Powers of x Fig. 12

variable x and any real numbers, we obtain the *rational integral functions* or *polynomials*:

$$y = a_0 + a_1 x + \ldots + a_n x^n.$$

The polynomials are the simplest and, in a sense, the basic functions of analysis.

If we now form the quotients of such functions, that is, expressions of the form

$$y = \frac{a_0 + a_1 x + \ldots + a_n x^n}{b_0 + b_1 x + \ldots + b_m x^m},$$

we obtain the *general* or *fractional rational functions*, which are defined at all points where the denominator differs from zero.

The simplest rational integral function is the *linear function*

$$y = ax + b.$$

It is represented graphically by a straight line. Every *quadratic function* of the form

$$y = ax^2 + bx + c$$

is represented by a parabola. The curves which represent rational integral functions of the third degree,

$$y = ax^3 + bx^2 + cx + d,$$

are occasionally called parabolas of the third order, and so on.

As examples, we give the graphs of the function $y = x^n$ for the indices $n = 1, 2, 3, 4$ in fig. 11. We see that for even values of n the function $y = x^n$ satisfies the equation $f(-x) = f(x)$, and is therefore an even function, while for odd values of n the function satisfies the condition $f(-x) = -f(x)$, and is therefore an odd function.

The simplest example of a rational function which is not a polynomial is the function $y = 1/x$ mentioned on p. 18; its graph is a rectangular hyperbola. Another is the function $y = 1/x^2$ (cf. fig. 12).

2. Algebraic Functions.

We are at once led away from the domain of rational functions by the problem of forming their inverses. The most important example of this is the introduction of the function $\sqrt[n]{x}$. We start with the function $y = x^n$, which for $x \geqq 0$ is monotonic. It therefore has a single-valued inverse, which we denote by the symbol $x = \sqrt[n]{y}$, or, interchanging the letters used for the dependent and independent variables,

$$y = \sqrt[n]{x} = x^{1/n}.$$

In accordance with the definition this root is always non-negative. In the case of odd values of n the function x^n is monotonic for all values of x, including negative values. Consequently for odd values of n we can also define $\sqrt[n]{x}$ uniquely for all values of x; in this case $\sqrt[n]{x}$ is negative for negative values of x.

More generally we may consider

$$y = \sqrt[n]{R(x)},$$

where $R(x)$ is a rational function. We arrive at further functions of similar type by applying rational operations to one or more of these special functions. Thus for example we may form the functions

$$y = \sqrt[n]{x} + \sqrt[n]{(x^2 + 1)}, \quad y = x + \sqrt{(x^2 + 1)}.$$

These functions are special cases of *algebraic functions*. (The general concept of an algebraic function cannot be defined here; see Chapter X.)

3. The Trigonometric Functions.

While the rational functions and the algebraic functions just considered are defined directly in terms of the elementary opera·tions of calculation, geometry is the source from which we first draw our knowledge of the other functions, the so-called *transcendental functions.** We shall here consider the *elementary transcendental functions*, namely, the trigonometric functions, the exponential function, and the logarithm.

In all higher analytical investigations where angles occur it is customary to measure these angles not in degrees, minutes,

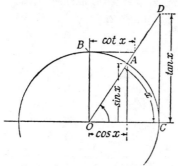

and seconds, but in *radians.* We place the angle to be measured with its vertex at the centre of a circle of radius 1, and measure the size of the angle by the length of the arc of the circumference which the angle cuts out. Thus an angle of 180° is the same as an angle of π radians (has radian measure π), an angle of 90° has radian measure $\pi/2$, an angle

Fig. 13.—The trigonometric functions

of 45° has radian measure $\pi/4$, an angle of 360° has radian measure 2π. Conversely, an angle of 1 radian expressed in degrees is

$$\frac{180°}{\pi}, \text{ or approximately } 57° \ 17' \ 45''.$$

Henceforward, whenever we speak of an angle x, we shall mean an angle whose radian measure is x.

After these preliminary remarks we may briefly remind the reader of the meanings of the trigonometric functions $\sin x$, $\cos x$, $\tan x$, $\cot x$.† These are shown in fig. 13, in which the angle x is measured from the arm OC (of length 1), angles being reckoned positive in the counter-clockwise direction. The rectangular

---

* The word " transcendental " does not mean anything particularly deep or mysterious; it merely suggests the fact that definition of these functions by means of the elementary operations of calculations is not possible, " *quod algebrae vires transcendit* ".

† It is sometimes convenient to introduce the functions $\sec x = 1/\cos x$, $\operatorname{cosec} x = 1/\sin x$.

co-ordinates of the point A at once give us the functions $\cos x$ and $\sin x$. The graphs of the functions $\sin x$, $\cos x$, $\tan x$, $\cot x$ are given in figs. 14 and 15.

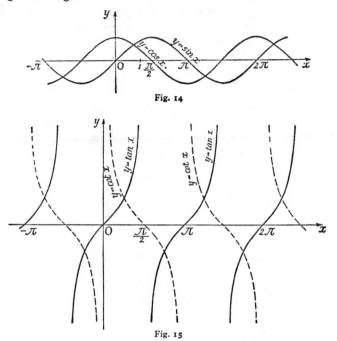

Fig. 14

Fig. 15

4. The Exponential Function and the Logarithm.

In addition to the trigonometric functions, the exponential function with the positive base a,

$$y = a^x,$$

and its inverse, the logarithm to the base a,

$$x = \log_a y,$$

are also regarded as elementary transcendental functions. In elementary mathematics it is customary to pass over certain inherent difficulties in the definition of these functions, and we too shall postpone the exact discussion of the functions until we have better methods at our disposal (cf. Chapter III, § 6, pp. 167–177, and also p. 191). We can, however, at least state the basis of the definitions here. If $x = p/q$ is a rational number (where p and q are positive integers), then— the number a being assumed positive—we define a^x as $\sqrt[q]{a^p} = a^{p/q}$, where the root, according to convention, is to be taken as positive. Since the rational values of x are everywhere dense, it is natural to extend this function a^x so as to make it a continuous function defined for irrational values of x also, giving values to a^x when x is irrational which are continuous with the values already defined when x is rational. This gives us a continuous function $y = a^x$, the " exponential function ", which for all *rational* values of x gives the value of a^x found above. That this extension is actually possible and can be carried out in only one way we meanwhile take for granted; but it must be borne in mind that we still have to *prove* that this is so.*

The function $$x = \log_a y$$

can then be defined for $y > 0$ as the inverse of the exponential function.

Examples

1. Plot the graph of $y = x^3$. From this, without further calculation, find the graph of $y = \sqrt[3]{x}$.

2. Sketch the following graphs, and state whether the functions are even or odd:

(a) $y = \sin 2x$.
(b) $y = 5 \cos x$.
(c) $y = \sin x + \cos x$.
(d) $y = 2 \sin x + \sin 2x$.
(e) $y = \sin(x + \pi)$.
(f) $y = 2 \cos\left(x + \dfrac{\pi}{3}\right)$.
(g) $y = \tan x - x$.

3. Sketch the graphs of the following functions, and state whether the functions are (1) monotonic or not, (2) even or odd:

(a) $y = x^2 \, (-\infty < x < \infty)$.
(b) $y = x^2 \, (0 \leq x \leq 1)$.
(c) $y = x \, (-1 \leq x \leq 1)$.
(d) $y = |x| \, (-1 \leq x \leq 1)$.
(e) $y = \sqrt{x^2} \, (-1 \leq x \leq 1)$.
(f) $y = |x - 1| \, (-\infty < x < \infty)$.
(g) $y = |x^2 + 4x + 2| \, (-4 \leq x \leq 3)$.
(h) $y = [x] \, (-\infty < x < \infty)$, where $[x]$ means the greatest integer which does not exceed x; that is, $[x] \leq x < [x] + 1$.

* Cf. pp. 70 and 173.

(i) $y = x - [x]\ (-\infty < x < \infty)$.

(j) $y = \sqrt{x - [x]}\ (-\infty < x < \infty)$.

(k) $y = x + \sqrt{x - [x]}\ (-\infty < x < \infty)$.

(l) $y = |x - 1| + |x + 1| - 2\ (-5 \leqq x \leqq 5)$.

(m) $y = |x - 1| - 2|x| + |x + 1|\ (-\infty < x < \infty)$.

Which two of these functions are identical?

4. A body dropped from rest falls approximately $16t^2$ ft. in t sec. If a ball falls from a window 25 ft. above ground, plot its height above ground as a function of t for the first 4 sec. after it starts to fall.

4. Functions of an Integral Variable. Sequences of Numbers

Hitherto we have considered the independent variable as a continuous variable, that is, as varying over a complete interval. However, numerous cases occur in mathematics in which a quantity depends only on an integer, a number n which can take the values 1, 2, 3, ... Such a function we call a function of an integral variable. This idea will most easily be grasped by means of examples.

1. The sum of the first n integers,

$$S_1(n) = 1 + 2 + 3 + 4 + \ldots + n = \tfrac{1}{2}n(n + 1),$$

is a function of n. Similarly, the sum of the first n squares,

$$S_2(n) = 1^2 + 2^2 + 3^2 + \ldots + n^2,$$

is a function * of the integer n.

* This last sum may easily be represented as a simple rational expression in n in the following way. We begin with the formula

$$(\nu + 1)^3 - \nu^3 = 3\nu^2 + 3\nu + 1,$$

write down this equation for the values $\nu = 0, 1, 2, \ldots, n$, and add. We thus obtain

$$(n + 1)^3 = 3S_2 + 3S_1 + n + 1;$$

on substituting the formula just given for S_1, this becomes

$$3S_2 = (n + 1)\left\{(n + 1)^2 - 1 - \tfrac{3}{2}n\right\} = (n + 1)\left\{n^2 + \tfrac{1}{2}n\right\},$$

so that

$$S_2 = \tfrac{1}{6}n(n + 1)(2n + 1).$$

By a similar process the functions

$$S_3(n) = 1^3 + 2^3 + \ldots + n^3,$$
$$S_4(n) = 1^4 + 2^4 + \ldots + n^4,$$

. . . . . .

can be represented as rational functions of n.

2. Other simple functions of integers are the expression

$$n! = 1 \cdot 2 \cdot 3 \dots n$$

and the binomial coefficients

$$\binom{n}{k} = \frac{n(n-1)\dots(n-k+1)}{k!} = \frac{n!}{k!\,(n-k)!}$$

for a fixed value of k.

3. Every whole number $n > 1$ which is not a prime number is divisible by more than two positive integers, while the prime numbers are divisible only by themselves and by 1. We can obviously consider the number $T(n)$ of divisors of n as a function of the number n itself. For the first few numbers this function is given by the following table:

$n =$	1	2	3	4	5	6	7	8	9	10	11	12
$T(n) =$	1	2	2	3	2	4	2	4	3	4	2	6

4. A function of this type which is of great importance in the theory of numbers is $\pi(n)$, the number of primes which are less than the number n. Its detailed investigation is one of the most interesting and attractive problems in the theory of numbers. Here we merely mention the principal result of these investigations: the number $\pi(n)$ is given approximately, for large values of n, by the function * $n/\log n$, where by $\log n$ we mean the logarithm to the " natural base " e, to be defined later (pp. 168, 174).

Functions of an integral variable usually occur in the form of so-called *sequences of numbers*. By a *sequence of numbers* we understand an *ordered array of infinitely many numbers* a_1, a_2, a_3, $\dots$, a_n, $\dots$, (*not necessarily all different*), *determined by any law whatever*. In other words, we are dealing simply with a function a of the integral variable n; the only difference is that we are using the index notation a_n instead of the symbol $a(n)$.

EXAMPLES

1. Prove that $1^3 + 2^3 + \dots + n^3 = (1 + 2 + \dots + n)^2$.

2. From the formula for $1^2 + 2^2 + \dots + n^2$, find a formula for $1^2 + 3^2 + 5^2 + \dots + (2n + 1)^2$.

3. Prove the following properties of the binomial coefficients:

(a) $\binom{n}{k} = \binom{n}{n-k} (k \leqq n)$. (b) $\binom{n}{k-1} + \binom{n}{k} = \binom{n+1}{k}$ (for $k > 0$).

(c) $1 + \binom{n}{1} + \binom{n}{2} + \dots + \binom{n}{n-1} + \binom{n}{n} = 2^n$.

* That is, the quotient of the number $\pi(n)$ by the number $n/\log n$ differs arbitrarily little from 1, provided only that n is large enough.

4. Evaluate the following sums:

(a) $1.2 + 2.3 + \ldots + n(n+1)$.

(b) $\dfrac{1}{1.2} + \dfrac{1}{2.3} + \ldots + \dfrac{1}{n(n+1)}$.

(c) $\dfrac{3}{1^2.2^2} + \dfrac{5}{2^2.3^2} + \ldots + \dfrac{2n+1}{n^2(n+1)^2}$.

5. A sequence is called an arithmetic progression of the first order if the differences of successive terms are constant. It is called an arithmetic progression of the second order if the differences of successive terms form an arithmetic progression of the first order; and in general, it is called an arithmetic progression of order k if the differences of successive terms form an arithmetic progression of order $(k-1)$.

The numbers 4, 6, 13, 27, 50, 84 are the first six terms of an arithmetic progression. What is its order? What is the eighth term?

6. Prove that the n-th term of an arithmetic progression of the second order can be written in the form $an^2 + bn + c$, where a, b, c are independent of n.

7.* Prove that the n-th term of an arithmetic progression of order k can be written in the form $an^k + bn^{k-1} + \ldots + pn + q$, where $a, b, \ldots, p, q$ are independent of n.
Find the n-th term of the progression in Ex. 5.

5. THE CONCEPT OF THE LIMIT OF A SEQUENCE

The fundamental concept on which the whole of analysis ultimately rests is that of the *limit* of a sequence. We shall first make the position clear by considering some examples.

1. $a_n = \dfrac{1}{n}$.

We consider the sequence

$$a_1 = 1, \quad a_2 = \frac{1}{2}, \quad a_3 = \frac{1}{3}, \ldots, \quad a_n = \frac{1}{n}, \ldots .$$

No number of this sequence is zero; but we see that the larger the number n is, the closer to zero is the number a_n. If, therefore, we mark off around the point 0 an interval as small as we please, then from a definite index onward all the numbers a_n will fall in this interval. This state of affairs we express by saying that as n increases the numbers a_n *tend to* 0, or that they possess the *limit* 0, or that the sequence $a_1, a_2, a_3 \ldots$ *converges* to 0.

If the numbers are represented as points on a line this means that the points $1/n$ crowd closer and closer to the point 0 as n increases.

The situation is similar in the case of the sequence

$$a_1 = 1, \quad a_2 = -\frac{1}{2}, \quad a_3 = \frac{1}{3}, \quad a_4 = -\frac{1}{4}, \ldots, \quad a_n = \frac{(-1)^{n-1}}{n}, \ldots$$

Here, too, the numbers a_n tend to zero as n increases; the only difference is that the numbers a_n are sometimes greater and sometimes less than the limit 0; as we say, they *oscillate* about the limit.

The convergence of the sequence to 0 is usually expressed symbolically by the equation

$$\lim_{n \to \infty} a_n = 0,$$

or occasionally by the abbreviation

$$a_n \to 0.$$

2. $a_{2m} = \dfrac{1}{m}; \quad a_{2m-1} = \dfrac{1}{2m}.$

In the preceding examples, the absolute value of the difference between a_n and the limit steadily becomes smaller as n increases. This is not necessarily the case, as is shown by the sequence

$$a_1 = \frac{1}{2}, \quad a_2 = 1, \quad a_3 = \frac{1}{4}, \quad a_4 = \frac{1}{2}, \quad a_5 = \frac{1}{6}, \quad a_6 = \frac{1}{3}, \ldots;$$

that is, in general, for even values $n = 2m$, $a_n = a_{2m} = 1/m$, for odd values $n = 2m - 1$, $a_n = a_{2m-1} = 1/2m$. This sequence also has a limit, namely, zero; for every interval about the origin, no matter how small, will contain all the numbers a_n from a certain value of n onward; but it is not true that every number lies nearer to the limit zero than the preceding one.

3. $a_n = \dfrac{n}{n+1}.$

We consider the sequence

$$a_1 = \frac{1}{2}, \quad a_2 = \frac{2}{3}, \ldots, \quad a_n = \frac{n}{n+1}, \ldots$$

where the integral index n takes all the values $1, 2, 3, \ldots$. If we write $a_n = 1 - \dfrac{1}{n+1}$ we see at once that as n increases the number a_n will approach closer and closer to the number 1, in the sense that if we mark off any interval about the point 1 all the numbers a_n following a certain a_N must fall in that interval. We write

$$\lim_{n \to \infty} a_n = 1.$$

The sequence $$a_n = \frac{n^2 - 1}{n^2 + n + 1}$$

behaves in a similar way. This sequence also tends to a limit as n increases, to the limit 1, in fact; in symbols, $\lim\limits_{n \to \infty} a_n = 1$. We see this most readily if we write

$$a_n = 1 - \frac{n + 2}{n^2 + n + 1} = 1 - r_n;$$

here we need only show that the numbers r_n tend to 0 as n increases. Now for all values of n greater than 2 we have $n + 2 < 2n$ and $n^2 + n + 1 > n^2$. Hence for the remainder r_n, we have

$$0 < r_n < \frac{2n}{n^2} = \frac{2}{n} \ (n > 2),$$

from which we see at once that r_n tends to 0 as n increases. Our discussion at the same time gives an estimate of the amount by which the number a_n (for $n > 2$) can at most differ from the limit 1; this difference certainly cannot exceed $2/n$.

The example just considered illustrates the fact, which we should naturally expect, that for large values of n the terms with the highest indices in the numerator and denominator of the fraction for a_n predominate and that they determine the limit.

4. $a_n = \sqrt[n]{p}$.

Let p be any fixed positive number. We consider the sequence $a_1, a_2, a_3, \ldots, a_n, \ldots$, where

$$a_n = \sqrt[n]{p}.$$

We assert that $$\lim\limits_{n \to \infty} a_n \equiv \lim\limits_{n \to \infty} \sqrt[n]{p} = 1.$$

We can prove this very easily by using a *lemma* which we shall find useful for other purposes also.

If $1 + h$ *is a positive number (that is, if* h > -1*), and* n *is an integer greater than* 1, *then*

$$(1 + h)^n > 1 + nh. \quad \ldots \ldots \quad (1)$$

Let us suppose that the inequality (1) is already proved for a certain value $m > 1$; we multiply both sides by $(1 + h)$ and obtain

$$(1 + h)^{m+1} > (1 + mh)(1 + h) = 1 + (m + 1)h + mh^2.$$

If, on the right, we omit the positive term mh^2 the inequality remains valid. We thus obtain

$$(1 + h)^{m+1} > 1 + (m + 1)h.$$

This, however, is our inequality for the index $m + 1$. It follows therefore that if the inequality holds for the index m it holds for the index $m + 1$

also. Since it holds for $m = 2$, it holds also for $m = 3$, hence for $m = 4$, and so on; therefore it holds for every index. This is a simple example of a proof by *mathematical induction*, a type of proof which is often useful.

Returning to our sequence, we distinguish between the case $p > 1$ and the case $p < 1$ (if $p = 1$, then $\sqrt[n]{p}$ is also equal to 1 for every n, and our statement becomes trivial).

If $p > 1$, then $\sqrt[n]{p}$ will also be greater than 1; we put $\sqrt[n]{p} = 1 + h_n$, where h_n is a positive quantity depending on n, and by the inequality (1) we have

$$p = (1 + h_n)^n > 1 + n h_n,$$

from which it at once follows that

$$0 < h_n < \frac{p - 1}{n}.$$

We therefore see that as n increases the number h_n must tend to 0, which proves that the numbers a_n converge to the limit 1, as stated. At the same time we have a means for estimating how close any a_n is to the limit 1; the difference between a_n and 1 is certainly not greater than $(p - 1)/n$.

If $p < 1$, then $\sqrt[n]{p}$ will likewise be less than 1 and therefore may be taken equal to $1/(1 + h_n)$, where h_n is a positive number. From this it follows, using the inequality (1), that

$$p = \frac{1}{(1 + h_n)^n} < \frac{1}{1 + n h_n}.$$

(By making the denominator smaller we increase the fraction.) It follows that

$$1 + n h_n < \frac{1}{p},$$

and therefore

$$h_n < \frac{1/p - 1}{n}.$$

From this we see that h_n tends to 0 as n increases. As the reciprocal of a quantity tending to 1, $\sqrt[n]{p}$ itself tends to 1.

5. $a_n = \alpha^n$.

We consider the sequence $a_n = \alpha^n$, where α is fixed and n runs through the sequence of positive integers.

First, let α be a positive number less than 1. We may then put $\alpha = 1/(1 + h)$, where h is positive, and inequality (1) gives

$$a_n = \frac{1}{(1 + h)^n} < \frac{1}{1 + nh} < \frac{1}{nh}.$$

Since the number h, and consequently $1/h$, depends only on α, and does not change as n increases, we see that as n increases α^n tends to 0:

$$\lim_{n \to \infty} \alpha^n = 0 \qquad (0 < \alpha < 1).$$

The same relationship holds when α is zero, or negative but greater than -1. This is immediately obvious, since in any case $\lim\limits_{n \to \infty} |\alpha|^n = 0$.

If $\alpha = 1$, then α^n will obviously be always equal to 1, and we shall have to regard the number 1 as the limit of α^n.

If $\alpha > 1$, we put $\alpha = 1 + h$, where h is positive, and at once see from our inequality that as n increases α^n does not tend to any definite limit, but increases beyond all bounds. We express this state of affairs by saying that α^n *tends to infinity* as n increases, or that α^n *becomes infinite*; in symbols,

$$\lim_{n \to \infty} \alpha^n = \infty \qquad (\alpha > 1).$$

Nevertheless, as we must explicitly emphasize, *the symbol ∞ does not denote a number with which we can calculate as with any other number*; equations or statements which express that a quantity is or becomes infinite never have the same sense as an equation between definite quantities. In spite of this, such modes of expression and the use of the symbol ∞ are extremely convenient, as we shall often see in the following pages.

If $\alpha = -1$, the values of α^n will not tend to any limit, but as n runs through the sequence of positive integers it will take the values $+1$ and -1 alternately. Similarly, if $\alpha < -1$ the value of α^n will increase numerically beyond all bounds, but its sign will be alternately positive and negative.

6. Geometrical Illustration of the Limits of α^n and $\sqrt[n]{p}$.

If we consider the curves $y = x^n$ and $y = x^{1/n} = \sqrt[n]{x}$ and restrict ourselves for the sake of convenience to non-negative values of x, the preceding limits are illustrated by figs. 16 and 17 respectively. In the case of the curves $y = x^n$ we see that in the interval from 0 to 1 they approach closer and closer to the x-axis as n increases, while outside that interval they climb more and more steeply and draw in closer and closer to a line parallel to the y-axis. All the curves pass through the point with co-ordinates $x = 1$, $y = 1$ and through the origin.

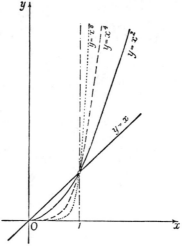

Fig. 16.—x^n as n increases

In the case of the functions $y = x^{1/n} = +\sqrt[n]{x}$, the curves approach closer and closer to the line parallel to the x-axis and at a distance 1 above it. On the other hand, all the curves must pass through the origin. Hence in the limit the curves approach the broken line consisting of the part of the y-axis between the points $y = 0$ and $y = 1$ and of the parallel to the x-axis $y = 1$. Moreover, it is clear that the two figures are closely related, as one would expect from the fact that the functions $y = \sqrt[n]{x}$ are actually the inverse functions of the n-th powers, from which we infer that each figure is transformed into the other on reflection in the line $y = x$.

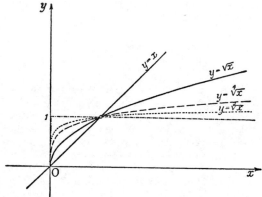

Fig. 17.—$x^{1/n}$ as n increases

7. The Geometric Series.

An example of a limit which is more or less familiar from elementary mathematics is the *geometric series*

$$1 + q + q^2 + \ldots + q^{n-1} = S_n;$$

the number q is called the *common ratio* of the series. The value of this sum may, as is well known, be expressed in the form

$$S_n = \frac{1 - q^n}{1 - q}$$

provided that $q \neq 1$; we can derive this expression by multiplying the sum S_n by q and subtracting the equation thus obtained from the original equation, or we may verify the formula by division.

The question now arises, what happens to the sum S_n when n increases indefinitely? The answer is this: the sum S_n has a definite limit S if q lies between -1 and $+1$, these end values being excluded, and it is then true that

$$S = \lim_{n \to \infty} S_n = \frac{1}{1 - q}.$$

In order to verify this statement we write the numbers S_n in the form $S_n = \dfrac{1 - q^n}{1 - q} = \dfrac{1}{1 - q} - \dfrac{q^n}{1 - q}$. We have already shown that provided

$|q| < 1$ the quantity q^n, and with it $\dfrac{q^n}{1-q}$, tends to 0 as n increases: hence with the above assumption the number S_n tends, as was stated, to the limit $\dfrac{1}{1-q}$ as n increases.

The passage to the limit $\lim\limits_{n \to \infty} (1 + q + q^2 + \ldots + q^{n-1}) = \dfrac{1}{1-q}$ is usually expressed by saying that when $|q| < 1$ the geometric series can be extended to infinity and that *the sum of the infinite geometric series is the expression* $\dfrac{1}{1-q}$.

The sums S_n of the finite geometric series are also called the *partial sums* of the infinite geometric series $1 + q + q^2 + \ldots$. (We must draw a sharp distinction between the *sequence* of numbers $S_1, S_2, \ldots, S_n, \ldots$ and the geometric *series*.)

The fact that the partial sums S_n of the geometric series tend to the limit $S = \dfrac{1}{1-q}$ as n increases may also be expressed by saying that the infinite geometric series $1 + q + q^2 + \ldots$ *converges* to the sum $S = \dfrac{1}{1-q}$ when $|q| < 1$.

8. $a_n = \sqrt[n]{n}$.

We shall show that the sequence of numbers

$$a_1 = 1, \quad a_2 = \sqrt{2}, \quad a_3 = \sqrt[3]{3}, \quad \ldots, \quad a_n = \sqrt[n]{n}, \quad \ldots$$

tends to 1 as n increases, i.e. that

$$\lim_{n \to \infty} \sqrt[n]{n} = 1.$$

Here we make use of a slight artifice. Instead of the sequence $a_n = \sqrt[n]{n}$ we first consider the sequence $b_n = \sqrt{a_n} = \sqrt{\sqrt[n]{n}} = \sqrt[n]{\sqrt{n}}$. When $n > 1$ the term b_n is also greater than 1. We can therefore put $b_n = 1 + h_n$, where h_n is positive and depends on n. By inequality (1), p. 31, we therefore have

$$\sqrt{n} = (b_n)^n = (1 + h_n)^n \geqq 1 + nh_n,$$

so that

$$h_n \leqq \frac{\sqrt{n} - 1}{n} \leqq \frac{\sqrt{n}}{n} = \frac{1}{\sqrt{n}}.$$

We now have

$$1 \leqq a_n = b_n^2 = 1 + 2h_n + h_n^2 \leqq 1 + \frac{2}{\sqrt{n}} + \frac{1}{n}.$$

The right-hand side of this inequality obviously tends to 1, and therefore so does a_n.

9. $a_n = \sqrt{n+1} - \sqrt{n}.$

We assert that $\qquad \lim\limits_{n \to \infty} (\sqrt{n+1} - \sqrt{n}) = 0.$

To prove this we need only write the expression under discussion in the form

$$\sqrt{n+1} - \sqrt{n} = \frac{(\sqrt{n+1} - \sqrt{n})(\sqrt{n+1} + \sqrt{n})}{\sqrt{n+1} + \sqrt{n}} = \frac{1}{\sqrt{n+1} + \sqrt{n}};$$

we see at once that this expression tends to 0 as n increases.

10. $a_n = \dfrac{n}{\alpha^n}.$

Let α be a number greater than 1. We assert that as n increases the sequence of numbers $a_n = \dfrac{n}{\alpha^n}$ tends to the limit 0.

As in the case of $\sqrt[n]{n}$ above we consider the sequence

$$\sqrt{a_n} = \frac{\sqrt{n}}{(\sqrt{\alpha})^n}.$$

We put $\sqrt{\alpha} = 1 + h.$ Here $h > 0$, since α and hence $\sqrt{\alpha}$ is greater than 1. By inequality (1), p. 31, we have

$$\sqrt{\alpha^n} = (1 + h)^n > 1 + nh,$$

so that $\qquad \sqrt{a_n} = \dfrac{\sqrt{n}}{(1+h)^n} \leqq \dfrac{\sqrt{n}}{1+nh} \leqq \dfrac{\sqrt{n}}{nh} = \dfrac{1}{h\sqrt{n}}.$

Hence $\qquad\qquad\qquad\qquad a_n \leqq \dfrac{1}{nh^2}.$

Since a_n is positive and the right-hand side of this equation tends to 0. we see that a_n must also tend to 0.

Examples

1. Prove that $\lim\limits_{n \to \infty} \dfrac{n^2 + n - 1}{3n^2 + 1} = \dfrac{1}{3}.$ Find an N such that for $n > N$ the difference between $\dfrac{n^2 + n - 1}{3n^2 + 1}$ and $\dfrac{1}{3}$ is (a) less than $\frac{1}{10}$, (b) less than $\frac{1}{1000}$, (c) less than $\frac{1}{1000000}$.

2. Find the limits of the following expressions as $n \to \infty$:

(a) $\dfrac{n^5 + 3n + 1}{n^6 + 7n^2 + 2}.$ (b) $\dfrac{n^6 + 3n + 1}{n^5 + 7n^2 + 2}.$

(c) $\dfrac{6n^3 + 2n + 1}{n^3 + n^2}.$ (d) $\dfrac{a_0 n^k + a_1 n^{k-1} + \ldots + a_k}{b_0 n^k + b_1 n^{k-1} + \ldots + b_k}.$

(e) $\dfrac{\sum\limits_{k=1}^{n} k^2}{n^3}.$

3. Prove that $\lim\limits_{n \to \infty} \sqrt[n]{n^2} = 1.$

4. Prove that $\lim\limits_{n \to \infty} \dfrac{n^2}{2^n} = 0.$ Find an N such that $\dfrac{n^2}{2^n} < \dfrac{1}{1000}$ whenever $n > N.$

5. Find numbers N_1, N_2, N_3 such that:

(a) $\dfrac{n}{2^n} < \dfrac{1}{10}$ for every $n > N_1$;

(b) $\dfrac{n}{2^n} < \dfrac{1}{100}$ for every $n > N_2$;

(c) $\dfrac{n}{2^n} < \dfrac{1}{1000}$ for every $n > N_3$.

6. Do the same thing for the sequence $a_n = \sqrt{n+1} - \sqrt{n}.$

7. Prove that $\lim\limits_{n \to \infty} (\sqrt{n+1} - \sqrt{n})(\sqrt{n + \tfrac{1}{2}}) = \tfrac{1}{2}.$

8. Prove that $\lim\limits_{n \to \infty} (\sqrt[3]{n+1} - \sqrt[3]{n}) = 0.$

9. Let $a_n = \dfrac{10^n}{n!}.$ (a) To what limit does a_n converge? (b) Is the sequence monotonic? (c) Is it monotonic from a certain n onwards? (d) Give an estimate of the difference between a_n and the limit. (e) From what value of n onwards is this difference less than $\tfrac{1}{100}$?

10. Prove that $\lim\limits_{n \to \infty} \dfrac{n!}{n^n} = 0.$

11. Prove that $\lim\limits_{n \to \infty} \left(\dfrac{1}{n^2} + \dfrac{2}{n^2} + \ldots + \dfrac{n}{n^2} \right) = \dfrac{1}{2}.$

12. Prove that $\lim\limits_{n \to \infty} \left(\dfrac{1}{n^2} + \dfrac{1}{(n+1)^2} + \ldots + \dfrac{1}{(2n)^2} \right) = 0.$

13. Prove that $\lim\limits_{n \to \infty} \left(\dfrac{1}{\sqrt{n}} + \dfrac{1}{\sqrt{n+1}} + \ldots + \dfrac{1}{\sqrt{2n}} \right) = \infty.$

14.* Prove that

$$\lim\limits_{n \to \infty} \left(\dfrac{1}{\sqrt{n^2 + 1}} + \dfrac{1}{\sqrt{n^2 + 2}} + \ldots + \dfrac{1}{\sqrt{n^2 + n}} \right) = 1.$$

15. Prove that if a and $b \leq a$ are positive, the sequence $\sqrt[n]{a^n + b^n}$ converges to a. Similarly, for any k fixed positive numbers $a_1, a_2, \ldots, a_k$ prove that $\sqrt[n]{a_1{}^n + a_2{}^n + \ldots + a_k{}^n}$ converges, and find its limit.

16. Prove that the sequence $\sqrt{2}$, $\sqrt{2\sqrt{2}}$, $\sqrt{2\sqrt{2\sqrt{2}}}$, ..., converges. Find its limit.

17.* If $\nu(n)$ is the number of distinct prime factors of n, prove that

$$\lim_{n \to \infty} \frac{\nu(n)}{n} = 0.$$

6. Further Discussion of the Concept of Limit

1. First Definition of Convergence.

From the cases discussed in the last section we are led to form the following general concept of limit:

If an infinite sequence of numbers a_1, a_2, a_3, ..., a_n, ... *is given and if there is a number ι such that every interval, no matter how small, marked off about the point l contains all the points* a_n *except for a finite number at most, we say that the number l is the limit of the sequence* a_1, a_2, ..., *or that the sequence* a_1, a_2, ... *converges to l*: in symbols, $\lim_{n \to \infty} a_n = l$. Here we expressly remark that this includes the trivial case in which all the numbers a_n are equal to one another and hence also coincide with the limit.

Instead of the above we may use the following equivalent statement:

If any positive number ϵ *be assigned—no matter how small— a whole number* $N = N(\epsilon)$ *can be found such that from the index N onward (i.e. for* $n > N(\epsilon)$*) it is always true that* $| a_n - l | < \epsilon$. Of course it is as a rule true that the bound $N(\epsilon)$ will have to be chosen larger and larger as smaller and smaller values of ϵ are chosen; in other words, $N(\epsilon)$ will increase beyond all bounds as ϵ tends to 0.

It is important to remember that *every convergent sequence is bounded*; that is, to every sequence a_1, a_2, a_3, ... for which a limit l exists there corresponds a positive number M, independent of n, such that for all the terms a_n of the sequence the inequality $| a_n | < M$ is valid.

This theorem readily follows from our definition. We choose ϵ equal to 1; then there is an index N such that for $n > N$ it is true that $| a_n - l | < 1$. Amongst the numbers

$$| a_1 - l |, \ | a_2 - l |, \ \ldots, \ | a_N - l |$$

let A be the largest. We can then put $M = | l | + A + 1$. For by the definition of A the inequality $|a_n - l| < A + 1$ certainly holds for $n = 1, 2, \ldots, N$, while for $n > N$

$$|a_n - l| < 1 \leqq A + 1.$$

A sequence which does not converge is said to be *divergent*. If as n increases the numbers a_n increase beyond all bounds we say that the sequence diverges to $+\infty$, and, as we have already done occasionally, we write $\lim_{n \to \infty} a_n = \infty$. Similarly, we write $\lim_{n \to \infty} a_n = -\infty$ if as n increases the numbers $-a_n$ increase beyond all bounds in the positive direction. But divergence may manifest itself in other ways, as, for example, in the case of the sequence $a_1 = -1$, $a_2 = +1$, $a_3 = -1$, $a_4 = +1$, $\ldots$, whose terms swing to and fro between two different values.*

In all the examples given above it has happened that the limit of the sequence considered is a *known* number. If the concept of limit yielded nothing more than the recognition that certain known numbers can be approximated to as closely as we like by certain sequences of other known numbers, we should have gained very little from it. The fruitfulness of the concept of limit in analysis rests essentially on the fact that limits of sequences of known numbers provide a means of dealing with *other* numbers which are not directly known or expressible.

The whole of higher analysis consists of a succession of examples of this fact, which will become steadily clearer to us in the following chapters. The representation of the irrational numbers as limits of rational numbers may be regarded as a first example. In this section we shall become acquainted with further examples. Before we take up this subject, however, we shall make a few preliminary general remarks.

2. Second (Intrinsic) Definition of Convergence.

How can we tell that a given sequence of numbers a_1, a_2, a_3, $\ldots$, a_n, $\ldots$ converges to a limit, even when we do not know beforehand what that limit is? This important question is answered for us by Cauchy's *convergence test*.†

---

* Another useful remark: the behaviour of a sequence as regards convergence is unaltered if we omit a finite number of the terms a_n. In what follows we shall frequently make use of this, speaking of the convergence or divergence of series in which the term a_n is undefined for a finite number of values of n.

† Sometimes referred to as the *general principle of convergence*.

We say that a sequence of numbers $a_1, a_2, \ldots, a_n, \ldots$ is *intrinsically* convergent if to every arbitrarily small positive number ϵ there corresponds a number $N = N(\epsilon)$, usually depending on ϵ, such that $|a_n - a_m| < \epsilon$, provided that n and m are both at least equal to $N(\epsilon)$. Cauchy's convergence test can then be expressed as follows:

Every intrinsically convergent sequence of numbers possesses a limit.

The importance of Cauchy's test lies in the fact that it allows us to speak of the limit of a sequence after considering the sequence itself, without any further information about the limit.

The converse of Cauchy's test is very easy to prove. For if the sequence $a_1, a_2, \ldots$ tends to the limit l, then by the definition of convergence we have

$$|l - a_n| < \frac{\epsilon}{2} \quad \text{and} \quad |l - a_m| < \frac{\epsilon}{2},$$

where ϵ is a positive quantity as small as we please, provided only that m and n are both large enough; therefore

$$|a_n - a_m| = |(l - a_m) - (l - a_n)| \leqq |l - a_m| + |l - a_n| < \epsilon.$$

Since ϵ can be chosen as small as we please, this inequality expresses our statement.

Cauchy's test itself becomes intuitively obvious if we think of the numbers as represented on the number axis. It then states that a sequence certainly has a limit if after a certain point N all the terms of the sequence are restricted to an interval which can be made arbitrarily small by choosing N large enough.

In the appendix we shall show how Cauchy's test can be proved by purely analytical methods. For the time being we accept it as a postulate.

3. Monotonic Sequences.

The question whether a given sequence converges to a limit is particularly easy to answer when the sequence is a so-called *monotonic sequence*; that is, if either every number of the sequence is larger than the preceding number (monotonic increasing sequence) or else every number is smaller than the preceding number (monotonic decreasing sequence). We have the following theorem:

Every monotonic increasing sequence whose terms are bounded above (that is, lie below a fixed number) possesses a limit; similarly, every monotonic decreasing sequence whose terms never fall below a certain fixed bound possesses a limit. For the present we shall regard these results as obvious, merely referring the student to the rigorous proof in the appendix (p. 61). A convergent monotonic increasing sequence must, of course, tend to a limit which is greater than any term of the sequence, while in the case of a convergent monotonic decreasing sequence the numbers tend to a limit which is smaller than any number of the sequence. Thus, for example, the numbers $1/n$ form a monotonic decreasing sequence with the limit 0, while the numbers $1 - 1/n$ form a monotonic increasing sequence with the limit 1.

In many cases it is convenient to replace the condition that a sequence shall increase monotonically by the weaker condition that the terms of the sequence shall never decrease; in other words, to allow successive terms of the sequence to be equal to one another. We then speak of a *monotonic nondecreasing* sequence, or of a *monotonic increasing sequence in the wider sense*. Our theorem on limits remains true for such sequences, and also for sequences which are *monotonic nonincreasing* or *monotonic decreasing in the wider sense*.

4. Operations with Limits.

We conclude with a remark concerning calculations with limits. From the definition of limit it follows almost at once that we can perform the elementary operations of addition, multiplication, subtraction, and division according to the following rules:

If $a_1, a_2, \ldots$ is a sequence with the limit a and $b_1, b_2, \ldots$ is a sequence with the limit b, then the sequence of numbers $c_n = a_n + b_n$ also has a limit, and

$$\lim_{n \to \infty} c_n = a + b.$$

The sequence of numbers $c_n = a_n b_n$ likewise converges, and

$$\lim_{n \to \infty} c_n = ab.$$

Similarly, the sequence $c_n = a_n - b_n$ converges, and

$$\lim_{n \to \infty} c_n = a - b.$$

Provided the limit b differs from 0, the numbers $c_n = \dfrac{a_n}{b_n}$ also converge, and have the limit

$$\lim_{n \to \infty} c_n = \frac{a}{b}$$

In words: we can *interchange* the rational operations of calculation with the process of forming the limit; that is, we obtain the same result whether we first perform a passage to the limit and then a rational operation or vice versa.

For the proof of these simple rules it is sufficient to give one example; using this for a model, the reader can establish the other statements for himself. We consider e.g. the multiplication of limits. The relations $a_n \to a$ and $b_n \to b$ amount to the following: if we choose any positive number ϵ, we need only take n greater than N, where $N = N(\epsilon)$ is a sufficiently large number depending on ϵ, in order to have both

$$|a - a_n| < \epsilon \quad \text{and} \quad |b - b_n| < \epsilon.$$

If we write $ab - a_n b_n = b(a - a_n) + a_n(b - b_n)$ and recall that there is a positive bound M, independent of n, such that $|a_n| < M$, we obtain

$$|ab - a_n b_n| \leqq |b||a - a_n| + |a_n||b - b_n| < (|b| + M)\epsilon.$$

Since the quantity $(|b| + M)\epsilon$ can be made arbitrarily small by choosing ϵ small enough, we see that the difference between ab and $a_n b_n$ actually becomes as small as we please for all sufficiently large values of n, which is precisely the statement made in the equation

$$ab = \lim_{n \to \infty} a_n b_n.$$

By means of these rules many limits can be evaluated very easily; for example, we have

$$\lim_{n \to \infty} \frac{n^2 - 1}{n^2 + n + 1} = \lim_{n \to \infty} \frac{1 - \dfrac{1}{n^2}}{1 + \dfrac{1}{n} + \dfrac{1}{n^2}} = 1,$$

since in the second expression the passages to the limit in numerator and denominator can be made directly.

Another simple and obvious rule is worth stating. *If* $\lim a_n = a$ *and* $\lim b_n = b$, *and if in addition* $a_n > b_n$ *for every* n, *then* $a \geqq b$. However, we are by no means entitled to expect that in general a will be *greater* than b, as is shown by the case of the sequences $a_n = 1/n$, $b_n = 1/2n$, for which $a = 0 = b$

5. The Number *e*.

As a first example of the generation of a number, which cannot be stated in advance, as the limit of a sequence of known numbers, we consider the sums

$$S_n = 1 + \frac{1}{1!} + \frac{1}{2!} + \ldots + \frac{1}{n!}.$$

We assert that as n increases these numbers S_n tend to a definite limit.

In order to prove the existence of the limit we observe that as n increases the numbers S_n increase monotonically. For all values of n we also have

$$S_n \leqq 1 + 1 + \frac{1}{2} + \frac{1}{2^2} + \ldots + \frac{1}{2^{n-1}} = 1 + \frac{1 - \dfrac{1}{2^n}}{1 - \dfrac{1}{2}} < 3.$$

The numbers S_n therefore have the upper bound 3 and, being a monotonic increasing sequence, they possess a limit, which we denote by e:

$$e = \lim_{n \to \infty} S_n.$$

Further, we assert that the number e *defined* as the above limit is also the limit of the sequence

$$T_n = \left(1 + \frac{1}{n}\right)^n.$$

The proof is simple and at the same time an instructive example of operations with limits. According to the binomial theorem, which we shall here assume,

$$T_n = \left(1 + \frac{1}{n}\right)^n$$

$$= 1 + n\frac{1}{n} + \frac{n(n-1)}{2!}\frac{1}{n^2} + \ldots + \frac{n(n-1)(n-2)\ldots 1}{n!}\frac{1}{n^n}$$

$$= 1 + 1 + \frac{1}{2!}\left(1 - \frac{1}{n}\right) + \ldots$$

$$+ \frac{1}{n!}\left(1 - \frac{1}{n}\right)\left(1 - \frac{2}{n}\right) \ldots \left(1 - \frac{n-1}{n}\right).$$

From this we see at once (1) that $T_n \leqq S_n$, and (2) that the T_n's also form a monotonic increasing sequence,* whence the existence of the limit $\lim\limits_{n \to \infty} T_n = T$ follows. In order to prove that $T = e$, we observe that

$$T_m > 1 + 1 + \frac{1}{2!}\left(1 - \frac{1}{m}\right) + \ldots + \frac{1}{n!}\left(1 - \frac{1}{m}\right) \ldots \left(1 - \frac{n-1}{m}\right),$$

provided that $m > n$. If we now keep n fixed and let m increase beyond all bounds, we obtain on the left the number T and on the right the expression S_n, so that $T \geqq S_n$. We have thus established the relationship $T \geqq S_n \geqq T_n$, for every value of n. We can now let n increase, so that T_n tends to T; from the double inequality it follows that $T = \lim\limits_{n \to \infty} S_n = e$. This was the statement to be proved.

We shall later (Chapter III, § 6, p. 172) reach this number e again from still another point of view.

6. The Number π as a Limit.

A limiting process which in essence goes back to classical antiquity (Archimedes) is that by which the number π is defined. Geometrically π means the area of the circle of radius 1. We therefore accept the existence of this number π as intuitive, regarding it as obvious that this area can be expressed by a (rational or irrational) number, which we then simply denote by π. However, this definition is not of much help to us if we wish to calculate the number with any accuracy. We have then no choice but to represent the number by means of a limiting process, namely, as the limit of a sequence of known and easily calculated numbers. Archimedes himself used this process in his method of exhaustions, where he steadily approximated to the circle by means of regular polygons with an increasing number of sides fitting it more and more closely. If we denote the area of the regular m-gon (polygon of m sides) inscribed in

---

* We obtain T_{n+1} from T_n by replacing the factors $1 - 1/n$, $1 - 2/n$, $\ldots$ by the larger factors $1 - \dfrac{1}{n+1}$, $1 - \dfrac{2}{n+1}$, $\ldots$ and finally adding a positive term.

the circle by f_m, the area of the inscribed $2m$-gon is given by the formula (proved by elementary geometry)

$$f_{2m} = \frac{m}{2} \sqrt{2 - 2\sqrt{1 - \left(\frac{2f_m}{m}\right)^2}}.$$

We now let m run, not through the sequence of all positive integers, but through the sequence of powers of 2, that is, $m = 2^n$; in other words, we form those regular polygons whose vertices are obtained by repeated bisection of the circumference. The area of the circle is then given by the limit

$$\pi = \lim_{n \to \infty} f_{2^n}.$$

This representation of π as a limit actually serves as a basis for numerical computations; for, starting with the value $f_4 = 2$, we can calculate in order the terms of our sequence tending to π. An estimate of the accuracy with which any term f_{2^n} represents π can be obtained by constructing the lines touching the circle and parallel to the sides of the inscribed 2^n-gon. These lines form a circumscribed polygon similar to the inscribed 2^n-gon, and having dimensions greater in the ratio $1 : \cos \frac{\pi}{2^{n-1}}$. Hence the area F_{2^n} of the circumscribed polygon is given by

$$\frac{f_{2^n}}{F_{2^n}} = \left(\cos \frac{\pi}{2^{n-1}}\right)^2.$$

Since the area of the circumscribed polygon is evidently greater than that of the circle, we have

$$f_{2^n} < \pi < F_{2^n} = \frac{f_{2^n}}{\left(\cos \frac{\pi}{2^{n-1}}\right)^2}.$$

These are matters with which the reader will be more or less familiar. What we wish to point out here is that the calculation of areas by means of exhaustion by rectilinear figures whose areas can be calculated easily forms the basis for the concept of integral, which will be introduced in the next chapter (p. 76).

Examples

1.* (a) Replace the statement "the sequence a_n is not absolutely bounded" by an equivalent statement not involving any form of the words "bounded" or "unbounded".

(b) Replace the statement "the sequence a_n is divergent" by an equivalent statement not involving any form of the words "convergent" or "divergent".

2.* Let a_1 and b_1 be any two positive numbers, and let $a_1 < b_1$. Let a_2 and b_2 be defined by the equations

$$a_2 = \sqrt{a_1 b_1}, \quad b_2 = \frac{a_1 + b_1}{2}.$$

Similarly, let
$$a_3 = \sqrt{a_2 b_2}, \quad b_3 = \frac{a_2 + b_2}{2},$$

and in general
$$a_n = \sqrt{a_{n-1} b_{n-1}}, \quad b_n = \frac{a_{n-1} + b_{n-1}}{2}.$$

Prove (a) that the sequence $a_1, a_2, \ldots$, converges, (b) that the sequence $b_1, b_2, \ldots$, converges, (c) that the two sequences have the same limit. (This limit is called the *arithmetic-geometric mean* of a_1 and b_1.)

3.* Prove that if $\lim\limits_{n \to \infty} a_n = \xi$, then $\lim\limits_{n \to \infty} \sigma_n = \xi$, where σ_n is the arithmetic mean $(a_1 + a_2 + \ldots + a_n)/n$.

4. If $\lim\limits_{n \to \infty} a_n = \xi$, show that the arithmetic means of the arithmetic means σ_n tend towards ξ.

5. Find the error involved in using $S_n = 1 + \dfrac{1}{1!} + \ldots + \dfrac{1}{n!}$ as an approximation to e. Calculate e accurately to 5 decimal places.

7. The Concept of Limit where the Variable is Continuous

Hitherto we have considered limits of sequences, that is, of functions of an integral variable n. The notion of limit, however, frequently occurs in connexion with the concepts of a continuous variable x and of a function $f(x)$.

We say that the value of the function $f(x)$ tends to a limit l as x tends to ξ, or in symbols

$$\lim_{x \to \xi} f(x) = l,$$

if all values of the function $f(x)$ for which x lies near enough to ξ differ arbitrarily little from l. Expressed more precisely, the condition is as follows:

If an arbitrarily small positive quantity ϵ is assigned, we can mark off about ξ an interval $|\,x - \xi\,| < \delta$ so small that for every point x in this interval different from ξ itself the inequality $|\,f(x) - l\,| < \epsilon$ holds.

Here we expressly exclude the equality of x and ξ. This is done purely for reasons of expediency, so as to have the definition in a form more convenient for application, e.g. in the case where the function $f(x)$ is undefined at the point ξ, although it is defined for all other points in a neighbourhood (p. 159) of ξ.

If our function is defined or considered in a given interval only, e.g. $\sqrt{1-x^2}$ in $-1 \leq x \leq 1$, we shall restrict the values of x to this interval. Thus if ξ denotes an end-point of the interval, x is made to approach ξ by values on one side of ξ only (limit from the interior of the interval or one-sided limit).

As an immediate consequence of this definition, we have the following fact: if $\lim\limits_{x \to \xi} f(x) = l$, and $x_1,\ x_2,\ x_3, \ldots,\ x_n, \ldots$ is a sequence of numbers all different from ξ but approaching ξ as a limit, then $\lim\limits_{n \to \infty} f(x_n) = l$.

For let ϵ be any positive number; we wish to show that for all values of n greater than a certain n_0 the inequality

$$| f(x_n) - l | < \epsilon$$

holds. By definition, there exists a $\delta > 0$ such that whenever $| x - \xi | < \delta$ the inequality

$$| f(x) - l | < \epsilon$$

is true. Since $x_n \to \xi$, the relation $| x_n - \xi | < \delta$ is satisfied for all sufficiently large values of n; and for such values it follows that $| f(x_n) - l | < \epsilon$, as was to be proved.

We shall now attempt to clarify this abstract definition by means of simple examples. Let us first consider the function

$$f(x) = \frac{\sin x}{x},$$

defined for $x \neq 0$. We state that

$$\lim_{x \to 0} \frac{\sin x}{x} = 1.$$

We cannot prove this statement simply by carrying out the passage to the limit in numerator and denominator separately, for the numerator and denominator vanish when $x = 0$, and the symbol $0/0$ has no meaning. We arrive at the proof in the following way.

From fig. 18 we find by comparing the areas of the triangles OAB and OAC and the sector OAB that if $0 < x < \pi/2$

$$\sin x < x < \tan x.$$

From this it follows that if $0 < |x| < \pi/2$,

$$1 < \frac{x}{\sin x} < \frac{1}{\cos x}.$$

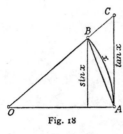

Fig. 18

Hence the quotient $\dfrac{\sin x}{x}$ lies between the numbers 1 and $\cos x$. We know that $\cos x$ tends to 1 as $x \to 0$, and from this it follows that the quotient $\dfrac{\sin x}{x}$ can differ only arbitrarily little from 1, provided that x is near enough to 0. This is exactly what is meant by the equation which was to be proved.

From the result just proved it follows that

$$\lim_{x \to 0} \frac{\tan x}{x} = \lim_{x \to 0} \frac{\sin x}{x} \lim_{x \to 0} \frac{1}{\cos x} = 1,$$

and also

$$\lim_{x \to 0} \frac{1 - \cos x}{x} = 0.$$

This last follows from the formula, valid for $0 < |x| < \dfrac{\pi}{2}$,

$$\frac{1 - \cos x}{x} = \frac{(1 - \cos x)(1 + \cos x)}{x(1 + \cos x)} = \frac{1 - \cos^2 x}{x(1 + \cos x)}$$

$$= \frac{\sin x}{x} \cdot \frac{1}{1 + \cos x} \cdot \sin x.$$

As $x \to 0$ the first factor on the right tends to 1, the second to $\frac{1}{2}$ and the third to 0; the product therefore tends to 0, as was stated.

From the same formula, dividing by x, we obtain

$$\frac{1 - \cos x}{x^2} = \left(\frac{\sin x}{x}\right)^2 \frac{1}{1 + \cos x},$$

whence

$$\lim_{x \to 0} \frac{1 - \cos x}{x^2} = \frac{1}{2}.$$

Finally, let us consider the function $\sqrt{x^2}$, defined for all values of x. This function is never negative; it is equal to x for $x \geqq 0$ and to $-x$ for $x < 0$. In other words, $\sqrt{x^2} = |x|$. Consequently the function $\sqrt{x^2}/x$, which is defined for all non-zero values of x, has the value $+1$ when $x > 0$ and -1 when $x < 0$. It is therefore impossible for the limit

$\lim\limits_{x \to 0} \sqrt{x^2}/x$ to exist, since arbitrarily near to 0 we can find values of x for which the quotient is $+1$ and other values for which it is -1.

In concluding this discussion on limits in connexion with continuous variables we remark that it is of course possible to consider limiting processes in which the continuous variable x increases beyond all bounds. For example, the meaning of the equation

$$\lim_{x \to \infty} \frac{x^2 + 1}{x^2 - 1} = \lim_{x \to \infty} \frac{1 + \dfrac{1}{x^2}}{1 - \dfrac{1}{x^3}} = 1$$

is clear without further discussion. It signifies that the function on the left differs arbitrarily little from 1, provided only that x is sufficiently large.

In these examples we have proceeded as if operations with limits obeyed the same laws in the case of continuous variables as in the case of sequences. That this is actually true the reader can verify for himself; the proofs are essentially the same as for limits of sequences.

EXAMPLES

1. Find the following limits, giving at each step the theorem on limits which justifies it:

(a) $\lim\limits_{x \to 2} 3x.$

(b) $\lim\limits_{x \to 3} 4x + 3.$

(c) $\lim\limits_{x \to 1} \dfrac{x^2 + 2x - 1}{2x + 2}.$

(d) $\lim\limits_{x \to 2} \sqrt{5 + \sqrt[3]{2x^5}}.$

2. Prove that

(a) $\lim\limits_{x \to 1} \dfrac{x^n - 1}{x - 1} = n;$ (b) $\lim\limits_{x \to \pi} \dfrac{\sin x}{\pi - x} = 1;$ (c) $\lim\limits_{x \to 0} \dfrac{\sin(x^2)}{x} = 0.$

3. Find whether or not the following limits exist, and if they do exist find their values:

(a) $\lim\limits_{x \to 0} \dfrac{\sqrt{1 - x}}{x};$ (b) $\lim\limits_{x \to 0} \dfrac{\sqrt{1 + x}}{x};$ (c) $\lim\limits_{x \to 0} \dfrac{\sqrt{1 + x} - \sqrt{1 - x}}{x}.$

8. THE CONCEPT OF CONTINUITY

1. Definitions.

We have already illustrated the notion of continuity in § 2 (p. 19) by means of examples. Now, with the help of the idea of a limit, we are in a position to make the concept of continuity precise.

We thought of the graph of a function which is continuous in an interval as a curve consisting of one unbroken piece; we also stated that the change in the function y must remain arbitrarily small provided only that the change of the independent variable x is restricted to a sufficiently small interval. This state of affairs is usually formulated as follows, with greater prolixity but increased precision. A function $f(x)$ is said to be *continuous* at the point ξ if it possesses the following property: at the point ξ the value of the function $f(\xi)$ is approximated to within an arbitrary pre-assigned degree of accuracy ϵ by all functional

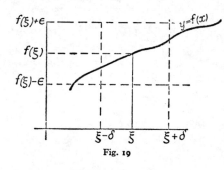

Fig. 19

values $f(x)$ for which x is near enough to ξ. In other words, $f(x)$ is continuous at ξ if for every positive number ϵ, no matter how small, there can be determined another positive number $\delta = \delta(\epsilon)$ such that $\mid f(x) - f(\xi) \mid < \epsilon$ (cf. fig. 19) for all points x for which $\mid x - \xi \mid < \delta$.

Or again: the condition of continuity requires that for the point ξ the limit equation

$$\lim_{x \to \xi} f(x) = f(\xi)$$

shall be true. The value of the function at the point ξ is the same as the limit of the functional values $f(x_n)$ for any arbitrary sequence x_n of numbers converging to ξ.

It is important to observe that our condition involves two different things: (1) the *existence* of the limit $\lim_{x \to \xi} f(x)$, and (2) the *coincidence* of this limit with $f(\xi)$, the value of the function at the point ξ.

Having now defined continuity of a function $f(x)$ at a point ξ, we proceed to state what we mean by the *continuity of a function* f(x) *in an interval*. This may be defined simply in the following way: the function $f(x)$ is continuous in an interval if it is continuous at each point of that interval. Stated fully, this requires that if a positive number ϵ be assigned, then for each point x of the interval there is a positive number δ, depending as a rule on ε and on x, such that

$$|f(\bar{x}) - f(x)| < \epsilon \text{ if } |\bar{x} - x| < \delta,$$

and $\bar{x}$ lies in the interval $a \leq \bar{x} \leq b$.

Closely related to this is another concept, that of *uniform continuity*. The function $f(x)$ is *uniformly* continuous in the interval $a \leq x \leq b$ if for every positive number ϵ there is a corresponding positive number δ such that for every pair of points x_1, x_2 in the interval whose distance apart, $|x_1 - x_2|$, is less than δ the inequality $|f(x_1) - f(x_2)| < \epsilon$ holds. This differs from the definition stated above in that the δ in the definition of uniform continuity does not depend on x, but is equally effective for all values of x—hence the name uniform continuity.

It is quite obvious that a uniformly continuous function is necessarily continuous. Conversely, it can be shown that every function $f(x)$ which is continuous in a *closed* interval $a \leq x \leq b$ is also uniformly continuous. The proof of this we leave for the appendix (p. 64). Even though the reader may not desire to read the proof at present he will find it helpful to study the examples given at the beginning of Appendix I, § 2, No. 2 (p. 65). But until the student has worked through this proof he may assume that whenever a function is said to be continuous in a closed interval uniform continuity is meant.

2. Points of Discontinuity.

We can understand the concept of continuity better if we study its opposite, the concept of discontinuity. The simplest type of discontinuity occurs at those points at which the function makes a *jump*; that is, at which the function has a definite limit as x tends to the point from the right and a definite limit as x tends to the point from the left, but these two limits are different. Whether or how the function is defined at the point of discontinuity itself does not matter.

For example, the function $f(x)$ defined by the equations

$$f(x) = 0 \text{ for } x^2 > 1, \quad f(x) = 1 \text{ for } x^2 < 1, \quad f(x) = \tfrac{1}{2} \text{ for } x^2 = 1$$

has discontinuities at the points $\xi = 1$ and $\xi = -1$. The limits on approaching these points from the right and from the left differ by 1, and the values of the function at these points agree with neither limit, but are equal to the arithmetic mean of the two limits.

It may be noted in passing that our function can be represented, using the idea of a limit, by the expression

$$f(x) = \lim_{n \to \infty} \frac{1}{1 + x^{2n}}.$$

For, if $x^2 < 1$, that is, if x lies in the interval $-1 < x < 1$, the numbers x^{2n} will have the limit 0, and the function will have the value 1. If, how-

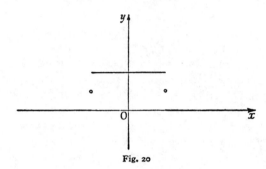

Fig. 20

ever, $x^2 > 1$, as n increases x^{2n} will increase beyond all bounds; our function will then have the value 0. Finally for $x^2 = 1$, that is, for $x = +1$ and $x = -1$, the value of the function is plainly $\frac{1}{2}$ (cf. fig. 20).

Other curves with jumps are sketched in figs. 21a and 21b; they represent functions having obvious discontinuities.

In the case of discontinuities of this kind the limit from the right and the limit from the left both exist. We now pass on to the consideration of discontinuities in which this is not the case. The most important of such discontinuities are the *infinite discontinuities* or *infinities*. These

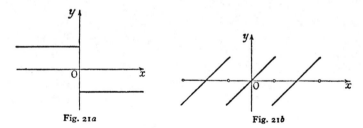

Fig. 21a   Fig. 21b

are discontinuities such as are exhibited by the functions $1/x$ or $1/x^2$ at the point $\xi = 0$; as $x \to \xi$ the absolute value $|f(x)|$ of the function increases beyond all bounds. In the case of $1/x$ the function increases numerically beyond all bounds through positive and through negative values respectively as x approaches the origin from the right and from the left. On the other hand, the function $1/x^2$ has for $x = 0$ an infinite discontinuity at which the value of the function becomes positively infinite from both sides (cf. fig. 6, p. 18, and fig. 12, p. 22). The function $y = \dfrac{1}{x^2 - 1}$, shown in fig. 22, has infinite discontinuities both at $x = 1$ and at $x = -1$.

Finally, we shall illustrate by an example another type of discontinuity in which no limit from the right or from the left exists. We consider the function

$$y = \sin \frac{1}{x},$$

defined for all non-zero values of x. This function takes all values

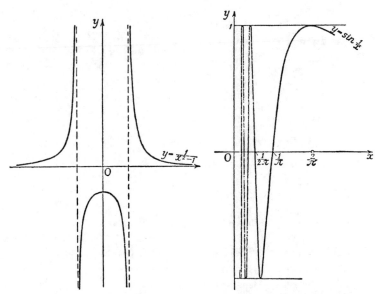

Fig. 22.—Function with infinite
discontinuities

Fig. 23.—Oscillating function with
discontinuity

between -1 and $+1$ when the number $1/x$ runs through the values from $(2n - \frac{1}{2})\pi$ to $(2n + \frac{1}{2})\pi$, no matter what value the integer n has. At the points $x = \dfrac{2}{(4n - 1)\pi}$ the function will have the value -1, at the points $x = \dfrac{2}{(4n + 1)\pi}$ it will have the value $+1$. From this we see that the function swings backwards and forwards more and more rapidly between the values $+1$ and -1 as x approaches nearer and nearer to the point $x = 0$, and that in the immediate neighbourhood of the point $x = 0$ an infinite number of such oscillations occur (cf. fig. 23).

It is interesting to observe that in contrast to the above example the function $y = x \sin 1/x$ (cf. fig. 24) remains continuous at the point $x = 0$ if we assign to it the value 0 at that point. This continuity is due to the fact that as the origin is approached the factor x damps the oscillations of the sine. Yet in the neighbourhood of the origin the function $y = x \sin 1/x$ does not change from monotonic increasing to monotonic decreasing a *finite* number of times. On the contrary, it oscillates backwards and forwards an infinite number of times, the magnitude of these oscillations becoming as small as we please as the origin is approached. This example shows us that even the simple idea of continuity permits of all sorts of remarkable possibilities foreign to our naïve intuitions.

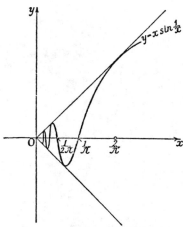

Fig. 24.—Continuous oscillating function

There is one important fact which must be taken into consideration if we are to give our ideas greater precision. It may happen that at a certain point a function is not defined by the original law, as for example at the point $x = 0$ in the last two examples discussed. We have then the right to extend the definition of the function by assigning to it any desired value at such a point. In the last example, however, we can extend the definition in such a way that the function *remains continuous* at that point also, namely, by putting $y = 0$ when $x = 0$. This can be done whenever the limits from the left and from the right both exist and are equal to one another; then we need only make the value of the function at the point in question equal to these limits in order to make the function continuous there. In the case of the function $y = \sin 1/x$ this is not possible.

3. Theorems on Continuous Functions.

In conclusion we quote the following important general theorems, whose proofs follow immediately from the remarks on operations with limits (p. 41):

The sum, difference, and product of two continuous functions are themselves continuous. The quotient of two continuous functions is continuous at every point at which the denominator does not vanish.

In particular, it follows that all polynomials and all rational functions are continuous except at the points where the denominator vanishes. The fact that the other elementary functions, such as the trigonometric functions, are continuous will follow naturally from later considerations (cf. pp. 69, 97).

EXAMPLES

1. Prove that
$$\lim_{x \to 0} \frac{x^2 \sin \dfrac{1}{x}}{\sin x} = 0.$$

2. Prove that

(a) $\lim\limits_{x \to a} \dfrac{\sin (x - \alpha)}{x^2 - \alpha^2} = \dfrac{1}{2\alpha};$ (b) $\lim\limits_{x \to \infty} \dfrac{x + \cos x}{x + 1} = 1;$

(c) $\lim\limits_{x \to \infty} \cos 1/x = 1.$

3. (a) Let $f(x)$ be defined by the equation $y = 6x$. Find a δ, depending on ξ, so small that $|f(x) - f(\xi)| < \varepsilon$ whenever $|x - \xi| < \delta$, where (1) $\varepsilon = \frac{1}{10}$; (2) $\varepsilon = \frac{1}{100}$; (3) $\varepsilon = \frac{1}{1000}$.

Do the same for   (b) $f(x) = x^2 - 2x;$

(c) $f(x) = 3x^4 + x^2 - 7;$

(d) $f(x) = \sqrt{x}, x \geq 0;$

(e) $f(x) = \sqrt{x^2}.$

4. (a) Let $f(x) = 6x$ in the interval $0 \leq x \leq 10$. Find a δ so small that $|f(x_1) - f(x_2)| < \varepsilon$ whenever $|x_1 - x_2| < \delta$, where (1) $\varepsilon = \frac{1}{100}$; (2) ε is arbitrary, > 0.

Do the same for (b) $f(x) = x^2 - 2x, -1 \leq x \leq 1;$

(c) $f(x) = 3x^4 + x^2 - 7, 2 \leq x \leq 4;$

(d) $f(x) = \sqrt{x}, 0 \leq x \leq 4;$

(e) $f(x) = \sqrt{x^2}, -2 \leq x \leq 2.$

5. Determine which of the following functions are continuous. For those which are discontinuous, find the points of discontinuity.

(a) $x^2 \sin x.$

(b) $x \sin^2 (x^2).$

(c) $\dfrac{1}{x} \sin x.$

(d) $\dfrac{\sin x}{\sqrt{x}}.$

(e) $\dfrac{x^3 + 3x + 7}{x^2 - 6x + 8}.$

(f) $\dfrac{x^3 + 3x + 7}{x^2 - 6x + 9}.$

(g) $\dfrac{x^3 + 3x + 7}{x^2 - 6x + 10}.$

(h) $\tan x.$

(i) $\dfrac{1}{\sin x}.$

(j) $\cot x.$

(k) $\dfrac{1}{\cos x}.$

(l) $x \cot x.$

(m) $(\pi - x) \tan x.$

Appendix I to Chapter I

PRELIMINARY REMARKS

In Greek mathematics we find an extensive working-out of the principle that all theorems are to be proved in a logically coherent way by reducing them to a system of axioms, as few in number as possible and not themselves to be proved. This axiomatic method of presentation, which at the same time served as a test for the accuracy of the investigation, was at the beginning of the modern era regarded as a model for other branches of knowledge. For example, in philosophy such men as Descartes and Spinoza believed that they had made their investigations more convincing by presenting them axiomatically, or, as they called it, " geometrically ".

But it was a different matter with modern mathematics, which began to develop at about the same time as the new philosophy. In mathematics the principle of reduction of the material to axioms was frequently abandoned. Intuitive evidence in each separate case became a favourite method of proof. Even in the case of scientists of the first rank we find operations with the new concepts based chiefly on a feeling for the right result and not always free from mystical associations—particularly in the case of the ominous " infinitely small quantities " or " infinitesimals ". Blind faith in the omnipotence of the new methods carried the investigator away along paths which he could never have travelled if subject to the limitations of complete rigour. It is no wonder that only the sure instinct of a great master could guard against gross errors.

It is fortunate that this was so, and that the critical countercurrents which sprang up in the eighteenth century and rose to their full strength in the nineteenth century did not come in time to check the development of modern mathematics, but only in time to establish and extend its results. But the need for critical investigation and consolidation of the advances made gradually increased to such an extent that its satisfaction is rightly regarded as one of the most important mathematical achievements of the nineteenth century.

In the differential and integral calculus the critical work of Cauchy is particularly important. By formulating the fundamental concepts in a clear and satisfactory way, Cauchy in many directions rounded off the work, begun in the eighteenth century, of presenting higher analysis in an intelligible manner, free from the vagueness due to the use of infinitesimals.

The principal thing which remained to be done was to replace intuitive considerations in proofs and discussions by considerations of pure analysis, depending only on numbers and on the operations which can be performed with numbers—as we say, to " arithmetize " analysis. As a matter of fact, the critically-trained mind feels there is something unsatisfactory about appeals to intuition in proofs in analysis. We need not go into the question of the accuracy or inaccuracy of intuition or of the existence of a " pure *a priori* intuition " in Kant's sense in order to recognize that naïve intuitive thinking includes much vagueness which hinders the approach to completely rigorous proofs in analysis. In the following chapters this will strike us more and more clearly. Even here we may mention, for example, that the concept of a continuous curve is very difficult to grasp intuitively. A continuous curve need not by any means possess a definite direction at every point. In fact, there actually exist continuous curves which at *no* point possess a direction, and continuous curves to which no length can be assigned. In the face of such facts, even the beginner will admit the need for arithmetizing analysis.*

Yet we must not allow ourselves to forget that a century of brilliant and fruitful development of mathematics was possible before these requirements were fulfilled. In spite of all its defects intuition still remains the most important driving force for mathematical discovery, and intuition alone can bridge the gap between theory and application.

We shall now follow Bolzano and Weierstrass in developing those lines of thought which yield the rigorous and complete proofs of the theorems which we formulated by intuitive means in the first chapter.

---

* Rigorous mathematical concepts are always very highly idealized forms of the ideas which arise intuitively. Hence it is absolutely impossible to dispose of problems relating to the ultimate foundations of mathematics by appealing to naïve intuition.

1. The Principle of the Point of Accumulation and its Applications

1. The Principle of the Point of Accumulation.

In the rigorous discussion of the fundamentals of analysis the leading part is played by Weierstrass's *principle of the point of accumulation*. From the intuitive point of view this principle is merely the statement of a triviality; but just because it summarizes a state of affairs which occurs frequently it is as useful as small change is in daily life. The principle is as follows:

If infinitely many numbers are given in a finite interval, these numbers possess at least one point of accumulation; that is, there is at least one point ξ such that in every interval, no matter how small, about the point ξ there lie infinitely many of the given numbers.

In order to prove the principle of the point of accumulation arithmetically, we assume to begin with that the given interval is the interval from 0 to 1. We now divide this into ten equal parts by means of the points $0{\cdot}1, 0{\cdot}2, \ldots, 0{\cdot}9$. At least one of these sub-intervals must contain infinitely many points. Let us suppose that the interval beginning with the number $0{\cdot}a_1$ is that interval (or one of those intervals if there are several). We now subdivide this interval into ten parts by means of the points of division $0{\cdot}a_1 1, 0{\cdot}a_1 2, \ldots, 0{\cdot}a_1 9$. Again, it is true that at least one of these sub-intervals must contain infinitely many points; let it be the sub-interval beginning with the number $0{\cdot}a_1 a_2$. We again subdivide into ten parts—notice that one of these parts must contain infinitely many points—and continue the process. We thus arrive at a sequence of digits, $a_1, a_2, a_3, \ldots$, each having one of the values $0, 1, 2, \ldots, 9$. We now consider the decimal

$$\xi = 0{\cdot}a_1 a_2 a_3 \ldots .$$

It is clear that this is a point of accumulation of our set of numbers. For every interval, no matter how small, in whose interior the point ξ lies, contains the sub-intervals of our system of subdivision from a certain degree of fineness onward, and these sub-intervals contain infinitely many numbers of the set.

If the interval under consideration, instead of being the interval from 0 to 1, is, say, the interval from a to $a + h$, nothing essential in the above argument is changed. The point of accumulation is then represented simply by a number of the form

$$a + h \times 0 \cdot a_1 a_2 a_3 \ldots .$$

2. Limits of Sequences.

The considerations above throw new light upon the concept of the limit of an infinite sequence of numbers $a_1, a_2, a_3, \ldots, a_n, \ldots$. We first consider the exceptional case in which infinitely many numbers of the sequence are equal to one another, and extend our definition by applying the name "point of accumulation" to this point (or these points) also. If there are infinitely many different numbers in the sequence, and if we assume that the numbers a_n of this sequence are "bounded", i.e. that there is a number M such that the inequality $|a_n| < M$ holds for all values of n, the numbers of the sequence form an infinite set of numbers in a finite interval, since they all lie between $-M$ and M. They must, therefore, possess at least one point of accumulation (ξ). If there is only *one* point of accumulation, it is easy to show that the sequence converges, and that its limit is the number ξ. For let us mark off any small interval about the point ξ. If infinitely many points of the sequence were outside this interval, they would have a limit point other than ξ, contrary to hypothesis. Hence only a finite number of the numbers of the sequence are exterior to the interval, and thus by definition the sequence approaches ξ. *If, on the other hand, there are several points of accumulation, the sequence approaches no limit.* The existence of a limit and the uniqueness of the point of accumulation of a bounded sequence of numbers are therefore equivalent ideas.

The case of the non-existence of a limit is to be regarded as the rule rather than the exception. For example, the sequence with the terms $a_{2n} = 1/n$, $a_{2n-1} = 1 - 1/n$ $(n = 1, 2, \ldots)$ has the two points of accumulation 0 and 1.

The aggregate of the positive rational numbers may be regarded as a sequence of numbers, in which the ordering by magnitude is, of course, completely destroyed. We arrive most easily at such an arrangement in a sequence if we first write down the rational numbers as shown on p. 60 and then run through this array as shown by the arrows, disregarding those numbers which have already been encountered (such as 2/4). The system of rational numbers obviously has all rational and irrational points as points of accumulation. It therefore forms a simple example of a sequence with an infinite number of points of accumulation.

By means of the concept of convergence we can state the principle of the point of accumulation in a remarkable form which is often convenient for applications.

$$\frac{1}{1} \longrightarrow \frac{1}{2} \quad \frac{1}{3} \longrightarrow \frac{1}{4} \quad \frac{1}{5} \longrightarrow \frac{1}{6} \quad \frac{1}{7} \longrightarrow \cdot$$

$$\frac{2}{1} \quad \frac{2}{2} \quad \frac{2}{3} \quad \frac{2}{4} \quad \frac{2}{5} \quad \frac{2}{6} \quad \cdot \quad \cdot$$

$$\frac{3}{1} \quad \frac{3}{2} \quad \frac{3}{3} \quad \frac{3}{4} \quad \frac{3}{5} \quad \cdot \quad \cdot$$

$$\frac{4}{1} \quad \frac{4}{2} \quad \frac{4}{3} \quad \frac{4}{4} \quad \cdot \quad \cdot$$

$$\frac{5}{1} \quad \frac{5}{2} \quad \frac{5}{3} \quad \cdot \quad \cdot$$

$$\frac{6}{1} \quad \frac{6}{2} \quad \cdot \quad \cdot$$

$$\frac{7}{1} \quad \cdot \quad \cdot$$

$$\cdot$$

Enumeration of the rational numbers

From every bounded infinite set of numbers it is possible to choose an infinite sequence a_1, a_2, a_3, ... *which converges to a definite limit* ξ. For this purpose we have only to take a point of accumulation ξ of the given set of numbers, then to select a number a_1 of the set whose distance from ξ is less than 1/10, then a second number a_2 of the set whose distance from ξ is less than 1/100, then a third number a_3 whose distance from ξ is less than 1/1000, and so on. We see at once that this sequence actually converges to the limit ξ.

3. Proof of Cauchy's Convergence Test.

Let us now return to convergent sequences, i.e. to bounded sequences with only one point of accumulation. Cauchy's convergence test, stated in § 6 (p. 40), now reduces almost to a triviality. For let us assume that $|a_m - a_n|$ is arbitrarily small when m and n are sufficiently large. Then the numbers a_n all lie in a finite interval, and therefore possess at least one point of accumulation ξ. If now there were a second point of accumulation η, the distance of this point from ξ would be $|\xi - \eta| = a$, a positive quantity. Within an arbitrarily small distance from ξ, say within a distance less than $a/3$ from ξ, there must be infinitely many numbers a_n, and hence, in particular, infinitely many numbers a_n for which $n > N$, however large N is chosen. Similarly, within an arbitrarily small distance from the point η, say within a distance less than $a/3$ from η, there are infinitely many numbers a_m of the sequence; in particular, infinitely many numbers a_m for which $m > N$. For these values a_n and a_m it is true that $| a_m - a_n | > a/3$, and this relation is incompatible with the hypothesis that for sufficiently large values of N the difference $| a_n - a_m |$ is arbitrarily small provided that n and m are both greater than N. Consequently there are not two distinct points of accumulation, and Cauchy's test is proved.

4. The Existence of Limits of Bounded Monotonic Sequences.

It is equally easy to see that *a bounded monotonic increasing or monotonic decreasing sequence of numbers must possess a limit.* For suppose that the sequence is monotonic increasing, and let ξ be a point of accumulation of the sequence; such a point of accumulation must certainly exist. Then ξ must be greater than any number of the sequence. For if a number a_l of the sequence were equal to or greater than ξ, every number a_n for which $n > l + 1$ would satisfy the inequality $a_n > a_{l+1} > a_l \geqq \xi$. Hence all numbers of the sequence, except the first $(l + 1)$ at most, would lie outside the interval of length $2(a_{l+1} - \xi)$ whose mid-point is at the point ξ. This, however, contradicts the assumption that ξ is a point of accumulation. Hence no numbers of the sequence, and *a fortiori* no points of accumulation, lie above ξ. So if another point of accumulation η exists we must have $\eta < \xi$. But if we repeat the above argument with η in place of ξ we obtain $\xi < \eta$, which is a contradiction. Hence only one point of accumulation can exist, and the convergence is proved. An argument exactly analogous to this of course applies to monotonic decreasing sequences.

As on p. 41, we can extend our statements about monotonic sequences by including the limiting case in which successive numbers of the sequence are equal to one another. It is in this case better to speak of monotonic non-decreasing and monotonic non-increasing sequences respectively. The theorem about the existence of a limit remains valid for such sequences.

5. Upper and Lower Points of Accumulation; Upper and Lower Bounds of a Set of Numbers.

In the construction on p. 58 which led us to a point of accumulation ξ we had at each step to make the choice of a sub-interval containing infinitely many points of the set. Had we always chosen the last sub-interval which contained an infinite number of points, we should have been led to a certain definite point of accumulation β. This point of accumulation β is called the *upper point of accumulation* or *upper limit* of the set of numbers, and is designated by the abbreviation $\overline{\lim}$. It is that point of accumulation of the sequence which lies farthest to the right; i.e. it is quite possible that an infinite number of points of the sequence lie above β, but no matter how small the positive number ϵ may be, there are *not* an infinite number above $\beta + \epsilon$.

If, in the construction on p. 58, we had always chosen the first of the intervals containing an infinite number of points of the set, we should again have arrived at a certain definite point of accumulation α. This point α is called the *lower point of accumulation* or *lower limit* of the set, and is denoted by $\underline{\lim}$. There may be infinitely many numbers of the set below α, but no matter how small the positive number ϵ may be there are only a finite number below $\alpha - \epsilon$. The proofs of these facts can be left to the reader.

Neither the upper limit β nor the lower limit α need belong to the set. For example, for the set of numbers $a_{2n} = 1/n$, $a_{2n-1} = 2 - 1/n$ these limits are respectively $\alpha = 0$ and $\beta = 2$, but the numbers 0 and 2 do not themselves occur in the set.

In this example there is no number of the set above $\beta = 2$. In this case we say that $\beta = 2$ is also the upper bound M of the set, according to the following definition: M is called the *least upper bound*, or simply *the* upper bound, of a set of numbers if (1) there is no number of the set greater than M, but (2) for every positive number ϵ there is a number of the set greater than $M - \epsilon$. The upper bound may coincide with the upper limit, as the above example shows. But the set $a_n = 1 + 1/n$ ($n = 1, 2, \ldots$) shows that this is not necessarily the case. Here $M = 2$ and $\beta = 1$.

Every bounded set of numbers has a least upper bound. For let β be the upper limit of the set. Either there are no numbers of the set greater than β, or there are such numbers. In the first case β is the least upper bound, since no numbers are above β and there are numbers arbitrarily close to β below it. In the second case let a be a number of the set greater than β. There are only a finite number of numbers of the set equal to or greater than a, since otherwise there would be an accumulation point above β, which is impossible. We therefore need only choose the greatest of these numbers; it will be the upper bound of the set.

We see that in any case $M \geqq \beta$, and we recognize the following fact:

If the upper bound of the set does not coincide with the upper limit, it must belong to the set, and is an " isolated " point of the set.

Corresponding statements hold for the lower bound m; it is always equal to or less than a, and if m and a do not coincide, m belongs to the set and is an isolated point of the set.

2. THEOREMS ON CONTINUOUS FUNCTIONS

1. Greatest and Least Values of Continuous Functions.

A bounded infinite set of numbers must possess a least upper bound M and a greatest lower bound m. But, as we have seen, these numbers M and m do not necessarily belong to the set; as we say, the set does not necessarily have a greatest or a least value.

In view of this fact the following theorem on continuous functions is by no means so obvious as it appears to simple intuition:

Every function f(x) *which is continuous in a closed interval* a $\leqq$ x $\leqq$ b *assumes a greatest value at least once and a least value at least once*, or, as we say, *it possesses a greatest and a least value.*

This may easily be proved in the following way. The values assumed by the continuous function $f(x)$ in the interval $a \leqq x \leqq b$ form a bounded set of numbers and therefore possess a least upper bound M. For otherwise a sequence of numbers $\xi_1, \xi_2, \ldots, \xi_n, \ldots$ in our interval would exist for which $f(\xi_n)$ increases beyond all bounds. This sequence would have at least one point of accumulation $\bar{\xi}$ in the interval. Then arbitrarily near to $\bar{\xi}$ there would always be numbers ξ_n of our sequence for which the expression $|f(\bar{\xi}) - f(\xi_n)|$ exceeds 1 (and in fact is arbitrarily large): that is, the function would be discontinuous at the point $\bar{\xi}$. Thus a least upper bound M exists and hence either there is a point ξ such that $f(\xi) = M$, which would prove the statement, or there is a sequence of numbers $x_1, x_2, \ldots, x_n, \ldots$ in the interval for which

$$\lim_{n \to \infty} f(x_n) = M.$$

According to the principle of the point of accumulation as formulated on p. 60 we can select a sub-sequence of the numbers x_n, which converges to a limit ξ. Let us call this sub-sequence $\xi_1, \xi_2, \ldots, \xi_n, \ldots$, so that

$$\lim_{n \to \infty} \xi_n = \xi.$$

It is then certain that

$$\lim_{n \to \infty} f(\xi_n) = M.$$

On the other hand, the function has been assumed continuous in the interval, and hence, in particular, at ξ, so that

$$\lim_{n \to \infty} f(\xi_n) = f(\xi).$$

Hence $f(\xi) = M$. The value M is therefore assumed by the function at a definite point ξ in the interior or on the boundary of the interval, as was stated. An exactly similar discussion applies to the least value.

The theorem about the greatest and least values of continuous functions does not remain true in general unless we expressly assume the interval to be closed, that is, unless we make the hypothesis of continuity refer to the end-points also. For example, the function $y = 1/x$ is continuous in the open interval $0 < x < \infty$. It has, however, no greatest value, but has arbitrarily large values near $x = 0$. Similarly, it has no least value, but becomes arbitrarily near 0 for sufficiently large values of x, without ever assuming the value 0.

2. The Uniformity of Continuity.

As we have already seen (cf. p. 54), and as we shall further see, the continuity of a function $f(x)$ in a closed interval $a \leq x \leq b$ leaves room for a variety of possibilities which do not suggest themselves intuitively. For this reason we shall give logically rigorous proofs of certain consequences of the idea of continuity which from a naïve point of view seem quite obvious. The definition of continuity simply states that from the relation $\lim_{n \to \infty} x_n = \xi$ the relation $\lim_{n \to \infty} f(x_n) = f(\xi)$ follows. We can also express this in the following way: for each point ξ there corresponds to every $\epsilon > 0$ a number $\delta > 0$ such that $|f(x) - f(\xi)| < \epsilon$ whenever $|x - \xi| < \delta$, provided that all the numbers x considered lie in the interval $a \leq x \leq b$.

For example, in the case of the function $y = cx$ (where $c \neq 0$) such a number δ is given by the relation $\delta = \epsilon/|c|$. For the function $y = x^2$ we can find such a number in the following way. We assume that $a = 0$ and $b = 1$, and ask ourselves how near to ξ the number x must lie in order that the expression $|x^2 - \xi^2|$ may be less than ϵ. For this purpose we write $|x^2 - \xi^2| = |x - \xi| |x + \xi| \leq |x - \xi| (1 + \xi)$. If, therefore, we choose $\delta \leq \epsilon/(1 + \xi)$ we can be sure that $|x^2 - \xi^2| < \epsilon$. We see in this example that the number δ found in this way depends not only on ϵ, but also on the point of the interval at which we are investigating the continuity of the function. But if we give up the attempt to make the best possible choice of δ for each ξ, we can eliminate this dependence of δ on ξ. For we need only replace ξ on the right by the number 1, thus obtaining for δ the expression $\epsilon/2$, which is smaller than the previous expression for δ but serves equally well for all points ξ.

The question now arises whether something similar does not hold for every function which is continuous in a closed interval. That is, we inquire whether it may not be possible to determine for each ϵ a $\delta = \delta(\epsilon)$ depending on ϵ *only* and *not* on ξ, such that the inequality

$$|f(x) - f(\xi)| < \epsilon$$

is true, provided $|x - \xi| < \delta$, for all values of ξ at the same time (or, better expressed, uniformly with respect to ξ). As a matter of fact, this is possible merely as a consequence of the general definition of continuity, without any additional hypotheses. This fact, which first attracted attention late in the nineteenth century, is called the *theorem of the uniform continuity of continuous functions*.

We shall prove the theorem indirectly. That is, we shall show that the assumption that a function $f(x)$ exists which in a closed interval $a \leq x \leq b$ is continuous and yet not uniformly continuous leads us to a contradiction. Uniform continuity means that if we wish to make the difference $|f(u) - f(v)|$ less than an arbitrarily chosen positive number ϵ, the numbers u and v being chosen in the closed interval $a \leq x \leq b$, we need only choose u and v near enough to one another, namely, at a distance apart less than $\delta = \delta(\epsilon)$; it is immaterial *where* in the interval the pair of numbers u, v is chosen. Now, if $f(x)$ were not uniformly continuous, there would exist a positive (perhaps very small) number a with the following property: to every number δ_n of an arbitrary sequence $\delta_1, \delta_2, \ldots$ of positive numbers tending to zero there corresponds a pair of values u_n, v_n of the interval for which $|u_n - v_n| < \delta_n$ and $|f(u_n) - f(v_n)| > a$. According to the principle of the point of accumulation the numbers u_n must have a point of accumulation ξ, and so the numbers v_n must have the same point of accumulation. If we mark off an arbitrarily small interval $|x - \xi| < \delta$ about this point ξ, an infinite number of the pairs u_n, v_n will lie in this interval. But this contradicts the assumed continuity of $f(x)$ at the point ξ; for that requires, by Cauchy's convergence test, that for points x_1 and x_2 near enough to ξ

$$|f(x_1) - f(x_2)| < a.$$

The uniformity of the continuity is thus proved.

In our proof we have made essential use of the fact that the interval is *closed*.* In fact, the theorem of the uniformity of continuity does not hold for intervals which are not closed.

For example, the function $1/x$ is continuous in the half-open interval $0 < x \leq 1$, but it is not uniformly continuous. For no matter how small the length $\delta\ (< 1)$ of an interval is chosen, the function will take values differing by any fixed number, say 1, in the interval, if only the interval lies near enough to the origin, say $\delta/2 \leq x \leq 3\delta/2$. The non-uniformity of continuity is of course due to the fact that in the closed interval $0 \leq x \leq 1$ the function possesses a discontinuity at the origin. If we had considered the example $y = x^2$ in the whole (open) interval $-\infty < x < \infty$ instead of in a closed interval, it would not have been uniformly continuous.

3. The Intermediate Value Theorem.

Another theorem which constantly recurs in analysis is the following:

A function f(x), *continuous in a closed interval* a $\leq$ x $\leq$ b, *which is negative for* x = a *and positive for* x = b, *(or conversely), assumes the value 0 at least once in the interval.*

---

* Otherwise the point of accumulation ξ need not belong to the interval.

Geometrically this theorem is trivial, since it merely states that a curve which begins below the x-axis and ends above it must cut the axis somewhere in between. Analytically the theorem is very easily proved. In the interval there are an infinite number of points for which $f(x) < 0$; on account of the continuity of the function, in fact, this is true for a whole *interval* beginning at the point a. The set consisting of those points x for which $f(x) < 0$ has a least upper bound ξ, which is greater than a. Since in every neighbourhood of ξ there are points x for which $f(x) < 0$, we must have $f(\xi) \leqq 0$ (whence in particular $\xi \neq b$). It is impossible, however, that $f(\xi) < 0$, for then $f(x)$ would be negative in a sufficiently small neighbourhood of ξ, including values x greater than ξ, in contradiction to the hypothesis that ξ is the upper bound of the values x for which $f(x) < 0$. Therefore $f(\xi) = 0$, and our assertion is proved.

A slight generalization of our theorem is:

If we assume that f(a) = α *and* f(b) = β, *and if* μ *is any value between* α *and* β, *the continuous function* f(x) *assumes the value* μ *at least once in the interval.* For the continuous function

$$\phi(x) = f(x) - \mu$$

will have different signs at the two ends of the interval, and will therefore assume the value 0 somewhere in the interval.

4. The Inverse of a Continuous Monotonic Function.

If the continuous function $y = f(x)$ is monotonic in the interval $a \leqq x \leqq b$, it will assume each value μ between $f(a)$ and $f(b)$ once and only once; hence if y describes the closed interval between the values $\alpha = f(a)$ and $\beta = f(b)$, to each value of y there will correspond exactly one value of x. We can therefore think of x as a single-valued function of y in this interval; that is, the function $y = f(x)$ has a unique inverse. We assert that this inverse function $x = \phi(y)$ is also a continuous monotonic function of y, as y varies within the interval between α and β.

The monotonic character of the inverse function $x = \phi(y)$ is obvious. In order to prove its continuity we observe that from the monotonic character of $f(x)$ it follows that

$$|f(x_2) - f(x_1)| = |y_2 - y_1| > 0,$$

provided that x_1 and x_2 are distinct numbers of the interval. If h is a positive number less than $b - a$, the function

$$|f(x + h) - f(x)|$$

is continuous in the closed interval $a \leq x \leq b - h$. At a point ξ it therefore has a least value $|f(\xi + h) - f(\xi)| = a(h)$, which by our preceding remark is not zero.* From this we conclude that if x_1 and x_2 are two points in the interval for which $|x_1 - x_2| \geq h$, then $|f(x_1) - f(x_2)| \geq a(h)$. But this implies the continuity of the inverse function. For if $|y_1 - y_2|$ falls below the positive number $a(h)$, then we must have $|x_1 - x_2| < h$; and hence if a positive number ϵ is given, we need only choose δ equal to $a(\epsilon)$ in order to ensure that for all values y for which $|y_1 - y| < \delta$ it is also true that $|\phi(y_1) - \phi(y)| < \epsilon$.

We have therefore established the following theorem: *If the function* $y = f(x)$ *is continuous and monotonic in the interval* $a \leq x \leq b$, *and* $f(a) = a$, $f(b) = \beta$, *then it has a single-valued inverse function* $x = \phi(y)$, $a \leq y \leq \beta$, *and this inverse function is also continuous and monotonic.*

5. Further Theorems on Continuous Functions.

We leave it to the reader to prove the following almost trivial fact: a continuous function of a continuous function is itself continuous. That is, if $\phi(x)$ is a function continuous in the interval $a \leq x \leq b$, and its functional values lie in the interval $a \leq \phi \leq \beta$, and if in addition $f(\phi)$ is a continuous function of ϕ in this last interval, then $f(\phi(x))$ is a continuous function of x in the interval $a \leq x \leq b$. (Theorem of the continuity of functions of continuous functions.)

It has already been mentioned on p. 54 that *the sum, difference, and product of continuous functions are themselves continuous, and that the quotient of continuous functions is continuous, provided that the denominator remains different from zero.*

3. Some Remarks on the Elementary Functions

In Chapter I we tacitly assumed that the elementary functions are continuous. The proof of this is very simple. First,

---

* On account of the continuity of $f(x)$, $a(h)$ itself of course tends to 0 as h does.

the function $f(x) = x$ is continuous; therefore $x^2 = x . x$ is continuous, being the product of two continuous functions, and every power of x is likewise continuous. Thus every polynomial is continuous, being the sum of continuous functions. Every rational fractional function, being a quotient of continuous functions, is likewise continuous in every interval in which the denominator does not vanish.

The function x^n is continuous and monotonic. Hence the n-th root, being the inverse function of the n-th power, is continuous. By the theorem of the continuity of functions of continuous functions, the n-th root of a rational function is continuous (except where the denominator vanishes).

The continuity of the trigonometric functions, with which the reader is familiar from elementary mathematics, could now readily be proved, using the concepts developed above. The discussion is not given here, since in Chapter II, § 3 (p. 97), this continuity will be seen to follow naturally as a consequence of differentiability.

We shall merely make a few remarks about the definition and continuity of the exponential function a^x, the general power function x^a, and the logarithm. We assume, as in § 3, pp. 25–26, that a is a positive number, say greater than 1, and if $r = p/q$ is a positive rational number (p and q being integers) we take $a^r = a^{p/q}$ as meaning the positive number whose q-th power is a^p. If a is any irrational number and $r_1, r_2, \ldots, r_m, \ldots$ is a sequence of rational numbers approaching a, we assert that $\lim\limits_{m \to \infty} a^{r_m}$ exists; we then call this limit a^a.

In order to prove the existence of this limit, by Cauchy's test we need only show that $| a^{r_n} - a^{r_m} |$ is arbitrarily small, provided that n and m are sufficiently large. We suppose for example that $r_n > r_m$, i.e. that $r_n - r_m = \delta$, where $\delta > 0$. Then

$$a^{r_n} - a^{r_m} = a^{r_m}(a^\delta - 1).$$

Since a^{r_m} remains bounded, we need only show that

$$| a^\delta - 1 | = a^\delta - 1$$

is arbitrarily small when the values of n and m are sufficiently large. But δ is a rational number, and certainly may be made as small as we please provided the values of n and m are sufficiently large. Hence if l is an arbitrarily large positive integer, $\delta < 1/l$ if n and m are large enough. Now the relations $\delta < 1/l$ and $a > 1$ give *

$$1 < a^\delta < a^{1/l},$$

and since $a^{1/l}$ tends to 1 as l increases (cf. p. 31) our assertion follows immediately.

It may be left to the reader to show that the function a^x extended to irrational values in this way is also continuous everywhere, and, moreover, that it is a monotonic function. For negative values of x this function is naturally defined by the equation

$$a^x = \frac{1}{a^{-x}}.$$

As x runs from $-\infty$ to $+\infty$, a^x takes all values between 0 and $+\infty$. Consequently it possesses a continuous and monotonic inverse function, which we call the *logarithm to the base a*. In like manner we can prove that the general power x^a is a continuous function of x, where a is any fixed rational or irrational number and x varies over the interval $0 < x < \infty$; also that x^a is monotonic if $a \neq 0$.

The "elementary" discussion of the exponential function, the logarithm, and the power x^a outlined here will later (Chapter III, § 6, p. 167) be replaced by another discussion which is in principle much simpler.

EXAMPLES

1. Give the upper and lower bounds and upper and lower limits for the following sequences, and state which belong to the sequence:

(a) $\dfrac{6^n}{n!}$, $n = 1, 2, \ldots$.

(b) $0, \dfrac{(-1)^n}{n!}$, $n = 1, 2, \ldots$.

(c) $\dfrac{(-1)^n}{n} + \dfrac{n}{2n+1}$, $n = 1, 2, \ldots$.

(d) $1 + \dfrac{(-1)^n}{n} + \dfrac{(-1)^n n}{2n+1}$, $n = 1, 2, \ldots$.

(e) $\dfrac{1}{m^2} + \dfrac{1}{n^2}$, $m, n = 1, 2, \ldots$.

* This statement follows from the fact that when $a > 1$ the power $a^{m/n}$ is greater than 1 if m/n is positive. This is clearly true. For if $a^{m/n}$ were less than 1, then $a^m = (a^{m/n})^n$ would be a product of n factors all less than 1, and would be less than 1. On the contrary, a^m is the product of m factors all greater than 1, and so is greater than 1.

2.* Prove that if $f(x)$ is continuous for $a \leq x \leq b$, then for every $\varepsilon > 0$ there exists a polygonal function $\varphi(x)$ (that is, a continuous function whose graph consists of a finite number of rectilinear segments meeting at corners) such that $|f(x) - \varphi(x)| < \varepsilon$ for every x in the interval.

3. Prove that every polygonal function $\varphi(x)$ can be represented by a sum $\varphi(x) = a + bx + \Sigma c_i |x - x_i|$, where the x_i's are the abscissæ of the corners.

Find a formula of this kind for the function $f(x)$ defined by the equations:

$$f(x) = 2x - 1 \ (0 \leq x \leq 2).$$
$$f(x) = 5 - x \ (2 \leq x \leq 3).$$
$$f(x) = x - 1 \ (3 \leq x \leq 5).$$
$$f(x) = 4 \ (5 \leq x \leq 7).$$

4. For the following functions $f(x)$ find as in § 1, No. 2, p. 65, a $\delta(\varepsilon)$ such that $|f(x_1) - f(x_2)| < \varepsilon$ whenever $|x_1 - x_2| < \delta(\varepsilon)$:

$$(a) \ f(x) = 2x^3, \ -1 \leq x \leq 1.$$
$$(b) \ f(x) = x^n, \ -a \leq x \leq a.$$
$$*(c) \ f(x) = \sqrt[3]{1 - x^2}, \ -1 \leq x \leq 1.$$

5.* The function $y = \sin 1/x$ has no discontinuities in the interval $0 < x < 1$. Prove that it is *not* uniformly continuous in that open interval.

6. A function $f(x)$ is defined for all values of x in the following manner:

$$f(x) = 0 \ \text{for all irrational values of } x;$$
$$f(x) = 1/q \ \text{for rational } x = p/q,$$

where p/q is a fraction in its lowest terms; (thus for $x = \frac{16}{29}$, $f(x) = \frac{1}{29}$).

Prove that $f(x)$ is continuous for all irrational values of x and discontinuous for all rational values of x.

Appendix II to Chapter I

1. Polar Co-ordinates

In Chapter I we have set the concept of function in the foreground and have represented functions geometrically by means of curves. It is, however, useful to recall * that analytical geometry follows the reverse procedure, beginning with a curve given

* See also p. 16.

by some geometrical property and representing this curve by a function, for example, by a function which expresses one of the co-ordinates of a point of the curve in terms of the other co-ordinate. This point of view naturally leads us to consider, apart from the rectangular co-ordinates to which we restricted ourselves in Chapter I, other systems of co-ordinates which may be better suited for the representation of curves given geometrically. The most important example

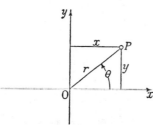

Fig. 25.—Polar co-ordinates

is that of *polar co-ordinates* r, θ, which are connected with the rectangular co-ordinates x, y of a point P by the equations

$$x = r\cos\theta, \quad y = r\sin\theta, \quad r^2 = x^2 + y^2, \quad \tan\theta = \frac{y}{x},$$

and whose geometrical interpretation is made clear in fig. 25.

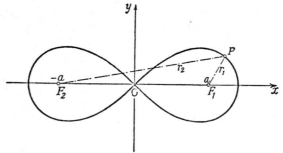

Fig. 26.—Lemniscate

Let us consider for example the *lemniscate*. This is geometrically defined as the locus of all points P for which the product of the distances r_1 and r_2 from the fixed points F_1 and F_2 with the rectangular co-ordinates $x = a$, $y = 0$ and $x = -a$, $y = 0$ respectively has the constant value a^2 (cf. fig. 26). Since

$$r_1{}^2 = (x - a)^2 + y^2, \quad r_2{}^2 = (x + a)^2 + y^2,$$

a simple calculation gives us the equation of the lemniscate in the form

$$(x^2 + y^2)^2 - 2a^2(x^2 - y^2) = 0.$$

If we now introduce polar co-ordinates, we obtain

$$r^4 - 2a^2r^2(\cos^2\theta - \sin^2\theta) = 0;$$

and if we divide by r^2 and use a simple trigonometrical formula this becomes

$$r^2 = 2a^2 \cos 2\theta.$$

Thus we see that the equation of the lemniscate is simpler in polar co-ordinates than in rectangular.

2. Remarks on Complex Numbers

Our studies will be based chiefly on the class of real numbers. Nevertheless, with a view to the discussions in Chapters VIII, IX, and XI we would remind the reader that the problems of algebra have led to a still wider extension of the concept of number, namely, to the introduction of *complex numbers*. The advance from the natural numbers to the class of all real numbers arose from the desire to eliminate exceptional phenomena and to make certain operations, such as subtraction, division, and correspondence between points and numbers, always possible. Similarly, we are compelled, by the requirement that every quadratic equation and in fact every algebraic equation shall have a solution, to introduce the complex numbers. If, for example, we wish the equation

$$x^2 + 1 = 0$$

to have roots, we are obliged to introduce new symbols i and $-i$ as the roots of this equation. (As is shown in algebra, this is sufficient to ensure that *every* algebraic equation shall have a solution.*)

If a and b are ordinary real numbers, the *complex number* $c = a + ib$ denotes a pair of numbers (a, b), calculations with such pairs of numbers being performed according to the following general rule: we add, multiply, and divide complex numbers (among which the real numbers are included as the special case $b = 0$), treating the symbol i as an undetermined quantity, and then simplify all expressions by using the equation $i^2 = -1$ to remove all powers of i higher than the first, thus leaving only an expression of the form $a + ib$.

We may assume that the reader already has a certain degree of familiarity with these complex numbers. We shall neverthe-

* That every algebraic equation possesses real or complex roots is the statement of the "fundamental theorem" of algebra.

less emphasize a particularly important relationship which we shall explain in connexion with the geometrical or trigonometrical representation of the complex numbers. If $c = x + iy$ is such a number, we represent it in a rectangular co-ordinate system by the point P with co-ordinates x and y. By means of the equations $x = r \cos \theta$, $y = r \sin \theta$, we now introduce the polar co-ordinates r and θ (cf. p. 72) instead of the rectangular co-ordinates x and y. Then $r^2 = \sqrt{x^2 + y^2}$ is the distance of the point P from the origin, and θ is the angle between the positive x-axis and the segment OP. The complex number c is now represented in the form

$$c = r(\cos \theta + i \sin \theta).$$

The angle θ is called the *amplitude* of the complex number c, the quantity r its *absolute value* or *modulus*, for which we also write $|c|$. To the "conjugate" complex number $\bar{c} = x - iy$ there obviously corresponds the same absolute value, but (except in the case of negative real values of c) the angle $-\theta$. Clearly

$$r^2 = |c|^2 = c\bar{c} = x^2 + y^2.$$

If we use this trigonometrical representation the multiplication of complex numbers takes a particularly simple form. For then

$$\begin{aligned} c \cdot c' &= r(\cos \theta + i \sin \theta) \cdot r'(\cos \theta' + i \sin \theta') \\ &= rr'(\cos \theta \cos \theta' - \sin \theta \sin \theta') \\ &\quad + i(\cos \theta \sin \theta' + \sin \theta \cos \theta'). \end{aligned}$$

If we recall the addition theorems for the trigonometric functions this becomes

$$c \cdot c' = rr'(\cos(\theta + \theta') + i \sin(\theta + \theta')).$$

We therefore multiply complex numbers by multiplying their absolute values and adding their amplitudes. The remarkable formula

$$(\cos \theta + i \sin \theta)(\cos \theta' + i \sin \theta') = \cos(\theta + \theta') + i \sin(\theta + \theta')$$

is usually called *De Moivre's theorem*. It leads us at once to the relation

$$(\cos \theta + i \sin \theta)^n = \cos n\theta + i \sin n\theta,$$

which e.g. at once enables us to solve the equation $x^n = 1$ for positive integers n, the roots (the so-called roots of unity) being

$$\epsilon_1 = \epsilon = \cos\frac{2\pi}{n} + i\sin\frac{2\pi}{n}, \quad \epsilon_2 = \epsilon^2 = \cos\frac{4\pi}{n} + i\sin\frac{4\pi}{n}, \quad \ldots,$$

$$\epsilon_{n-1} = \epsilon^{n-1} = \cos\frac{(n-1)\pi}{n} + i\sin\frac{(n-1)\pi}{n}, \quad \epsilon_n = \epsilon^n = 1.$$

Moreover, if we imagine the expression on the left-hand side of the equation $(\cos\theta + i\sin\theta)^n = \cos n\theta + i\sin n\theta$ expanded by the binomial theorem, we need only separate real and imaginary parts in order to obtain expressions for $\cos n\theta$ and $\sin n\theta$ in terms of powers and products of powers of $\sin\theta$ and $\cos\theta$.

EXAMPLES

1. Plot the graphs of the following functions:

$r = \sin\varphi.$ $\qquad\qquad$ $r = \cos 5\varphi.$

$r = \varphi.$

$r = \sin 6\varphi.$ $\qquad\qquad$ $r = \dfrac{1}{\cos(\varphi - \alpha)}$, α constant.

2. Find the polar equation of

(a) the circle of radius a with its centre at the origin;

(b) the circle of radius a with centre (a, φ_0);

(c) the general straight line.

3. Use De Moivre's theorem to express $\cos 2\theta$ and $\sin 2\theta$ in terms of $\sin\theta$ and $\cos\theta$. Similarly, for $\cos 3\theta$, $\sin 3\theta$, $\cos 5\theta$, $\sin 5\theta$.

Prove that $\cos n\theta$ is a polynomial in $\cos\theta$, and also that if n is odd $\sin n\theta$ is a polynomial in $\sin\theta$.

4. Work out the following expressions, and state the modulus and amplitude of each of the numbers involved and of the answers:

(a) $-3 \cdot 2i.$ $\qquad\qquad\qquad$ (f) $i^{1/2}.$

(b) $(4 + 4i)(\frac{1}{4} - \frac{1}{4}\sqrt{3}i).$ $\qquad$ (g) $(1 + i)^{1/2}.$

(c) $(1 + i)(1 - i).$ $\qquad\qquad$ (h) $(3 - 3i)^{2/3}.$

(d) $(\sqrt{3} - i)^2.$ $\qquad\qquad\quad$ (k) $1^{1/3}.$

(e) $1^{1/2}.$ $\qquad\qquad\qquad\quad$ (l) $(16i)^{1/4}.$

5.* Prove that if $\varepsilon = \cos\dfrac{2\pi}{n} + i\sin\dfrac{2\pi}{n}$, where n is an integer greater than 1,

$$\varepsilon^\nu + \varepsilon^{2\nu} + \varepsilon^{3\nu} + \ldots + \varepsilon^{n\nu} = \begin{cases} 0 \text{ if } n \text{ is not a factor of } \nu, \\ n \text{ if } n \text{ is a factor of } \nu. \end{cases}$$

CHAPTER II

The Fundamental Ideas of the Integral and Differential Calculus

Among the limiting processes of analysis there are two which play an especially important part, not only because they arise in many different connexions, but chiefly because of the very close reciprocal relation between them. Isolated examples of these two limiting processes, *differentiation* and *integration*, were considered even in classical times; but it is the recognition of their complementary nature and the resulting development of a new and methodical mathematical procedure that marks the beginning of the real systematic differential and integral calculus. The credit of initiating this development belongs equally to the two great geniuses of the seventeenth century, Newton and Leibnitz, who, as we know to-day, made their discoveries independently of one another. While Newton in his investigations may have succeeded in stating his concepts more clearly, Leibnitz's notation and methods of calculation are more highly developed; even to-day these formal portions of Leibnitz's work form an indispensable element in the theory.

1. The Definite Integral

We first encounter the integral in the problem of measuring the area of a plane region bounded by curved lines. More refined considerations then permit us to separate the notion of integral from the naïve intuitive idea of area, and to express it analytically in terms of the notion of number only. This analytical definition of the integral we shall find to be of great significance, not only because it alone enables us to attain complete clarity in our concepts, but also because its applications extend far beyond the calculation of areas.

We begin by considering the matter intuitively.

1. The Integral as an Area.

Let us suppose that we are given a function $f(x)$ which is continuous and positive in an interval, and that a and b $(a < b)$ are two values in that interval. We think of the function as represented by a curve, and consider the area of the region which is bounded above by the curve, at the sides by the straight lines $x = a$ and $x = b$, and below by the portion of the x-axis between the points a and b (fig. 1).

That there is a definite meaning in speaking of the area of this region is an assumption inspired by intuition, which we

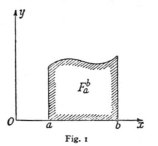

Fig. 1

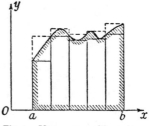

Fig. 2.—Upper sum and lower sum

here state expressly as a hypothesis. We call this area F_a^b the *definite integral of the function* f(x) *between the limits* a *and* b. When we actually seek to assign a numerical value to this area, we find that we are in general unable to measure areas with curved boundaries; but we can measure polygons with straight sides by dividing them into rectangles and triangles. Such a subdivision of our area is usually impossible. It is, however, only a short step to conceive of the area as the limiting value of a sum of areas of rectangles, in the following manner. We subdivide the part of the x-axis between a and b into n equal parts, and at each point of division we erect the ordinate up to the curve; the area is thus divided into n strips. We can no more calculate the area of such strips than we could that of the original surface; but if, as shown in fig. 2, we find first the least and then the greatest value of the function $f(x)$ in each subinterval, and then replace the corresponding strip (1) by a rectangle whose height is equal to the least value of the function, and (2) by a rectangle whose height is equal to the greatest value of the function, we obtain two step-shaped figures. (In fig. 2 the first of these is drawn with a solid line, the second being shown by dotted lines.) The first step-shaped figure obviously has an area which is at most equal to the area $F_a{}^b$ which we are trying to determine; the second has an area which is at least as large as $F_a{}^b$. If we denote the sum of the areas of the first set of rectangles by $\underline{F_n}$ (*lower sum*), and the sum of the areas of the second set by $\overline{F_n}$ (*upper sum*), we have the relation

$$\underline{F_n} \leqq F_a{}^b \leqq \overline{F_n}.$$

If we now make the subdivision finer and finer, i.e. let n increase without limit, intuition tells us that the quantities $\overline{F_n}$ and $\underline{F_n}$ approach closer and closer to each other and tend to the same limit $F_a{}^b$. We may therefore consider our integral as the limiting value

$$F_a{}^b = \lim_{n \to \infty} \underline{F_n} = \lim_{n \to \infty} \overline{F_n}.$$

Intuition also shows us the possibility of an immediate generalization. It is by no means necessary that the n sub-intervals should all be of the same length. They may, on the contrary, have different lengths, provided only that as n increases the length of the longest sub-interval tends to 0.

2. The Analytical Definition of the Integral.

In the above section we have considered the definite integral as a number given by an area, and hence to a certain extent previously known, and have subsequently represented it as a limiting value. We shall now reverse the procedure. We no longer take the point of view that we know by intuition how an area can be assigned to the region under a continuous curve, or, indeed, that this is possible; we shall, on the contrary, begin with sums formed in a purely analytical way, like the upper and lower sums defined previously, and shall then prove that these sums tend to a definite limit. We take this limiting value as the *definition* of the integral and of the area. We are naturally led to adopt the formal symbols which have been used in the integral calculus since Leibnitz's time.

Let $f(x)$ be a function which is positive and continuous in the interval $a \leq x \leq b$ (of length $b - a$). We think of the interval as divided by $(n - 1)$ points $x_1, x_2, \ldots, x_{n-1}$ into n equal or unequal sub-intervals, and in addition we put $x_0 = a$, $x_n = b$. In each interval we choose a perfectly arbitrary point, which may be within the interval or at either end; suppose that in the first interval we choose the point ξ_1, in the second the point $\xi_2, \ldots$, in the n-th the point ξ_n. Instead of the continuous function $f(x)$ we now consider a discontinuous function (step-function) which has the constant value $f(\xi_1)$ in the first sub-interval, the constant value $f(\xi_2)$ in the second sub-interval, ...,

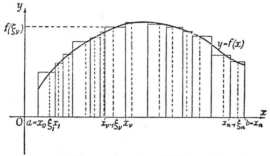

Fig. 3.—To illustrate the analytical definition of integral

the constant value $f(\xi_n)$ in the n-th sub-interval. As is shown in fig. 3, the graph of this step-function defines a series of rectangles, the sum of whose areas is given by the expression

$$F_n = (x_1 - x_0)f(\xi_1) + (x_2 - x_1)f(\xi_2) + \ldots + (x_n - x_{n-1})f(\xi_n).$$

This expression is usually shortened by using the summation sign Σ:

$$F_n = \sum_{\nu=1}^{n} (x_\nu - x_{\nu-1})f(\xi_\nu);$$

by introducing the symbol

$$\Delta x_\nu = x_\nu - x_{\nu-1}$$

we can abbreviate it still further:

$$F_n = \sum_{\nu=1}^{n} f(\xi_\nu)\Delta x_\nu.$$

(Here the symbol Δ is *not* a factor, but denotes " difference ". The whole inseparable symbol Δx_ν, by definition, means the length of the ν-th sub-interval.) Our basic assertion may now be stated as follows:

If we let the number of points of division increase without limit and at the same time let the length of the longest sub-interval tend to 0, then the above sum tends to a limit. This limit is independent of the particular manner in which the points of division x_1, x_2, . . . , x_{n-1} and the intermediate points ξ_1, ξ_2, . . . , ξ_n are chosen.

This limiting value we shall call the *definite integral* of the function $f(x)$, the *integrand*, between the limits a and b; as we have already mentioned, we shall consider it as the *definition* * of the area under the curve $y = f(x)$, for $a \leqq x \leqq b$. Our basic assertion may then be re-worded thus: If $f(x)$ is continuous in $a \leqq x \leqq b$ its definite integral between the limits a and b exists.

This theorem on the existence of the definite integral of a continuous function can be proved by purely analytical methods, without appealing to intuition. We shall nevertheless pass it over for the present and return to it in the Appendix to this chapter (p. 131), after the use of the concept of integral has stimulated the reader's interest in constructing a firm foundation for it. For the moment we shall content ourselves with the fact that the intuitive considerations on pp. 77–78 have made the theorem appear extremely plausible.

3. Extensions, Notation, Fundamental Rules.

The above definition of the integral as the limit of a sum led Leibnitz to express the integral by the following symbol:

$$\int_a^b f(x)\,dx.$$

The integral sign is a modification of a summation sign which had the shape of a long S. The passage to the limit from a sub-division of the interval into finite portions Δx_ν is suggested by the use of the letter d in place of Δ. We must, however, guard ourselves against thinking of dx as an " infinitely small quantity " or " infinitesimal ", or of the integral as the " sum of

---

* Of course we may also define the notion of area in a purely geometrical way, and then prove that such a definition is equivalent to the above limit-definition (cf. Chap. V, § 2, No. 1 (p. 268)).

an infinite number of infinitely small quantities ". Such a conception would be devoid of any clear meaning; it is only a naïve befogging of what we have previously carried out with precision.

In the above figures we have assumed (1) that the function $f(x)$ is positive throughout the interval, and (2) that $b > a$. The formula which defines the integral as the limit of a sum is, however, independent of any such assumptions. For if $f(x)$ is negative in all or part of our interval, the only effect is to make the corresponding factors $f(\xi_\nu)$ in our sum negative instead of positive. To the region bounded by the part of the curve below the x-axis we shall naturally assign a negative area, which is in agreement with the familiar convention of sign in analytical geometry. The total area bounded by a curve will thus in general be the sum of positive and negative terms, corresponding respectively to the portions of the curve above and below the x-axis.*

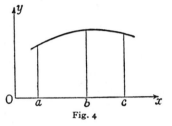

Fig. 4

If we also omit the condition $a < b$ and assume that $a > b$, we can still retain our arithmetical definition of integral; the only change is that when we traverse the interval from a to b the differences Δx_ν are negative. We are thus led to the relation

$$\int_a^b f(x)\,dx = -\int_b^a f(x)\,dx,$$

which holds for all values of a and b ($a \neq b$). In conformity with this we define $\int_a^a f(x)\,dx$ as equal to zero.

Our definition immediately gives the basic relation (see fig. 4):

$$\int_a^b f(x)\,dx + \int_b^c f(x)\,dx = \int_a^c f(x)\,dx$$

for $a < b < c$. By means of the preceding relations we at once find that this equation is also true for any position of the points a, b, c relative to one another.

We obtain a simple but important fundamental rule by

* For the area of regions bounded by arbitrary closed curves see Chap. V, § 2, p. 269.

4

considering the function $cf(x)$, where c is a constant. From the definition of the integral we immediately obtain

$$\int_a^b cf(x)\,dx = c\int_a^b f(x)\,dx.$$

Further, we assert the following addition rule: **If**

$$f(x) = \phi(x) + \psi(x),$$

then

$$\int_a^b f(x)\,dx = \int_a^b \phi(x)\,dx + \int_a^b \psi(x)\,dx.$$

The proof is quite simple.

We add a final remark, which is perfectly obvious, but very important in applications, about the " variable of integration ". We have written our integral in the form $\int_a^b f(x)\,dx$. For evaluating the integral it does not matter whether we use the letter x or any other letter to denote the abscissæ of the co-ordinate system, i.e. the independent variable. The particular symbol we use for the variable of integration is therefore a matter of complete indifference; instead of $\int_a^b f(x)\,dx$ we could equally well write $\int_a^b f(t)\,dt$ or $\int_a^b f(u)\,du$ or any similar expression.

2. Examples

We are now in a position to carry out the limiting process prescribed by our definition of the integral, and thus actually to calculate the area in question in a number of special cases; this we shall do in a series of examples, where (except in No. 5, p. 86) we shall make use of the upper or lower sum alone.*

1. Integration of a Linear Function.

We first consider the function $f(x) = x^n$, where n is an integer greater than or equal to 0. For $n = 0$, i.e. for $f(x) = 1$, the result is so obvious that we simply write it down:

$$\int_a^b 1\,dx = \int_a^b dx = b - a.$$

For the function $f(x) = x$ the integration is again a triviality from the geometrical point of view. The integral of the function $f(x) = x$,

$$\int_a^b x\,dx,$$

---

* We leave it as a useful exercise for the reader to prove that in the following examples we actually do arrive at the same result, whether we use the upper sum or the lower.

is simply the area of the trapezoid shown in fig. 5, which by an elementary formula has the value

$$\tfrac{1}{2}(b-a)(b+a) = \tfrac{1}{2}(b^2 - a^2).$$

We shall now verify that our limiting process leads to exactly the same result. In calculating the limit we can restrict ourselves to the discussion of upper sums or lower sums. We subdivide the interval from a to b into n equal parts by means of the points of division

$$a + h, \quad a + 2h, \quad \ldots, \quad a + (n-1)h,$$

where $h = (b-a)/n$. The integral must then be the limit of the following sum, which is an upper sum if $b < a$ and a lower sum if $b > a$:

$$h\{a + (a+h) + (a+2h) + \ldots + (a + \overline{n-1}\,h)\}$$
$$= h\{na + h + 2h + \ldots + (n-1)h\}.$$

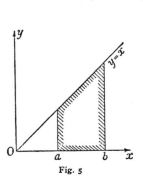

Fig. 5

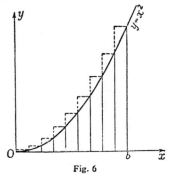

Fig. 6

By an elementary formula we have

$$1 + 2 + \ldots + (n-1) = \tfrac{1}{2}n(n-1),$$

and our expression may therefore be written in the form

$$nh\left(a + h\,\frac{n-1}{2}\right) = (b-a)\left(a + \frac{b-a}{2}\,\frac{n-1}{n}\right).$$

As n increases the right-hand side obviously tends to the limit

$$(b-a)\{a + \tfrac{1}{2}(b-a)\} = \tfrac{1}{2}(b^2 - a^2),$$

which was to be proved.

2. Integration of x^2.

A problem not quite so simple is that of integrating the function $f(x) = x^2$, or, in geometrical language, of determining the area of the region bounded by a segment of a parabola, a segment of the x-axis and two ordinates. We consider e.g. the integral

$$\int_0^b x^2\,dx,$$

where $b \geqq 0$ (see fig. 6), and divide the interval $0 \leqq x \leqq b$ into n equal parts of length $h = b/n$; the area which we wish to find is then the limit of the following expression (upper sum):

$$h(h^2 + 2^2h^2 + 3^2h^2 + \ldots + n^2h^2) = h^3(1^2 + 2^2 + \ldots + n^2)$$
$$= b^3(1^2 + 2^2 + \ldots + n^2)/n^3.$$

The sum enclosed in brackets, however, has already been found (see p. 27, footnote):

$$1^2 + 2^2 + \ldots + n^2 = \tfrac{1}{6}n(n + 1)(2n + 1).$$

If we substitute this expression and rewrite the result in a slightly different form, our sum becomes

$$\frac{b^3}{6}\left(1 + \frac{1}{n}\right)\left(2 + \frac{1}{n}\right).$$

As n increases beyond all bounds this tends to the limit $\tfrac{1}{3}b^3$, and we obtain the required integral formula

$$\int_0^b x^2 \, dx = \tfrac{1}{3}b^3.$$

From this, using the general relationships given above, we immediately derive the general formula

$$\int_a^b x^2 \, dx = \int_0^b x^2 \, dx - \int_0^a x^2 \, dx = \tfrac{1}{3}(b^3 - a^3).$$

3. Integration of x^α, where α is any Positive Integer.

As a third example we consider the integration of the function

$$y = f(x) = x^\alpha,$$

where α is any positive integer. For the computation of the integral

$$\int_a^b x^\alpha \, dx$$

(where we assume $0 < a < b$) it would be inconvenient to divide the interval into n equal parts.* The passage to the limit may, however, be accomplished very easily if we effect a subdivision in " geometric progression " in the following manner. We put $\sqrt[n]{b/a} = q$ and subdivide the interval by the points

$$a, \ aq, \ aq^2, \ \ldots, \ aq^{n-1}, \ aq^n = b.$$

* We should then be obliged to base the evaluation of the integral upon the calculation of the limit of $\dfrac{1}{n^{\alpha+1}}(1^\alpha + 2^\alpha + \ldots + n^\alpha)$ as $n \to \infty$; the reader may work this out for himself as indicated in the footnote on p. 27.

The required integral is then the limit of the following sum:

$$a^\alpha(aq - a) + (aq)^\alpha(aq^2 - aq) + (aq^2)^\alpha(aq^3 - aq^2) + \cdots$$
$$+ (aq^{n-1})^\alpha(aq^n - aq^{n-1})$$
$$= a^{\alpha+1}(q-1)\{1 + q^{\alpha+1} + q^{2(\alpha+1)} + q^{3(\alpha+1)} + \cdots + q^{(n-1)(\alpha+1)}\}.$$

The terms in the last bracket form a geometric progression with common ratio $q^{\alpha+1} \neq 1$. If we sum this progression, we obtain for the whole expression the value

$$a^{\alpha+1}(q-1)\frac{q^{n(\alpha+1)} - 1}{q^{\alpha+1} - 1}.$$

We now replace q by its value $(b/a)^{1/n}$; our sum then takes the form

$$(b^{\alpha+1} - a^{\alpha+1})\frac{q-1}{q^{n+1} - 1}.$$

If we now let n increase without limit, the first factor retains its value. Since $q \neq 1$ we can use the formula for the sum of a geometric progression and write the second factor in the form

$$\frac{1}{q^\alpha + q^{\alpha-1} + \cdots + 1};$$

and as the equation $q = (b/a)^{1/n}$ shows that q tends to 1 as $n \to \infty$, the second factor will have the limit $1/(\alpha + 1)$. Thus finally the value of our integral is given by the formula

$$\int_a^b x^\alpha\, dx = \frac{1}{\alpha + 1}(b^{\alpha+1} - a^{\alpha+1}).$$

The above calculation is simple in principle, but somewhat complicated in detail. We shall later find that it can be entirely avoided, once we are better acquainted with integration theory.

4. Integration of x^α, where α is any Rational Number other than -1.

The result obtained above may be generalized considerably without essentially complicating the method. Let $\alpha = r/s$ be a positive rational number, r and s being positive integers; then in the evaluation of the integral given above nothing is changed except the evaluation of the limit $\frac{q-1}{q^{\alpha+1} - 1}$ as q approaches 1. This expression is now simply $\frac{q-1}{q^{(r+s)/s} - 1}$. Let us put $q^{1/s} = \tau\ (\tau \neq 1)$: then as q tends to 1, τ will also tend to 1. We have therefore to find the limiting value of $\frac{\tau^s - 1}{\tau^{r+s} - 1}$ as τ approaches 1. If we divide both numerator and denominator by $\tau - 1$ and transform them as before the limit simply becomes

$$\lim_{\tau \to 1}\frac{\tau^{s-1} + \tau^{s-2} + \cdots + 1}{\tau^{r+s-1} + \tau^{r+s-2} + \cdots + 1}.$$

Since both numerator and denominator are continuous in τ this limit is at once determined if we substitute $\tau = 1$. We thus arrive at the limit $\dfrac{s}{r+s} = \dfrac{1}{\alpha+1}$; and so for every positive rational value of α we obtain the integral formula

$$\int_a^b x^\alpha \, dx = \frac{1}{\alpha+1} (b^{\alpha+1} - a^{\alpha+1}).$$

This formula remains valid for negative rational values of α, provided we exclude the value $\alpha = -1$ for which the formula used above for the sum of the geometric progression loses its meaning. We have now to investigate the limit of the expression $\dfrac{q-1}{q^{\alpha+1}-1}$ for negative values of α, say $\alpha = -r/s$. To do this we put $q^{-1/s} = \tau$; we obtain

$$q = \tau^{-s}, \quad q^{\alpha+1} = q^{-(r-s)/s} = \tau^{r-s}.$$

We accordingly seek to determine the limiting value of

$$\frac{\tau^{-s}-1}{\tau^{r-s}-1} = \frac{1-\tau^s}{\tau^r - \tau^s}.$$

We leave it to the reader to prove that this limit is again equal to $\dfrac{1}{\alpha+1}$; that is, that we have the integral formula

$$\int_a^b x^\alpha \, dx = \frac{1}{\alpha+1} (b^{\alpha+1} - a^{\alpha+1})$$

for the general case of rational values of α either positive or negative, with the exception of $\alpha = -1$.

The form of the right-hand side of this equation shows that the expression is not valid for $\alpha = -1$, since both numerator and denominator would then be zero.

It is natural to suppose that the range of validity of our last formula extends also to irrational values of α. We shall actually establish this in § 7 (p. 129) by a simple passage to the limit.

5. Integration of $\sin x$ and $\cos x$.

As a last example we consider the function $f(x) = \sin x$. This too we shall treat by means of a special device. We express the integral

$$\int_a^b \sin x \, dx$$

as the limit of the following sum:

$$S_h = h\{\sin(a+h) + \sin(a+2h) + \ldots + \sin(a+nh)\},$$

where $h = \dfrac{b-a}{n}$. We multiply the right-hand bracket by $2\sin\dfrac{h}{2}$, and recall the well-known trigonometrical formula

$$2 \sin u \sin v = \cos(u - v) - \cos(u + v);$$

provided h is not a multiple of 2π, we thus obtain the formula

$$S_h = \frac{h}{2 \sin \frac{h}{2}} \left\{ \cos\left(a + \frac{h}{2}\right) - \cos\left(a + \frac{3}{2}h\right) + \cos\left(a + \frac{3}{2}h\right) - \cos\left(a + \frac{5}{2}h\right) \right.$$
$$\left. + \ldots + \cos\left(a + \frac{2n-1}{2}h\right) - \cos\left(a + \frac{2n+1}{2}h\right) \right\}$$

$$= \frac{h}{2 \sin \frac{h}{2}} \left\{ \cos\left(a + \frac{h}{2}\right) - \cos\left(a + \frac{2n+1}{2}h\right) \right\}.$$

Since $a + nh = b$, the integral becomes the limit of

$$\frac{h}{2 \sin \frac{h}{2}} \left\{ \cos\left(a + \frac{h}{2}\right) - \cos\left(b + \frac{h}{2}\right) \right\} \text{ as } h \to 0.$$

Now we know from Chapter I (p. 47) that as h tends to 0, the expression $\frac{h}{2}/\sin\frac{h}{2}$ approaches the limit 1. The desired limit is then simply $\cos a - \cos b$, and we thus arrive at the integral formula

$$\int_a^b \sin x \, dx = -(\cos b - \cos a).$$

In the same way, as the reader may verify for himself, we obtain the formula

$$\int_a^b \cos x \, dx = \sin b - \sin a.$$

Almost every one of these examples has been attacked by means of some special method or particular device. The essential point of the systematic integral and differential calculus is the very fact that instead of such special devices we utilize considerations of a general character which lead us directly to the desired result. In order to arrive at these methods we must now turn our attention to the other fundamental concept of higher analysis, the derivative.

EXAMPLES

1. Find the area bounded by the parabola $y = 2x^2 + x + 1$, the ordinates $x = 1$ and $x = 3$, and the x-axis.

2. Find the area bounded by the parabola $y = \frac{1}{2}x^2 + 1$ and the straight line $y = 3 + x$.

3. Find the area bounded by the parabola $y^2 = 5x$ and the straight line $y = 1 + x$.

4. Find the area bounded by the parabola $y = x^2$ and the straight line $y = ax + b$.

5. Using the methods in the text, evaluate the integrals

$$(a) \int_a^b (x + 1)^\alpha \, dx, \quad (b) \int_a^b \sin \alpha x \, dx, \quad (c) \int_a^b \cos \alpha x \, dx,$$

where α is an arbitrary integer.

6. Use the formulæ obtained in Ex. 5, along with the identities $\sin^2 x = \frac{1}{2} - \frac{1}{2} \cos 2x$, $\cos^2 x = \frac{1}{2} + \frac{1}{2} \cos 2x$, to prove that

$$\int_a^b \cos^2 x \, dx = \frac{b - a}{2} + \frac{\sin 2b - \sin 2a}{4};$$

$$\int_a^b \sin^2 x \, dx = \frac{b - a}{2} - \frac{\sin 2b - \sin 2\alpha}{4}.$$

7. By use of Ex. 1, p. 28, evaluate $\int_a^b x^3 \, dx$, using division into equal sub-intervals.

8. Evaluate $\int_0^1 (1 - x)^n \, dx$ (where n is an integer) by expanding the bracket.

3. The Derivative

The concept of the derivative, like that of the integral, is of intuitive origin. Its sources are (1) the problem of constructing the *tangent* to a given curve at a given point and (2) the problem of finding a precise definition for the *velocity* in an arbitrary motion.

1. The Derivative and the Tangent.

We shall first take up the tangent problem. If P is a point on a given curve (see fig. 7), we shall, in conformity with naïve intuition, define the tangent to the curve at the point P by means of the following geometrical limiting process. In addition to the point P, we consider a second point P_1 on the curve. Through the two points P, P_1 we draw a straight line, a secant of the curve. If we now let the point P_1 move along the curve towards the point P, this secant will tend to a limiting position which is independent of the side from which P_1 approaches P. This limiting position of the secant is the tangent, and the statement that such a limiting position of the secant exists is equivalent to the assumption that the curve has a definite tangent or a definite direction at the point P. (We have used the word " assumption " because we have actually made one. The hypothesis that the tangent exists is valid for most simple curves, but is by no means true for all curves, or even for all continuous curves.)

Once we have represented our curve by means of a function $y = f(x)$ the problem arises of representing our geometrical limiting process analytically, using the function $f(x)$. We take the angle which a straight line l makes with the x-axis as being the angle through which the positive x-axis must be turned in the positive direction* in order to become for the first time parallel to the line l. Let a_1 be the angle which the secant PP_1 forms with the

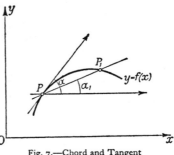

Fig. 7.—Chord and Tangent

positive x-axis (cf. fig. 7) and a the angle which the tangent forms with the positive x-axis. Then if we disregard the case of a perpendicular tangent we obviously have

$$\lim_{P_1 \to P} a_1 = a,$$

where the meaning of the symbols is perfectly clear. If $x, y (= f(x))$ and $x_1, y_1 (= f(x_1))$ are the co-ordinates of the points P and P_1 respectively, we immediately have †

$$\tan a_1 = \frac{y_1 - y}{x_1 - x} = \frac{f(x_1) - f(x)}{x_1 - x};$$

and thus our limiting process is represented by the equation

$$\lim_{x_1 \to x} \frac{f(x_1) - f(x)}{x_1 - x} = \tan a.$$

* That is, in such a direction that a rotation of $\pi/2$ brings it into coincidence with the positive y-axis; in other words, counter-clockwise.

† In order that this equation may have a meaning, we must assume that $0 < |x - x_1| < \delta$, δ being chosen sufficiently small. In what follows, corresponding assumptions will often be made tacitly in the steps leading up to limiting processes.

4*

The expression

$$\frac{f(x_1) - f(x)}{x_1 - x} = \frac{y_1 - y}{x_1 - x} = \frac{\Delta y}{\Delta x}$$

we call the *difference quotient* of the function $y = f(x)$, since the symbols Δy and Δx denote the differences of the function $y = f(x)$ and of the independent variable x. (Here, as on p. 79, the symbol Δ is an abbreviation for difference, and is *not* a factor.) The tangent of α, the direction angle of the curve,* is therefore equal to the limit to which the difference quotient of our function tends when x_1 tends to x.

We call this limit the *derivative* † of the function $y = f(x)$ at the point x and, as Lagrange did, use the symbol $y' = f'(x)$ to denote it, or, as Leibnitz did, the symbol ‡ $\dfrac{dy}{dx}$ or $\dfrac{df(x)}{dx}$ or $\dfrac{d}{dx} f(x)$. On p. 100 we shall discuss the meaning of Leibnitz's notation in more detail; here we would point out that the notation $f'(x)$ expresses the fact that the *derivative is itself a function of x*, since it has a definite value for each value of x in the interval which we are considering. This fact is sometimes emphasized by the use of the terms *derived function, derived curve* (see p. 99). We again quote the definition of the derivative:

$$f'(x) = \lim_{x_1 \to x} \frac{f(x_1) - f(x)}{x_1 - x},$$

or

$$\frac{dy}{dx} = \frac{df(x)}{dx} = f'(x) = \lim_{x_1 \to x} \frac{f(x_1) - f(x)}{x_1 - x} = \lim_{\Delta x \to 0} \frac{\Delta y}{\Delta x}$$

$$= \lim_{h \to 0} \frac{f(x + h) - f(x)}{h},$$

where in the last expression we have replaced x_1 by $x + h$.

It is impossible to find the derivative merely by putting $x_1 = x$ in the expression for the difference quotient, for then the numerator and denominator would both be equal to 0 and we

---

* The slope or gradient of the curve is given by $\tan \alpha$, and hence the term *gradient* is occasionally used for the derivative of the function represented by the curve.

† The term *differential coefficient* is also used, particularly in the older text-books.

‡ Cauchy's notation $Df(x)$ is also occasionally found in the literature.

should be led to the meaningless expression 0/0. On the contrary, the actual performance of the passage to the limit in each individual case depends upon certain preliminary steps (transformation of the difference quotient).

For example, for the function $f(x) = x^2$ we have *

$$\frac{f(x_1) - f(x)}{x_1 - x} = \frac{x_1^2 - x^2}{x_1 - x} = x_1 + x.$$

This function $x_1 + x$ is not the same function as $\dfrac{x_1^2 - x^2}{x_1 - x}$, for the function $x_1 + x$ is defined at one point where the quotient $\dfrac{x_1^2 - x^2}{x_1 - x}$ is undefined, namely, the point $x_1 = x$. For all other values of x_1 the two functions are equal to one another; hence in the above passage to the limit, in which we specifically required that $x_1 \neq x$, we obtain the same value for $\lim\limits_{x_1 \to x} \dfrac{x_1^2 - x^2}{x_1 - x}$ as for $\lim\limits_{x_1 \to x} (x_1 + x)$. But since the function $x_1 + x$ is defined and continuous at the point $x_1 = x$, we can do with it what we could not do with the quotient, namely, pass to the limit by simply putting $x_1 = x$. For the derivative we then obtain the expression

$$f'(x) = \frac{d(x^2)}{dx} = 2x.$$

The carrying out of such a process, i.e. the actual formation of the derivative, is called the *differentiation* of the function $f(x)$. We shall see later how this process of differentiation can actually be carried out in all important cases.

Now the fact that the problem of differentiating a given function has a definite meaning apart from the geometrical intuition of the tangent is of great significance. The reader will recall that in the case of the integral we freed ourselves from the geometrical intuition of area, and on the contrary based the notion of area on the definition of the integral. Now, independently of the geometrical representation of a function $y = f(x)$ by means of a curve, we shall define the derivative of the function $y = f(x)$ as being the new function $y' = f'(x)$ given by the equation above, provided always that the limit of the difference quotient exists. If this limit exists we say that the function $f(x)$ is *differentiable*. From now on we shall assume that every function dealt with is differentiable unless specific men-

* Cf. p. 89, second footnote.

tion is made to the contrary.* It should be observed that if the function $f(x)$ is to be differentiable at the point x the limit as $h \to 0$ of the quotient $\dfrac{f(x + h) - f(x)}{h}$ must exist *independently of the manner in which* h *tends to* 0, whether it be through positive values or through negative values or without restriction as to sign.

Once we have found the derivative $f'(x)$, we take the direction which makes an angle α with the positive x-axis given by the equation $\tan \alpha = f'(x)$ as the direction of the tangent to the curve at the point (x, y). We thus avoid the difficulties which arise out of the indefiniteness of the geometrical view, since we

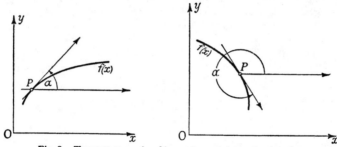

Fig. 8.—Tangents to graphs of increasing and decreasing functions

base the geometrical definition on the analytical and not vice versa.

Nevertheless, the visualization of the derivative as the tangent to the curve is an important aid to understanding, even in purely analytical discussions. Accordingly we shall at once accept the following statement based on geometrical intuition:

If f'(x) *is positive and the curve is traversed in the direction of increasing* x, *then the tangent slants upwards, and therefore at the point in question the curve rises as* x *increases; if, on the other hand,* f'(x) *is negative, the tangent slants downwards and the curve falls as* x *increases* (see fig. 8). Analytically this follows from the remark that the limit of $\dfrac{f(x + h) - f(x)}{h}$ cannot be positive unless the function is increasing at the point x, by which we

---

* Examples of cases in which this assumption is not satisfied will be given later (see p. 97).

mean that for all values of h sufficiently close to 0 the value of $f(x+h)$ is greater or smaller than $f(x)$ according as h is positive or negative. We can, of course, make a corresponding statement for the case where $f'(x)$ is negative.

2. The Derivative as a Velocity.

Just as naïve intuition led us to the notion of the direction of the tangent to a curve, so it causes us to assign a *velocity* to a *motion*. The definition of velocity leads us once again to exactly the same limiting process as we have already called differentiation.

Let us consider, for example, the motion of a point on a straight line, the position of the point being determined by a single co-ordinate y. This co-ordinate y is the distance, with its proper sign, of our moving point from a fixed point on the line. The motion is given if we know y as a function of the time t, $y = f(t)$. If this function is a linear function $f(t) = ct + b$, we call the motion a *uniform motion with the velocity c*, and for every pair of values t and t_1 which are not equal to one another we can write

$$c = \frac{f(t_1) - f(t)}{t_1 - t}.$$

The velocity is therefore the difference quotient of the function $ct + b$, and this difference quotient is completely independent of the particular pair of instants which we fix upon. But what are we to understand by the velocity of motion at an instant t if the motion is no longer uniform?

In order to arrive at this definition we consider the difference quotient $\frac{f(t_1) - f(t)}{t_1 - t}$, which we shall call the average velocity in the time interval between t_1 and t. If now this average velocity tends to a definite limit when we let the instant t_1 come closer and closer to t, we shall naturally define this limit as the velocity at the time t. In other words: *the velocity at the time t is the derivative*

$$f'(t) = \lim_{t_1 \to t} \frac{f(t_1) - f(t)}{t_1 - t}.$$

From this new meaning of the derivative, which in itself has nothing to do with the tangent problem, we see that it really is appropriate to define the limiting process of differentiation as a purely analytical operation independent of geometrical intuitions. Here again the differentiability of the position-function is an assumption which we shall always tacitly make, and which, in fact, is absolutely necessary if the notion of velocity is to have any meaning.

As a simple example of the connexion between motion and velocity we consider the case of a freely falling body. We begin with the experimentally established law that the distance traversed in time t by a freely falling body is proportional to t^2, and therefore can be represented by a function of the form

$$y = f(t) = at^2.$$

As on p. 91, we find immediately that the velocity is given by the expression $f'(t) = 2at$, which shows us that the velocity of a freely falling body increases in proportion to the time.

3. Examples.

We now proceed to work out a number of examples of the actual differentiation of functions. We begin with the function $y = f(x) = c$, where c is a constant. It is then always true that $f(x + h) - f(x) = c - c = 0$, so that $\lim\limits_{h \to 0} \dfrac{f(x + h) - f(x)}{h} = 0$; that is, *the derivative of a constant is zero.*

For a linear function $y = f(x) = cx + b$, we find that

$$\lim_{h \to 0} \frac{f(x + h) - f(x)}{h} = \lim_{h \to 0} \frac{ch}{h} = c.$$

Further, we shall differentiate the function

$$y = f(x) = x^\alpha,$$

at first assuming that α is a positive integer. Provided $x_1 \neq x$, we have

$$\frac{f(x_1) - f(x)}{x_1 - x} = \frac{x_1^\alpha - x^\alpha}{x_1 - x};$$

the right-hand side of this equation is equal to $x_1^{\alpha-1} + x_1^{\alpha-2}x + \ldots + x^{\alpha-1}$, as we see either by direct division or by using the formula for the sum of a geometric progression. The new expression for the right-hand side of the equation is a continuous function, and so we can carry out the passage to the limit ($x_1 \to x$) by simply replacing x_1 everywhere by x. Each term is then $x^{\alpha-1}$, and since the number of terms is exactly α, we obtain

$$y' = f'(x) = \frac{d(x^\alpha)}{dx} = \alpha x^{\alpha-1}.$$

We arrive at the same result if α is a negative integer $-\beta$; we must, however, assume that x is not zero. We then find that

$$\frac{f(x_1) - f(x)}{x_1 - x} = \frac{\frac{1}{x_1^\beta} - \frac{1}{x^\beta}}{x_1 - x} = -\frac{x^\beta - x_1^\beta}{x - x_1} \cdot \frac{1}{x^\beta x_1^\beta}$$

$$= -\frac{x^{\beta-1} + x^{\beta-2}x_1 + \ldots + x_1^{\beta-1}}{x_1^\beta x^\beta}.$$

Once again we can carry out the passage to the limit simply by replacing x_1 everywhere by x. Then just as above we obtain for the limit the expression

$$y' = -\beta \frac{x^{\beta-1}}{x^{2\beta}} = -\beta x^{-\beta-1}.$$

Hence for *negative* integral values of α the derivative is again given by

$$y' = \alpha x^{\alpha-1}.$$

Finally, we shall prove the same formula where x is positive and α any rational number. We suppose that $\alpha = p/q$, where p and q are both integers and, moreover, positive. (If one of them were negative no essential changes in the proof would be needed; for $\alpha = 0$ the result is already known, since x^α is then constant.) We now have

$$\frac{f(x_1) - f(x)}{x_1 - x} = \frac{x_1^{p/q} - x^{p/q}}{x_1 - x}.$$

If we now put $x^{1/q} = \xi$ and $x_1^{1/q} = \xi_1$, we obtain

$$\frac{f(x_1) - f(x)}{x_1 - x} = \frac{\xi_1^p - \xi^p}{\xi_1^q - \xi^q} = \frac{\xi_1^{p-1} + \xi_1^{p-2}\xi + \ldots + \xi^{p-1}}{\xi_1^{q-1} + \xi_1^{q-2}\xi + \ldots + \xi^{q-1}}.$$

After this last transformation we can immediately perform the passage to the limit ($x_1 \to x$ or, what amounts to the same thing, $\xi_1 \to \xi$), and thus obtain for the limiting value the expression

$$y = \frac{p}{q} \frac{\xi^{p-1}}{\xi^{q-1}} = \frac{p}{q} \xi^{p-q} = \frac{p}{q} x^{(p-q)/q} = \frac{p}{q} x^{p/q-1},$$

or finally

$$f'(x) = y' = \alpha x^{\alpha-1},$$

which is formally the same result as before. We leave it for the reader to prove for himself that the same differentiation formula holds for negative rational indices also. We shall come back (p. 130) to the differentiation of powers once we have developed the theory in a more connected form.

As a further example we finally consider the differentiation of the trigonometric functions, $\sin x$ and $\cos x$. We use the elementary trigonometrical formula

$$\frac{\sin(x+h) - \sin x}{h} = \frac{\sin x \cos h + \cos x \sin h - \sin x}{h}$$

$$= \sin x \cdot \frac{\cos h - 1}{h} + \cos x \frac{\sin h}{h}.$$

Now from Chapter I, § 7 (pp. 47–48). we know that

$$\lim_{h \to 0} \frac{\sin h}{h} = 1, \quad \lim_{h \to 0} \frac{\cos h - 1}{h} = 0.$$

For the required derivative we thus immediately obtain

$$y' = \frac{d(\sin x)}{dx} = \cos x.$$

The function $y = \cos x$ can be differentiated in exactly the same way. Starting with

$$\frac{\cos(x+h) - \cos x}{h} = \cos x \frac{\cos h - 1}{h} - \sin x \frac{\sin h}{h},$$

and taking the limit as $h \to 0$, we at once obtain the derivative

$$y' = \frac{d(\cos x)}{dx} = -\sin x.$$

4. Some Fundamental Rules for Differentiation.

Just as in the case of the integral, certain simple but fundamental rules for forming the derivative follow immediately from the definition. If $\phi(x) = f(x) + g(x)$, then $\phi'(x) = f'(x) + g'(x)$; again, if $\psi(x) = cf(x)$ (where c is a constant), then $\psi'(x) = cf'(x)$. For we have

$$\frac{\phi(x+h) - \phi(x)}{h} = \frac{f(x+h) - f(x)}{h} + \frac{g(x+h) - g(x)}{h}$$

and

$$\frac{\psi(x+h) - \psi(x)}{h} = c\frac{f(x+h) - f(x)}{h},$$

and our statements follow directly by passage to the limit.

According to these rules, for example, the derivative of the function $\phi(x) = f(x) + ax + b$ (where a and b are constants) is given by the equation

$$\phi'(x) = f'(x) + a.$$

5. Differentiability and Continuity of Functions.

It is useful to note that if we know that a function can be differentiated we need not give any special proof of its continuity.

If a function is differentiable, then it is necessarily continuous.

For if the difference quotient $\dfrac{f(x + h) - f(x)}{h}$ approaches a definite limit as h tends to zero, the numerator of the fraction, that is, $f(x + h) - f(x)$, must tend to zero with h; and this fact expresses the continuity of the function $f(x)$ at the point x.

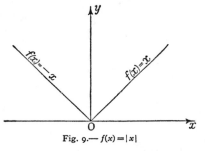

Fig. 9.— $f(x) = |x|$

The converse of this, however, is absolutely false; it is *not* true that every continuous function has a derivative at every point. The simplest example disproving this assumption is the function $f(x) = |x|$, i.e. $f(x) = -x$ for $x \leqq 0$ and $f(x) = x$ for $x \geqq 0$; its graph is shown in fig. 9. At the point $x = 0$ this function is continuous, but has no derivative. The limit of $\dfrac{f(x + h) - f(x)}{h}$ is equal to 1 if h tends to 0 through positive values, and is equal to -1 if h tends to zero through negative values; if we do not restrict the sign of h, no limit exists. We say that our function has different *right-hand* and *left-hand* derivatives at the point x, where by right-hand derivative and left-hand derivative we mean respectively the limiting values of $\dfrac{f(x + h) - f(x)}{h}$ as h approaches 0 through positive values only and negative values only. The *differentiability* of a function thus requires not merely that the right-hand and left-hand derivatives exist, but that they are equal. Geometrically the inequality of the two derivatives means that the curve has a sharp corner.

As further examples of points where a continuous function is not differentiable we consider the points where the derivative becomes infinite, i.e. the points at which there exists neither a right-hand nor a left-hand derivative, the difference quotient $\dfrac{f(x + h) - f(x)}{h}$ increasing beyond all bounds as $h \to 0$. For example, the function $y = f(x) = \sqrt[3]{x} = x^{\frac{1}{3}}$ is defined and continuous for all values of x. For all non-zero values of x its derivative is given (p. 95) by the formula $y' = \frac{1}{3} x^{-\frac{2}{3}}$. At the point $x = 0$ we have $\dfrac{f(x + h) - f(x)}{h} = \dfrac{h^{\frac{1}{3}}}{h} = h^{-\frac{2}{3}}$, and we see at once that as $h \to 0$ the expression has no limiting value, but, on the contrary, tends to ∞. This state of affairs is often briefly described by saying that the function possesses an infinite deri-

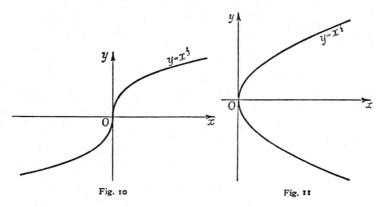

Fig. 10   Fig. 11

vative, or the derivative ∞, at the point in question; we should remember, however, that this merely means that as h tends to 0 the difference quotient increases beyond all bounds, and that the derivative in the sense in which we have defined it really does not exist. The geometrical meaning of an infinite derivative is that the tangent to the curve is vertical (cf. fig. 10).

The function $y = f(x) = \sqrt{x}$, which is defined and continuous for $x \geqq 0$, is also non-differentiable at the point $x = 0$. Since y is undefined for negative values of x, we here consider the right-hand derivative only. The equation $\dfrac{f(h) - f(0)}{h} = \dfrac{1}{\sqrt{h}}$ shows us that this derivative is infinite; the curve touches the y-axis at the origin (fig. 11).

Finally, in the function $y = \sqrt[3]{x^2} = x^{\frac{2}{3}}$ we have a case in which the right-hand derivative at the point $x = 0$ is positive

and infinite, while the left-hand derivative is negative and
infinite, as follows from the relation

$$\frac{f(h) - f(0)}{h} = \frac{1}{\sqrt[3]{h}}.$$

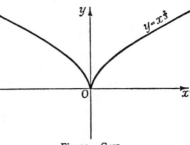

Fig. 12.—Cusp

As a matter of fact, the
continuous curve $y = x^{\frac{2}{3}}$,
the so-called *semi-cubical
parabola* or *Neil's parabola*, has at the origin a
cusp perpendicular to the
x-axis (cf. fig. 12).

6. Higher Derivatives and their Significance.

The derivative $f'(x)$ of a function is itself a function of x,
the graph of which we call the *derived curve* of the given curve.
For example, the derived curve of the parabola $y = x^2$ is a

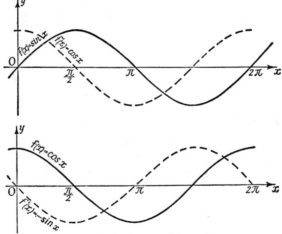

Fig. 13.—Derived curves of sin x and cos x

straight line, represented by the function $y = 2x$. The derived
curve of the sine curve $y = \sin x$ is the cosine curve $y = \cos x$;
similarly, the derived curve of the curve $y = \cos x$ is the curve
$y = -\sin x$. (Any of these latter curves can be obtained from
the others by translation in the direction of the x-axis, as is
shown in fig. 13.)

It is now quite a natural step to form the derived curves of the derived curves, i.e. to form the derivative of the function $f'(x) = \phi(x)$. This derivative

$$\phi'(x) = \lim_{h \to 0} \frac{f'(x + h) - f'(x)}{h},$$

provided that it really exists, we shall call the *second derivative* of the function $f(x)$, and we shall denote it by $f''(x)$.

Similarly, we may attempt to form the derivative of $f''(x)$, the so-called *third derivative* of $f(x)$, which we then denote by $f'''(x)$. In the case of most functions of importance there is nothing to hinder us from imagining this process repeated as many times as we like, and from thus defining an n-th *derivative* $f^{(n)}(x)$. Occasionally it will be convenient to call the function $f(x)$ its own 0-th derivative.*

If the independent variable is interpreted as the time t and the motion of a point is represented by means of the function $f(t)$, the *physical* meaning of the second derivative is found to be the velocity with which the velocity $f'(t)$ changes, or, as it is usually called, the *acceleration*. Later (pp. 158–159) we shall discuss the geometrical interpretation of the second derivative in detail. Here, however, we may note the following facts: at a point where $f''(x)$ is positive, $f'(x)$ increases as x increases; if, on the other hand, $f''(x)$ is negative, $f'(x)$ decreases as x increases.

7. The Derivative and the Difference Quotient.

The fact that in the limiting process which defines the derivative the difference Δx tends to 0 is sometimes expressed by saying that the quantity Δx *becomes infinitely small*. This expression indicates that the passage to the limit is regarded as a process during which the quantity Δx is never zero, yet approaches zero as closely as we please. In Leibnitz's notation the passage to the limit in the process of differentiation is symbolically expressed by replacing the symbol Δ by the symbol d, so that we can define Leibnitz's symbol for the derivative by the equation

$$\frac{dy}{dx} = \lim_{\Delta x \to 0} \frac{\Delta y}{\Delta x}.$$

* The terms *second, third,* . . . , *n-th differential coefficient* are also used; cf. second footnote, p. 90.

If, however, we wish to obtain a clear grasp of the meaning of the differential calculus we must beware of regarding the derivative as the quotient of two quantities which are actually "infinitely small". The *difference quotient* $\dfrac{\Delta y}{\Delta x}$ absolutely must be formed with differences Δx which are not equal to 0. *After* the forming of this difference quotient we must imagine the passage to the limit carried out by means of a transformation or some other device. We have no right to suppose that *first* Δx goes through something like a limiting process and reaches a value which is infinitesimally small but still not 0, so that Δx and Δy are replaced by "infinitely small quantities" or "infinitesimals" dx and dy, and that the quotient of these quantities is then formed. Such a conception of the derivative is incompatible with the clarity of ideas demanded in mathematics; in fact, it is entirely meaningless. For a great many simpleminded people it undoubtedly has a certain charm, the charm of mystery which is always associated with the word "infinite"; and in the early days of the differential calculus even Leibnitz himself was capable of combining these vague mystical ideas with a thoroughly clear understanding of the limiting process. It is true that this fog which hung round the foundations of the new science did not prevent Leibnitz or his great successors from finding the right path. But this does not release us from the duty of avoiding every such hazy idea in our building-up of the differential and integral calculus.

The notation of Leibnitz, however, is not merely attractive in itself, but is actually of great flexibility and the utmost usefulness. The reason is that in many calculations and formal transformations we can deal with the *symbols* dy *and* dx *in exactly the same way as if they were ordinary numbers.* They enable us to give neater expression to many calculations which can be carried out without their use. In the following pages we shall see this fact verified over and over again, and shall find ourselves justified in making free and repeated use of it, provided we do not lose sight of the symbolical character of the signs *dy* and *dx*.

For the second and higher derivatives also Leibnitz has devised a notation of great suggestiveness and practical utility. He thinks of the second derivative as the limit of the "second difference quotient " in the following manner. In addition to the variable x we consider $x_1 = x + h$ and $x_2 = x + 2h$. We then take the second difference quotient as meaning the first difference quotient of the first difference quotient, i.e. the expression

$$\frac{1}{h}\left(\frac{y_2 - y_1}{h} - \frac{y_1 - y}{h}\right) = \frac{1}{h^2}(y_2 - 2y_1 + y),$$

where $y = f(x)$, $y_1 = f(x_1)$, and $y_2 = f(x_2)$. If we also write $h = \Delta x$ and $y_2 - y_1 = \Delta y_1$, $y_1 - y = \Delta y$, we may appropriately call the expression in the last bracket the difference of the difference of y or the *second difference* of y and write symbolically *

$$y_2 - 2y_1 + y = \Delta y_1 - \Delta y = \Delta(\Delta y) = \Delta^2 y.$$

In this symbolic notation the second difference quotient is then written $\dfrac{\Delta^2 y}{(\Delta x)^2}$, where the denominator is really the square of Δx, while in the numerator the number 2 symbolically denotes the repetition of the difference process. This symbolism for the difference quotient † led Leibnitz to introduce the notation

$$y'' = f''(x) = \frac{d^2 y}{dx^2}, \ \ y''' = f'''(x) = \frac{d^3 y}{dx^3}, \ \ \&c.,$$

for the second and higher derivatives, and we shall find that this notation also stands the test of use.

8. The Mean Value Theorem.

Between the derivative $\dfrac{dy}{dx} = f'(x)$ and the difference quotient there exists a simple relation which is important for many purposes. This relation is known as the *mean value theorem*,

---

* Here $\Delta\Delta = \Delta^2$ is not a square, but merely a symbol for " difference of difference " or " second difference ".

† We must emphasize that the statement that the second derivative may be represented as the limit of the second difference quotient requires proof. For we previously defined the second derivative, not in this way, but as the limit of the first difference quotient of the first derivative. In actual fact, the two definitions are equivalent, provided the second derivative is continuous; the proof, however, is not given, as we have no particular need of it here.

and is obtained in the following way. We consider the difference quotient

$$\frac{f(x_1) - f(x_2)}{x_1 - x_2} = \frac{\Delta f}{\Delta x}$$

of a function $f(x)$, and assume that the derivative exists everywhere in the interval $x_1 \leqq x \leqq x_2$, so that the graph of the curve has a tangent everywhere. The difference quotient will be represented by the direction of the secant (see fig. 14); it is, in fact, the tangent of the angle α shown in the figure. Let us imagine this secant shifted parallel to itself. At least once it will reach a position in which it is a tangent to the curve at a point between x_1 and x_2, namely, at the point of the curve which is at the greatest distance from the secant.

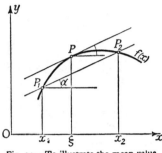

Fig. 14.—To illustrate the mean value theorem

Hence there will be an intermediate value ξ such that

$$\frac{f(x_1) - f(x_2)}{x_1 - x_2} = f'(\xi).$$

This statement is called *the mean value theorem of the differential calculus.* We can also express it somewhat differently by noticing that the number ξ may be written in the form

$$\xi = x_1 + \theta(x_2 - x_1),$$

where θ is a certain number between 0 and 1. In applications of the mean value theorem we shall often find that θ cannot be more accurately determined than this, but it will usually turn out that a more accurate value is not needed. When accurately formulated, then, the mean value theorem runs as follows:

If f(x) *is continuous in the closed interval* $x_1 \leqq x \leqq x_2$ *and differentiable at every point of the open interval* $x_1 < x < x_2$, *then there is at least one value* θ, *where* $0 < \theta < 1$, *such that*

$$\frac{f(x_2) - f(x_1)}{x_2 - x_1} = f'\{x_1 + \theta(x_2 - x_1)\}.$$

If we replace x_1 by x and x_2 by $x + h$, we can express the mean value theorem by the formula

$$\frac{f(x + h) - f(x)}{h} = f'(\xi) = f'(x + \theta h), \quad x < \xi < x + h.$$

We wish to emphasize that while it is essential that $f(x)$ should be continuous for all points of the interval, including the end-points, we need not assume that the derivative exists at the end-points. This apparently trivial remark is actually useful in many applications.

If at any point in the interior of the interval the derivative fails to exist, the mean value theorem is not necessarily true.

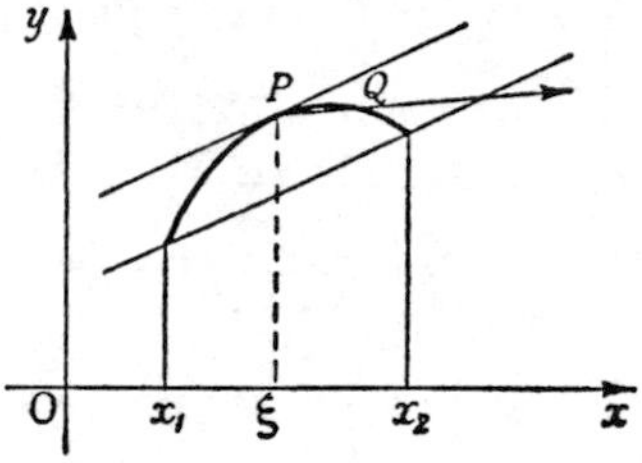

Fig. 15.—To illustrate the mean value theorem

This is shown by the example $f(x) = |x|$, p. 97.

We can complete our intuitive argument by the following considerations. There is at least one point P on the curve which has the greatest possible distance from the chord joining the points on the curve whose abscissæ are x_1 and x_2 (see fig. 15). At this point the curve by hypothesis has a definite tangent. We shall now prove that this tangent must be parallel to the chord. By definition the tangent is the limiting position of the secant and is obtained by joining P to a point Q on the curve and letting the point Q move towards P. Since by hypothesis Q is not farther from the chord than P, the line PQ produced in the direction P to Q must either cut the chord or run parallel to it; and this must be the case, no matter on which side of P the point Q lies. This, however, is only possible if the limiting position is parallel to the chord. If we denote the abscissa of the point P by the letter ξ, the slope $f'(\xi)$ of the tangent at P is then equal to the slope of the chord, $\dfrac{f(x_1) - f(x_2)}{x_1 - x_2}$; hence for the number ξ in the theorem we may simply take the abscissa of P.

The *rigorous* proof of the mean value theorem is usually developed in the following way. We first establish *Rolle's theorem*, which is a special case of the mean value theorem:

If a function $\phi(x)$ is continuous in the closed interval $x_1 \leqq x \leqq x_2$

and differentiable in the open interval $x_1 < x < x_2$, *and if in addition* $\phi(x_1) = 0$ *and* $\phi(x_2) = 0$, *then there exists at least one point* ξ *in the interior of the interval at which* $\phi'(\xi) = 0$.

In fact, there must be at least one point ξ, interior to the interval, at which the function $\phi(x)$ takes on its greatest or its least value (cf. Chap. I, Appendix I, § 2, p. 63); to be specific, we assume that ξ is a point where $\phi(\xi)$ is a maximum, so that for every x in the interval $\phi(x) \leq \phi(\xi)$. Then for every number h whose absolute value $|h|$ is small enough it is certainly true that $\phi(\xi) - \phi(\xi + h) \geq 0$. If h is positive,

$$\frac{\phi(\xi + h) - \phi(\xi)}{h} \leq 0;$$

we now let h tend to zero through positive values and obtain $\phi'(\xi) \leq 0$. If, on the other hand, h is negative, $\frac{\phi(\xi + h) - \phi(\xi)}{h} \geq 0$, and thus by letting h tend to zero through negative values we obtain $\phi'(\xi) \geq 0$; comparing this with the preceding inequality, we see that $\phi'(\xi) = 0$, which establishes our theorem.

We now apply Rolle's theorem to the function *

$$\phi(x) = f(x) - f(x_1) - \frac{x - x_1}{x_2 - x_1} \{ f(x_2) - f(x_1) \}.$$

This function obviously satisfies the condition $\phi(x_1) = \phi(x_2) = 0$, and is of the form $\phi(x) = f(x) + ax + b$ with constant coefficients $a = -\frac{f(x_2) - f(x_1)}{x_2 - x_1}$ and b. By p. 96, we know that

$$\phi'(x) = f'(x) + a,$$

and thus by Rolle's theorem we have

$$0 = \phi'(\xi) = f'(\xi) + a$$

for a suitably chosen intermediate value ξ; hence

$$f'(\xi) = -a = \frac{f(x_2) - f(x_1)}{x_2 - x_1},$$

and the mean value theorem is proved.

---

* This function, apart from a factor independent of x, is the distance of the point $(x, f(x))$ of the curve from the secant; the reader can easily verify this for himself.

As the first of many applications of the mean value theorem we shall prove the following. *Let the function* f(x) *be continuous in the closed interval* $a \leqq x \leqq b$ *and have a derivative* f'(x) *at every point of the open interval* $a < x < b$. *Then if* f'(x) *is positive everywhere in* $a < x < b$, *the function* f(x) *is monotonic increasing in the interval* $a \leqq x \leqq b$; *and likewise if* f'(x) *is negative in* $a < x < b$, *then* f(x) *is monotonic decreasing.*

We shall prove the first statement; the second can be proved in a similar way. Suppose that $f'(x) > 0$, and let x_1 and $x_2 > x_1$ be any two values of x in the closed interval. Then by the mean value theorem

$$f(x_2) - f(x_1) = (x_2 - x_1) f'(\xi),$$

where $x_1 < \xi < x_2$. Since both factors on the right are positive, this proves that $f(x_2) > f(x_1)$; hence $f(x)$ is monotonic increasing.

9. The Approximate Representation of Arbitrary Functions by Linear Functions. Differentials.

The equation $\lim\limits_{h \to 0} \dfrac{f(x+h) - f(x)}{h} = f'(x)$ defining the derivative is equivalent to the equations

$$f(x+h) - f(x) = h f'(x) + \epsilon h$$

or $\qquad y + \Delta y = f(x + \Delta x) = f(x) + f'(x) \Delta x + \epsilon \Delta x,$

where ϵ is a quantity which tends to zero with $h = \Delta x$. If for the moment we think of the point x as fixed and the increment Δx as variable, then by this formula the increment of the function, that is, the quantity Δy, consists of two terms, namely, a part $h f'(x)$ which is proportional to h, and an " error " which can be made as small as we please relative to h by making h itself small enough. Thus the smaller the interval about the point x which we consider, the more accurately is the function $f(x+h)$ (which is a function of h) represented by its linear part $f(x) + h f'(x)$. This approximate representation of the function $f(x+h)$ by a linear function of h is expressed geometrically by the substitution for the curve of its tangent at the point x. Later, in Chapter VII, we shall consider the practical application of these ideas to the performance of approximate calculations.

Here we merely remark in passing that it is possible to use this approximate representation of the increment Δy by the linear expression $h f'(x)$ to construct a logically satisfactory definition of the notion of a " differential ", as was done by Cauchy in particular.

While the idea of the differential as an infinitely small quantity has no meaning, and it is accordingly futile to define the derivative as the quotient of two such quantities, we may still try to assign a sense to the equation $f'(x) = dy/dx$ in such a way that the expression dy/dx need not be thought of as purely symbolic, but as the actual quotient of two quantities dy and dx. For this purpose we first define the derivative $f'(x)$ by our limiting process, then think of x as fixed and consider the increment $h = \Delta x$ as the independent variable. This quantity h we *call* the *differential of x*, and write $h = dx$. We now define the expression $dy = y'dx = hf'(x)$ as the *differential of the function y*; dy is therefore a number which has nothing to do with infinitely small quantities. So the derivative $y' = f'(x)$ is now really the *quotient of the differentials dy and dx*; but in this statement there is nothing remarkable; it is, in fact, merely a tautology, a restatement of the verbal definition. The differential dy is accordingly the linear part of the increment Δy (see fig. 16).

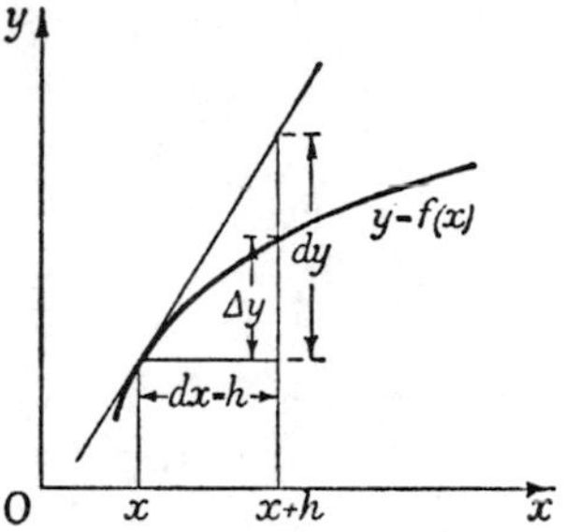

Fig. 16.—The differential *dy*

We shall not make any immediate use of these differentials. Nevertheless, it may be pointed out for the sake of completeness that we may also form second and higher differentials. For if we think of h as chosen in any manner, but always the same for every value of x, then $dy = hf'(x)$ is a function of x, of which we can again form the differential. The result will be called the *second differential of y*, and will be denoted by the symbol $d^2y = d^2f(x)$. The increment of $hf'(x)$ being $h\{f'(x + h) - f'(x)\}$, the second differential is obtained by replacing the quantity in brackets by its linear part $hf''(x)$, so that $d^2y = h^2f''(x)$. We may naturally proceed further along the same lines, obtaining third, fourth, ... differentials of y, &c., which can be defined by the expressions $h^3f'''(x)$, $h^4f^{iv}(x)$, and so on.

10. Remarks on Applications to the Natural Sciences.

In the applications of mathematics to natural phenomena we never have to deal with sharply defined quantities. Whether a length is *exactly* a metre is a question which cannot be decided by any experiment and which consequently has no "physical meaning". Again, there is no immediate physical meaning in saying that the length of a material rod is rational or irrational; we can always measure it with any desired degree of accuracy in rational numbers, and the real matter of interest is whether or not we can manage to perform such a measurement using rational numbers with relatively small denominators. Just as the question of rationality or irrationality in the rigorous sense of "exact mathematics" has no physical meaning, so the actual carrying out of limiting processes in applications will usually be nothing more than a mathematical idealization.

The practical significance of such idealizations lies chiefly in the fact that if they are used all analytical expressions become essentially simpler and more manageable. For example, it is vastly simpler and more convenient to work with the notion of instantaneous velocity, which is a function of only *one* definite time-instant, than with the notion of average velocity between two different instants. Without such idealization every rational investigation of nature would be condemned to hopeless complications and would break down at the very outset.

We do not intend, however, to enter into a discussion of the relationship of mathematics to reality. We merely wish to emphasize, for the sake of our better understanding of the theory, that in applications we have the right to replace a derivative by a difference quotient and vice versa, provided only that the differences are small enough to guarantee a sufficiently close approximation. The physicist, the biologist, the engineer, or anyone else who has to deal with these ideas in practice, will therefore have the right to identify the difference quotient with the derivative within his limits of accuracy. The smaller the increment $h = dx$ of the independent variable, the more accurately can he represent the increment $\Delta y = f(x + h) - f(x)$ by the differential $dy = hf'(x)$. So long as he keeps within the limits of accuracy required by the problem, he is accustomed to speak of the quantities $dx = h$ and $dy = hf'(x)$ as "infinitesimals". These "physically infinitesimal" quantities have a precise meaning. They are finite quantities, not equal to zero, which are chosen small enough for the given investigation, e.g. smaller than a fractional part of a wave-length or smaller than the distance between two electrons in an atom; in general, smaller than the degree of accuracy required.

EXAMPLES

1.* Replace the statement " At the point $x = \xi$ the function $f(x)$ is not differentiable " by an equivalent statement not using any form of the word " differentiable ".

2. Differentiate the following functions directly by using the definition of the derivative:

$$(a) \ \frac{1}{x+1}. \qquad (b) \ \frac{1}{x^2+2}. \qquad (c) \ \frac{1}{2x^3+1}. \qquad (d) \ \frac{1}{\sin x}.$$

$$(e) \ \sin 3x. \qquad (f) \ \cos ax. \qquad (g) \ \sin^2 x. \qquad (h) \ \cos^2 x.$$

3. Find the intermediate value ξ of the mean value theorem for the following functions, and illustrate graphically:

$$(a) \ 2x. \quad (b) \ x^3. \quad (c) \ 5x^3 + 2x. \quad (d) \ 1/(x^2+1). \quad (e) \ x^{1/3}.$$

4. Show that the mean value theorem fails for the following functions when the two points are taken with opposite signs, e.g. $x_1 = -1$, $x_2 = 1$:

$$(a) \ 1/x. \quad (b) \ |x|. \quad (c) \ x^{2/3}.$$

Illustrate graphically, and compare with the previous example.

4. The Indefinite Integral, the Primitive Function, and the Fundamental Theorems of the Differential and Integral Calculus.

As we have already mentioned above, the connexion between the problem of integration and the problem of differentiation is the corner-stone of the differential and integral calculus. This connexion we will now study.

1. The Integral as a Function of the Upper Limit.

The value of the definite integral of a function $f(x)$ depends on the choice of the two limits of integration a and b. It is a function of the lower limit a as well as of the upper limit b. In order to study this dependence more closely we imagine the lower limit a to be a definite fixed number, denote the variable of integration no longer by x but by u (cf. p. 82), and denote the upper limit by x instead of b in order to suggest that we shall let the upper limit vary and that we wish to investigate the value of the integral as a function of the upper limit. Accordingly, we write

$$\int_a^x f(u)\,du = \Phi(x).$$

We call this function $\Phi(x)$ *an indefinite integral* of the function

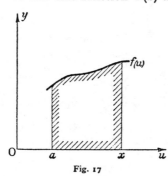

Fig. 17

$f(x)$. When we speak of *an* and not of *the* indefinite integral, we suggest that instead of the lower limit a any other could be chosen, in which case we should ordinarily obtain a different value for the integral. Geometrically the indefinite integral for each value of x will be given by the area (shown by shading in fig. 17) under the curve $y = f(u)$ and bounded by the ordinates $u = a$ and $u = x$, the sign being determined by rules given earlier (p. 81).

If we choose another lower limit a in place of the lower limit a, we obtain the indefinite integral

$$\Psi(x) = \int_a^x f(u)\,du.$$

The difference $\Psi(x) - \Phi(x)$ will obviously be given by

$$\int_a^a f(u)\,du,$$

which is a constant, since a and a are each taken as fixed given numbers. Therefore

$$\Psi(x) = \Phi(x) + \text{const.};$$

Different indefinite integrals of the same function differ only by an additive constant.

We may likewise regard the integral as a function of the lower limit, and introduce the function

$$\phi(x) = \int_x^b f(u)\,du,$$

in which b is a fixed number. Here again two such integrals with different upper limits b and β differ only by an additive constant $\int_b^\beta f(u)\,du$.

2. The Derivative of the Indefinite Integral.

We will now differentiate the indefinite integral $\Phi(x)$ with respect to the variable x. The result is the following theorem:

The indefinite integral

$$\Phi(x) = \int_a^x f(u)\,du$$

of a continuous function f(x) *always possesses a derivative* $\Phi'(x)$, *and, moreover,*

$$\Phi'(x) = f(x);$$

that is, differentiation of the indefinite integral of a given continuous function always gives us back that same function.

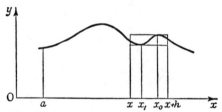

Fig. 18.—Differentiation of the indefinite integral

This is the root idea of the whole of the differential and integral calculus. The proof follows extremely simply from the interpretation of the integral as an area. We form the difference quotient

$$\frac{\Phi(x+h) - \Phi(x)}{h},$$

and observe that the numerator

$$\Phi(x+h) - \Phi(x) = \int_a^{x+h} f(u)\,du - \int_a^x f(u)\,du = \int_x^{x+h} f(u)\,du$$

is the area between the ordinate corresponding to x and the ordinate corresponding to $x + h$.

Now let x_0 be a point in the interval between x and $x + h$ at which the function $f(x)$ takes its greatest value, and x_1 a point at which it takes its least value in that interval (cf. fig. 18).

Then the area in question will lie between the values $hf(x_0)$ and $hf(x_1)$, which represent the areas of rectangles with the interval from x to $x + h$ as base and $f(x_0)$ and $f(x_1)$ respectively as altitudes. Expressed analytically,

$$f(x_0) \geqq \frac{\Phi(x + h) - \Phi(x)}{h} \geqq f(x_1).$$

This can also be proved directly from the definition of the integral without appealing to the geometrical interpretation.* To do this we write

$$\int_x^{x+h} f(u)\,du = \lim_{n \to \infty} \sum_{\nu=1}^{n} f(u_\nu)\,\Delta u_\nu,$$

where $u_0 = x$, u_1, u_2, $\ldots$, $u_n = x + h$ are points of division of the interval from x to $x + h$, and the greatest of the absolute values of the differences $\Delta u_\nu = u_\nu - u_{\nu-1}$ tends to zero as n increases. Then $\Delta u_\nu / h$ is certainly positive, no matter whether h is positive or negative. Since we know that $f(x_0) \geqq f(u_\nu) \geqq f(x_1)$ and since the sum of the quantities Δu_ν is equal to h, it follows that

$$f(x_0) \geqq \frac{1}{h} \Sigma f(u_\nu)\,\Delta u_\nu \geqq f(x_1);$$

and thus if we let n tend to infinity we obtain the inequalities stated above for

$$\frac{1}{h}\int_x^{x+h} f(u)\,du \quad \text{or} \quad \frac{\Phi(x + h) - \Phi(x)}{h}.$$

If h now tends to zero, both $f(x_0)$ and $f(x_1)$ must tend to the limit $f(x)$, owing to the continuity of the function. We therefore see at once that

$$\Phi'(x) = \lim_{h \to 0} \frac{\Phi(x + h) - \Phi(x)}{h} = f(x),$$

as stated by our theorem.

Owing to the differentiability of $\Phi(x)$, we have the following theorem, by § 3, No. 5, p. 97:

The integral of a continuous function f(x) is itself a continuous function of the upper limit.

* Compare also the later discussion on p. 127.

For the sake of completeness we would point out that if we regard the definite integral not as a function of its upper limit but as a function of its lower limit, the derivative is not equal to $f(x)$, but is instead equal to $-f(x)$. In symbols: if we put

$$\phi(x) = \int_x^b f(u)\, du,$$

then $$\phi'(x) = -f(x).$$

The proof follows immediately from the remark that

$$\int_x^b f(u)\, du = -\int_b^x f(u)\, du.$$

3. The Primitive Function; General Definition of the Indefinite Integral.

The theorem which we have just proved shows us that the indefinite integral $\Phi(x)$ at once gives the solution of the following problem: *given a function* f(x), *to determine a function* F(x) *such that*

$$F'(x) = f(x).$$

This problem requires us to *reverse the process of differentiation*. It is a typical inverse problem such as occurs in many parts of mathematics and such as we have already found to be a fruitful mathematical method for generating new functions. (For example, the first extension of the idea of natural numbers was made under the pressure of the necessity for reversing certain elementary processes of calculation. The formation of inverse functions has led and will lead us to new kinds of functions.)

A function $F(x)$ such that $F'(x) = f(x)$ is called a *primitive function of $f(x)$*, or simply a *primitive of $f(x)$*; this terminology suggests that the function $f(x)$ arises from $F(x)$ by differentiation.

This problem of the inversion of differentiation or of the finding of a primitive function is at first sight of quite a different character from the problem of integration. From p. 111, however, we know that:

Every indefinite integral $\Phi(x)$ of the function f(x) *is a primitive of* f(x).

Yet this result does not completely solve the problem of finding primitive functions. For we do not yet know if we have found *all* the solutions of the problem. The question about the group of all primitive functions is answered by the following theorem, sometimes referred to as the fundamental theorem of the differential and integral calculus:

The difference of two primitives $F_1(x)$ *and* $F_2(x)$ *of the same function* f(x) *is always a constant:*

$$F_1(x) - F_2(x) = c.$$

Thus from any one primitive function F(x) *we can obtain all the others in the form*

$$F(x) + c$$

by suitable choice of the constant c. *Conversely, for every value of the constant* c *the expression* $F_1(x) = F(x) + c$ *represents a primitive function of* f(x).

It is clear that for any value of the constant c the function $F(x) + c$ is a primitive, provided that $F(x)$ itself is one. For (cf. p. 96) we have

$$\frac{\{F(x+h)+c\} - \{F(x)+c\}}{h} = \frac{F(x+h) - F(x)}{h},$$

and since by hypothesis the right-hand side tends to $f(x)$ as $h \to 0$, so does the left-hand side, and therefore

$$\frac{d}{dx}\{F(x) + c\} = f(x) = F'(x).$$

Thus to complete the proof of our theorem it only remains to show that the difference of two primitive functions $F_1(x)$ and $F_2(x)$ is always a constant. For this purpose we consider the difference

$$F_1(x) - F_2(x) = G(x)$$

and form the derivative

$$G'(x) = \lim_{h \to 0} \left\{ \frac{F_1(x+h) - F_1(x)}{h} - \frac{F_2(x+h) - F_2(x)}{h} \right\}.$$

Both the expressions on the right-hand side, by hypothesis, have the same limit $f(x)$ as $h \to 0$; thus for every value of x we have $G'(x) = 0$. But a function whose derivative is everywhere zero must have a graph whose tangent is everywhere parallel to the x-axis, i.e. must be a constant; and therefore we have $G(x) = c$, as we stated above. We can prove this last fact without relying upon intuition, by using the mean value theorem. Applying the mean value theorem to $G(x)$, we have

$$G(x_2) - G(x_1) = (x_2 - x_1) G'(\xi); \ x_1 < \xi < x_2.$$

But we have seen that the derivative $G'(x)$ is equal to 0 for every value of x, and hence in particular for the value ξ; hence it follows immediately that $G(x_1) = G(x_2)$. Since x_1 and x_2 are arbitrary values of x in the given interval, $G(x)$ must be a constant.

Combining the theorem just proved with the result of No. 2 (p. 111), we can now make the following statement:

Every primitive function F(x) *of a given function* f(x) *can be represented in the form*

$$F(x) = c + \Phi(x) = c + \int_a^x f(u) \, du,$$

where c *and* a *are constants, and conversely, for any constant values of* a *and* c *chosen arbitrarily this expression always represents a primitive function.*

It may readily be guessed that the constant c can as a rule be omitted, since by changing the lower limit a we change the primitive function by an additive constant. In many cases, however, we should not obtain *all* the primitive functions if we omitted the c, as the example $f(x) = 0$ shows. For this function the indefinite integral of No. 1 (p. 110) is always 0, independently of the lower limit; yet any arbitrary constant is a primitive function of $f(x) = 0$. A second example is the function $f(x) = \sqrt{x}$, which is defined for non-negative values of x only. The indefinite integral is

$$\Phi(x) = \tfrac{2}{3} x^{3/2} - \tfrac{2}{3} a^{3/2},$$

and we see that no matter how we choose the lower limit a the indefinite integral $\Phi(x)$ is always obtained from $\tfrac{2}{3}(x)^{3/2}$ by addition of a constant which is less than or equal to zero, namely, the constant $-\tfrac{2}{3} a^{3/2}$; yet such a function as $\tfrac{2}{3} x^{3/2} + 1$ is also a primitive function for $\sqrt{x}$. Thus in the general expression for the primitive function we cannot dispense with the additive constant. The relationship which we have found suggests an extension of the idea of the indefinite integral. We shall henceforth call every expression of the form $c + \Phi(x) = c + \int_a^x f(u) \, du$ an indefinite integral of $f(x)$. In other words, *we shall no longer make any distinction between the primitive function and the indefinite integral.* Nevertheless, if the reader is to have a proper understanding of the interrelations of these concepts it is absolutely necessary that he should clearly bear in mind that in the first instance integration and inversion of differentiation are two entirely different things, and that it is only the knowledge of the relationship between them that gives us the right to apply the term "indefinite integral" to the primitive function also.

It is customary to represent the indefinite integral by a notation which in itself is perhaps not perfectly clear. We write

$$F(x) = c + \int_a^x f(u)\,du = \int f(x)\,dx;$$

that is, we omit the upper limit x and the lower limit a and also the additive constant c and use the letter x for the variable of integration. It would really be more consistent to avoid this last change, in order to prevent confusion with the upper limit x which is the independent variable in $F(x)$. In using the notation $\int f(x)\,dx$ we must never lose sight of the indeterminacy connected with it, i.e. the fact that that symbol always denotes *an* indefinite integral only.

4. The Use of the Primitive Function in the Evaluation of Definite Integrals.

Suppose that we know any one primitive function $F(x) = \int f(x)\,dx$ for the function $f(x)$ and that we wish to evaluate the definite integral $\int_a^b f(u)\,du$. We know that the indefinite integral

$$\Phi(x) = \int_a^x f(u)\,du,$$

being also a primitive of $f(x)$, can only differ from $F(x)$ by an additive constant. Consequently

$$\Phi(x) = F(x) + c,$$

and the additive constant c is at once determined if we recollect that the indefinite integral $\Phi(x) = \int_a^x f(u)\,du$ must take the value 0 when $x = a$. We thus obtain $0 = \Phi(a) = F(a) + c,$

whence $c = -F(a)$ and $\Phi(x) = F(x) - F(a)$. In particular, for the value $x = b$ we have

$$\int_a^b f(u)\,du = F(b) - F(a),$$

which gives us the important rule:

If F(x) *is any primitive of the function* f(x) *whatsoever, the definite integral of* f(x) *between the limits* a *and* b *is equal to the difference* F(b) — F(a).

If we use the relation $F'(x) = f(x)$ this may be written in the form

$$F(b) - F(a) = \int_a^b F'(x)\,dx = \int_a^b \frac{dF(x)}{dx}\,dx.$$

This formula can easily be proved and understood directly. We divide the interval $a \leqq x \leqq b$ into sub-intervals $\Delta x_1, \Delta x_2, \ldots, \Delta x_n$, and consider the sum $\Sigma \dfrac{\Delta F}{\Delta x_\nu} \Delta x_\nu$. On the one hand, this sum is simply $\Sigma \Delta F = F(b) - F(a)$, independently of the particular subdivision; hence its limit is $F(b) - F(a)$. On the other hand, its limit is also equal to $\int_a^b F'(x)\,dx$, as follows from the mean value theorem. For $\Delta F / \Delta x_\nu = F'(\xi_\nu)$, where ξ_ν is a point intermediate between the ends $x_{\nu-1}$ and x_ν of the interval Δx_ν. The sum is therefore equal to $\Sigma \Delta x_\nu\, F'(\xi_\nu)$; and by the definition of integral this tends to the limit $\int_a^b F'(x)\,dx$ as the subdivision is made finer, which establishes our formula.

In applying our rule we often use the symbol $|$ to denote the difference $F(b) - F(a)$; i.e. we write

$$\int_a^b f(x)\,dx = F(b) - F(a) = F(x)\,\Big|_a^b,$$

meaning by the vertical line that in the preceding expression first the value b, and then the value a, is to be substituted for x, and finally the difference of the resulting numbers is to be found.

5. Examples.

We are now in a position to illustrate by a series of simple examples the relationships between the definite integral, the indefinite integral, and the derivative, which we have just investigated. In virtue of the theorem on p. 111, from each of the integration formulæ directly proved in § 2 (p. 82) we can derive a differentiation formula.

On p. 86 we obtained the integration formula

$$\int_a^b x^a dx = \frac{1}{\alpha + 1}(b^{a+1} - a^{a+1})$$

for every rational number $\alpha \neq 1$ and all positive values of a and b; if we replace the variable of integration by u and the upper limit by x, this may be written

$$\int_a^x u^a \, du = \frac{1}{\alpha + 1}(x^{a+1} - a^{a+1}).$$

From this it follows by the fundamental theorem that the right-hand side is a primitive function of the integrand, i.e. the differentiation formula

$$\frac{d}{dx} x^{a+1} = (\alpha + 1)x^a$$

is valid for every rational value of $\alpha \neq -1$ and all positive values of x. By direct substitution we find that this *last* formula is also true for $\alpha = -1$, if $x > 0$. The result obtained exactly agrees with what we have already found (p. 95) by direct differentiation. Thus by using the fundamental theorem after we had carried out the integration, we could have saved ourselves the trouble of that differentiation.

Further, from the integration formula

$$\int_a^x \cos u \, du = \sin x - \sin a$$

given on p. 87 it follows that $\dfrac{d}{dx} \sin x = \cos x$, in agreement with the result found on p. 96.

Conversely, however, we may regard every directly-proved differentiation formula $F'(x) = f(x)$ as a connexion between a primitive function $F(x)$ and a derived function $f(x)$, that is, we may regard it as a formula for indefinite integration and then obtain from it the definite integral of $f(x)$ as on p. 117. This very method is frequently made use of, as we shall see in Chapter IV (p. 205). In particular, we may start from the results of § 3 (p. 94) and obtain the integral formulæ of § 2, p. 82, in virtue of the fundamental theorem. For example, from p. 95 we know that $\dfrac{d}{dx} x^{a+1} = (\alpha + 1)x^a$. Therefore $\dfrac{x^{a+1}}{\alpha + 1}$ is a primitive function or indefinite integral of x^a, provided that $\alpha \neq -1$, and thus by p. 117 we again arrive at the integral formula above.

1. From the differentiations performed in Examples 2, 3 on p. 109, set up the corresponding integrations.

2. Evaluate (a) $\int_0^1 \dfrac{dx}{(x+1)^2}.$ (b) $\int_0^1 \dfrac{2x\,dx}{(x^2+1)^2}.$

3. Using Example 2, prove from the definition of the definite integral that

(a) $\lim\limits_{n\to\infty} n\left[\dfrac{1}{(n+1)^2}+\dfrac{1}{(n+2)^2}+\cdots+\dfrac{1}{(2n)^2}\right]=\dfrac{1}{2}.$

(b) $\lim\limits_{n\to\infty} n^2\left[\dfrac{1}{(n^2+1)^2}+\dfrac{2}{(n^2+2^2)^2}+\cdots+\dfrac{n}{(n^2+n^2)^2}\right]=\dfrac{1}{4}.$

5. SIMPLE METHODS OF GRAPHICAL INTEGRATION

An indefinite integral or primitive function of $f(x)$ is a function $y = F(x)$ which not only can be visualized as an area, but like any other function can be represented graphically by a curve. Our definition immediately suggests the possibility of constructing this curve approximately and thus obtaining a graph of the integral function. To begin with we must remember that this last curve is not unique, but on account of the additive constant can be shifted parallel to itself in the direction of the y-axis. We can therefore require that the integral curve shall pass through

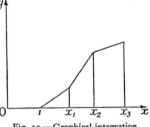

Fig. 19.—Graphical integration

an arbitrarily selected point, e.g. if $x = 1$ belongs to the interval of definition of $f(x)$, through the point with the co-ordinates $x = 1$, $y = 0$. The curve is thereafter determined by the requirement that for each value of x its direction is given by the corresponding value of $f(x)$. To obtain an approximate construction of a curve which satisfies these conditions, we seek to construct not the curve $y = F(x)$ itself, but a polygonal path (broken line) whose corners lie vertically above previously assigned points of division of the x-axis and whose segments have approximately the same direction as the portion of the integral curve between the same points of division. For this purpose we divide our interval of the x-axis by means of the points $x = 1$, $x_1, x_2, \ldots$ into a certain number of parts, not necessarily all

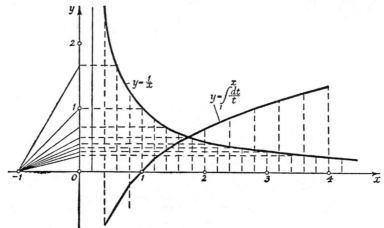

Fig. 20.—Graphical integration of $\frac{1}{x}$

of the same length, and at each point of division we erect the parallel to the y-axis. Then (fig. 19) we draw through the point $x = 1$, $y = 0$ the straight line whose slope is equal to $f(1)$;

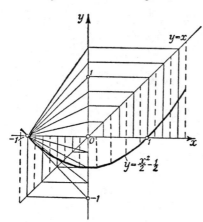

Fig. 21.—Graphical integration of x

through the intersection of this line with the line $x = x_1$ we draw the line with the slope $f(x_1)$; through the intersection of this line with $x = x_2$ we draw the line with slope $f(x_2)$, and so on. In the actual practical construction of these lines, we erect at each point of division the ordinate to the curve $y = f(x)$, and project these ordinates on to any parallel to the y-axis; to be specific, let us suppose that they are projected on to the y-axis itself. We then obtain the direction of the integral curve by joining the point with co-ordinates $x = 0$ and $y = f(x)$ to the point $x = -1$, $y = 0$. By carrying over these directions parallel to themselves we obtain a polygonal path whose corners lie vertically above the given points of division of the x-axis and whose direction agrees with the direction of the integral curve at the initial point of each interval. This polygonal path can be made to represent the integral curve with any desired degree of accuracy by making the subdivision of the interval fine enough. We can frequently improve the accuracy of the construction by choosing for the direction of each segment of the polygon that direction which belongs not to the beginning but to the mid-point of the corresponding interval (cf. figs. 20 and 21).*

In fig. 21 the construction described above is carried out for the function $f(x) = x$. By graphical integration we obtain an approximation to the integral curve, which is the parabola $y = \frac{1}{2}x^2 - \frac{1}{2}$. In addition, fig. 20 shows an approximation to the integral function of the function $f(x) = 1/x$. We shall study this integral later in greater detail—it will turn out to be the logarithmic function. Finally, the reader would be well advised to work out some other examples for himself, e.g. the graphical integration of the functions $\sin x$ and $\cos x$.

EXAMPLES

1. By graphical integration with the interval $h = \frac{1}{10}$ construct the following integral curves:

(a) $\displaystyle\int_0^x x^2\, dx\ (0 \leqq x \leqq 2).$ (b) $\displaystyle\int_1^x \frac{1}{x^2}\, dx\ (1 \leqq x \leqq 2).$

(c) $\displaystyle\int_0^x \frac{1}{1 + x^2}\, dx\ (0 \leqq x \leqq 1).$

In particular, evaluate $\displaystyle\int_0^1 \frac{1}{1 + x^2}\, dx.$

6. FURTHER REMARKS ON THE CONNEXION BETWEEN THE INTEGRAL AND THE DERIVATIVE

Before we begin to follow up the relationships found in § 4 (p. 109) systematically we shall illustrate them from another point of view, which is closely related to the intuitive idea of density and to other physical concepts.

---

* We may mention in passing that graphical integration (that is, the finding of the graph of a primitive $F(x)$ of a function $f(x)$ which is itself given by a graph) can also be performed by means of a mechanical device, the so-called integraph. In this mechanism a pointer is moved along the given curve and a pen automatically traces one of the curves $y = F(x)$ for which $F'(x) = f(x)$. The indeterminacy of the constant of integration is expressed by a certain arbitrariness in the initial position of the instrument. For integrating devices generally see B. Williamson, *Integral Calculus*, pp. 214–217 (Longmans); *Dictionary of Applied Physics*, Vol. III, pp. 450–457 (Macmillan, 1923).

1. Mass Distribution and Density; Total Quantity and Specific Quantity.

We suppose that any mass whatsoever is distributed along a straight line, the x-axis, the distribution being continuous but not necessarily uniform. We may, for example, think of a vertical column of air standing on a surface of area 1; as x-axis we take a line pointing vertically upwards and as origin the point on the earth's surface. The total mass between two abscissæ x_1 and x_2 is then determined in the following way by means of a so-called sum-function $F(x)$. We measure distance along the line from the initial point of the mass-distribution, $x = 0$, and by $F(x)$ we mean the total mass between the abscissa 0 and the abscissa x. The increment of mass from the abscissa x_1 to the abscissa x_2 is then given simply by

$$F(x_2) - F(x_1);$$

a sign is thus assigned to the increment, and this sign changes if x_1 and x_2 are interchanged.

The average mass per unit length in the interval x_1 to x_2 is

$$\frac{F(x_2) - F(x_1)}{x_2 - x_1}.$$

If we assume that the function $F(x)$ is differentiable, then as $x_2 \to x_1$ this value tends to the derivative $F'(x_1)$. This quantity is precisely what is usually called the *specific mass* or *density* of the distribution at the point x_1; as a rule, of course, its value depends on the particular point chosen. Between the density $f(x)$ and the sum-function $F(x)$ there accordingly exists the relation

$$F(x) = \int_0^x f(u)\, du; \quad f(x) = F'(x).$$

The sum-function is a primitive function of the density, or, what amounts to the same thing, *the mass is the integral of the density*; conversely, *the density is the derivative of the sum-function*.

Exactly the same relation is very frequently encountered in physics. For example, if by $Q(t)$ we denote the total amount of heat needed to raise unit mass of a substance from temperature t_0 to temperature t, then to raise the temperature from t_1 to t_2 an amount of heat equal to

$$Q(t_2) - Q(t_1)$$

is needed. Between t_1 and t_2 the average amount of heat used per unit increase in temperature is then

$$\frac{Q(t_2) - Q(t_1)}{t_2 - t_1}.$$

If we once again assume the differentiability of the function $Q(t)$ we obtain in the limit a function

$$q(t) = \lim_{t_1 \to t} \frac{Q(t) - Q(t_1)}{t - t_1},$$

which we call the *specific heat* of the substance. This specific heat is in general to be regarded as a function of the temperature.

Here again, between specific heat and total quantity of heat there exists the characteristic relation of integral and derivative,

$$\int_a^b q(t)\,dt = Q(b) - Q(a).$$

We shall meet with the same relations in all cases where total quantity and specific quantity are contrasted, e.g. electric charge as contrasted with density of charge, or the total force on a surface as contrasted with the force-density or pressure.

In nature it usually happens that what we know directly is not the density or specific quantity, but the total quantity; thus it is the integral which is primary (as the name " primitive " suggests) and the specific quantity is only arrived at after a limiting process, namely, differentiation.

Incidentally it may be noted that if the masses considered are by their nature positive, the sum-function $F(x)$ must be a monotonic increasing function of x, and consequently the specific quantity, the density $f(x)$, must be non-negative. Nothing hinders us, however, from considering negative quantities also (e.g. negative electricity); then our sum-functions $F(x)$ need no longer be monotonic.

2. The Question of Applications.

The relation of the primitive sum-function to the density of distribution perhaps becomes clearer when it is realized that from the point of view of physical facts the limiting processes of integration and differentiation represent an idealization, and that they do not express anything exact in nature. On the contrary, in the realm of physical actuality we can form in place of the integral only a sum of very many small quantities and in place of the derivative only a difference quotient of very small quantities. The quantities Δx remain different from 0; the passage to the limit $\Delta x \to 0$ is merely a mathematical simplification, in which the accuracy of the mathematical representation of the reality is not essentially impaired.

As an example we return to the vertical column of air. According to the atomic theory we find that we cannot think of the distribution of mass as a continuous function of x. On the contrary, we will assume (and this, too, is a simplifying idealization) that the mass is distributed along the x-axis in the form of a large number of point-molecules lying very close to one another. Then the sum-function $F(x)$ will not be a continuous function, but will have a constant value in the interval between two molecules and will take a sudden jump as the variable x passes the point occupied by a molecule. The amount of this jump will be equal to the mass of the molecule, while the average distance between molecules, according to results established in atomic theory, is of the order of 10^{-8} cm. If now we are performing upon this air column some measurement in which masses of the order 10^4 molecules are to be considered negligible, our function cannot be distinguished from a continuous function. For if we choose two values x and $x + \Delta x$ whose difference Δx is less than 10^{-4} cm., then the difference between $F(x)$ and $F(x + \Delta x)$ will be the mass of the molecules in the interval; since the number of these molecules is of the order of 10^4, the values of $F(x)$ and $F(x + \Delta x)$ are, so far as our experiment is concerned, equal. As density of distribution we consider simply the difference quotient $\dfrac{\Delta F(x)}{\Delta x} = \dfrac{F(x + \Delta x) - F(x)}{\Delta x}$; it is an important physical assumption that we do not obtain measurably different values for this quotient when Δx is allowed to vary between certain bounds, say between 10^{-4} and 10^{-5} cm. Now let us imagine that $F(x)$ is measured and plotted for a large number of points about 10^{-4} cm. apart, and that the points thus found are joined by straight lines; we obtain a polygon, and by rounding off the corners we finally obtain a curve with a continuously turning tangent. This curve is the graph of some function, say $F_1(x)$. This new function $F_1(x)$ cannot within the limits of experimental accuracy be distinguished from $F(x)$, and its derivative is within the same limits equal to $\Delta F/\Delta x$; we thus have found a continuous differentiable function which for the purposes of physics *is* the function $F(x)$.

It is perhaps appropriate to discuss yet another example of the concepts of sum-function and density of distribution. In statistics, e.g. in the kinetic theory of matter or in statistical biology, these concepts frequently occur in a form in which the nature of the mathematical idealization is particularly clear. Let us consider e.g. the molecules of a gas confined in a vessel and observe their velocities at a given instant of time. Let the number of molecules be N, and let the number of those with velocities less than x be $N\Phi(x)$. Then $\Phi(x)$ denotes the ratio of the number of molecules moving with velocities between 0 and x to the total number of molecules. This sum-function is, of course, not continuous, but is sectionally * constant and suddenly increases by $1/N$ when x as it increases passes a value which is equal to the velocity of some molecule.

The idealization which we shall make here is that we shall think of the number N as increasing beyond all bounds. We assume that in this passage to the limit $N \to \infty$ the sum-function $\Phi(x)$ tends to a definite continuous limit function $F(x)$. That this is really the case (i.e. that we can with sufficient accuracy replace $\Phi(x)$ by this continuous function $F(x)$) is obviously an important physical assumption; and it is another such assumption to suppose that this sum-function $F(x)$ possesses a derivative $F'(x) = f(x)$, which we then call the density of distribution. The sum-function is connected with the density of distribution by the equations

$$F(x) = \int_0^x f(u)\,du; \quad F(b) - F(a) = \int_a^b f(x)\,dx.$$

* Ger. *stückweise*; cf. Chap. IX, § 3.

This density of distribution is occasionally referred to as the *specific probability* that a molecule possesses the velocity x. The idealization we have just carried out plays a great part in the kinetic theory of gases originated by Maxwell; in exactly the same mathematical form it appears in many problems of mathematical statistics.

7. The Estimation of Integrals and the Mean Value Theorem of the Integral Calculus

We close this chapter with some considerations about a matter of general significance, the whole importance of which will not appear until somewhat later. The point in question is the estimation of integrals.

1. The Mean Value Theorem of the Integral Calculus.

The first and simplest of these estimation rules runs as follows: if in an interval $a \leq x \leq b$ the continuous function $f(x)$ is everywhere non-negative (is either positive or zero), then the definite integral

$$\int_a^b f(x)\,dx$$

is also non-negative. Similarly, the integral is not positive if the function is positive nowhere in the interval. The proof of this theorem follows directly from the definition of the integral.

From this the following theorem arises:. if

$$f(x) \geqq g(x)$$

everywhere in the interval $a \leq x \leq b$, then

$$\int_a^b f(x)\,dx \geqq \int_a^b g(x)\,dx$$

also. For by our first remark the integral of the difference $f(x) - g(x)$ is non-negative and by our addition rule (cf. p. 82)

$$\int_a^b (f(x) - g(x))\,dx = \int_a^b f(x)\,dx - \int_a^b g(x)\,dx.$$

Let M be the greatest and m the least value of the function $f(x)$ in the interval ab. The function $M - f(x)$ is non-negative in the interval, and the same is true for the function $f(x) - m$.

From the above remark we immediately obtain the double inequality

$$\int_a^b m\,dx \leqq \int_a^b f(x)\,dx \leqq \int_a^b M\,dx.$$

But $\int_a^b m\,dx = m\int_a^b dx = m(b-a)$ and likewise $\int_a^b M\,dx = M(b-a)$,

whence $m(b-a) \leqq \int_a^b f(x)\,dx \leqq M(b-a)$. The integral under consideration can therefore be represented as the product of $(b-a)$ and some number μ between m and M:

$$\int_a^b f(x)\,dx = \mu(b-a), \qquad m \leqq \mu \leqq M.$$

As a rule there is no need to state the exact value of this mean value μ. We may, however, state that it will be assumed by the function at one point ξ of the interval $a \leqq \xi \leqq b$ at least, since in its interval of definition a continuous function assumes all values between its greatest value and its least. As in the case of the mean value theorem of the differential calculus, the exact statement of the value ξ is in many cases unimportant. We may therefore put $\mu = f(\xi)$, where ξ is an intermediate value of x, and we then have

$$\int_a^b f(x)\,dx = (b-a)f(\xi), \qquad a \leqq \xi \leqq b.$$

This last formula is called the *mean value theorem of the integral calculus*.

We can generalize the theorem somewhat by considering, instead of the integrand $f(x)$, an integrand of the form $f(x)\,p(x)$, where $p(x)$ *is an arbitrary non-negative function, which, like* f(x), *is to be assumed continuous*. Since $mp(x) \leqq f(x)\,p(x) \leqq Mp(x)$, we immediately obtain the relation

$$m\int_a^b p(x)\,dx \leqq \int_a^b f(x)\,p(x)\,dx \leqq M\int_a^b p(x)\,dx,$$

or, in a single equation,

$$\int_a^b f(x)\,p(x)\,dx = f(\xi)\int_a^b p(x)\,dx,$$

where ξ is again a number intermediate between a *and* b.

We have thus proved the following theorem:

If f(x) *and* p(x) *are continuous functions in* a $\leqq$ x $\leqq$ b, *and* p(x) $\geqq$ 0, *then*

$$\int_a^b f(x) p(x) dx = f(\xi) \int_a^b p(x) dx,$$

where a $\leqq \xi \leqq$ b.

2. Applications.   The Integration of x^α for any Irrational Value of α.

The mean value theorem and the equivalent integral estimates immediately afford us an insight into an intuitive and easily

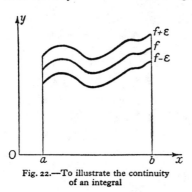

Fig. 22.—To illustrate the continuity of an integral

apprehended fact: the value of an integral changes very little if the function itself is everywhere changed very little. In precise language: if in the whole interval $a \leqq x \leqq b$ the absolute value of the difference of two functions $f(x)$ and $g(x)$ is less than a number ϵ, then the difference of their integrals is in absolute value less than $\epsilon(b - a)$. In symbols: if throughout the interval $a \leqq x \leqq b$ we have $|f(x) - g(x)| < \epsilon$, then

$$\left| \int_a^b f(x) dx - \int_a^b g(x) dx \right| < \epsilon(b - a)$$

or, otherwise expressed,

$$- \epsilon(b - a) + \int_a^b g(x) dx < \int_a^b f(x) dx < \int_a^b g(x) dx + \epsilon(b - a).$$

Fig. 22 illustrates this theorem very clearly. For the curve $y = f(x)$ we draw the "parallel curves" $y = f(x) + \epsilon$ and $y = f(x) - \epsilon$. By hypothesis the function $g(x)$ keeps within the strip bounded by these "parallel curves". It is clear from this that the areas which are bounded by the curves $f(x)$ and $g(x)$ differ from one another by less than half the area of the strip, and the area of the strip is just

$$\int_a^b \{f(x) + \epsilon\} dx - \int_a^b \{f(x) - \epsilon\} dx = 2\epsilon(b - a).$$

No appeal to intuition is needed. Since

$$-\epsilon + g(x) < f(x) < \epsilon + g(x);$$

it follows, by considerations analogous to those on p. 126, that

$$\int_a^b \{-\epsilon + g(x)\} dx < \int_a^b f(x) dx < \int_a^b \{g(x) + \epsilon\} dx,$$

which as the result of the fundamental rules of integration takes the form

$$-\epsilon(b - a) + \int_a^b g(x) dx < \int_a^b f(x) dx < \int_a^b g(x) dx + \epsilon(b - a);$$

here we have merely replaced the integral of a sum by the corresponding sum of integrals, and have noticed that

$$\int_a^b \epsilon \, dx = \epsilon(b - a).$$

As an indication of the importance of this theorem, we shall show that with its help we are able to integrate the function x^α for any irrational value of α, or, more exactly, to calculate the definite integral $\int_a^b x^\alpha \, dx$. Here we assume that $0 < a < b$.

We represent the index α as the limit of a sequence of rational numbers $\alpha_1, \alpha_2, \ldots, \alpha_n, \ldots$, so that $\alpha = \lim_{n \to \infty} \alpha_n$; here we can assume that none of the values α_n is equal to -1, since α itself is different from -1. For the power x^α we then use the definition

$$x^\alpha = \lim_{n \to \infty} x^{\alpha_n}$$

and notice the following: no matter how small a positive number ε we choose, we can always find an n so large that in the whole interval * $\imath \leqq x \leqq b$ we have $|x^\alpha - x^{\alpha_n}| < \varepsilon$.

---

* This can be proved quite simply as follows. (Cf. Appendix I, § 3, p. 69). Remembering that x^α is monotonic, and putting $\delta_n = a_n - a$, we have

$$|x^\alpha - x^{\alpha_n}| = x^\alpha |1 - x^{\delta_n}| \leqq (a^a + b^a)(|1 - a^{\delta_n}| + |1 - b^{\delta_n}|);$$

for x^α lies between a^a and b^a, so that $x^a \leqq a^a + b^a$, and likewise $1 - x^{\delta_n}$ lies between $1 - a^{\delta_n}$ and $1 - b^{\delta_n}$, so that $|1 - x^{\delta_n}| \leqq (|1 - a^{\delta_n}| + |1 - b^{\delta_n}|)$. From $\lim_{n \to \infty} a^{\delta_n} = \lim_{n \to \infty} b^{\delta_n} = 1$, it follows that

$$\lim_{n \to \infty} |1 - a^{\delta_n}| = \lim_{n \to \infty} |1 - b^{\delta_n}| = 0;$$

so if n is chosen large enough the right-hand side of the inequality is less than ϵ. This gives us $|x^{\alpha_n} - x^\alpha| < \epsilon$ simultaneously for all values of x in the interval $a \leqq x \leqq b$.

Now we need only apply the relationship mentioned above to the functions $f(x) = x^\alpha$ and $g(x) = x^{\alpha_n}$, obtaining

$$-\varepsilon(b - a) + \int_a^b x^{\alpha_n}\, dx < \int_a^b x^\alpha\, dx < \int_a^b x^{\alpha_n}\, dx + \varepsilon(b - a).$$

The integrals on the right and on the left, however, may be evaluated in accordance with the result on p. 85, giving

$$-\varepsilon(b - a) + \frac{1}{\alpha_n + 1}\,(b^{\alpha_n+1} - a^{\alpha_n+1})$$

$$< \int_a^b x^\alpha\, dx < \frac{1}{\alpha_n + 1}\,(b^{\alpha_n+1} - a^{\alpha_n+1}) + \varepsilon(b - a).$$

If we now let the number ε steadily decrease and tend to 0, the corresponding values of n increase beyond all bounds; the numbers α_n, a^{α_n}, and b^{α_n} must then converge to α, a^α, and b^α respectively, and we immediately obtain the result

$$\int_a^b x^\alpha\, dx = \frac{1}{\alpha + 1}\,(b^{\alpha+1} - a^{\alpha+1}).$$

In other words, the integration formula that holds for rational values of α holds also for irrational values of α.

From this it follows in virtue of the fundamental theorem of p. 111 that for positive values of x the differentiation formula

$$\frac{d}{dx}\,x^{\alpha+1} = (\alpha + 1)x^\alpha$$

already obtained for rational values of α remains valid for irrational values of α also.

Examples

1. Find the intermediate value ξ of the mean value theorem of the integral calculus for the following, and interpret geometrically:

$$(a)\ \int_a^b 1\, dx. \qquad\qquad (b)\ \int_a^b x\, dx.$$

$$(c)\ \int_a^b x^n\, dx. \qquad\qquad (d)\ \int_a^b \frac{dx}{x^2}.$$

2. Let $f(x)$ be continuous. Prove, from the mean value theorem of the integral calculus, that the derivative of the indefinite integral of $f(x)$ is equal to $f(x)$.

3. (a) Evaluate $I_n = \int_0^a x^{1/n}\, dx$. What is $\lim\limits_{n \to \infty} I_n$? Interpret geometrically. (b) Do the same for $I_n = \int_0^a x^n\, dx$.

4.* Let the function $f(\xi)$ be continuous for all values of ξ, and let $F(x)$ be defined by the equation

$$F(x) = \frac{1}{2\delta} \int_{-\delta}^{\delta} f(x+t)\,dt,$$

where δ is an arbitrary positive number. Prove that:

(a) the function $F(x)$ possesses a continuous derivative for all values of x;

(b) in any fixed interval $a \leq x \leq b$ we can make $|F(x) - f(x)| < \varepsilon$, where ε is an arbitrary pre-assigned positive number, by choosing δ small enough.

5.* *Schwarz's inequality for integrals.*

Prove that for all continuous functions $f(x)$, $g(x)$

$$\int_a^b (f(x))^2\,dx \int_a^b (g(x))^2\,dx \geqq \left(\int_a^b f(x)g(x)\,dx \right)^2.$$

Appendix to Chapter II

1. The Existence of the Definite Integral of a Continuous Function

We have still to give a proof of the fact that the definite integral of a continuous function between the limits a and b ($a < b$) always exists. For this purpose we recall the notation of § 1 (p. 79), and consider the sum

$$F_n = \sum_{\nu=1}^{n} f(\xi_\nu)\Delta x_\nu.$$

It is certainly true that

$$\underline{F_n} = \sum_{\nu=1}^{n} f(v_\nu)\Delta x_\nu \leqq F_n \leqq \sum_{\nu=1}^{n} f(u_\nu)\Delta x_\nu = \overline{F_n},$$

where $f(v_\nu)$ denotes the least and $f(u_\nu)$ the greatest value of the function in the ν-th sub-interval. The problem is to prove that F_n tends to a definite limit independent of the particular manner of subdivision and of the particular choice of the quantities ξ_ν, provided that as n increases the length of the longest sub-interval tends to zero. To establish this it is obviously necessary and sufficient to show that the two expressions $\underline{F_n}$ and $\overline{F_n}$ converge to one and the same limit.

No matter how small the positive number ϵ is chosen, we know by the uniform continuity of $f(x)$ that in every sufficiently small interval the "oscillation" $|f(u_\nu) - f(v_\nu)|$ is less than ϵ; so that if the subdivision is fine enough we certainly must have

$$0 \leq \overline{F}_n - \underline{F}_n = \sum_{\nu=1}^{n} \Delta x_\nu \{ f(u_\nu) - f(v_\nu) \} < \epsilon(b-a).$$

We therefore see that as n increases this difference must tend to zero, and so we can content ourselves with proving that one of the sums, say $\overline{F}_n$, converges. This convergence will be proved as soon as we show that $|\overline{F}_n - \overline{F}_m|$ can be made as small as desired by requiring that the corresponding subdivisions (which we shall refer to as "subdivision n" and "subdivision m" respectively) go beyond a certain degree of fineness. This degree of fineness is characterized by the property that for both subdivisions the oscillation of the function in each sub-interval is less than ϵ ($\epsilon > 0$). We pass to a third subdivision whose points of division consist of all the points of subdivision n and of subdivision m taken together. This new subdivision, which has say l points of division, we denote by the suffix l, and we consider the corresponding upper sum $\overline{F}_l$. We shall now estimate the value of $|\overline{F}_n - \overline{F}_m|$ by first obtaining estimates for the expressions $|\overline{F}_n - \overline{F}_l|$ and $|\overline{F}_m - \overline{F}_l|$. We assert that the following two relationships hold:

$$\underline{F}_n \leq \overline{F}_l \leq \overline{F}_n \quad \text{and} \quad \underline{F}_m \leq \overline{F}_l \leq \overline{F}_m.$$

The proof follows at once from the meaning of our expressions. Let us consider say the ν-th sub-interval of the subdivision n. This sub-interval will consist of one or several sub-intervals of the subdivision l; the terms corresponding to these intervals will each consist of two factors, one of which is a difference Δx and the other of which is certainly not greater than $f(u_\nu)$ and not less than $f(v_\nu)$. The sum of the lengths Δx of those intervals of the subdivision l which lie in the ν-th sub-interval of the coarser subdivision n is, however, exactly Δx_ν. We therefore see that the corresponding contribution to the sum $\overline{F}_l$ must lie between the limits $f(u_\nu)\Delta x_\nu$ and $f(v_\nu)\Delta x_\nu$. If we now sum over all the n sub-intervals we obtain the first of the above inequalitics; the second is obtained in exactly the same way if we consider the subdivision m instead of the subdivision n.

We have already seen that $\overline{F_n} - F_n < \epsilon(b - a)$; it is likewise true that $\overline{F_m} - F_m < \epsilon(b - a)$. From the inequalities for $\overline{F_l}$ proved above, it therefore follows that

$$0 \leqq \overline{F_n} - \overline{F_l} < \epsilon(b - a) \quad \text{and} \quad 0 \leqq \overline{F_m} - \overline{F_l} < \epsilon(b - a).$$

Thus it is also certain that

$$|\overline{F_n} - \overline{F_m}| = |(\overline{F_n} - \overline{F_l}) - (\overline{F_m} - \overline{F_l})| < 2\epsilon(b - a).$$

Since ϵ can be chosen as small as we please, this relation shows us by Cauchy's convergence test (p. 40) that the sequence of numbers $\overline{F_n}$ actually converges. At the same time we see at once from our argument that the limiting value is completely independent of the manner of subdivision.

The proof of the existence of the definite integral of a continuous function is thus complete.

Our method of proof teaches us still more. It shows us that in many cases we are also led to the integral by a somewhat more general limiting process. If, for example, $f(x) = \phi(x)\psi(x)$ and the interval from a to b is subdivided into n parts by the points of division x_ν, we consider instead of the sum $\Sigma f(\xi_\nu)\Delta x_\nu$ the more general sum

$$\Sigma \phi(\xi_\nu')\psi(\xi_\nu'')\Delta x_\nu,$$

where ξ_ν' and ξ_ν'' are two not necessarily coincident points of the ν-th sub-interval. This sum will also tend to the integral

$$\int_a^b f(x)\,dx = \int_a^b \phi(x)\psi(x)\,dx$$

as n increases, provided that the length of the longest sub-interval tends to zero.

A corresponding statement holds for all sums formed in an analogous way; for example, the sum

$$\sum_{\nu=1}^n \sqrt{\{\phi(\xi_\nu')^2 + \psi(\xi_\nu'')^2\}}\,\Delta x_\nu$$

tends to the integral

$$\int_a^b \sqrt{\{\phi(x)^2 + \psi(x)^2\}}\,dx.$$

The proof of these facts follows lines exactly similar to the above and hence need not be worked out in detail.

2. THE RELATION BETWEEN THE MEAN VALUE THEOREM OF THE DIFFERENTIAL CALCULUS AND THE MEAN VALUE THEOREM OF THE INTEGRAL CALCULUS.

Between the mean value theorem of the differential calculus and that of the integral calculus there exists a simple relation which is arrived at by way of the fundamental theorem (p. 111) and which we give as an instructive example of the use of that theorem. We take the mean value theorem of the integral calculus in its more special form,

$$\int_a^b f(x)\,dx = (b-a)f(\xi).$$

If we put $\int f(x)\,dx = F(x)$, so that $f(x) = F'(x)$, the theorem just written takes the form

$$F(b) - F(a) = (b-a)F'(\xi)$$

or

$$\frac{F(b) - F(a)}{b-a} = F'(\xi).$$

Here we can obviously choose for $F(x)$ any function whose first derivative $F'(x) = f(x)$ is continuous, and thus for such functions the mean value theorem of the differential calculus is proved.

If we consider the more general form of the mean value theorem of the integral calculus,

$$\int_a^b f(x)p(x)\,dx = f(\xi)\int_a^b p(x)\,dx,$$

where $p(x)$ is a function which in our interval is continuous and positive and $f(x)$ is an arbitrary continuous function, we are led to a correspondingly more general mean value theorem of the differential calculus. We put

$$\int f(x)\,p(x)\,dx = F(x), \quad \text{i.e. } f(x)p(x) = F'(x),$$

and

$$\int p(x)\,dx = G(x), \quad \text{i.e. } p(x) = G'(x);$$

the above mean value formula then takes the form

$$F(b) - F(a) = \{G(b) - G(a)\}f(\xi),$$

or, since $f(x) = \dfrac{F'(x)}{G'(x)}$,

$$\frac{F(b) - F(a)}{G(b) - G(a)} = \frac{F'(\xi)}{G'(\xi)},$$

where $a \neq b$.

This formula, in which ξ once again denotes a number intermediate between a and b, is called the *generalized mean value theorem* of the differential calculus. For this to be valid it is obviously sufficient to assume that $F(x)$ and $G(x)$ are continuous functions with continuous first derivatives and that in addition $G'(x)$ is everywhere positive (or everywhere negative). For with these assumptions the whole process can be reversed.

Finally, it should be observed that in the present discussion of the mean value theorem of the differential calculus we have had to make assumptions more stringent than the theorems in themselves require. (Cf. § 3, No. 8, p. 103, and later p. 203.)

EXAMPLE

1. Show that if $f(x)$ has a continuous derivative in the interval $a \leq x \leq b$, then $f(x)$ can be represented as the difference of two monotonic functions.

CHAPTER III

Differentiation and Integration of the Elementary Functions

1. THE SIMPLEST RULES FOR DIFFERENTIATION AND THEIR APPLICATIONS

In higher analysis and its applications it is usually the case that the problems of integration are more important than those of differentiation, but that differentiation offers less difficulty than integration. Consequently the natural method of building up the integral and differential calculus is first to learn to differentiate the widest possible classes of functions and then by virtue of the fundamental theorem (Chap. II, § 4, p. 116) to make the results thus obtained available for the solution of integration problems. In the following sections it will be our task to carry out this programme. To a certain extent we shall make a fresh start, since we shall work out the most important differentiations and integrations systematically without calling upon the results of last chapter. In this development of the subject certain rules for differentiation, with the first of which we are already acquainted (p. 96), will play an important part.

1. Rules for Differentiation.

We assume that in the interval which we are considering the functions $f(x)$ and $g(x)$ are differentiable; our rules then run as follows:

Rule 1. *Multiplication by a constant.*

If c is a constant and $\phi(x) = cf(x)$, then $\phi(x)$ is bifferentiable, and

$$\phi'(x) = cf'(x).$$

This follows immediately from the relation

$$\frac{\phi(x+h)-\phi(x)}{h} = c\,\frac{f(x+h)-f(x)}{h}$$

if we take the limits as $h \to 0$.

Rule 2. *Derivative of a sum.*

If $\phi(x) = f(x) + g(x)$, then $\phi(x)$ is differentiable, and

$$\phi'(x) = f'(x) + g'(x);$$

that is, the processes of differentiation and addition are inter-changeable. The same holds for the sum of any finite number (n) of terms

$$\phi(x) = \sum_{\nu=1}^{n} f_\nu(x),$$

for which we obtain

$$\phi'(x) = \sum_{\nu=1}^{n} f_\nu{}'(x).$$

We may pass over the proof, which after Chap. II, § 3 (p. 88) is fairly obvious.

Rule 3. *Derivative of a product.*

If $\phi(x) = f(x)g(x)$, then $\phi(x)$ is differentiable, and

$$\phi'(x) = f(x)g'(x) + g(x)f'(x).$$

The proof follows from the equation

$$\frac{\phi(x+h)-\phi(x)}{h} = \frac{f(x+h)g(x+h)-f(x)g(x)}{h}$$

$$= \frac{f(x+h)g(x+h)-f(x+h)g(x)+f(x+h)g(x)-f(x)g(x)}{h}$$

$$= f(x+h)\frac{g(x+h)-g(x)}{h} + g(x)\frac{f(x+h)-f(x)}{h}.$$

In this last expression the passage to the limit $h \to 0$ can be directly carried out, yielding the formula stated.

This formula takes a still more elegant form if we divide * throughout by $\phi(x) = f(x)g(x)$. We then obtain

$$\frac{\phi'(x)}{\phi(x)} = \frac{f'(x)}{f(x)} + \frac{g'(x)}{g(x)}.$$

* We must, of course. assume that $\phi(x)$ is nowhere equal to zero.

By repeated application of this product formula we obtain by induction for the derivative of a product of n factors an expression consisting of n terms, each of which consists of the derivative of one factor multiplied by all the other factors of the original product.   In symbols:

$$\phi'(x) = \frac{d}{dx} \{ f_1(x) f_2(x) \ldots f_n(x) \}$$

$$= f_1'(x) f_2(x) \ldots f_n(x) + f_1(x) f_2'(x) f_3(x) \ldots f_n(x)$$
$$+ \ldots + f_1(x) f_2(x) \ldots f_n'(x)$$

$$= \sum_{\nu=1}^{n} f_\nu'(x) \frac{\phi(x)}{f_\nu(x)},$$

or on division * by $\phi(x) = f_1(x) f_2(x) \ldots f_n(x)$

$$\frac{\phi'(x)}{\phi(x)} = \frac{f_1'(x)}{f_1(x)} + \frac{f_2'(x)}{f_2(x)} + \ldots + \frac{f_n'(x)}{f_n(x)} = \sum_{\nu=1}^{n} \frac{f_\nu'(x)}{f_\nu(x)}.$$

Rule 4. *Derivative of a quotient.*

For a quotient

$$\phi(x) = \frac{f(x)}{g(x)}$$

the following rule holds: the function $\phi(x)$ is differentiable at every point at which $g(x)$ does not vanish, and

$$\phi'(x) = \frac{g(x) f'(x) - g'(x) f(x)}{\{ g(x) \}^2}.$$

If $\phi(x) \neq 0$, this can be written

$$\frac{\phi'(x)}{\phi(x)} = \frac{f'(x)}{f(x)} - \frac{g'(x)}{g(x)}.$$

If we accept the differentiability of $\phi(x)$ as a hypothesis, we can apply the product rule to $f(x) = \phi(x) g(x)$ and conclude that

$$f'(x) = \phi(x) g'(x) + g(x) \phi'(x).$$

By substituting $\dfrac{f(x)}{g(x)}$ for $\phi(x)$ on the right and solving for $\phi'(x)$ we obtain the rule stated above.   In order to prove the differen-

---

* We must, of course, assume that $\phi(x)$ is nowhere equal to zero.

tiability of $\phi(x)$ as well as the rule we use the following method. We write

$$\frac{\phi(x+h)-\phi(x)}{h} = \frac{\dfrac{f(x+h)}{g(x+h)} - \dfrac{f(x)}{g(x)}}{h}$$

$$= \frac{g(x)\dfrac{f(x+h)-f(x)}{h} - \dfrac{g(x+h)-g(x)}{h}f(x)}{g(x)g(x+h)}.$$

If we now let h tend to 0, we arrive at the result stated; for by hypothesis the two terms obtained by performing the division on the right have definite limits, which are respectively $\dfrac{g(x)f'(x)}{\{g(x)\}^2}$ and $\dfrac{g'(x)f(x)}{\{g(x)\}^2}$. This at once proves both the existence of the limit on the left-hand side and the differentiation formula.

2. Differentiation of the Rational Functions.

To begin with, we shall again deduce the differentiation formula

$$\frac{d}{dx} x^n = nx^{n-1}$$

for every positive integer n, basing the proof on the rule for differentiating a product. We think of x^n as a product of n factors, $x^n = x \ldots x$, and thence obtain

$$\frac{d}{dx} x^n = 1 \cdot x^{n-1} + 1 \cdot x^{n-1} + \ldots + 1 \cdot x^{n-1} = nx^{n-1}.$$

The second derivative of the function x^n results if we use the above formula and the first rule of differentiation:

$$\frac{d^2}{dx^2} x^n = n(n-1)x^{n-2}.$$

Continuing the process, we obtain

$$\frac{d^3}{dx^3} x^n = n(n-1)(n-2)x^{n-3}$$

$$\bullet \quad \bullet \quad \bullet \quad \bullet \quad \bullet \quad \bullet \quad \bullet \quad \bullet \quad \bullet \quad \bullet \quad \bullet$$

$$\frac{d^n}{dx^n} x^n = 1 \cdot 2 \ldots n = n!$$

From the last of these it is clear that the $(n+1)$-th derivative of x^n vanishes everywhere.

In virtue of our first two rules, a knowledge of the differentiation of powers at once enables us to differentiate any polynomial

$$y = a_0 + a_1 x + a_2 x^2 + \ldots + a_n x^n.$$

We have simply

$$y' = a_1 + 2a_2 x + 3a_3 x^2 + \ldots + na_n x^{n-1},$$

and further

$$y'' = 2a_2 + 3 \cdot 2 a_3 x + 4 \cdot 3 a_4 x^2 + \ldots + n(n-1)a_n x^{n-2},$$

and so on.

The differentiation of any rational function now follows with the help of the quotient rule. In particular, we shall again deduce the differentiation formula for the function x^n, where $n = -m$ is a negative integer. The application of the quotient rule, together with the fact that the derivative of a constant is equal to zero, gives us the result

$$\frac{d}{dx}\left(\frac{1}{x^m}\right) = -\frac{mx^{m-1}}{x^{2m}} = -\frac{m}{x^{m+1}},$$

or, if we take $m = -n$,

$$\frac{d}{dx} x^n = nx^{n-1},$$

which agrees formally with the result for positive values of n and with the results given earlier (p. 95).

3. Differentiation of the Trigonometric Functions.

For the trigonometric functions $\sin x$ and $\cos x$ we have already (p. 96) obtained the differentiation formulæ

$$\frac{d}{dx} \sin x = \cos x \quad \text{and} \quad \frac{d}{dx} \cos x = -\sin x.$$

The quotient rule now enables us to differentiate the functions

$$y = \tan x = \frac{\sin x}{\cos x} \quad \text{and} \quad y = \cot x = \frac{\cos x}{\sin x}.$$

According to the rule, the derivative of the first of these functions is

$$y' = \frac{\cos^2 x + \sin^2 x}{\cos^2 x} = \frac{1}{\cos^2 x},$$

and we obtain the result

$$\frac{d}{dx} \tan x = \frac{1}{\cos^2 x} = \sec^2 x = 1 + \tan^2 x.$$

Similarly, we obtain

$$\frac{d}{dx} \cot x = -\frac{1}{\sin^2 x} = -\operatorname{cosec}^2 x = -(1 + \cot^2 x).$$

2. The Corresponding Integral Formulæ

1. General Rules for Integration.

The fundamental theorem of p. 116 and the definition of the indefinite integral reveal to us the possibility of writing down an integral formula corresponding to each differentiation formula. The following rules of integration (of which the first two have already been mentioned on p. 82) are completely equivalent to the first three rules of differentiation.

Multiplication by a constant: If c is a constant, then

$$\int cf(x)\,dx = c\int f(x)\,dx.$$

Integration of a sum: It is always true that

$$\int \{f(x) + g(x)\}\,dx = \int f(x)\,dx + \int g(x)\,dx.$$

To the third rule of differentiation corresponds the rule for the *integration of a product*, or, as it is usually called, the rule for *integration by parts*. On integration the product rule gives

$$\int \{f(x)g(x)\}'\,dx = \int f(x)g'(x)\,dx + \int g(x)f'(x)\,dx.$$

The indefinite integral on the left is obviously $f(x)g(x)$ (except possibly for an additive constant), and we can therefore write the rule for integration by parts in the following form:

$$\int f(x)g'(x)\,dx = f(x)g(x) - \int g(x)f'(x)\,dx.$$

This last integration formula, the counterpart of the rule for the differentiation of a product, has been given here only for the sake of completeness; it will not become important for us until the next chapter (p. 218).

2. Integration of the Simplest Functions.

Corresponding to the differentiation formulæ for special functions which we have recently found, we now set down the equivalent integration formulæ. The formula

$$\frac{d}{dx} x^n = nx^{n-1}$$

when expressed as an integration formula becomes

$$\int x^{n-1} dx = \frac{x^n}{n}, \quad n \neq 0.$$

For this formula merely means that the derivative of the right-hand side is equal to the expression under the integral sign on the left. If we replace n by $n + 1$, we obtain the integral formula

$$\int x^n dx = \frac{1}{n+1} x^{n+1}, \quad n \neq -1.$$

This formula holds for every integral index n (where $n < 0$ it of course holds only if $x \neq 0$) with the exception of $n = -1$, for which the denominator $n + 1$ would vanish. Later (p. 167) this exceptional case will be studied in detail.

The fundamental theorem of the integral calculus at once permits us to use our integral formulæ for the determination of areas, that is, of definite integrals. By p. 117 we immediately obtain

$$\int_a^b x^n dx = \frac{1}{n+1} (b^{n+1} - a^{n+1}), \quad n \neq -1,$$

where if n is negative we assume that a and b are of the same sign, since otherwise the integrand would be discontinuous in the interval of integration.

To the differentiation formulæ for $\sin x$, $\cos x$, $\tan x$, and $\cot x$ correspond the following integration formulæ:

$$\int \cos x\,dx = \sin x, \qquad \int \sin x\,dx = -\cos x,$$

$$\int \frac{1}{\cos^2 x}\,dx = \tan x, \qquad \int \frac{1}{\sin^2 x}\,dx = -\cot x.$$

From these formulæ we obtain by way of the fundamental rule of Chap. II, § 4 (p. 117) the value of the definite integral between any limits, the only restriction being that when the last two formulæ are used the interval of integration must not contain any point of discontinuity of the integrand. For example,

$$\int_a^b \cos x\,dx = \sin x \Big|_a^b = \sin b - \sin a.$$

It scarcely needs to be emphasized that with the help of the first two rules of integration we are now in a position to integrate any polynomial in x, and, in fact, any linear combination with arbitrary constant coefficients of the functions integrated here. The following point, however, should be noted. Rules of integration and rules of differentiation must according to the fundamental theorem be equivalent to one another; it is therefore possible first to prove the general integration rules of this section and then to read off the differentiation rules of the preceding section. The reader would be well advised to carry out this suggestion for himself.

EXAMPLES

1. Find the numerical values of all the derivatives of $x^5 - x^4$ at $x = 1$.

2. What is the numerical value of the eleventh derivative of
$$317x^9 - 202x^7 + 76 \quad \text{at} \quad x = 13\tfrac{1}{2}?$$

3. Differentiate the following functions and write down the corresponding integral formulæ:

(a) $ax + b$.

(b) $25cx^7$.

(c) $a + 2bx + cx^2$.

(d) $\dfrac{ax + b}{cx + d}$.

(e) $\dfrac{ax^2 + 2bx + c}{\alpha x^2 + 2\beta x + \gamma}$.

(f) $\dfrac{1}{1 - x^2} - \dfrac{1}{1 + x^2}$.

(g) $\dfrac{(x^8 - \sqrt{8}x^4 + 4)(x^8 + \sqrt{8}x^4 + 4)}{x^{16} + 16}$.

4. Let $P(x) = a_0 + a_1 x + a_2 x^2 + \ldots + a_n x^n$.
(a) Calculate the polynomial $F(x)$ from the equation $F(x) - F'(x) = P(x)$.
(b)* Calculate $F(x)$ from the equation $c_0 F(x) + c_1 F'(x) + c_2 F''(x) = P(x)$.

5. Differentiate the following functions and write down the corresponding integral formulæ:

(a) $2 \sin x \cos x$.

(b) $\dfrac{1}{1 + \tan x}$.

(c) $x \tan x$.

(d) $\dfrac{\sin x + \cos x}{\sin x - \cos x}$.

(e) $\dfrac{\sin x}{x}$.

Recalling that $\sec x = \dfrac{1}{\cos x}$, $\operatorname{cosec} x = \dfrac{1}{\sin x}$, find the derivatives indicated in Ex. 6–9:

6. $\dfrac{d^2}{dx^2} \sec x$.

7. $\dfrac{d^3}{dx^3} \sec x \tan x$.

8. $\dfrac{d^3}{dx^3} \operatorname{cosec} x$.

9. $\dfrac{d^4}{dx^4} \tan x \sin x$.

10. Find the limit as $n \to \infty$ of the absolute value of the n-th derivative of $\dfrac{1}{x}$ at the point $x = 2$.

Evaluate:

11. $\displaystyle\int (ax + b)\, dx$.

12. $\displaystyle\int (ax^2 + 2bx + c)\, dx$.

13. $\displaystyle\int (9x^8 + 7x^6 + 5x^4 + 3x^2 + 1)\, dx$.

14. $\displaystyle\int \left(\dfrac{1}{x^2} + \dfrac{1}{x^3} + \dfrac{1}{x^4} \right) dx$.

15. $\displaystyle\int \left(x^2 + \dfrac{1}{x^2} \right) dx$.

16. $\displaystyle\int \left(a \cos x + \dfrac{b}{\sin^2 x} \right) dx$.

17. $\displaystyle\int \left(3x + 7 \sin x + \dfrac{5}{x^3} - \dfrac{9}{\cos^2 x} \right) dx$.

18. $\displaystyle\int \sec x \tan x\, dx$.

3. THE INVERSE FUNCTION AND ITS DERIVATIVE

1. The General Formula for Differentiation.

We have seen earlier (pp. 21 and 67) that a continuous function $y = f(x)$ has a continuous inverse in every interval in which it is monotonic. More exactly:

If $a \leqq x \leqq b$ *is an interval in which the continuous function* $y = f(x)$ *is monotonic, and if* $f(a) = \alpha$ *and* $f(b) = \beta$, *then* x *is a function of* y *which in the interval between* α *and* β *is one-valued, continuous, and monotonic.*

As we have already shown on p. 92, the concept of the derivative gives us a simple means for recognizing that a function is monotonic and therefore has an inverse. For a differentiable function is certainly always monotonic increasing if $f'(x)$ is greater than zero throughout the corresponding interval, and similarly is monotonic decreasing if $f'(x)$ is everywhere less than zero in the interval.

We shall now prove the following theorem:

If in the interval $a < x < b$ *the function* $y = f(x)$ *is differentiable, and in that interval either* $f'(x) > 0$ *everywhere or else* $f'(x) < 0$ *everywhere, then the inverse function* $x = \phi(y)$ *also possesses a derivative at every point of its interval of definition, and between the derivative of the given function* $y = f(x)$ *and that of the inverse function* $x = \phi(y)$ *there exists for corresponding values of* x *and* y *the relationship* $f'(x) \cdot \phi'(y) = 1$, *which we can also write in the form*

$$\frac{dy}{dx} = \frac{1}{\dfrac{dx}{dy}}.$$

In this last formula we again observe the flexibility of Leibnitz's notation. It is just *as if* the symbols dy and dx were quantities which could be operated with like actual numbers. The proof of this formula is correspondingly simple if we regard the derivative as the limit of the difference quotient,

$$y' = f'(x) = \lim_{\Delta x \to 0} \frac{\Delta y}{\Delta x} = \lim_{x_1 \to x} \frac{y_1 - y}{x_1 - x},$$

where x and $y = f(x)$, and x_1 and $y_1 = f(x_1)$, respectively denote pairs of corresponding values. By hypothesis the first of these limiting values is not equal to zero. On account of the continuity of $y = f(x)$ and $x = \phi(y)$ the equation $\lim \Delta x = 0$ is equivalent to $\lim \Delta y = 0$, and consequently the relations $y_1 \to y$ and $x_1 \to x$ are also equivalent. Therefore the limiting value

$$\lim_{x_1 \to x} \frac{x_1 - x}{y_1 - y} = \lim_{y_1 \to y} \frac{x_1 - x}{y_1 - y},$$

exists and is equal to $\dfrac{1}{f'(x)}$. On the other hand, the limiting value is by definition the derivative $\phi'(y)$ of the inverse function $\phi(y)$, and thus our formula is proved.

This formula has a simple geometrical meaning, which is clearly shown in fig. 1. The tangent to the curve $y = f(x)$ or $x = \phi(y)$ forms with the

6   (E 798)

positive x-axis an angle α, with the positive y-axis an angle β, and from the geometrical meaning of the derivative

$$f'(x) = \tan \alpha, \quad \varphi'(y) = \tan \beta.$$

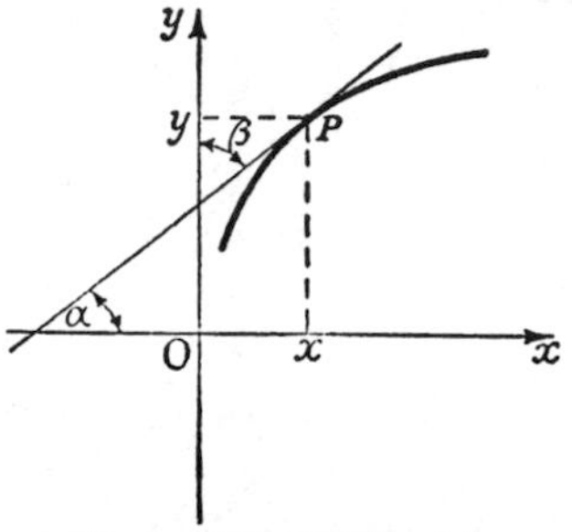

Since, however, the sum of the angles α and β is $\pi/2$, $\tan \alpha \, \tan \beta = 1$, and this relationship is exactly equivalent to our differentiation formula.

We have hitherto expressly assumed that either $f'(x) > 0$ or $f'(x) < 0$, i.e. that $f'(x)$ is never zero. What, then, happens if $f'(x) = 0$? If $f'(x) = 0$ everywhere in an interval the function is constant there, and consequently has no inverse, since the same value of y must correspond to all values of x in the interval. If the equation $f'(x) = 0$ is true only at isolated points, and if for the sake of simplicity $f'(x)$ is assumed continuous, then we must distinguish whether on passing through these points $f'(x)$

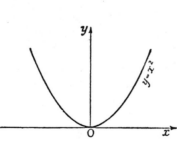

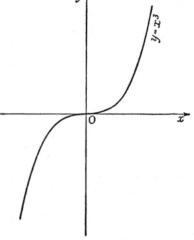

Fig. 1.—Differentiation of the inverse function

Fig. 2.—Parabola

Fig. 3.—Cubical parabola

changes sign or not. In the first case this point separates a point where the function is monotonic increasing from another where it is monotonic decreasing. In the neighbourhood of such a point there can be no single-valued inverse function. In the second case the vanishing of the derivative does not destroy the monotonic character of the function $y = f(x)$, so that a single-

valued inverse exists.   But the inverse function will no longer
be differentiable at the corresponding point; in fact, its derivative
will be infinite there.   The functions $y = x^2$ and $y = x^3$ at the
point $x = 0$ offer examples of the two types.   Figs. 2 and 3
illustrate the behaviour of the two functions where they pass
through the origin and at the same time show that one of the
functions, namely $y = x^3$, has a single-valued inverse, but that
the other function, $y = x^2$, has not.

2. The Inverse of the Power Function.

The simplest example of an inverse function is offered by the
functions $y = x^n$ for positive integers n and, as we at first assume, positive
values of x.   Under these conditions y' is always positive, so that for
all positive values of y we can form a unique positive inverse function

$$x = \sqrt[n]{y} = y^{1/n}.$$

The derivative of this inverse function is immediately obtained in accordance with the above general rule by the following calculations:

$$\frac{d(y^{1/n})}{dy} = \frac{dx}{dy} = \frac{1}{\dfrac{dy}{dx}} = \frac{1}{nx^{n-1}} = \frac{1}{n}\frac{1}{y^{(n-1)/n}} = \frac{1}{n}y^{1/n-1},$$

and if we now denote the independent variable by x, we may finally
write

$$\frac{d\sqrt[n]{x}}{dx} = \frac{d}{dx}(x^{1/n}) = \frac{1}{n}x^{1/n-1},$$

which agrees with the result obtained directly on p. 94.
    The point $x = 0$ requires special consideration.   If x approaches 0
through positive values, $d(x^{1/n})/dx$, where $n > 1$, will obviously increase
beyond all bounds; this corresponds to the fact that for $n > 1$ the derivative of the n-th power $f(x) = x^n$ vanishes at the origin. Geometrically
this means that the curves $y = x^{1/n}$, $n > 1$, touch the y-axis at the origin
(cf. fig. 17, p. 34).
    For the sake of completeness it should be noted that for odd values of n
the assumption that $x > 0$ can be omitted and the function $y = x^n$ can be
considered for all values of x without loss of its monotonic character or of
the uniqueness of its inverse. The differentiation formula $\dfrac{d}{dy}(y^{1/n}) = \dfrac{1}{n}y^{1/n-1}$
still holds for negative values of y; for $x = 0$, $n > 1$, we have $\dfrac{d(x^n)}{dx} = 0$,
which corresponds to an infinite derivative (dx/dy) of the inverse function
at the point $y = 0$.

3. The Inverse Trigonometric Functions.

In order to form the inverses of the trigonometric functions we once again consider the graphs of $\sin x$, $\cos x$, $\tan x$, and $\cot x$. We at once see from figs. 14 and 15, p. 25, that for each of these functions it is necessary to select a definite interval if we are to speak of a unique inverse; for the lines $y = c$ parallel to the x-axis cut the curves in an infinite number of points, if at all.

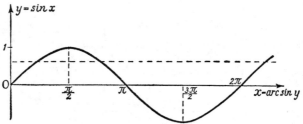

Fig. 4.—The inverse sine function

For the function $y = \sin x$ the derivative $y' = \cos x$ will e.g. be positive in the interval $-\pi/2 < x < \pi/2$. In this interval the sine accordingly has an inverse function; we write the inverse function of the sine in the form *

$$x = \text{arc } \sin y$$

(read arc sine y; this means the angle whose sine has the value y). This function runs monotonically from $-\pi/2$ to $+\pi/2$ as y traverses the interval -1 to $+1$. If we especially wish to emphasize that we are considering the inverse function of the sine for this very interval, we speak of the *principal value* of the arc sine. If we form the inverse function for some other interval in which $\sin x$ is monotonic, e.g. the interval $+\pi/2 < x < 3\pi/2$, we obtain " another *branch* " of the arc sine; without the exact statement of the interval in which the values of the function must lie the arc sine is a *multiple-valued function,* and in fact has an *infinite* number of values.

In general, the fact that arc $\sin y$ is multiple-valued is expressed by the statement that to any one value y of the sine there corresponds not only the angle x but also the angle $2k\pi + x$, as well as the angle $(2k + 1)\pi - x$, where k is any integer (cf. fig. 4).

* The notation $x = \sin^{-1} y$ is also used in English books.

The differentiation of the function $x = \text{arc} \sin y$ is performed in virtue of our general rule by the following short calculation:

$$\frac{dx}{dy} = \frac{1}{y'} = \frac{1}{\cos x} = \frac{1}{\pm \sqrt{(1 - \sin^2 x)}} = \frac{1}{\pm \sqrt{(1 - y^2)}},$$

where the square root is to be taken as positive if we confine ourselves to the first interval mentioned.*

If the independent variable is finally changed back from y to x, the differentiation formula for the function arc $\sin x$ is obtained in the following form:

$$\frac{d}{dx} \text{arc} \sin x = \frac{1}{\sqrt{(1 - x^2)}}.$$

Here it is assumed that the arc sine lies between $-\pi/2$ and $+\pi/2$, and the square root sign is chosen positive.

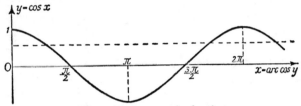

Fig. 5.—The inverse cosine function

For the inverse function of $y = \cos x$, denoted by arc $\cos x$, we obtain the differentiation formula

$$\frac{d}{dx} \text{arc} \cos x = - \frac{1}{\sqrt{(1 - x^2)}}$$

in exactly the same way. Here we take the positive sign of the root if the value of arc $\cos x$ is taken in the interval between 0 and π (not, as in the case of arc $\sin x$, between $-\pi/2$ and $+\pi/2$); cf. fig. 5.

A word remains to be said about the end-points $x = -1$ and $x = +1$. The derivatives become infinite on approaching these end-points, corresponding to the fact that the graphs of the

---

* If instead of this we had chosen the interval $\pi/2 < x < 3\pi/2$, corresponding to the substitution of $x + \pi$ for x, we should have had to use the negative square root, since $\cos x$ is negative in this interval.

inverse sine and inverse cosine must possess vertical tangents at these points.

We can deal with the inverse functions of the tangent and cotangent in an analogous way. The function $y = \tan x$, whose de-

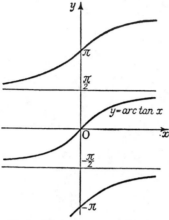

rivative $1/\cos^2 x$ for $x \neq \pi/2 + k\pi$ is everywhere positive, has a unique inverse in the interval $-\pi/2 < x < \pi/2$. We call this inverse function $x = \arctan y$ or (by interchange of the letters x and y) $y = \arctan x$. We see at once from fig. 6 that the original many-valuedness of the inverse —i.e. the many-valuedness which occurs if the interval of the values of the function is not fixed—is expressed by the fact that for each x we could have chosen instead of y any of the values $y + k\pi$ (where k is an

Fig. 6.—The inverse tangent function

integer). For the function $y = \cot x$ the inverse $x = \text{arc cot} y$, or (by interchange of x and y) $y = \text{arc cot} x$, is uniquely determined if we require that its value shall lie in the interval from 0 to π; the many-valuedness of arc cot x is otherwise the same as for arc tan x.

The differentiation formulæ may be found as follows:

$$x = \arctan y, \quad \frac{dx}{dy} = \frac{1}{\dfrac{dy}{dx}} = \cos^2 x = \frac{1}{1 + \tan^2 x} = \frac{1}{1 + y^2};$$

$$x = \text{arc cot} y, \quad \frac{dx}{dy} = -\sin^2 x = -\frac{1}{1 + \cot^2 x} = -\frac{1}{1 + y^2};$$

or finally, if we denote the independent variable by x,

$$\frac{d}{dx} \arctan x = \frac{1}{1 + x^2},$$

$$\frac{d}{dx} \text{arc cot} x = -\frac{1}{1 + x^2}.$$

4. The Corresponding Integral Formulæ.

Expressed in the language of the indefinite integral, the formulæ which we have just derived read as follows:

$$\int \frac{1}{\sqrt{(1-x^2)}}\, dx = \text{arc sin}\, x, \quad \int \frac{1}{\sqrt{(1-x^2)}}\, dx = -\text{arc cos}\, x,$$

$$\int \frac{1}{1+x^2}\, dx = \text{arc tan}\, x, \quad \int \frac{1}{1+x^2}\, dx = -\text{arc cot}\, x.$$

Between the pair of formulæ on the left and that on the right, which express each indefinite integral in the form of two functions which appear entirely different, no contradiction exists. We must remember that in the case of the indefinite integral an arbitrary additive constant remains at our disposal. If we choose these constants so that they differ by $\pi/2$ and recall that $\pi/2 - \text{arc cos}\, x = \text{arc sin}\, x$ and likewise $\pi/2 - \text{arc cot}\, x = \text{arc tan}\, x$ this formal disagreement is immediately cleared up. The indefiniteness simply depends on the fact that the indefinite integral is not a *single* definite function, but a whole family of functions which differ from one another by arbitrary additive constants. The equation for an indefinite integral specifies not *the* value, but only *a* value, of it. As we have already remarked, it would be more correct to express this fact by always including the undetermined constant, thus writing, not

$$\int f(x)\, dx = F(x),$$

but
$$\int f(x)\, dx = F(x) + c.$$

For convenience, however, it is usual to avoid this more detailed form; the reader should therefore be all the more careful to bear in mind the indefiniteness which is always associated with the shorter form (see also p. 116).

From the formulæ for indefinite integration there immediately follow formulæ for definite integration, as on p. 117. In particular,

$$\int_a^b \frac{dx}{1+x^2} = \text{arc tan}\, x \,\Big|_a^b = \text{arc tan}\, b - \text{arc tan}\, a.$$

If we put $a = 0$, $b = 1$ and recall that $\tan 0 = 0$ and $\tan \pi/4 = 1$, we obtain the remarkable formula

$$\frac{\pi}{4} = \int_0^1 \frac{1}{1 + x^2}\, dx.$$

The number π, which originally arose from the consideration of the circle, is by this formula brought into a very simple relation-

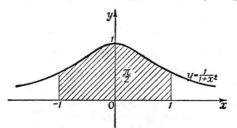

Fig. 7.—$\pi/2$ illustrated by an area

ship with the rational function $\dfrac{1}{1 + x^2}$, *and is expressed by the area defined as shown in fig. 7.*

EXAMPLES

1. If $y = \dfrac{x^2}{4}$, $y = 16$ corresponds to $x = 8$. Find $\dfrac{dy}{dx}$ for $x = 8$; solve $y = \dfrac{x^2}{4}$ for x and find $\dfrac{dx}{dy}$ for $y = 16$, and show that the values of these derivatives are consistent with the rule for inverse functions.

2. Prove that (a) $\operatorname{arc} \sin \alpha + \operatorname{arc} \sin \beta = \operatorname{arc} \sin(\alpha\sqrt{1 - \beta^2} + \beta\sqrt{1 - \alpha^2})$;

      (b) $\operatorname{arc} \sin \alpha + \operatorname{arc} \sin \beta = \operatorname{arc} \cos(\sqrt{1 - \alpha^2}\sqrt{1 - \beta^2} - \alpha\beta)$;

      (c) $\operatorname{arc} \tan \alpha + \operatorname{arc} \tan \beta = \operatorname{arc} \tan \dfrac{\alpha + \beta}{1 - \alpha\beta}$.

Differentiate the expressions in Ex. 3–10 and write down the corresponding integral formulæ:

3. $\dfrac{\sqrt{x}}{1 + x}$.

4. $\sqrt{x}\cos^2 x$.

5. $\dfrac{1 + \sqrt{x}}{1 - \sqrt{x}}$.

6. $\dfrac{\sqrt{x}}{1 - \tan x}$.

7. $\operatorname{arc} \sin x \cdot \operatorname{arc} \cos x$.

8. $\dfrac{1 + \operatorname{arc} \tan x}{1 - \operatorname{arc} \tan x}$.

9. $\dfrac{\operatorname{arc} \sin x}{\operatorname{arc} \tan x}$.

10. $5 \operatorname{arc} \cot x + \dfrac{1}{\operatorname{arc} \cos x}$.

11. Using graph paper, plot $y = \dfrac{1}{1 + x^2}$ on a large scale. By counting squares, find $\displaystyle\int_0^1 \dfrac{1}{1 + x^2}\, dx$, thus obtaining an estimate for $\dfrac{\pi}{4}$ (cf. Ex. 1, p. 121).

4. DIFFERENTIATION OF A FUNCTION OF A FUNCTION

1. The Chain Rule.

The preceding rules for differentiation enable us to differentiate every function which can be expressed as a rational expression whose terms are functions with known derivatives. We can, however, take yet another important step forward and differentiate all those functions obtained by *compounding* functions with known derivatives. Let $\phi(x)$ be a function which is differentiable in an interval $a \leq x \leq b$ and assumes all values in the interval $\alpha \leq \phi \leq \beta$. We now wish to consider a second differentiable function $g(\phi)$ of the independent variable ϕ, in which the variable ϕ ranges over the interval from α to β. We can now regard the function $g(\phi) = g\{\phi(x)\} = f(x)$ as a function of x in the interval $a \leq x \leq b$. The function $f(x) = g\{\phi(x)\}$ will then be called a function of x *compounded* from the functions g and ϕ, or a *function of a function*.

If, for example, $\varphi(x) = 1 - x^2$ and $g(\varphi) = \sqrt{\varphi}$, this compound function is simply $f(x) = \sqrt{(1 - x^2)}$. For the interval $a \leq x \leq b$ we here take the interval $0 \leq x \leq 1$. The values of the function $\phi(x)$ exactly fill up the interval $0 \leq \varphi \leq 1$; the compound function $f(x) = \sqrt{(1 - x^2)}$ is therefore defined in the interval $0 \leq x \leq 1$.

Another example of the compounding of functions is the function $f(x) = \sqrt{(1 + x^2)}$, where the compounding process may be indicated by the equations

$$\varphi(x) = 1 + x^2, \quad g(\varphi) = \sqrt{\varphi}$$

and where the value of the function $\varphi(x)$ runs through all positive numbers ≥ 1, so that the function $f(x) = g\{\varphi(x)\}$ can be formed for all values of x.

In compounding functions in this way we must naturally be careful to restrict ourselves to intervals $a \leq x \leq b$ for which the compound function is defined. For example, the compound function $\sqrt{(1 - x^2)}$ is defined only for values of x in the region $-1 \leq x \leq 1$, and not in the region $1 < x \leq 2$, for when x is in this last interval the values of the function $\varphi(x)$ consist of negative numbers, for which the function $g(\varphi)$ is not defined.

Just as we can compound two functions with one another, we can and

6*   (1793)

must consider functions in which the compounding process is performed more than once. Such a function is

$$\sqrt{(1 + \text{arc tan } x^2)}$$

which can be built up by the compounding process

$$\varphi(x) = x^2, \quad \psi(\varphi) = 1 + \text{arc tan } \varphi, \quad g(\psi) = \sqrt{\psi(\varphi)} = f(x).$$

For the differentiation of compound functions we have the following fundamental theorem, the *chain rule of the differential calculus*:

The function f(x) = g{ϕ(x)} *is differentiable, and its derivative is given by the equation*

$$f'(x) = g'(\phi) \cdot \phi'(x),$$

or, in Leibnitz's notation,

$$\frac{dy}{dx} = \frac{dy}{d\phi} \cdot \frac{d\phi}{dx}.$$

In words: *the derivative of the compound function is the product of the derivatives of the constituent functions.*

The proof of this formula follows very easily if we recall the meaning of the derivative. For any arbitrary $\Delta x \neq 0$ and corresponding values of $\Delta \phi$ and Δg there exist two quantities ϵ and η, tending to 0 with Δx, such that

$$\Delta g = g'(\phi)\Delta \phi + \epsilon \Delta \phi \quad \text{and} \quad \Delta \phi = \phi'(x)\Delta x + \eta \Delta x;$$

we have only to calculate η from the second equation and, where $\Delta \phi \neq 0$, ϵ from the first equation, while if $\Delta \phi = 0$, we put $\epsilon = 0$. If in the first of these equations we now substitute the value of $\Delta \phi$ from the second equation, we obtain

$$\Delta g = g'(\phi)\phi'(x)\Delta x + \{\eta g'(\phi) + \epsilon \phi'(x) + \epsilon \eta\}\Delta x,$$

or
$$\frac{\Delta g}{\Delta x} = g'(\phi)\phi'(x) + \{\eta g'(\phi) + \epsilon \phi'(x) + \epsilon \eta\}.$$

In this equation, however, we can let Δx tend to 0, and at once obtain the result stated, since the bracket on the right tends to zero with Δx. Consequently the left-hand side of our equation has a limit $f'(x)$, and this limit is equal to the first term on the right-hand side, as was stated.*

* We could also have proved the rule by carrying out the passage to the limit $\Delta x \to 0$, and consequently $\Delta \phi \to 0$, in the equation $\dfrac{\Delta g}{\Delta x} = \dfrac{\Delta g}{\Delta \phi} \cdot \dfrac{\Delta \phi}{\Delta x}$. The

By successive application of our formula we can immediately extend it to functions which arise from the *compounding of more than two functions*. If, for example,

$$y = g(u), \quad u = \phi(v), \quad v = \psi(x),$$

we can think of $y = f(x)$ as a function of x; its derivative is given by the rule

$$\frac{dy}{dx} = y' = g'(u)\,\phi'(v)\,\psi'(x) = \frac{dy}{du} \cdot \frac{du}{dv} \cdot \frac{dv}{dx}.$$

The case of a function compounded of an arbitrary number of functions is essentially similar. The proof may be left to the reader.

2. Examples.

As a very simple example we consider the function $y = x^\alpha$, where we put $\alpha = p/q$, q being a positive integer and p a positive or negative integer, so that α is an arbitrary positive or negative rational number. Let x be positive. By the chain rule with

$$y = \phi^p, \quad \phi = x^{1/q}$$

we have the formula

$$y' = p\phi^{p-1} \cdot \frac{1}{q}\, x^{(1-q)/q} = \frac{p}{q}\, x^{p/q-1},$$

so that for arbitrary rational values of α we obtain the differentiation formula

$$\frac{d}{dx}\, x^\alpha = \alpha x^{\alpha-1},$$

in agreement with the result already found in another way in Chap. II, § 3 (p. 94).

As a second example, we consider

$$y = \sqrt{(1 - x^2)} \quad \text{or} \quad y = \sqrt{\phi},$$

where $\phi = 1 - x^2$ and $-1 < x < 1$. The chain rule gives us

$$y' = \frac{1}{2\sqrt{\phi}} \cdot (-2x) = -\frac{x}{\sqrt{(1 - x^2)}}.$$

Further examples are given in the following brief calculations:

1. $y = \arcsin \sqrt{(1 - x^2)}$,

$$\frac{dy}{dx} = \frac{1}{\sqrt{\{1 - (1 - x^2)\}}} \cdot \frac{d\sqrt{(1 - x^2)}}{dx}$$

$$= \frac{1}{|x|} \cdot \frac{-x}{\sqrt{(1 - x^2)}} = \mp \frac{1}{\sqrt{(1 - x^2)}}.$$

---

method in the text is, however, to be preferred, since it avoids the necessity for considering the case $\phi'(x) = 0$ specially.

2. $$y = \sqrt{\left(\frac{1+x}{1-x}\right)},$$

$$\frac{dy}{dx} = \frac{1}{2\sqrt{\left(\frac{1+x}{1-x}\right)}} \cdot \frac{d\left(\frac{1+x}{1-x}\right)}{dx}$$

$$= \frac{\sqrt{(1-x)}}{2\sqrt{(1+x)}} \cdot \frac{2}{(1-x)^2} = \frac{1}{(1+x)^{1/2}(1-x)^{3/2}}.$$

The chain rule for differentiation can also be expressed in the form of an integration formula, in agreement with the fact that to each differentiation formula there corresponds a completely equivalent integration formula. Nevertheless, we will pass over this formula for the present, since we have no immediate need of it here and, moreover, it is discussed in detail later (Chap. IV, § 2, p. 207).

3. Further Remarks on the Integration and Differentiation of x^α when α is Irrational.

In view of the elementary definition of the power x^α by the equation

$$x^\alpha = \lim x^{r_n},$$

where the numbers r_n form a sequence of rational numbers with the limit α, we might be tempted to effect the differentiation of x^α by direct passage to the limit in the differentiation formula

$$\frac{d}{dx} x^{r_n} = r_n x^{r_n - 1}.$$

We are not entitled to do this unless we have the right to conclude that from the relation $x^{r_n} \to x^\alpha$ there follows the relation $\frac{d}{dx} x^{r_n} \to \frac{d}{dx} x^\alpha$. There

is, however, a very serious objection to such a passage to the limit. For in any arbitrarily small neighbourhood of a given curve other curves may be drawn whose direction at arbitrarily selected points differs from the direction of the original curve by any desired amount; for example, we may approximate to a straight line by a wave lying ar-

Fig. 8.—Approximation to a straight line by wavy curves

bitrarily near it, the angle between the wave and the line reaching a value as high as $45°$ (see fig. 8). In other words, the above example shows us that *from the fact that two functions differ only very little from one another, we cannot immediately conclude that their derivatives also are everywhere nearly equal to one another.* This objection forbids us to perform the apparently obvious passage to the limit, in the absence of further justification.

In this respect, however, the integral behaves quite differently from the derivative. We have already observed on p. 128 that if two functions differ by less than ε throughout the interval from a to b, their integrals must differ by less than $\varepsilon(b-a)$. We there used this result to establish the validity of the differentiation formula

$$\frac{1}{\alpha+1}\frac{d}{dx}x^{\alpha+1}=x^{\alpha},$$

or, replacing $\alpha+1$ by α,

$$\frac{d}{dx}x^{\alpha}=\alpha x^{\alpha-1}.$$

In this indirect way, therefore, the relation $\frac{d}{dx}x^{r_n}\to\frac{d}{dx}x^{\alpha}$ given above is verified.

The above discussion is a characteristic example of the interrelations of the differential calculus and the integral calculus. Yet in principle it is preferable to replace (as we shall do on p. 173 *et seq.*) the elementary definition of x^{α} by another, essentially simpler, definition which will lead us once more to the same result, and this time directly.

EXAMPLES

Differentiate the following functions:

1. $(x+1)^3$.

2. $(3x+5)^2$.

3. $(x^9-3x^6-x^3)^5$.

4. $\dfrac{1}{1+x}$.

5. $\dfrac{1}{1-x^2}$.

6. $(ax+b)^n$ (n an integer).

7. $\dfrac{1}{x+\sqrt{(x^2-1)}}$.

8. $\sqrt{\left(\dfrac{ax^2+bx+c}{lx^2+mx+n}\right)}$.

9. $(\sqrt{(1-x)^{2/3}})^5$.

10. $\sin^2 x$.

11. $\sin(x^2)$.

12. $\sqrt{(1+\sin^2 x)}$.

13. $x^2\sin\dfrac{1}{x^2}$.

14. $\tan\dfrac{1+x}{1-x}$.

15. $\sin(x^2+3x+2)$.

16. $\arcsin(3+x^3)$.

17. $\arcsin(\cos x)$.

18. $\sin(\arccos\sqrt{(1-x^2)})$.

19. $x^{\sqrt{2}}-x^{-\sqrt{2}}$.

20. $[\sin(x+7)]^{\sqrt[3]{5}}$.

21. $[\arcsin(a\cos x+b)]^x$.

5. Maxima and Minima

Now that we have attained a certain mastery of the problem of differentiating the elementary functions and the functions compounded from them, we are in a position to make a variety of applications. Here we shall consider the simplest of these applications, the theory of maxima and minima of a function, in conjunction with a geometrical discussion of the second derivative, and then in the next section we shall again take up the thread of the general theory.

1. Convexity or Concavity of Curves.

By definition the derivative $\dfrac{d}{dx} f(x)$ of a function $f(x)$ gives the slope of the curve $y = f(x)$. This slope can itself be repre-

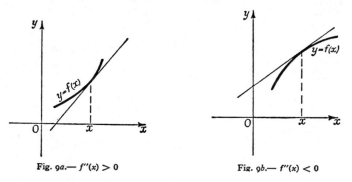

Fig. 9a.— $f''(x) > 0$ Fig. 9b.— $f''(x) < 0$

sented by a curve $y = \dfrac{d}{dx} f(x) = f'(x)$, the *derived* curve of the given curve. The slope of this last curve will be given by the derivative $\dfrac{d}{dx} f'(x) = \dfrac{d^2}{dx^2} f(x) = f''(x)$, the second derivative of $f(x)$, and so on. If the second derivative $f''(x)$ is positive at a point x—so that owing to continuity (which we here assume) it is positive in a certain neighbourhood of the point x—then the derivative $f'(x)$ must increase as it passes this point in the direction of increasing values of x. Hence the curve $y = f(x)$ turns its convex side towards the direction of decreasing values of y. The opposite is true if $f''(x)$ is negative. In the first case, therefore, the curve in the neighbourhood of the point lies above the tangent, in the second case below the tangent (see figs. 9a and b).

Special consideration is required only in the case of points where $f''(x) = 0$. On passing through such a point the second

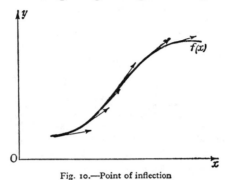

Fig. 10.—Point of inflection

derivative $f''(x)$ will, as a rule, change its sign. Such a point will then be a point of transition between the two cases indicated above; that is, the tangent will on one side be above the curve, and on the other side below it, so that besides touching the curve it will also cross it (see fig. 10). Such a point is called a *point of inflection* of the curve, and the corresponding tangent is called an inflectional tangent.

The simplest example is given by the function $y = x^3$, the cubical parabola, for which the x-axis itself is an inflectional tangent at the point $x = 0$. Another example is given by the function $f(x) = \sin x$, for which $f'(x) = d(\sin x)/dx = \cos x$ and $f''(x) = d^2(\sin x)/dx^2 = -\sin x$. Consequently $f'(0) = 1$ and $f''(0) = 0$; since the sign of $f''(x)$ changes at $x = 0$, the sine curve has at the origin an inflectional tangent inclined at an angle of 45° to the x-axis.

It must, however, be noted that points can exist where $f''(x) = 0$ although the tangent does not cut the curve, but remains entirely on one side of it. For example, the curve $y = x^4$ lies entirely above the x-axis, although the second derivative $f''(x)$ vanishes for $x = 0$.

2. Maxima and Minima.

We say that a continuous function or a curve $y = f(x)$ has a *maximum* (*minimum*) at a point ξ if in at least some neighbourhood of the point $x = \xi$ the values of the function $f(x)$ for $x \neq \xi$ are all less than $f(\xi)$ (greater than $f(\xi)$). By a *neighbourhood* of a point we mean an interval $\alpha \leqq x \leqq \beta$ which contains the point ξ in its interior. Geometrically speaking, such maxima and minima are respectively the wave-crests and wave-troughs of the curve. A glance at fig. 11 shows us that the value of the

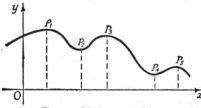

Fig. 11.—Maxima and minima

maximum at one point P_5 may very well be less than the value of the minimum at another point P_2; thus the concept of maximum and minimum is always to some extent relative, on account of the restriction to a certain neighbourhood.

If we wish to fix upon the actual greatest or least value of the function we must employ special means for deciding how this value is to be selected from among the maxima or minima.

The point for us at present is to find the (relative) maxima or minima, or, to use a word that covers both maxima and minima, the *relative extreme values* (*extrema*) * of a given function or curve. This problem, which is very frequently encountered in geometry, mechanics, and physics and which occurs in many other applications, formed one of the principal incentives for the development of the differential and integral calculus in the seventeenth century.

We see at once that if the function is assumed to be differentiable, the tangent to the curve at an extreme value ξ must be horizontal. Hence the condition

$$f'(\xi) = 0$$

is a *necessary* condition for an extreme value; by solving this equation for the unknown ξ we obtain the points at which an extreme value *may possibly* occur. Our condition, however, is by no means a *sufficient* condition for an extreme value; there may be points at which the derivative vanishes, i.e. at which the tangent is horizontal, although the curve has neither a maximum nor a minimum there. This occurs if at the given point the curve has a horizontal inflectional tangent cutting it, as in the above example of the function $y = x^3$ at the point $x = 0$.

* The expressions *turning value, turning point*, are also used. On the other hand, the terms *stationary value, stationary point*, include inflections as well as maxima and minima.

If, however, we have found a point at which $f'(x)$ vanishes, we may immediately conclude that the function has a maximum at that point if $f''(\xi) < 0$, a minimum if $f''(\xi) > 0$. For in the first case the curve in the neighbourhood of this point lies completely below the tangent, in the second case completely above the tangent.

Instead of basing the deduction of our necessary condition on intuition we could, of course, have given an easy proof by purely analytical methods (cf. the exactly analogous considerations for Rolle's theorem, p. 105). If the function $f(x)$ has a maximum at the point ξ, then for all sufficiently small values of h different from 0 the expression $f(\xi) - f(\xi + h)$ must be positive. Therefore the quotient $\dfrac{f(\xi + h) - f(\xi)}{h}$ will be positive or negative, according as h is negative or positive. Thus if h tends to zero through negative values the limit of this quotient cannot be negative, while if h tends to zero through positive values the limit cannot be positive. But since we have assumed that the derivative exists these two limits must be equal to one another and, in fact, to $f'(\xi)$, which therefore can only have the value zero; we must have $f'(\xi) = 0$. A similar proof holds for the case of a minimum.

We can also formulate, and prove analytically, conditions which are *necessary and sufficient* for the occurrence of a *maximum or a minimum*, without involving the second derivative. We suppose that the function $f(x)$ is continuous and has a continuous derivative $f'(x)$ which vanishes only at a finite number of points.

Then f(x) *has a maximum or a minimum at the point* x $= \xi$ *if, and only if, the derivative* f'(x) *changes sign on passing through this point; in particular, the function has a minimum if the derivative is negative to the left of ξ and positive to the right, while in the contrary case it has a maximum.*

We prove this by using the mean value theorem. First, we observe that to the left and right of ξ there exist intervals $\xi_1 < x < \xi$ and $\xi < x < \xi_2$ (extending to the nearest points at which $f'(x) = 0$) in each of which $f'(x)$ has only one sign. If the signs of $f'(x)$ in these two intervals are different, then $f(\xi + h) - f(\xi) = hf'(\xi + \theta h)$ has the same sign for all numerically small values of h, whether h is positive or negative, so that

$f(\xi)$ is an extreme value. If $f'(x)$ has the same sign in both intervals, then $hf'(\xi + \theta h)$ changes sign when h does, so that $f(\xi + h)$ is greater than $f(\xi)$ on one side and less than $f(\xi)$ on the other side, and there is no extreme value. Our theorem is thus proved.

At the same time we see that the value $f(\xi)$ is the greatest or least value of the function in every interval, containing the point ξ, in which the only change of sign of $f'(x)$ occurs at ξ itself.

The mean value theorem on which this proof is based can still be used even if $f(x)$ is not differentiable at an end-point of the interval in which it is applied, provided that $f(x)$ is differentiable at all the other points of the interval; for example, the above proof still holds if $f'(x)$ does not exist at $x = \xi$. This leads us to the following more general result: if the function $f(x)$ is continuous in an interval containing the point ξ, and everywhere in this interval, with the possible exception of ξ itself, has a derivative $f'(x)$ which vanishes at not more than a finite number of points, then $f(x)$ has an extreme value at the point $x = \xi$ if, and only if, the point ξ separates two intervals in which $f'(x)$ has different signs. For example, the function $y = |x|$ has a minimum at $x = 0$, since $y' > 0$ for $x > 0$ and $y' < 0$ for $x < 0$ (cf. fig. 9, p. 97). The function $y = \sqrt[3]{x^2}$ likewise has a minimum at the point $x = 0$, even though its derivative $\frac{2}{3}x^{-\frac{1}{3}}$ is infinite there (cf. fig. 12, p. 99).

In addition we make the following remark on the general theory of maxima and minima: the finding of maxima and minima is not directly equivalent to the finding of the greatest and least values of a function in a closed interval. In the case of a monotonic function these greatest and least values will be assumed at the ends of the interval and are therefore not maxima and minima in our sense; for this latter concept refers to a *complete neighbourhood* of the place in question. Thus for example the function $f(x) = x$ in the interval $0 \leq x \leq 1$ assumes its greatest value at the point $x = 1$, and its least value at $x = 0$, and a corresponding statement holds for every monotonic function. The function $y = \arctan x$, whose derivative is $1/(1 + x^2)$, is monotonic for $-\infty < x < +\infty$, and in that open interval possesses neither a maximum nor a minimum, nor a greatest or a least value.

If after finding the zeros of $f'(x)$ we wish to make sure that we have thereby found the points at which the function has its greatest or least values, we can often make use of the following criterion:

A point ξ at which f'(x) vanishes gives the least or greatest value of the function f(x) in a whole interval, if throughout that interval f''(x) > 0 or f''(x) < 0 respectively.

For if ξ and $\xi + h$ both belong to the interval

$$f'(\xi + h) = f'(\xi + h) - f'(\xi) = hf''(\xi + \theta h),$$

by the mean value theorem. Hence at the point $x = \xi + h$ the derivative $f'(x)$ has the same sign as h or the opposite sign, according as $f''(x) > 0$ or $f''(x) < 0$; the statement then follows from the remark following the theorem at the top of p. 162.

3. Examples of Maxima and Minima.

Ex. 1. Of all rectangles of given area, to find that with the least perimeter.

Let a^2 be the area of the rectangle and x the length of one side (here we must consider x as ranging over the interval $0 < x < \infty$); then the length of the other side is a^2/x, and half the perimeter is given by

$$f(x) = x + \frac{a^2}{x}.$$

We have
$$f'(x) = 1 - \frac{a^2}{x^2}, \qquad f''(x) = \frac{2a^2}{x^3}.$$

The equation $f'(\xi) = 0$ has the single positive root $\xi = a$. For this value $f''(x)$ is positive (as it is for any positive value of x); it therefore gives the required least value, and we obtain the very plausible result that of all rectangles of given area the square has the smallest perimeter.

Ex. 2. Of all triangles with given base and given area, to find that with the least perimeter.

To solve this problem, we take the x-axis along the given base AB and the middle point of AB as the origin. If C is the vertex of the triangle, h its altitude (which is fixed), and (x, h) are the co-ordinates of the vertex, then the sum of the two sides of the triangle AC and BC which are to be determined will be given by

$$f(x) = \sqrt{\{(x + a)^2 + h^2\}} + \sqrt{\{(x - a)^2 + h^2\}}$$

where $2a$ is the length of the base. From this we obtain

$$f'(x) = \frac{x + a}{\sqrt{\{(x + a)^2 + h^2\}}} + \frac{x - a}{\sqrt{\{(x - a)^2 + h^2\}}},$$

$$f''(x) = \frac{-(x+a)^2}{\sqrt{\{(x+a)^2+h^2\}^3}} + \frac{1}{\sqrt{\{(x+a)^2+h^2\}}} + \frac{-(x-a)^2}{\sqrt{\{(x-a)^2+h^2\}^3}}$$

$$+ \frac{1}{\sqrt{\{(x-a)^2+h^2\}}}$$

$$= \frac{h^2}{\sqrt{\{(x+a)^2+h^2\}^3}} + \frac{h^2}{\sqrt{\{(x-a)^2+h^2\}^3}}.$$

We see at once (1) that $f'(0)$ vanishes, (2) that $f''(x)$ is always positive; hence at $x = 0$ there is a least value. For since $f''(x) > 0$ the first derivative $f'(x)$ always increases and therefore cannot be equal to zero at any other point, so that the point $x = 0$ must really give the least value of $f(x)$. This least value is accordingly given by the isosceles triangle.

Similarly, we find that of all triangles with given perimeter and given base the isosceles triangle has the greatest area.

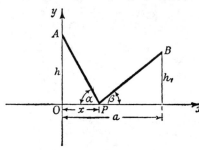

Fig. 12.—Law of reflection

Ex. 3. To find a point on a given straight line such that the sum of its distances from two given fixed points is a minimum.

Let there be given a straight line and two fixed points A and B on the same side of the line. We wish to find a point P on the straight line such that the distance $PA + PB$ has the least possible value.

We take the given line as the x-axis and use the notation of fig. 12. Then the distance in question is given by

$$f(x) = \sqrt{(x^2+h^2)} + \sqrt{\{(x-a)^2+h_1^2\}},$$

and we obtain

$$f'(x) = \frac{x}{\sqrt{(x^2+h^2)}} + \frac{x-a}{\sqrt{\{(x-a)^2+h_1^2\}}},$$

$$f''(x) = \frac{-x^2}{\sqrt{(x^2+h^2)^3}} + \frac{1}{\sqrt{(x^2+h^2)}} + \frac{-(x-a)^2}{\sqrt{\{(x-a)^2+h_1^2\}^3}}$$

$$+ \frac{1}{\sqrt{\{(x-a)^2+h_1^2\}}}$$

$$= \frac{h^2}{\sqrt{(x^2+h^2)^3}} + \frac{h_1^2}{\sqrt{\{(x-a)^2+h_1^2\}^3}}.$$

The equation $f'(\xi) = 0$ accordingly gives us

$$\frac{\xi}{\sqrt{(\xi^2+h^2)}} = \frac{a-\xi}{\sqrt{\{(\xi-a)^2+h_1^2\}}},$$

or

$$\cos\alpha = \cos\beta,$$

which means that the two lines PA and PB must form equal angles with the given line. The positive sign of $f''(x)$ shows us that we really have a least value.

The solution of this problem is closely connected with the optical law of reflection. By an important principle of optics, known as Fermat's *principle of least time*, the path of a light ray is determined by the property that the time that the light takes to go from a point A to a point B under known conditions must be the least possible. If the condition is imposed that a ray of light shall on its way from A to B pass through some point on a given straight line (say on a mirror), we see that the shortest time will be taken along the ray for which the " angle of incidence " is equal to the " angle of reflection ".

Ex. 4. The Law of Refraction.—Let there be given two points A and B on opposite sides of the x-axis. Which path from A to B corresponds to the shortest possible time if the velocity on one side of the x-axis is c_1 and on the other side c_2?

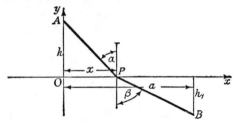

Fig. 13.—Law of refraction

It is clear that this shortest path must lie along two portions of straight lines meeting one another at a point P on the x-axis. Using the notation of fig. 13, we obtain the two expressions $\sqrt{(h^2 + x^2)}$ and $\sqrt{\{h_1^2 + (a - x)^2\}}$ for the lengths PA, PB respectively, and we find the time of passage along this path by dividing the lengths of the two segments by the corresponding velocities and adding. This gives us

$$f(x) = \frac{1}{c_1} \sqrt{(h^2 + x^2)} + \frac{1}{c_2} \sqrt{\{h_1^2 + (a - x)^2\}},$$

for the time taken.

By differentiation, we obtain

$$f'(x) = \frac{1}{c_1} \frac{x}{\sqrt{(h^2 + x^2)}} - \frac{1}{c_2} \frac{a - x}{\sqrt{\{h_1^2 + (a - x)^2\}}},$$

$$f''(x) = \frac{1}{c_1} \frac{h^2}{\sqrt{(h^2 + x^2)^3}} + \frac{1}{c_2} \frac{h_1^2}{\sqrt{\{h_1^2 + (a - x)^2\}^3}}.$$

As we readily see from the figure, the equation $f'(x) = 0$, i.e.

$$\frac{1}{c_1} \frac{x}{\sqrt{(h^2 + x^2)}} = \frac{1}{c_2} \frac{a - x}{\sqrt{\{h_1^2 + (a - x)^2\}}},$$

is equivalent to the condition $\dfrac{1}{c_1} \sin \alpha = \dfrac{1}{c_2} \sin \beta$, or

$$\frac{\sin \alpha}{\sin \beta} = \frac{c_1}{c_2}.$$

We leave it to the reader to prove that there is only *one* point which satisfies this condition and that this point actually yields the required least value. The physical meaning of our example is again given by the optical principle of least time. A ray of light travelling between two points describes the path of shortest time. If c_1 and c_2 are the velocities of light on either side of the boundary of two optical media, the path of the light will be that given by our result, which accordingly gives Snell's *law of refraction.*

Examples

1. Find the maxima, minima, and points of inflection of the following functions. Graph them, and determine the regions of increase and decrease, and of convexity and concavity:

(a) $x^3 - 6x + 2$. (b) $x^{2/3}(1 - x)$. (c) $2x/(1 + x^2)$.
(d) $x^3/(x^4 + 1)$. (e) $\sin^2 x$.

2. Determine the maxima, minima, and points of inflection of $x^3 + 3px + q$. Discuss the nature of the roots of $x^3 + 3px + q = 0$.

3. Which point of the hyperbola $y^2 - \tfrac{1}{2}x^2 = 1$ is nearest to the point $x = 0$, $y = 3$?

4. Let P be a fixed point with co-ordinates x_0, y_0 in the first quadrant of a rectangular co-ordinate system. Find the equation of the line through P such that the length intercepted between the axes is a minimum.

5. A statue 12 ft. high stands on a pillar 15 ft. high. At what distance must a man 6 ft. high stand in order that the statue may subtend the greatest possible angle at his eye?

6. Two sources of light, of intensities a and b, are at a distance d apart. At which point of the line joining them is the illumination least? (Assume that the illumination is proportional to the intensity and inversely proportional to the square of the distance.)

7. Of all rectangles with a given area, find

(a) the one with the smallest perimeter;
(b) the one with the shortest diagonal.

8. In the ellipse $\dfrac{x^2}{a^2} + \dfrac{y^2}{b^2} = 1$ inscribe the rectangle of greatest area.

9. Two sides of a triangle are a and b. Determine the third side so that the area is a maximum.

10. A circle of radius r is divided into two segments by a line g at a distance h from the centre. In the smaller of these segments inscribe the rectangle of greatest possible area.

11. Of all circular cylinders with a given volume, find the one with the least area.

12. Given the parabola $y^2 = 2px$, $p > 0$, and a point $P(x = \xi, y = \eta)$ within it ($\eta^2 < 2p\xi$), find the shortest path (consisting of two line segments) leading from P to a point Q on the parabola and then to the focus $F(x = \frac{1}{2}p, y = 0)$ of the parabola. Show that the angle FQP is bisected by the normal to the parabola, and that QP is parallel to the axis of the parabola. (Principle of the parabolic mirror.)

13.* A prism deflects a beam of light travelling in a plane perpendicular to the edge of the prism. What must the relative position of prism and beam be for the deflection to be a minimum?

14. Given n fixed numbers $a_1, \ldots, a_n$, determine x so that $\sum\limits_{i=1}^{n} (a_i - x)^2$ is a minimum.

15. Prove that if $p > 1$ and $x > 0$, $x^p - 1 \geqq p(x - 1)$.

16. Prove the inequality $1 \geqq \dfrac{\sin x}{x} \geqq \dfrac{2}{\pi}$, $0 \leqq x \leqq \dfrac{\pi}{2}$.

17. Prove that (a) $\tan x \geqq x$, $0 \leqq x \leqq \dfrac{\pi}{2}$.

(b) $\cos x \geqq 1 - \dfrac{x^2}{2}$.

18.* Given $a_1 > 0$, $a_2 > 0, \ldots, a_n > 0$, determine the minimum of

$$\frac{\dfrac{a_1 + \ldots + a_{n-1} + x}{n}}{\sqrt[n]{a_1 a_2 \ldots a_{n-1} x}}$$

for $x > 0$. Use the result to prove by mathematical induction that

$$\sqrt[n]{a_1 a_2 \ldots a_n} \leqq \frac{a_1 + \ldots + a_n}{n}.$$

6. The Logarithm and the Exponential Function

The systematic relations between the differential calculus and the integral calculus lead naturally to a convenient method of approach to the exponential function and the logarithm. Although we have already (pp. 25, 69) investigated these functions, we now define them afresh and develop their theory again without making any use of our previous definition and the results based on it. We begin with the logarithm, and then obtain the exponential function as its inverse.

1. Definition of the Logarithm. The Differentiation Formula.

We have seen that indefinite integration of the power x^n for integral indices n in general leads to a power of x. The only exception is the function $1/x$, which does not appear as the derivative of any of the functions which we have dealt with so far. It is natural to suppose that the indefinite integral of the function $1/x$ represents a new sort of function; so, following up this idea, we will proceed to investigate the function

$$y = \int_1^x \frac{d\xi}{\xi} = f(x)$$

for $x > 0$. We call it the *logarithm of x*, or, more accurately, the *natural logarithm of x*, and write it $y = \log x$ or $y = \text{nat} \log x$. We have denoted the variable of integration by ξ in order to avoid confusion with the upper limit x.

The choice of the number 1 as lower limit is an arbitrary one, which, however, will soon prove its convenience.

In the course of the following argument it will appear that the logarithm defined here is the same as the logarithm which we

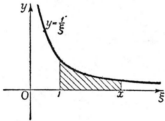

previously (p. 70) defined in an "elementary way". But, as we once more emphasize, the results of the following investigations are independent of those obtained earlier.

Fig. 14.—Log x illustrated by an area

Geometrically our logarithmic function means the area shown shaded in fig. 14, which is bounded above by the rectangular hyperbola $y = 1/\xi$, below by the ξ-axis, and at the sides by the lines $\xi = 1$ and $\xi = x$. This area is to be reckoned positive if $x > 1$, negative if $x < 1$. For $x = 1$ the area vanishes, and we therefore have $\log 1 = 0$.

According to the above definition the derivative of the logarithm is given by the formula

$$\frac{d(\log x)}{dx} = \frac{1}{x}.$$

Here let us expressly emphasize that we assume throughout that the argument x is positive; the logarithm of 0 or of any negative value cannot be formed in accordance with the formula above, for the integrand $1/\xi$ becomes infinite when $\xi = 0$. On the other hand, if we choose some negative number, say -1, for the lower limit, we can form the integral with a negative upper limit x, i.e. we can consider the expression

$$\int_{-1}^{x}\frac{d\xi}{\xi} \qquad (x<0).$$

Owing to the significance of the integral as the limit of a sum or as an area, we see that for $x<0$

$$\int_{-1}^{x}\frac{d\xi}{\xi}=\int_{1}^{-x}\frac{d\xi}{\xi}=\int_{1}^{|x|}\frac{d\xi}{\xi}=\log|x|.$$

In conformity with this we can in general write the formula for indefinite integration as

$$\int\frac{dx}{x}=\log|x|.$$

The logarithm can, of course, be represented by means of a graph. This graph, the logarithmic curve, is shown in fig. 15. We have already seen (p. 119 *et seq.*) how to construct it.

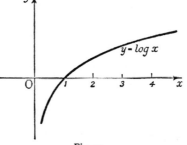

Fig. 15

2. The Addition Theorem.

The logarithm defined as above obeys the following fundamental law:

$$\log(ab)=\log a+\log b.$$

The proof of this *addition theorem* follows directly from the differentiation formula. For, writing $z=\log(ax)$, and applying the chain rule, we have

$$\frac{dz}{dx}=\frac{1}{ax}\cdot a=\frac{1}{x}.$$

But

$$\frac{d}{dx}\log x=\frac{1}{x};$$

and since the functions z and $\log x$ have the same derivative they differ only by a constant, so that $z=\log x+c$ or

$$\log ax=\log x+c.$$

This being true for all positive values of x, we first put $x = 1$ to find c; since $\log 1 = 0$, this yields

$$\log a = c.$$

Substituting this value for c, we have

$$\log ax = \log x + \log a,$$

whence, for $x = b$,

$$\log ab = \log a + \log b,$$

which was to be proved.

For arbitrary positive numbers $a_1, a_2, \ldots, a_n$ the equation

$$\log(a_1 a_2 \ldots a_n) = \log a_1 + \log a_2 + \ldots + \log a_n$$

follows from the addition theorem for the logarithm.

In particular, if all the numbers $a_1, a_2, \ldots, a_n$ are equal to one and the same number a, we have

$$\log a^n = n \log a.$$

Similarly, it follows that

$$\log a + \log \frac{1}{a} = \log 1 = 0,$$

so that

$$\log a = -\log \frac{1}{a}.$$

If, further, we put $\sqrt[n]{a} = \alpha$ it follows that $\log a = n \log \alpha$, or

$$\log \sqrt[n]{a} = \log a^{1/n} = \frac{1}{n} \log a.$$

From this by repeated use of the addition theorem we find that, when m is a positive integer,

$$\frac{m}{n} \log a = \log \sqrt[n]{a^m} = \log a^{m/n}.$$

The equation $\qquad \log a^r = r \log a$

is thus proved for all positive rational values of r, and for $r = 0$ it is obviously correct. For negative rational values of r it is also valid, for then

$$\log a^r = \log \frac{1}{a^{-r}} = -\log a^{-r} = r \log a.$$

3. Monotonic Character and Values of the Logarithm.

The value of the logarithm obviously increases when x increases, and decreases when x decreases; the logarithm is therefore a monotonic function.

Since the derivative $1/x$ becomes smaller and smaller as x increases, the function increases more and more slowly as x increases. Nevertheless, as x increases beyond all bounds the function $\log x$ does not tend to a positive limit, but becomes infinite; that is to say, for every positive number A, no matter how large, there are values of x for which $\log x > A$. This fact follows very readily from the addition theorem. For $\log 2^n = n \log 2$, and since $\log 2$ is a positive number, by taking $x = 2^n$ with sufficiently large values of n we can make $\log x$ as large as we please.

Since $\log(1/2^n) = -n \log 2$, we see that as x tends to zero through positive values $\log x$ is negative and increases numerically beyond all bounds.

Summing up these results:

The function $\log x$ is a monotonic function which assumes all values between $-\infty$ and $+\infty$ as the independent variable x ranges over the continuum of positive numbers.

4. The Inverse Function of the Logarithm (the Exponential Function).

Since the function $y = \log x$ $(x > 0)$ is a monotonic function of x which assumes all real values, its inverse function, which we shall at first denote by $x = E(y)$, must be a single-valued monotonic function defined for every real value of y; it is differentiable, since $\log x$ itself is differentiable. We interchange the notation for the dependent and independent variables, and proceed to study the function $E(x)$ in detail. In the first place, it must clearly be positive for every value of x. Further, we must have

$$E(0) = 1;$$

for this equation is equivalent to the statement that $\log 1 = 0$.

Secondly, from the addition theorem for the logarithm there immediately follows the *multiplication theorem*

$$E(a)E(\beta) = E(a + \beta).$$

To prove this we need only notice that the equations

$$E(\alpha) = a, \quad E(\beta) = b, \quad E(\alpha + \beta) = c$$

are equivalent to

$$a = \log a, \quad \beta = \log b, \quad \alpha + \beta = \log c.$$

Since by the addition theorem for the logarithm $\alpha + \beta = \log ab$, it must be true that $c = ab$, which proves the multiplication theorem.

From this theorem we derive a fundamental property of the function $y = E(x)$, which gives us the right to call our function the *exponential function* and to write it symbolically in the form

$$y = e^x.$$

In order to obtain this property we observe that there must be a number—which we shall call * e—for which

$$\log e = 1.$$

This is equivalent to the definition

$$E(1) = e.$$

Using the multiplication theorem for the function $E(x)$, we have

$$E(n) = e^n,$$

and, in the same way, for positive integers m and n,

$$E\left(\frac{m}{n}\right) = e^{m/n},$$

which we could also have found directly from the addition theorem for the logarithm.

The equation $E(r) = e^r$ thus proved for positive rational numbers r holds also for negative rational numbers in virtue of the equation

$$E(r)E(-r) = E(0) = 1.$$

The function $E(x)$ is therefore a function which is continuous for *all* values of x, and which for *rational* values of x coincides with e^x. These facts give us the right to call our function e^x for

---

* Its identity with the number e considered on p. 43 will be proved in No. 6 (p. 175).

arbitrary irrational values of x also.* (It should be noticed that here the continuity of e^x is an immediate consequence of its definition as the inverse function of a continuous monotonic function, while if the elementary definition is adopted the continuity must be *proved*.)

The exponential function is differentiated according to the formula

$$\frac{d}{dx} e^x = e^x \quad \text{or} \quad y' = y.$$

This formula expresses the important fact that *the derivative of the exponential function is the function itself.*

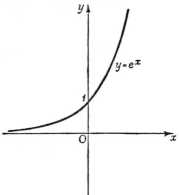

The proof is extremely simple. For we have $x = \log y$, whence, by the formula for the differentiation of the logarithm, we have $\dfrac{dx}{dy} = \dfrac{1}{y}$, and then by the rule for inverse functions

$$\frac{dy}{dx} = y = e^x,$$

as was stated.

Fig. 16.—The exponential function

The graph of the exponential function e^x, the so-called exponential curve, is obtained by reflection of the logarithmic curve in the line which bisects the first quadrant. It is shown in fig. 16.

5. The General Exponential Function a^x and the General Power x^a.

The exponential function a^x for an arbitrary positive base a is now simply *defined* by the equation

$$y = a^x = e^{x \log a},$$

---

* If we anticipate the fact, which will be proved on p. 175, that our number e is identical with the number so denoted previously, we have now proved that the definition given here yields the same exponential function with base e as was formerly defined by the process of raising to powers. For, according to that elementary definition, we defined the values of e^x for irrational x's as the limit of the expressions e^{x_n}, where x_n takes on a sequence of rational values with the limit x.

which agrees with the earlier definition in virtue of the relation

$$e^{\log a} = a.$$

Using the chain rule we immediately obtain

$$\frac{d}{dx}\, a^x = \frac{d}{dx}\, e^{x \log a} = e^{x \log a} \cdot \log a,$$

$$= a^x \log a.$$

The inverse function of the exponential function $y = a^x$ is called the *logarithm to the base a* and is written

$$x = \log_a y,$$

while the logarithmic function previously introduced, when a distinction is necessary, is spoken of as the natural logarithm, or logarithm to the base e.

From the definition it follows immediately that

$$\log y = x \log a = \log_a y \cdot \log a,$$

which shows us that the logarithm of y to an arbitrary positive base $a \neq 1$ is obtained by multiplying the natural logarithm of y by the reciprocal of the natural logarithm of a, the *modulus* of the system of logarithms to the base * a.

Instead of our previous definition of the *general power* $x^a (x > 0)$ we shall now define this power by the equation

$$x^a = e^{a \log x}.$$

The rule for differentiating the power x^a follows immediately from the definition, using the chain rule; for

$$\frac{d}{dx}\, x^a = e^{a \log x} \cdot \frac{a}{x} = ax^{a-1},$$

in agreement with our previous result (cf. p. 155).

---

* If we take $a = 10$ we obtain the ordinary " Briggian " logarithms, which have already been met with in elementary mathematics and which are advantageous for use in numerical calculations.

6. The Exponential Function and the Logarithm represented as Limits.

We are now in a position to state important limiting relations referring to the quantities introduced above. We begin with the formula for differentiating the function $f(x) = \log x$,

$$\frac{1}{x} = f'(x) = \lim_{h \to 0} \frac{f(x+h) - f(x)}{h} = \lim_{h \to 0} \frac{\log(x+h) - \log x}{h}$$

$$= \lim_{h \to 0} \frac{1}{h} \log\left(1 + \frac{h}{x}\right).$$

If we put $\dfrac{1}{x} = z$, this becomes

$$\lim_{h \to 0} \frac{1}{h} \log(1 + zh) = z.$$

Since the function e^x is continuous for all values of x, this implies that

$$e^z = \lim_{h \to 0} e^{\{\log(1+zh)/h\}} = \lim_{h \to 0} (1 + zh)^{1/h}. \quad . \quad . \quad (a)$$

If in particular we give h the sequence of values $1, \dfrac{1}{2}, \dfrac{1}{3}, \ldots, \dfrac{1}{n}, \ldots,$ we have

$$\lim_{n \to \infty} \left(1 + \frac{z}{n}\right)^n = e^z. \quad . \quad . \quad . \quad (b)$$

If to z we assign the value 1, formula (a) gives the following important fact:

As h *tends to zero, the expression* $(1 + h)^{1/h}$ *tends to the number* e:

$$\lim_{h \to 0} (1 + h)^{1/h} = e.$$

Formula (b) gives

$$\lim_{n \to \infty} \left(1 + \frac{1}{n}\right)^n = e,$$

which proves that the number e is the same as the number denoted by the symbol e on p. 43.

From the differentiation formula for a^x,

$$a^x \log a = \lim_{h \to 0} \frac{a^{x+h} - a^x}{h},$$

it follows for $x = 0$ that

$$\log a = \lim_{h \to 0} \frac{a^h - 1}{h},$$

a formula which expresses the logarithm of a directly as a limit.

To this equation we append the remark that by its means we can complete the relation

$$\int_a^b x^a \, dx = \frac{1}{a+1} (b^{a+1} - a^{a+1})$$

established earlier. We have always been obliged to exclude the case $a = -1$. Now, however, we can trace what happens when the number a tends to the *limit* -1. If we put $a = 1$ the left-hand side will by our definition of the logarithm have the limit *

$$\int_1^b \frac{dx}{x} = \log b;$$

the right-hand side therefore has the same limit when $a \to -1$. This fact, moreover, is in accordance with the formula

$$\log b = \lim_{h \to 0} \frac{b^h - 1}{h};$$

we need only write $a + 1 = h$.

We have thus cleared up the exceptional case $a = -1$ in the integration formula which we have so often used. The formula above is still meaningless when $a = -1$, but as a limit formula it retains its significance as $a \to -1$.

7. Final Remarks.

Here we briefly review the train of thought followed out in this section. We first defined the natural logarithm $y = \log x$ for $x > 0$ by means of an integral, whence we immediately deduced the differentiation formula, the addition theorem, and the existence of an inverse. We then investigated the inverse function $y = e^x$, where the number e was seen to be the number

---

* We have here carried out the passage to the limit $a \to -1$ under the integral sign without further investigation; cf. the discussion on p. 128 *et seq.*

whose logarithm is 1, and we derived its differentiation formula, as well as limit expressions for it and for the logarithm. The introduction of the functions $y = x^a = e^{a \log x}$ and $y = a^x = e^{x \log a}$ followed naturally.

In the discussion given here, as contrasted with the " elementary " treatment, the question of continuity causes no difficulty, since the logarithm is defined as an integral and therefore as a continuous and differentiable function, whose inverse function is also continuous.

EXAMPLES

1. Sketch the function $y = \dfrac{1}{x}$ $(1 \leq x \leq 2)$ on a large scale, using graph paper, and find $\log_e 2$ by counting squares.

Differentiate the functions in Ex. 2–5:

2. $x (\log x - 1)$.

3. $\log \log x$.

4. $\log \{x + \sqrt{(1 + x^2)}\}$.

5. $\log \{ \sqrt{(1 + \log x)} - \sin x\}$.

6. Differentiate $\log \dfrac{\sqrt{(x^2 + 1)}}{\sqrt[3]{(2 + x)}}$; (a) by using the chain rule and the quotient rule, without preliminary simplification; (b) first simplifying by means of the theorems on logarithms.

7. (a) Differentiate $y = \dfrac{\sqrt[3]{(7x^2 + 1)}}{\sqrt[4]{(x - 2)} \sqrt{(x^4 + 1)}}$.

(b) Differentiate the same function, first taking logarithms and simplifying.

8.* Given $\lim\limits_{n \to \infty} \varepsilon_n = 0$, prove that $\lim\limits_{n \to \infty} \left(1 + \varepsilon_n \cdot \dfrac{x}{n}\right)^n = 1$.

9. Show that the function $y = e^{-\alpha x} (a \cos x + b \sin x)$ satisfies the equation

$$y'' + 2\alpha y' + (\alpha^2 + 1)y = 0$$

for all values of a and b.

10.* Show that $\dfrac{d^n}{dx^n} (e^{-1/x^2}) = \dfrac{P_n(x)}{x^{3n}} e^{-1/x^2}$, when $x \neq 0$, where $P_n(x)$ is a polynomial of degree $2n - 2$. Establish the " recurrence formula "

$$P_{n+1}(x) = (2 - 3nx^2) P_n(x) + x^3 P_n'(x).$$

11. Find the maximum of $y = x \lambda^a e^{-\lambda x}$, where λ and α are constants. Find the locus of this maximum when λ is allowed to vary.

12. Differentiate $a^{(a^x)}$ $(a > 0)$.

13. Differentiate $a^{\sin x (\log x)^2}$.

7

(E 798)

7. Some Applications of the Exponential Function

In this section we shall consider some miscellaneous problems involving the exponential function, and we shall thus gain an insight into the fundamental importance of this function in all sorts of applications.

1. Definition of the Exponential Function by Means of a Differential Equation.

We can define the exponential function by a simple theorem, whose use will save us many detailed investigations of particular cases.

If a function $\mathrm{y} = \mathrm{f}(\mathrm{x})$ *satisfies an equation of the form*

$$y' = ay$$

where a is a constant other than zero, then y *has the form*

$$y = f(x) = ce^{ax},$$

where c is also a constant; and conversely, every function of the form ce^{ax} *satisfies the equation* $\mathrm{y}' = a\mathrm{y}$. The latter is usually briefly referred to as a *differential equation*, since it expresses a relation between the function and its derivative.

In order to make the theorem clear, we notice first of all that in the simplest case $a = 1$ the above equation becomes $y' = y$. We know that $y = e^x$ satisfies this equation, and it is clear that the same is also true of $y = ce^x$, if c is an arbitrary constant. Conversely, we can easily see that no other function satisfies the differential equation. For if y is such a function, we consider the function $u = ye^{-x}$. We must then have

$$u' = y'e^{-x} - ye^{-x} = e^{-x}(y' - y).$$

But the right-hand side vanishes, since we have assumed that $y' = y$; hence $u' = 0$, so that by p. 114 *et seq.* u is a constant c and $y = ce^x$, as we wished to prove.

The case of any non-zero value of a can be treated in exactly the same way as the special case $a = 1$. If we introduce the function $u = ye^{-ax}$, we obtain the equation $u' = y'e^{-ax} - aye^{-ax}$. Hence from the assumed differential equation we find that $u' = 0$, so that $u = c$ and $y = ce^{ax}$. The converse is clear.

We will now apply this theorem to a number of examples and thus make it more intelligible.

2. Interest Compounded Continuously. Radioactive Disintegration.

A capital sum, or principal, which has its interest added to it at regular periods of time, increases by jumps at these interest periods in the following manner. If 100α is the rate of interest per cent, and if further the interest accrued is added to the principal at the end of each year, then after x years the accumulated amount of an original principal of 1 will be

$$(1 + \alpha)^x.$$

If, however, the principal had the interest added to it not at the end of each year, but at the end of each n-th part of a year, then after x years the principal would amount to

$$\left(1 + \frac{\alpha}{n}\right)^{nx}.$$

Taking $x = 1$ for the sake of simplicity, i.e. reckoning the interest at 100α per cent for one year, we find that if the interest is computed in this latter way the principal 1 amounts after one year to

$$\left(1 + \frac{\alpha}{n}\right)^n.$$

If we now let n increase beyond all bounds, i.e. if we let the interest be calculated at shorter and shorter intervals, the limiting case will signify in a sense that the interest is compounded continuously, at each instant; and we see that the total amount after one year will be e^α times the original principal. Similarly, if the interest is calculated in this manner, an original principal of 1 will have grown after x years to an amount $e^{\alpha x}$; here x may be any number, integral or otherwise.

The discussion in No. 1 (p. 178) forms a framework within which examples of this type are readily understood. We consider a quantity, given by the number y, which increases (or decreases) with the time. Let the rate at which this quantity increases or decreases be proportional to the total quantity. Then if we take the time as the independent variable x, we obtain a law of the form $y' = \alpha y$ for the rate of increase, where α, the factor of proportionality, is positive or negative according as the quantity is increasing or decreasing. Then in accordance with No. 1 the quantity y itself will be given by a formula

$$y = ce^{\alpha x},$$

where the meaning of the constant c is immediately obvious if we consider the instant $x = 0$. At that instant $e^{\alpha x} = 1$, and we find that $c = y_0$ is the quantity at the beginning of the time considered, so that we may write

$$y = y_0 e^{\alpha x}.$$

A characteristic example of the use of these ideas is the case of *radio-active disintegration*. The rate at which the total quantity y of the radio-active substance is diminishing at any instant is proportional to the total quantity present at that instant; this is *a priori* plausible, as each portion of the substance decreases as rapidly as every other portion. Therefore the quantity y of the substance expressed as a function of the time satisfies a relation of the form $y' = -ky$, where k is to be taken as positive since we are dealing with a diminishing quantity. The quantity of substance is thus expressed as a function of the time by $y = y_0 e^{-kx}$, where y_0 is the amount of the substance at the beginning of the time considered (time $x = 0$).

After a certain time τ the radioactive substance will have diminished to half its original quantity. This so-called *half-value period* is given by the equation

$$\tfrac{1}{2}y_0 = y_0 e^{-k\tau},$$

whence we immediately obtain $\tau = \dfrac{\log 2}{k}$.

3. Cooling or Heating of a Body by a Surrounding Medium.

Another typical example of the occurrence of the exponential function is offered by the cooling of a body, e.g. a metal plate, which is immersed in a very large bath of given temperature. In considering this cooling we assume that the surrounding bath is so large that its temperature is unaffected by the cooling process. We further assume that at each instant all parts of the immersed body are at the same temperature, and that the rate at which the temperature changes is proportional to the difference between the temperature of the body and that of the surrounding medium (Newton's law of cooling).

If we denote the time by x and the temperature difference by $y = y(x)$, this law of cooling is expressed by the equation

$$y' = -ky,$$

where k is a positive constant whose value depends on the body itself. From this instantaneous relationship, which expresses the effect of the cooling process at a given instant, we now wish to derive an "integral law" which will allow us to find the temperature at an arbitrary time x from the temperature at an initial time $x = 0$. The theorem of No. 1 (p. 178) immediately gives us this integral law in the form

$$y = ce^{-kx},$$

where k is the above-mentioned constant depending on the body. This shows that the temperature decreases "exponentially" and tends to become equal to the external temperature. The rapidity with which this happens is expressed by the number k. As before, we find the meaning of the constant c by considering the instant $x = 0$; this gives us $y_0 = c$, so that our law of cooling can finally be written in the form

$$y = y_0 e^{-kx}.$$

It is obvious that the same discussion will also apply to the heating of a body. The only difference is that the initial difference of temperature y_0 is in this case negative instead of positive.

4. Variation of the Atmospheric Pressure with the Height above the Surface of the Earth.

As a further example of the occurrence of the exponential formula we shall deduce the law according to which the atmospheric pressure varies with height. We here make use (1) of the physical fact that the atmospheric pressure is equal to the weight of the column of air vertically above a surface of area 1, and (2) of Boyle's law, according to which the pressure of the air (p) at a given constant temperature is proportional to the density of the air (σ). Boyle's law, expressed in symbols, is $p = a\sigma$, where a is a constant which depends on a specific physical property of the air and in addition is proportional to the absolute temperature—here we are not concerned with this, as we shall assume that the temperature is constant. Our problem is to determine $p = f(h)$ as a function of the height (h) above the surface of the earth.

If by p_0 we denote the atmospheric pressure at the surface of the earth, i.e. the total weight of the air column supported by a unit area, and by $\sigma(\lambda)$ the density of the air at the height λ above the earth, the weight of the column up to the height h will be given by the integral $\int_0^h \sigma(\lambda)\,d\lambda$. The pressure at height h will therefore be

$$p = f(h) = p_0 - \int_0^h \sigma(\lambda)\,d\lambda.$$

By differentiation this yields the following relation between the pressure $p = f(h)$ and the density $\sigma(h)$:

$$\sigma(h) = -f'(h) = -p'.$$

We now use Boyle's law to eliminate the quantity σ from this equation, thus obtaining an equation

$$p' = -\frac{1}{a}p$$

which involves the unknown pressure function only. From p. 178 it follows that

$$p = f(h) = ce^{-h/a}.$$

If as above we denote the pressure at the earth's surface, i.e. $f(0)$, by p_0, it follows immediately that $c = p_0$, and consequently

$$p = f(h) = p_0 e^{-h/a}.$$

Changing to logarithms, we obtain

$$h = a \log \frac{p_0}{p}.$$

These two formulæ find frequent application. For example, if the constant a is known they enable us to find the height of a place from the barometric pressure, or to find the difference in height of two places by measuring the atmospheric pressure at each place. Again, if the atmospheric pressure and the height h are known we can determine the constant a, which is of great importance in gas theory.

5. Advance of a Chemical Reaction.

We now consider an example from chemistry, namely, the so-called *unimolecular reaction*. We suppose that a substance is dissolved in a relatively large amount of solvent, say a quantity of cane sugar in water. If a chemical reaction takes place, the chemical law of mass action in this simple case states that the rate of reaction is proportional to the quantity of reacting substance present. If we suppose that the cane sugar is being transformed by catalytic action into invert sugar, and if by $u(x)$ we denote the quantity of cane sugar which at time x is still unchanged, the velocity of reaction will be $-du/dx$, and in accordance with the law of mass action an equation of the form

$$\frac{du}{dx} = -ku$$

holds, where k is a constant depending on the substance reacting. From this instantaneous law we immediately obtain, as on p. 178, an integral law, which gives us the amount of cane sugar as a function of the time:

$$u(x) = ae^{-kx}.$$

This formula clearly shows us how the chemical reaction tends asymptotically to its final state $u = 0$, that is, complete transformation of the reacting substance. The constant a is obviously the quantity present at time $x = 0$.

6. Making and Breaking an Electric Circuit.

As a final example we consider the growth of a (direct) electric current when a circuit is completed (or its decay when the circuit is broken). If R is the resistance of the circuit and E the impressed electromotive force (voltage), the current I will gradually increase from its original value 0 to the steady final value E/R. We have therefore to consider I as a function of the time. The growth of the current depends on the self-induction of the circuit; the circuit has a characteristic constant L, the coefficient of self-induction, of such a nature that as the current increases an electromotive force of magnitude LdI/dx, opposed to the external electromotive force E, is developed. From Ohm's law, according to which the product of the resistance and the current is at each instant equal to the actual effective voltage, we obtain the relation

$$IR = E - L\frac{dI}{dx}.$$

Here we put

$$f(x) = I(x) - \frac{E}{R};$$

we immediately find that $f'(x) = -\frac{R}{L} f(x)$, so that by the theorem on p. 178 $f(x) = f(0)e^{-Rx/L}$. Recalling that $I(0) = 0$, we see that $f(0) = -\frac{E}{R}$, and thus we obtain the expression

$$I = f(x) + \frac{E}{R} = \frac{E}{R}(1 - e^{-Rx/L})$$

for the current as a function of the time.

From this expression we see that when the circuit is closed the current tends asymptotically to its steady value E/R.

EXAMPLES

1. The function $f(x)$ satisfies the equation

$$f(x + y) = f(x)f(y).$$

(a) If $f(x)$ is differentiable, either $f(x) \equiv 0$ or else $f(x) = e^{\alpha x}$.

(b)* If $f(x)$ is continuous, either $f(x) \equiv 0$ or else $f(x) = e^{\alpha x}$.

2. If a differentiable function $f(x)$ satisfies the equation

$$f(xy) = f(x) + f(y),$$

then $f(x) = \alpha \log x$.

3. A quantity of radium weighs 1 gm. at time $t = 0$. At time $t = 10$ (years) it has diminished to ·997 gm. After what time will it have diminished to ·5 gm.?

4. Solve the following differential equations:

(a) $y' = \alpha(y - \beta)$.   (c) $y' - \alpha y = \beta e^{\alpha x}$.

(b) $y' - \alpha y = \beta$.   (d) $y' - \alpha y = \beta e^{\gamma x}$.

8. THE HYPERBOLIC FUNCTIONS

1. Analytical Definition.

In many applications the exponential function does not enter alone, but in combinations of the form

$$\frac{1}{2}(e^x + e^{-x}) \quad \text{or} \quad \frac{1}{2}(e^x - e^{-x}).$$

It is convenient to introduce these and similar combinations as special functions; we denote them as follows:

$$\sinh x = \frac{e^x - e^{-x}}{2}, \quad \cosh x = \frac{e^x + e^{-x}}{2},$$

$$\tanh x = \frac{e^x - e^{-x}}{e^x + e^{-x}}, \quad \coth x = \frac{e^x + e^{-x}}{e^x - e^{-x}},$$

and we call them the *hyperbolic sine, hyperbolic cosine, hyperbolic tangent,* and *hyperbolic cotangent* respectively. The functions $\sinh x$, $\cosh x$, and $\tanh x$ are defined for all values of x, while in the case of $\coth x$ the point $x = 0$ must be excluded. This notation is designed to express a certain analogy with the trigonometric functions; it is this analogy, which we are about to study in detail, that justifies special consideration of our new functions. In figs. 17, 18, and 19 the graphs of the hyperbolic functions are shown; the dotted lines in fig. 17 are the graphs of $y = \frac{1}{2}e^x$ and $y = \frac{1}{2}e^{-x}$, from which the graphs of $\sinh x$ and $\cosh x$ may easily be constructed.

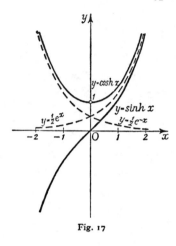

Fig. 17

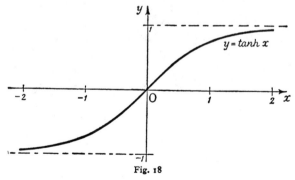

Fig. 18

We see that $\cosh x$ is an even function, i.e. a function which remains unchanged when x is replaced by $-x$, while $\sinh x$ is an odd function, i.e. a function that changes sign when x is replaced by $-x$. (Cf. p. 20.)

The function
$$\cosh x = \frac{e^x + e^{-x}}{2}$$

is, by its definition, positive for all values of x. It has its least value when $x = 0$; $\cosh 0 = 1$.

Between $\cosh x$ and $\sinh x$ there exists the fundamental relation

$$\cosh^2 x - \sinh^2 x = 1,$$

which follows immediately from the definitions of these functions. If we now denote the independent variable by t instead of x and write

$$x = \cosh t, \quad y = \sinh t,$$

we have

$$x^2 - y^2 = 1;$$

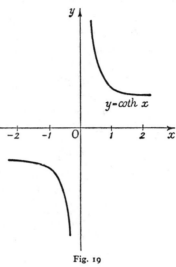

Fig. 19

that is, the point with the coordinates $x = \cosh t, y = \sinh t$ moves along the rectangular hyperbola $x^2 - y^2 = 1$ as t runs through the whole scale of values from $-\infty$ to $+\infty$.

According to the defining equation, $x \geqq 1$, and we may easily convince ourselves that y runs through the whole scale of values $-\infty$ to $+\infty$ as t does; for if t tends to infinity so does e^t, while e^{-t} tends to zero. We may therefore state more exactly that as t runs from $-\infty$ to $+\infty$, the equations $x = \cosh t, y = \sinh t$ give us one branch, namely, the right-hand one, of the rectangular hyperbola.

2. Addition Theorems and Formulæ for Differentiation.

From the definitions of our functions there follow the formulæ, known as addition theorems:

$$\cosh (a + b) = \cosh a \cosh b + \sinh a \sinh b,$$
$$\sinh (a + b) = \sinh a \cosh b + \cosh a \sinh b.$$

The proofs are obtained at once if we write

$$\cosh (a + b) = \frac{e^a e^b + e^{-a} e^{-b}}{2}, \quad \sinh (a + b) = \frac{e^a e^b - e^{-a} e^{-b}}{2},$$

7* (E 798)

and in these equations put

$$e^a = \cosh a + \sinh a, \quad e^{-a} = \cosh a - \sinh a,$$
$$e^b = \cosh b + \sinh b, \quad e^{-b} = \cosh b - \sinh b.$$

The analogy between these formulæ and the corresponding trigonometrical formulæ is clear. The only difference in the addition theorems is one sign in the first formula.

A corresponding analogy holds for the differentiation formulæ. Remembering that $d(e^x)/dx = e^x$, we readily find that *

$$\frac{d}{dx} \cosh x = \sinh x, \quad \frac{d}{dx} \sinh x = \cosh x,$$

$$\frac{d}{dx} \tanh x = \frac{1}{\cosh^2 x}, \quad \frac{d}{dx} \coth x = -\frac{1}{\sinh^2 x}.$$

3. The Inverse Hyperbolic Functions.

To the hyperbolic functions $x = \cosh t$, $y = \sinh t$, there correspond inverse functions, which we denote † by

$$t = \operatorname{ar} \cosh x, \quad t = \operatorname{ar} \sinh y.$$

Since the function $\sinh t$ is monotonic increasing throughout the interval $-\infty < t < +\infty$, its inverse function is uniquely determined for all values of y; on the other hand, we learn from a glance at the graph (cf. fig. 17, p. 184) that $t = \operatorname{ar} \cosh x$ is not uniquely determined, but has an ambiguity of sign, for to a given value of x corresponds not only the number t but also the number $-t$. Since $\cosh t \geq 1$ for all values of t, its inverse $\operatorname{ar} \cosh x$ is defined only for $x \geq 1$.

We can express these inverse functions very easily in terms of the logarithm, by regarding the quantity $e^t = u$ in the definitions

$$x = \frac{e^t + e^{-t}}{2}, \quad y = \frac{e^t - e^{-t}}{2}$$

as unknown and solving these (quadratic) equations for u. Then

$$u = x \pm \sqrt{(x^2 - 1)}, \quad u = y + \sqrt{(y^2 + 1)};$$

since $u = e^t$ can have only positive values the square root in

---

* It is sometimes convenient to introduce the functions $\operatorname{sech} x = 1/\cosh x$, $\operatorname{cosech} x = 1/\sinh x$.

† The notation $\cosh^{-1} x$, &c., is also used; cf. footnote. p. 148.

the second equation must be taken with the positive sign, while in the first either sign is possible. In the logarithmic form,

$$t = \log(x \pm \sqrt{(x^2 - 1)}) = \text{ar}\cosh x,$$
$$t = \log(y + \sqrt{(y^2 + 1)}) = \text{ar}\sinh y.$$

In the case of ar cosh x the variable x is restricted to the interval $x \geq 1$, while ar sinh y is defined for all values of y.

The formula gives us two values, $\log\{x + \sqrt{(x^2 - 1)}\}$ and $\log\{x - \sqrt{(x^2 - 1)}\}$, for ar cosh x, corresponding to the two branches of ar cosh x. Since

$$\{x + \sqrt{(x^2 - 1)}\} \{x - \sqrt{(x^2 - 1)}\} = 1$$

the sum of these two values of ar cosh x is zero, which agrees with a remark made above.

The inverses of the hyperbolic tangent and hyperbolic cotangent can be defined analogously, and can also be expressed in terms of logarithms. These functions we denote by ar tanh x and ar coth x; and, expressing the independent variable everywhere by x, we readily obtain

$$\text{ar}\tanh x = \frac{1}{2}\log\frac{1+x}{1-x} \text{ in the interval } -1 < x < 1,$$

$$\text{ar}\coth x = \frac{1}{2}\log\frac{x+1}{x-1} \text{ in the intervals } x < -1,\, x > 1.$$

The differentiation of these inverse functions may be carried out by the reader himself; here he may make use of either the rule for differentiating an inverse function or the chain rule in conjunction with the above expressions for the inverse functions in terms of logarithms. If x is the independent variable, the results are

$$\frac{d}{dx}\text{ar}\cosh x = \pm\frac{1}{\sqrt{(x^2 - 1)}}, \quad \frac{d}{dx}\text{ar}\sinh x = \frac{1}{\sqrt{(x^2 + 1)}},$$

$$\frac{d}{dx}\text{ar}\tanh x = \frac{1}{1 - x^2}, \quad \frac{d}{dx}\text{ar}\coth x = \frac{1}{1 - x^2}.$$

The last two formulæ do not contradict each other, since the first holds only for $-1 < x < 1$ and the second only for $x < -1$ and $1 < x$. The two values of $\dfrac{d}{dx}$ ar cosh x, expressed by the sign $\pm$ in the first formula, correspond to the two different branches of the curve $y = \text{ar cosh}\, x = \log\{x \pm \sqrt{(x^2 - 1)}\}$.

4. Further Analogies.

In the above representation of the rectangular hyperbola by the quantity t we did not attempt to bring out any geometrical meaning of the "parameter" t itself. We shall now return to this matter, thus gaining still more insight into the analogy between the trigonometric functions and the hyperbolic functions. If we represent the circle with equation $x^2 + y^2 = 1$ by means of a parameter t in the form $x = \cos t$, $y = \sin t$, we can interpret the quantity t as an angle or as a length of arc measured along the circumference; we may, however, also regard t as twice the area of the circular sector corresponding to that angle, the area being reckoned positive or negative according as the angle is positive or negative.

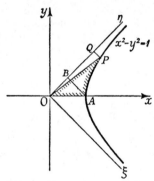

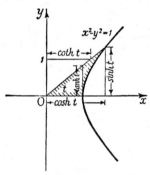

Fig. 20.—Parametric representation   Fig. 21.—To illustrate the hyperbolic
of the hyperbola   functions

We now make the analogous statement that for the hyperbolic functions the quantity t is twice the area of the hyperbolic sector* shown shaded in fig. 20. The proof is obtained without difficulty if we refer the hyperbola to its asymptotes as axes by means of the transformation of co-ordinates

$$x - y = \sqrt{2}\, \xi, \quad x + y = \sqrt{2}\, \eta,$$

or

$$x = \frac{1}{\sqrt{2}}(\xi + \eta), \quad y = \frac{1}{\sqrt{2}}(\eta - \xi);$$

with these new co-ordinates the equation of the hyperbola is $\xi\eta = \frac{1}{2}$. We thus see immediately that the area in question is equal to the area of the figure $ABQP$; for the two right-angled triangles OPQ and OAB

* Just as the notation $t = \text{arc cos}\, x$ recalls that t is an *arc* of a circle of reference, so $t = \text{ar cosh}\, x$ recalls that t is a certain *area* connected with a rectangular hyperbola.

have the same area, according to the equation of the hyperbola. The two points A and P obviously have the co-ordinates

$$\xi = \frac{1}{\sqrt{2}}, \quad \eta = \frac{1}{\sqrt{2}} \quad \text{and} \quad \xi = \frac{x-y}{\sqrt{2}}, \quad \eta = \frac{x+y}{\sqrt{2}}$$

respectively, and for double the area of our figure we thus obtain

$$2 \int_{1/\sqrt{2}}^{(x+y)/\sqrt{2}} (1/2\eta) \, d\eta = \log(x+y) = \log\{x \pm \sqrt{(x^2-1)}\}.$$

Comparison of this expression with the formula of p. 187 for the inverse function $t = \text{ar cosh} \, x$ shows us that our statement about the quantity t is true.

In conclusion, it may be pointed out that, as shown in fig. 21, the hyperbolic functions can be diagrammatically represented on the hyperbola, just as the trigonometric functions can be represented on the circle.*

EXAMPLES

1. Prove the formula

$$\sinh a + \sinh b = 2 \sinh\left(\frac{a+b}{2}\right) \cosh\left(\frac{a-b}{2}\right).$$

Obtain similar formulæ for $\sinh a - \sinh b$, $\cosh a + \cosh b$, $\cosh a - \cosh b$.

2. Express $\tanh(a \pm b)$ in terms of $\tanh a$ and $\tanh b$.
Express $\coth(a \pm b)$ in terms of $\coth a$ and $\coth b$.
Express $\sinh \frac{1}{2} a$ and $\cosh \frac{1}{2} a$ in terms of $\cosh a$.

3. Differentiate

(a) $\cosh x + \sinh x$; (b) $e^{\tanh x + \coth x}$; (c) $\log \sinh(x + \cosh^2 x)$;

(d) $\text{ar cosh} \, x + \text{ar sinh} \, x$; (e) $\text{ar sinh}(\alpha \cosh x)$; (f) $\text{ar tanh} \dfrac{2x}{1+x^2}$.

4. Calculate the area bounded by the catenary $y = \cosh x$, the ordinates $x = a$ and $x = b$, and the x-axis.

9. THE ORDER OF MAGNITUDE OF FUNCTIONS

The various functions that we have met in this chapter exhibit very important differences as regards their behaviour for large values of the argument or, as we also say, in the *order*

---

* The numerical values of the hyperbolic functions, which are useful in a variety of calculations, are to be found in many tables. We may mention the following: J. B. Dale, *Five-figure Tables of Mathematical Functions* (Arnold, 1918); K. Hayashi, *Fünfstellige Tafeln der Kreis- und Hyperbelfunktionen* (Berlin, 1930); E. Jahnke and F. Emde, *Funktionentafeln mit Formeln und Kurven* (German and English, Leipzig, 1933).

of magnitude of their increase. On account of the great importance of this we shall here discuss the matter briefly, even although it is not directly connected with the idea of the integral or of the derivative.

1. The Concept of Order of Magnitude. The Simplest Cases.

If the variable x increases beyond all bounds, then when $a > 0$ the functions x^a, $\log x$, e^x, e^{ax} will also increase beyond all bounds. As regards the manner of this increase, however, we can immediately point out an essential difference between them. For example, the function x^3 will become infinite to a higher order than x^2; we mean thereby that as x increases the quotient x^3/x^2 itself increases beyond all bounds. Similarly, we shall say that the function x^a becomes infinite to a higher order than x^β if $a > \beta > 0$, and so on.

Quite generally, we shall say of two functions $f(x)$ and $g(x)$ whose absolute values increase with x beyond all bounds that *$f(x)$ becomes infinite of a higher order than $g(x)$*, if as x increases the quotient $\left| \dfrac{f(x)}{g(x)} \right|$ increases beyond all bounds; we shall say that $f(x)$ becomes infinite of a lower order than $g(x)$ if the quotient $\left| \dfrac{f(x)}{g(x)} \right|$ tends to zero as x increases; and we shall say that the two functions become infinite of the same order of magnitude if as x increases the quotient $\left| \dfrac{f(x)}{g(x)} \right|$ possesses a limit different from 0 or at least remains between two fixed positive bounds. For example, the function $ax^3 + bx^2 + c = f(x)$, where $a \neq 0$, will be of the same order of magnitude as the function $x^3 = g(x)$; for the quotient $\left| \dfrac{f(x)}{g(x)} \right| = \left| \dfrac{ax^3 + bx^2 + c}{x^3} \right|$ has the limit $|a|$. On the other hand, the function $x^3 + x + 1$ becomes infinite of a higher order of magnitude than the function $x^2 + x + 1$.

A sum of two functions $f(x)$ and $\phi(x)$, where $f(x)$ is of higher order of magnitude than $\phi(x)$, has the same order of magnitude as $f(x)$. For $\left| \dfrac{f(x) + \phi(x)}{f(x)} \right| = \left| 1 + \dfrac{\phi(x)}{f(x)} \right|$, and by hypothesis this expression tends to 1 as x increases.

We might be tempted to measure the order of magnitude of functions by a scale, assigning to the quantity x the order of magnitude 1 and to the power x^a ($a > 0$) the order of magnitude a. A polynomial of the n-th degree then obviously has the order of magnitude n; a rational function, the degree of whose numerator is higher by h than that of the denominator, has the order of magnitude h.

2. The Order of Magnitude of the Exponential Function and of the Logarithm.

It turns out, however, that any attempt to fix the order of magnitude of arbitrary functions by the above scale must end in failure. For there are functions that become infinite of higher order than the power x^a of x, no matter how large a is chosen; again, there are functions which become infinite of lower order than the power x^a, no matter how small the positive number a is chosen. These functions therefore will not fit in anywhere in our scale.

Without entering into a detailed theory of the order of magnitude we shall prove the following theorem:

If a is an arbitrary number greater than 1, then the quotient $\dfrac{a^x}{x}$ tends to infinity as x increases.

To prove this we construct the function

$$\phi(x) = \log \frac{a^x}{x} = x \log a - \log x;$$

it is obviously sufficient to show that this increases beyond all bounds if x tends to $+\infty$. For this purpose we consider the derivative

$$\phi'(x) = \log a - \frac{1}{x}$$

and notice that for $x \geq c = \dfrac{2}{\log a}$ this is not less than the positive number $\dfrac{1}{2} \log a$. Hence it follows that for $x \geq c$

$$\phi(x) - \phi(c) = \int_c^x \phi'(t)\,dt \geq \int_c^x \tfrac{1}{2} \log a \, dt \geq \tfrac{1}{2}(x - c) \log a,$$

$$\phi(x) \geq \phi(c) + \tfrac{1}{2}(x - c) \log a,$$

and the right-hand side becomes infinite as x increases.

We shall give a second proof of this important theorem. If we write $\sqrt{a} = b = 1 + h$, we have $b > 1$ and $h > 0$. Let n be the integer such that $n \leqq x < n + 1$; we may take $x > 1$, so that $n \geqq 1$. Applying the lemma of p. 31, we have

$$\sqrt{\left(\frac{a^x}{x}\right)} = \frac{b^x}{\sqrt{x}} = \frac{(1+h)^x}{\sqrt{x}} > \frac{(1+h)^n}{\sqrt{(n+1)}} > \frac{1+nh}{\sqrt{(n+1)}} > \frac{nh}{\sqrt{2n}} = \frac{h}{\sqrt{2}}\sqrt{n};$$

so that

$$\frac{a^x}{x} > \frac{h^2}{2} \cdot n,$$

and therefore tends to infinity with x.

From the fact just proved many others follow. For example, for every positive index α and every number $a > 1$ the quotient a^x/x^α tends to infinity as x increases; that is:

The exponential function becomes infinite of a higher order of magnitude than any power of x.

In order to see this, we need only show that the α-th root of the expression, that is,

$$\frac{a^{x/\alpha}}{x} = \frac{1}{\alpha} \cdot \frac{a^{x/\alpha}}{x/\alpha} = \frac{1}{\alpha} \cdot \frac{a^y}{y} \qquad \left(y = \frac{x}{\alpha}\right),$$

tends to infinity. This, however, follows immediately from the preceding theorem, when x is replaced by $y = x/\alpha$.

We can prove the following theorem in a similar fashion. For every positive value of α the quotient $(\log x)/x^\alpha$ tends to zero when x tends to infinity; that is:

The logarithm becomes infinite of a lower order of magnitude than any arbitrarily small positive power of x.

The proof follows immediately if we put $\log x = y$, by which our quotient is transformed into $y/e^{\alpha y}$. We then put $e^\alpha = a$; then a is a number > 1, and our quotient y/a^y approaches 0 as y increases. Since y approaches infinity as x does our theorem is proved.*

---

* Another very simple proof may be suggested: for $x > 1$ and $\epsilon > 0$

$$\log x = \int_1^x \frac{d\xi}{\xi} < \int_1^x \xi^{\epsilon-1} d\xi = \frac{1}{\epsilon}(x^\epsilon - 1);$$

if we choose ϵ smaller than α and divide both members of this inequality by x^α, then as $x \to \infty$ it follows that $(\log x)/x^\alpha \to 0$.

With these results as a basis we can construct functions of an order of magnitude far higher than that of the exponential function and other functions of an order of magnitude far lower than that of the logarithm. For example, the function e^{e^x} is of a higher order than the exponential function, and the function $\log \log x$ of a lower order than the logarithm; and we can obviously repeat these iteration processes as often as we like, piling up the symbols e or log to any extent we please.

3. General Remarks.

These considerations show that it is not possible by systematic reasoning to assign to all functions definite numbers as orders of magnitude in such a way that when two functions are compared the function of the higher order of magnitude has the higher number. If, for example, the function x is of the order of magnitude 1 and the function $x^{1+\epsilon}$ of the order of magnitude $1 + \epsilon$, then the function $x \log x$ must be of an order of magnitude that is greater than 1 and less than $1 + \epsilon$ no matter how small ϵ is chosen. But there is no such number. Apart from this, however, it is easy to see that functions need not possess a clearly defined order of magnitude. For example, the function $\dfrac{x^2(\sin x)^2 + x + 1}{x^2(\cos x)^2 + x}$ approaches no definite limits as x increase; on the contrary, for $x = n\pi$ (where n is an integer) the value is

$\dfrac{1}{n\pi}$, while for $x = \left(n + \dfrac{1}{2}\right)\pi$ it is $\left(n + \dfrac{1}{2}\right)\pi + 1 + \dfrac{1}{(n + \frac{1}{2})\pi}$.

Although the numerator and denominator both become infinite, the quotient neither remains between positive bounds, nor tends to zero, nor tends to infinity. The numerator, therefore, is neither of the same order as the denominator, nor of lower order, nor of higher order. This apparently startling situation merely means that our definitions are not designed in such a way that we can compare any pair of functions. This is not a defect; we have no desire to compare the orders of such functions as the numerator and denominator above, since knowledge of the value of one of them gives us no useful information about the other.

4. The Order of Magnitude of a Function in the Neighbourhood of an Arbitrary Point.

Just as we can inquire into the behaviour of a function when x increases without limit, we may also ask ourselves whether and how functions that become infinite at the point $x = \xi$ may be distinguished as regards their behaviour at that point. We further state that the function $f(x) = \dfrac{1}{|x - \xi|}$ becomes infinite of the first order at the point $x = \xi$, and correspondingly that the function $\dfrac{1}{|x - \xi|^a}$ becomes infinite of the order a, provided that a is positive.

We then recognize that the function $e^{1/|x-\xi|}$ becomes infinite of higher order, and the function $\log|x - \xi|$ infinite of lower order, than all these powers; i.e. that the limiting relations

$$\lim_{x \to \xi}(|x - \xi|^a \cdot e^{1/|x-\xi|}) = \infty \quad \text{and} \quad \lim_{x \to \xi}(|x - \xi|^a \cdot \log|x - \xi|) = 0$$

hold.

In order to see this we merely put $\dfrac{1}{|x - \xi|} = y$; our statements then reduce to the known theorem on p. 192, since $|x - \xi|^a \cdot e^{1/|x-\xi|} = e^y/y^a$ and $|x - \xi|^a \cdot \log|x - \xi| = -(\log y)/y^a$ and y increases beyond all bounds as x tends to ξ. The method of reducing the behaviour at a finite point to the behaviour at infinity by the substitution $\dfrac{1}{|x - \xi|} = y$ frequently proves useful.

5. The Order of Magnitude of a Function tending to Zero.

Just as we seek to describe the approach of a function to infinity more definitely by means of the concept of order of magnitude, we may also specify the way in which a function approaches zero. We say that as $x \to \infty$ the quantity $1/x$ vanishes to the first order, the quantity x^{-a}, where a is positive, to the order a. We find once again that *the function $1/\log x$ vanishes to a lower order than an arbitrary power* x^{-a}, that is, for every positive a the relation

$$\lim_{x \to 0}(x^{-a} \cdot \log x) = 0$$

holds.

In the same way we say that for $x = \xi$ the quantity $x - \xi$ vanishes to the first order, the quantity $| x - \xi |^a$ to the order a. With the above results it is easy to prove the relations

$$\lim_{x \to 0} (\, | x |^a . \log | x | \,) = 0, \quad \lim_{x \to 0} (\, | x |^{-a} . e^{-1/|x|}) = 0$$

which are usually expressed as follows:

The function $\dfrac{1}{\log | x |}$ *vanishes as* $x \to 0$ *to a lower order than any power of* x; *the exponential function* $e^{-1/|x|}$ *vanishes to a higher order than any power of* x.

EXAMPLES

1. Compare the following functions with powers of x as regards their order of magnitude as $x \to \infty$:

(a) $e^{x^\beta} - 1$.

(b) $(\log x)^\beta$.

(c) $\sin x$.

(d) $\sinh x$.

(e) $x^{1/2} \sin x . \arctan x$.

(f) $x^{1/2} \sin x + \dfrac{x^2 \cos^3 x}{x^2 + 1}$.

(g) $\dfrac{e^{-1/x}}{1 - e^{-1/x}}$.

(h) $x^x - 1$.

(j) $\log (x \log x)$.

2. Compare the functions of Ex. 1 with e^{ax}, e^{x^a}, $(\log x)^a$.

3. Compare the functions of Ex. 1 with powers of x as $x \to 0$.

4. Does the limit $\lim_{x \to \infty} e^{x^n} e^{(-e^x)}$ exist?

5. What are the limits, as $x \to \infty$, of $e^{(-e^x)}$ and $e^{(e^{-x})}$?

6. Let $f(x)$ be a continuous function vanishing, together with its first derivative, for $x = 0$. Show that $f(x)$ vanishes to a higher order than x as $x \to 0$.

7. Show that $f(x) = \dfrac{a_0 x^n + a_1 x^{n-1} + \ldots + a_n}{b_0 x^m + b_1 x^{m-1} + \ldots + b_m}$, when a_0, $b_0 \neq 0$, is of the same order of magnitude as x^{n-m}, when $x \to \infty$.

8.* Prove that e^x is not a rational function.

9.* Prove that e^x cannot satisfy an algebraic equation with polynomials in x as coefficients.

Appendix to Chapter III

1. Some Special Functions

From time to time we have made it clear by examples that the general concept of function contains many possibilities foreign to naïve intuition. As a rule these examples were not given in terms of single analytical expressions. Here, therefore, we wish to show that it is possible to represent various typical discontinuities and abnormal phenomena by means of very simple expressions built up from the elementary functions. We begin, however, with an example in which no discontinuity is present.

1. The Function $y = e^{-1/x^2}$.

This function (cf. fig. 22), which is defined in the first instance only for values of x other than zero, obviously has the limit zero as $x \to 0$. For by the transformation $1/x^2 = \xi$ our function becomes $y = e^{-\xi}$ and $\lim_{\xi \to \infty} e^{-\xi} = 0$.

Hence in order to extend our function so that it is continuous for $x = 0$

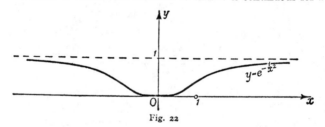

$$y = e^{-\frac{1}{x^2}}$$

Fig. 22

we define the value of the function at the point $x = 0$ by the equation $y(0) = 0$.

By the chain rule the derivative of our function for $x \neq 0$ is $y' = \dfrac{2}{x^3} e^{-1/x^2}$.

If x tends to 0, this derivative will also have the limit 0, as we find immediately from p. 194 *et seq.* At the point $x = 0$ itself the derivative

$$y'(0) = \lim_{h \to 0} \frac{y(h) - y(0)}{h} = \lim_{h \to 0} \frac{e^{-1/h^2}}{h}$$

is also zero.

If we form the higher derivatives for $x \neq 0$, we shall obviously always obtain the product of the function e^{-1/x^2} and a polynomial in $1/x$, and the passage to the limit $x \to 0$ will always yield the limit 0. All the higher derivatives will likewise vanish, like y', at the point $x = 0$.

Thus we see that our function is continuous everywhere and differentiable as many times as we please, and yet at the point $x = 0$ it vanishes with all its derivatives. We shall later realize (Chap. VI, Appendix, p. 336) how remarkable this behaviour really is.

2. The Function $y = e^{-1/x}$.

We may readily convince ourselves that for positive values of x this function behaves in the same way as the function just dealt with; if x tends to 0 through positive values the function tends to 0, and the same is true of all its derivatives. If we define the value of the function at

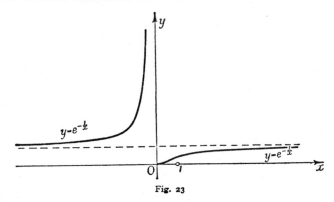

Fig. 23

$x = 0$ as $y(0) = 0$, all the right-hand derivatives at the point $x = 0$ will have the value 0. It is quite another matter when x tends to 0 through negative values; for then the function and all its derivatives become infinite, and left-hand derivatives at the point $x = 0$ do not exist. At the point $x = 0$, therefore, the function has a remarkable sort of discontinuity, quite unlike the infinite discontinuities of rational functions considered on pp. 22, 53 (cf. fig. 23).

3. The Function $y = \tanh \dfrac{1}{x}$.

We have already seen on pp. 33, 52 that functions with "jump" discontinuities can be obtained from simple functions by a passage to the limit. The exponential function defined on p. 171 and the principle of compounding of functions give us another method for constructing functions with such discontinuities from elementary functions, without any further limiting process. An example of this is the function

$$y = \tanh \frac{1}{x} = \frac{e^{1/x} - e^{-1/x}}{e^{1/x} + e^{-1/x}}$$

and its behaviour at the point $x = 0$. The function is in the first instance not defined at this point. If we approach the point $x = 0$ through positive values of x, we obviously obtain the limit 1; if, on the other hand, we approach the point $x = 0$ through negative values, we obtain the limit -1.

The point $x = 0$ is therefore a point of discontinuity; as x increases through 0 the value of the function jumps by 2 (cf. fig. 24). On the other hand, the derivative

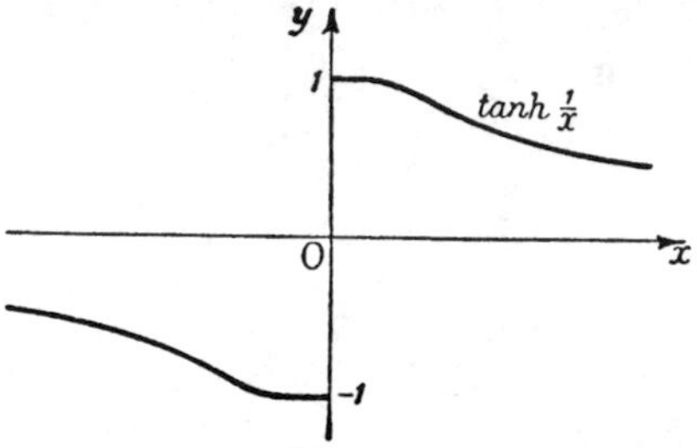

$$y' = -\frac{1}{\cosh^2(1/x)}\frac{1}{x^2}$$

$$= -\frac{1}{x^2}\frac{4}{(e^{1/x} + e^{-1/x})^2}$$

Fig. 24

approaches the limit 0 from both sides, as follows readily from * § 9, p. 194.

4. The Function $y = x\tanh\dfrac{1}{x}$.

In the case of the function

$$y = x\tanh\frac{1}{x} = x\frac{e^{1/x} - e^{-1/x}}{e^{1/x} + e^{-1/x}}$$

the above discontinuity is removed by the factor x. This function has the limit 0 as $x \to 0$ from either side, so that we can again appropriately

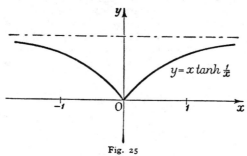

Fig. 25

define $y(0)$ as equal to 0. Our function is then continuous at $x = 0$, but its first derivative

$$y' = \tanh\frac{1}{x} - \frac{1}{x}\frac{1}{\cosh^2(1/x)}$$

* Another example of the occurrence of a " jump " discontinuity is given by the function $y = \arctan\dfrac{1}{x}$ as $x \to 0$.

has just the same kind of discontinuity as the preceding example. The graph of the function is a curve with a corner (cf. fig. 25); at the point $x = 0$ the function has no actual derivative, but a right-hand derivative with the value $+1$ and a left-hand derivative with the value -1.

5. The Function $y = x \sin \dfrac{1}{x}$, $\ y(0) = 0$.

We have already seen that this function is not composed of a finite number of monotonic pieces—as we may say, it is not " sectionally * monotonic "—but that it is nevertheless continuous (p. 54). Its first derivative

$$y' = \sin\frac{1}{x} - \frac{1}{x}\cos\frac{1}{x}, \qquad (x \neq 0)$$

on the contrary, has a discontinuity at $x = 0$; for as x tends to 0 this derivative oscillates continually between bounding curves, one positive and one negative, which themselves tend to $+\infty$ and $-\infty$ respectively. At the actual point $x = 0$ the difference quotient is $\dfrac{y(h) - y(0)}{h} = \sin\dfrac{1}{h}$; since as $h \to 0$ this swings backwards and forwards between 1 and -1 an infinite number of times, the function possesses neither a right-hand derivative nor a left-hand derivative.

2. Remarks on the Differentiability of Functions

The derivative of a function which is continuous and has a derivative at every point need not be continuous.

As the simplest example of this we consider the function

$$y = f(x) = x^2 \sin\frac{1}{x}.$$

This function is in the first instance not defined at $x = 0$; we shall define $f(0)$, its value there, as 0, so that the function is now defined and continuous everywhere. For all values of x different from zero the derivative is given by the expression

$$f'(x) = -x^2 \cos\frac{1}{x} \cdot \frac{1}{x^2} + 2x \sin\frac{1}{x} = -\cos\frac{1}{x} + 2x \sin\frac{1}{x}.$$

When x tends to 0, $f'(x)$ has no limit. If, on the other hand, we form the difference quotient $\dfrac{f(h) - f(0)}{h} = \left(h^2 \sin\dfrac{1}{h}\right)/h = h \sin\dfrac{1}{h}$, we see at once that this tends to 0 as h does. The derivative therefore exists for $x = 0$, and

* Ger. *stückweise*; cf. p. 438, footnote.

has the value 0. In order to grasp intuitively the reason for this paradoxical behaviour we represent the function graphically (cf. fig. 26). It swings backwards and forwards between the curves $y = x^2$ and $y = -x^2$, which it touches alternately. Thus the ratio of the heights of the wave-crests of our curve and their distances from the origin steadily becomes smaller. Yet these waves do not become flatter; for their slope is given by the derivative $f'(x) = 2x \sin\dfrac{1}{x} - \cos\dfrac{1}{x}$; at the points $x = \dfrac{1}{2n\pi}$ where $\cos\dfrac{1}{x} = 1$ this is equal to -1, and at the points $x = \dfrac{1}{(2n+1)\pi}$ where $\cos\dfrac{1}{x} = -1$ it is equal to $+1$.

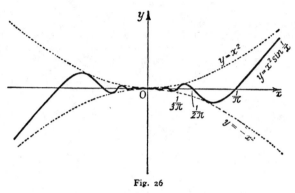

Fig. 26

In contrast to the possibility illustrated here, that a derivative may exist everywhere and yet not be continuous, we state the following simple theorem, which throws light on a whole series of earlier examples and discussions: if we know that in a neighbourhood of a point $x = a$ the function $f(x)$ is continuous and has a derivative $f'(x)$ everywhere, except that we do not know whether $f'(a)$ exists, and if in addition the equation $\lim\limits_{x \to a} f'(x) = b$ holds, then the derivative $f'(x)$ exists at the point a also, and $f'(a) = b$. The proof follows immediately from the mean value theorem. For we have $\dfrac{f(a+h) - f(a)}{h} = f'(\xi)$, where ξ is a value intermediate between a and $a + h$. If h now tends to 0, by hypothesis $f'(\xi)$ tends to b, and our statement follows at once.

A companion theorem to this is the following, which may be proved in a similar way: if the function $f(x)$ is continuous in $a \leqq x \leqq b$ and for $a < x < b$ possesses a derivative which increases beyond all bounds as x tends to a, then the right-hand difference quotient $\dfrac{f(a + h) - f(a)}{h}$ also increases beyond all bounds as h tends to 0, so that no finite right-hand derivative exists at $x = a$. The geometrical meaning of this state of affairs is that at the point with the (finite) co-ordinates $(a, f(a))$ the curve has a vertical tangent.

3. Some Special Formulæ

1. Proof of the Binomial Theorem.

Our rules for differentiation enable us to give a simple proof of the binomial theorem; this proof will be introduced here as an example of the *method of undetermined coefficients* which we shall find important later. We wish to expand the quantity $(1 + x)^n$ in powers of x for all positive integral values of n. We see at once that the function $(1 + x)^n$ must be a polynomial of degree n, i.e. it must be of the form

$$(1 + x)^n = a_0 + a_1 x + a_2 x^2 + \ldots + a_n x^n,$$

and the problem now is to determine the coefficients a_ν. If we put $x = 0$, we at once obtain $a_0 = 1$. If we differentiate both sides of the equation once, twice, three times, &c., we obtain the equations

$$n(1 + x)^{n-1} = a_1 + 2a_2 x + \ldots + na_n x^{n-1},$$
$$n(n - 1)(1 + x)^{n-2} = 2a_2 + 3 \cdot 2a_3 x + \ldots + n(n - 1)a_n x^{n-2},$$
$$\cdot \quad \cdot \quad \cdot \quad \cdot \quad \cdot \quad \cdot \quad \cdot$$

Since these equations hold for all values of x, we can put $x = 0$ in each of them and thus obtain for the coefficients $a_1, a_2, \ldots$ the expressions

$$a_1 = n, \quad a_2 = \frac{n(n - 1)}{1 \cdot 2}, \quad a_3 = \frac{n(n - 1)(n - 2)}{1 \cdot 2 \cdot 3}, \quad \ldots$$

$$a_k = \frac{n(n - 1)(n - 2) \ldots (n - k + 1)}{k!} = \binom{n}{k}.$$

We thus finally obtain the binomial theorem in the form

$$(1 + x)^n = 1 + nx + \binom{n}{2}x^2 + \ldots + \binom{n}{k}x^k + \ldots + x^n.$$

2. Successive Differentiation.   Leibnitz's Rule.

In connexion with the above we leave the reader to prove as an exercise that the successive differentiation of a product may be performed according to the following rule (*Leibnitz's rule*):

$$\frac{d^n}{dx^n}(fg) = \frac{d^nf}{dx^n}g + \binom{n}{1}\frac{d^{n-1}f}{dx^{n-1}}\frac{dg}{dx} + \binom{n}{2}\frac{d^{n-2}f}{dx^{n-2}}\frac{d^2g}{dx^2} + \cdots$$

$$+ \binom{n}{n-1}\frac{df}{dx}\frac{d^{n-1}g}{dx^{n-1}} + f\frac{d^ng}{dx^n}.$$

The repeated differentiation of a compound function $y = f\{\phi(x)\}$, however, follows no such easily remembered law. From the rules for differentiation in last chapter (the product rule and the chain rule) we have

$$\frac{dy}{dx} = \frac{df}{d\phi}\frac{d\phi}{dx} = f'\phi',$$

$$\frac{d^2y}{dx^2} = f''\phi'^2 + f'\phi'',$$

$$\frac{d^3y}{dx^3} = f'''\phi'^3 + 3f''\phi'\phi'' + f'\phi''',$$

. . . .

3. Further Examples of the Use of the Chain Rule.   The Differentiation of $f(x)^{g(x)}$.   The Generalized Mean Value Theorem.

To form the derivative of the function x^x we write $x^x = e^{x\log x}$, whence we obtain

$$\frac{d}{dx}x^x = x^x(\log x + 1)$$

by the chain rule. Similarly, we can carry out the differentiation of the more general expression $f(x)^{f(x)} = e^{f(x)\log f(x)}$ by means of the chain rule, in the following way:

$$\frac{d}{dx}\{f(x)^{f(x)}\} = f(x)^{f(x)} \cdot f'(x)\{\log f(x) + 1\}.$$

As a further application of the chain rule we here give a proof of the theorem which we have already called the generalized mean value theorem of the differential calculus (p. 135), the theorem being established here under less stringent conditions. Let $G(x) = u$ be a function which in the closed interval $a \leqq x \leqq b$ is continuous and monotonic, and which in the open interval $a < x < b$ has a derivative which is nowhere equal to 0, and let $F(x)$ be a function which is also continuous for $a \leqq x \leqq b$ and differentiable for $a < x < b$. By means of the inverse function $x = \Phi(u)$ of $G(x)$ we introduce the new independent variable u instead of x in $F(x)$, thus obtaining the compound function $f(u) = F(\Phi(u))$; according to the chain rule,

$$f'(u) = F'(x)\Phi'(u) = \frac{F'(x)}{G'(x)}.$$

The ordinary mean value theorem, applied to the function $f(u)$ and to the interval between $u_1 = G(a)$ and $u_2 = G(b)$, shows us that for an intermediate value ω

$$\frac{f(u_2) - f(u_1)}{u_2 - u_1} = f'(\omega) \quad \text{or} \quad \frac{F(b) - F(a)}{G(b) - G(a)} = \frac{F'(\xi)}{G'(\xi)},$$

where $\xi = \Phi(\omega)$ is a value intermediate between a and b.

EXAMPLES

1. Find the second derivative of $f[g\{h(x)\}]$.

2. Differentiate the following functions:

(a) $x^{\sin x}$.   (b) $(\cos x)^{\tan x}$.

(c) $\log_{v(x)} u(x)$, (that is, the logarithm of $u(x)$ to the base $v(x)$; $v(x) > 0$).

3. Prove Leibnitz's rule.

4. Find the n-th derivative of:

      (a) $x^3 e^{ax}$.   (d) $\cos mx \sin kx$.

      (b) $(\log x)^2$.   (e) $e^x \cos 2x$.

      (c) $\sin x \sin 2x$.   (f) $(1 + x)^6 e^x$.

5.* Find the n-th derivative of arc $\sin x$ at $x = 0$, and then that of $(\text{arc } \sin x)^2$ at $x = 0$.

6. Prove that $\sum\limits_{k=2}^{n} k(k-1)\binom{n}{k} = n(n-1)2^{n-2}$.

CHAPTER IV

Further Development of the Integral Calculus

The rules for differentiation which we formulated in the preceding chapter have given us extensive powers over the problem of *differentiating* given functions. Almost always, however, the inverse problem of integration greatly exceeds it in importance. Hence we must now study the art of integrating given functions.

The results attained by means of our differentiation formulæ may be summed up as follows:

*Every function which is formed from the elementary functions by means of a " closed expression " * can be differentiated, and its derivative is also a closed expression formed from the elementary functions.*

On the other hand, we have not met with any exactly corresponding fact applying to the integration of elementary functions. We do know that every elementary function, and, in fact, every continuous function, *can* be integrated, and we *have* integrated a large number of elementary functions either directly or by inversion of differentiation formulæ and have found their integrals to be expressions involving elementary functions only. But we are still far from being able to find a general solution of the following problem: given a function $f(x)$ which is expressed in terms of the elementary functions by any closed expression, to find an expression for its indefinite integral, $F(x) = \int f(x)\,dx$, which is itself a closed expression in terms of the elementary functions.

---

* By this we mean a function which can be built up from the elementary functions by repeated application of the rational operations and the processes of compounding and inversion.

In this connexion it should, however, be emphasized that the distinction between " elementary " functions and others is in itself quite arbitrary.

The fact is that this problem is in general insoluble; it is by no means true that every elementary function has an integral which itself is an elementary function. In spite of this, it is extremely important that we should be able actually to carry out such integrations when they *are* possible, and that we should acquire a certain amount of technical skill in the integration of given functions.

The first part of this chapter will be devoted to the development of devices useful for this purpose. In this connexion we would expressly warn the beginner against merely memorizing the many formulæ obtained by using these technical devices. The student should instead direct his efforts towards gaining a clear understanding of the *methods* of integration and learning how to apply them. Moreover, he should remember that even when integration by these devices is impossible the integral does exist (at least for all continuous functions), and can actually be calculated to as high a degree of accuracy as is desired by means of numerical methods which will be developed later (Chap. VII, p. 342).

In the latter part of the chapter we shall endeavour to deepen and extend our conceptions of integration and integral, quite apart from the problem of the technique of integration.

1. Elementary Integrals

First of all we repeat that to each of the differentiation formulæ proved earlier there corresponds an equivalent integration formula. Since these elementary integrals are used time and again as materials for the art of integration, we collect them in a table (p. 206). The right-hand column contains a number of elementary functions, the left-hand column the corresponding derivatives. If we read the table from left to right we obtain in the right-hand column an indefinite integral of the function in the left-hand column.

We would also remind the reader of the fundamental theorems of the differential and integral calculus, proved in Chap. II, § 4 (p. 117), in particular, of the fact that the definite integral is obtained from the indefinite integral $F(x)$ by the formula

$$\int_a^b f(x)\, dx = F(x) \Big|_a^b = F(b) - F(a).$$

$F'(x) = f(x)$	$F(x) = \int f(x)\,dx$
1. x^a ($a \neq -1$).	$\dfrac{x^{a+1}}{a+1}$.
2. $\dfrac{1}{x}$.	$\log \mid x \mid$.
3. e^x.	e^x.
4. a^x ($a \neq 1$).	$\dfrac{a^x}{\log a}$.
5. $\sin x$.	$-\cos x$.
6. $\cos x$.	$\sin x$.
7. $\dfrac{1}{\sin^2 x}$ ($\equiv \operatorname{cosec}^2 x$).	$-\cot x$.
8. $\dfrac{1}{\cos^2 x}$ ($\equiv \sec^2 x$).	$\tan x$.
9. $\sinh x$.	$\cosh x$.
10. $\cosh x$.	$\sinh x$.
11. $\dfrac{1}{\sinh^2 x}$ ($\equiv \operatorname{cosech}^2 x$).	$-\coth x$.
12. $\dfrac{1}{\cosh^2 x}$ ($\equiv \operatorname{sech}^2 x$).	$\tanh x$.
13. $\dfrac{1}{\sqrt{(1-x^2)}}$ ($\mid x \mid < 1$).	$\begin{cases} \operatorname{arc}\sin x. \\ -\operatorname{arc}\cos x. \end{cases}$
14. $\dfrac{1}{1+x^2}$.	$\begin{cases} \operatorname{arc}\tan x. \\ -\operatorname{arc}\cot x. \end{cases}$
15. $\dfrac{1}{\sqrt{(1+x^2)}}$.	$\operatorname{ar\,sinh} x \equiv \log\{x + \sqrt{(1+x^2)}\}$.
16. $\dfrac{1}{\pm\sqrt{(x^2-1)}}$ ($\mid x \mid > 1$).	$\operatorname{ar\,cosh} x \equiv \log\{x \pm \sqrt{(x^2-1)}\}$.
17. $\dfrac{1}{1-x^2}\begin{cases} \mid x \mid < 1. \\ \\ \mid x \mid > 1 \end{cases}$	$\operatorname{ar\,tanh} x \equiv \dfrac{1}{2}\log\dfrac{1+x}{1-x}$. $\operatorname{ar\,coth} x \equiv \dfrac{1}{2}\log\dfrac{x+1}{x-1}$.

Finally, for the technique of integration the reader should have the elementary rules of integration collected in Chap. II, § 1 (pp. 81–82) at his finger-tips.

In the following sections we shall attempt to reduce the calculation of integrals of given functions in some way or other to the elementary integrals collected in this table. Apart from devices which the beginner certainly could not acquire systematically, but which, on the contrary, occur only to those with long experience, this reduction is based essentially on two useful methods. Each of these methods enables us to transform a given integral in many ways; the object of such transformations is to reduce the given integral, in one step or in a sequence of steps, to one or more of the elementary integration formulæ given above..

2. The Method of Substitution

The first of these useful methods for attacking integration problems is the introduction of a new variable (i.e. the method of *substitution* or *transformation*). The corresponding integral formula is just the chain rule of the differential calculus expressed in the integral form.

1. The Substitution Formula.

We suppose that a new variable u is introduced into a function $F(x)$ by means of the equation $x = \phi(u)$, so that $F(x)$ becomes a function of u:

$$F(x) = F(\phi(u)) = G(u).$$

By the chain rule of the differential calculus

$$\frac{dG}{du} = \frac{dF}{dx} \, \phi'(u).$$

If we now write

$$F'(x) = f(x) \quad \text{and} \quad G'(u) = g(u),$$

or the equivalent expressions

$$F(x) = \int f(x)\,dx \quad \text{and} \quad G(u) = \int g(u)\,du,$$

then on the one hand the chain rule takes the form

$$g(u) = f(x)\,\phi'(u)$$

and on the other hand $G(u) = F(x)$ by definition, that is,

$$\int g(u)\,du = \int f(x)\,dx,$$

and we obtain the integral formula equivalent to the chain rule,

$$\int f(\phi(u))\,\phi'(u)\,du = \int f(x)\,dx, \qquad \{x = \phi(u)\}.$$

This is the basic formula for the substitution of a new variable in an integral. It means that if we wish to find an indefinite integral of a function of u, which is given in the special form $f(\phi(u))\,\phi'(u)$, then we can instead find the indefinite integral of the function $f(x)$ as a function of x and after integration return to the variable u by putting $x = \phi(u)$.

If, for example, we apply the formula to the integrand $\dfrac{\phi'(u)}{\phi(u)}$ we obtain

$$\int \frac{\phi'(u)}{\phi(u)}\,du = \int \frac{dx}{x} = \log|\,x\,| = \log|\,\phi(u)\,|$$

or, replacing u by x,

$$\int \frac{\phi'(x)}{\phi(x)}\,dx = \log|\,\phi(x)\,|.$$

If in this important formula we substitute particular functions, such as $\varphi(x) = \log x$ or $\varphi(x) = \sin x$ or $\varphi(x) = \cos x$, then we obtain *

$$\int \frac{dx}{x \log x} = \log|\,\log x\,|,$$

$$\int \cot x\,dx = \log|\,\sin x\,|, \qquad \int \tan x\,dx = -\log|\,\cos x\,|.$$

A further example is

$$\int \varphi(u)\,\varphi'(u)\,du = \int x\,dx = \frac{1}{2}x^2 = \frac{1}{2}[\varphi(u)]^2,$$

where $f(x) = x$. This yields for $\varphi(u) = \log u$

$$\int \frac{\log u}{u}\,du = \frac{1}{2}(\log u)^2.$$

---

* These and the following formulæ are verified by showing that differentiation of the result gives us back the integrand. The formulæ, moreover, are of course only asserted as true in so far as the expressions occurring in them have a meaning.

We finally consider the example

$$\int \sin^n u \cos u\, du.$$

Here $x = \sin u = \varphi(u)$, and hence

$$\int \sin^n u \cos u\, du = \int x^n\, dx = \frac{x^{n+1}}{n+1} = \frac{\sin^{n+1} u}{n+1}.$$

In many cases, however, we shall use the above formula in the reverse direction, starting with the right-hand side, the integral $\int f(x)\, dx$. We now have to evaluate or simplify a prescribed indefinite integral $F(x) = \int f(x)\, dx$ by introducing the new variable of integration u by means of the transformation formula $x = \phi(u)$, then working out the indefinite integral

$$G(u) = \int f(\phi(u))\, \phi'(u)\, du,$$

and finally replacing the variable u in this integral by x. In order to carry out this last step we must be certain that a definite value u does actually correspond to the value x, i.e. that the function $x = \phi(u)$ has an inverse. Accordingly we now make the following assumption, in which we regard x as the primary variable. In the interval under consideration $u = \psi(x)$ is a monotonic differentiable function whose derivative $\psi'(x)$ does not vanish anywhere in the interval. The inverse function—which under these conditions is definite and single-valued—we denote by $x = \phi(u)$; its derivative is then given by $\phi'(u) = 1/\psi'(x)$. As the *basic formula* for the substitution of a new variable u in an integral, we obtain

$$\int f(x)\, dx = \int f(\phi(u))\, \phi'(u)\, du \qquad (u = \psi(x)).$$

The indefinite integral $\int f(x)\, dx$ can be obtained by calculating the indefinite integral $\int f(\phi(u))\, \phi'(u)\, du$ and finally introducing x instead of u for the independent variable by means of the equation $u = \psi(x)$.

It is therefore not sufficient merely to express the old variable x in terms of the new one u, and then to integrate with respect to this new variable; before integrating we must multiply by the derivative of the original variable x with respect to the new variable u.

The corresponding formula for definite integration between two limits is

$$\int_a^b f(x)\,dx = \int_{\psi(a)}^{\psi(b)} f(\phi(u))\,\phi'(u)\,du.$$

In the new integral we have to choose those limits of integration which are obtained by subjecting the old integration limits to the transformation $x = \phi(u)$, $u = \psi(x)$.

In most applications the integrand $f(x)$ will appear at the outset as a function of a function, say $f(x) = h(u)$, where $u = \psi(x)$. It is then more convenient to write our integral formula in a slightly different form by identifying the expression $f\{\phi(u)\}$ with the expression $h(u)$. If for u we make the substitution $u = \psi(x)$, $x = \phi(u)$, then our transformation formula is simply

$$\int h\{\psi(x)\}\,dx = \int h(u)\,\frac{dx}{du}\,du.$$

As a first example we consider the integration of the function $f(x) = \sin 2x$, taking $u = \psi(x) = 2x$ and $h(u) = \sin u$. We have

$$\frac{du}{dx} = \psi'(x) = 2.$$

If we now introduce $u = 2x$ into the integral as the new variable, then it is transformed, *not* into $\int \sin u\,du$, but into

$$\frac{1}{2}\int \sin u\,du = -\frac{1}{2}\cos u = -\frac{1}{2}\cos 2x;$$

this may of course be verified at once by differentiating the right-hand side.

If we integrate for x between the limits 0 and $\pi/4$, the corresponding limits for u are 0 and $\pi/2$, and we obtain

$$\int_0^{\pi/4} \sin 2x\,dx = \frac{1}{2}\int_0^{\pi/2} \sin u\,du = -\frac{1}{2}\cos u \Big|_0^{\pi/2} = \frac{1}{2}.$$

Another simple example is the integral $\int_1^4 \frac{dx}{\sqrt{x}}$. Here we take $u = \psi(x) = \sqrt{x}$, whence $x = \varphi(u) = u^2$. Since $\varphi'(u) = 2u$, we have

$$\int_1^4 \frac{dx}{\sqrt{x}} = 2\int_1^2 \frac{u\,du}{u} = 2\int_1^2 du = 2.$$

2. Another Proof of the Substitution Formula.

Our integration formula can also be explained in another and more direct manner, by aiming at the formula for *definite integration* and basing the proof on the meaning of the definite integral as a limit of a sum. To calculate the integral

$$\int_a^b h(\psi(x))\,dx$$

(for the case $a < b$), we begin with an arbitrary subdivision of the interval $a \leqq x \leqq b$, and then make the subdivision finer and finer. We choose these subdivisions in the following way. If the function $u = \psi(x)$ is assumed to be monotonic increasing, there is a (1, 1) correspondence between the interval $a \leqq x \leqq b$ on the x-axis, and an interval $\alpha \leqq u \leqq \beta$ of the values of $u = \psi(x)$, where $\alpha = \psi(a)$, $\beta = \psi(b)$. We divide up this u-interval into n parts of length * Δu; there is a corresponding subdivision of the x-interval into sub-intervals which in general are not all of the same length. We denote the points of division of the x-interval by $x_0 = a, x_1, x_2, \ldots, x_n = b$ and the lengths of the corresponding sub-intervals by

$$\Delta x_1, \Delta x_2, \ldots, \Delta x_n.$$

The integral we are considering is then the limit † of the sum

$$\sum_{\nu=1}^{n} h\{\psi(\xi_\nu)\}\Delta x_\nu,$$

where the value ξ_ν is arbitrarily selected from the ν-th sub-interval of the x-subdivision. This sum we now write in the form $\sum_{\nu=1}^{n} h(u_\nu)\dfrac{\Delta x_\nu}{\Delta u}\Delta u$, where $u_\nu = \psi(\xi_\nu)$. By the mean value theorem of the differential calculus $\dfrac{\Delta x_\nu}{\Delta u} = \phi'(\eta_\nu)$, where η_ν is a suitably chosen intermediate value of the variable u in the ν-th sub-interval of the u-subdivision and $x = \phi(u)$ denotes the

---

* The assumption that these sub-intervals are all equal is by no means essential for the proof.

† This limit exists (for $\Delta u \to 0$) and is the integral, since on account of the uniform continuity of $x = \phi(u)$ the greatest of the lengths Δx tends to 0 with Δu.

inverse function of $u = \psi(x)$. If we now select the value ξ_ν in such a way that ξ_ν and η_ν coincide, i.e. $\xi_\nu = \phi(\eta_\nu)$, $\eta_\nu = \psi(\xi_\nu)$, then our sum takes the form

$$\sum_{\nu = 1}^{n} h(\eta_\nu)\,\phi'(\eta_\nu)\,\Delta u.$$

If we here make the passage to the limit we immediately obtain the expression

$$\int_a^\beta h(u)\frac{dx}{du}\,du,$$

as the limiting value, that is, as the value of the integral we are considering, in agreement with the formula given above.

We have therefore proved the following theorem:

Let h(u) *be a continuous function of* u *in the interval* $a \leqq u \leqq \beta$. *Then if the function* u = ψ(x) *is continuous and monotonic and has a continuous non-vanishing derivative* $\dfrac{du}{dx}$ *in* a $\leqq$ x $\leqq$ b, *and* ψ(a) = a, ψ(b) = β,

$$\int_a^b h\{\psi(x)\}dx = \int_a^b h(u)\,dx = \int_a^\beta h(u)\frac{dx}{du}\,du.$$

This formula exhibits the advantage of Leibnitz's notation. In order to carry out the substitution $u = \psi(x)$, we need only write $\dfrac{dx}{du}\,du$ in place of dx, changing the limits from the original values of x to the corresponding values of u.

3. Examples. Integration Formulæ.

With the help of the substitution rule we can in many cases evaluate a given integral $\int f(x)\,dx$ if we reduce it by means of a suitable substitution $x = \phi(u)$ to one of the elementary integrals in our table. Whether such substitutions exist and how to find them are questions to which no general answer can be given; this is rather a matter in which practice and ingenuity, in contrast to systematic method, come into their own.

As an example, we shall work out the integral $\int \dfrac{dx}{\sqrt{(a^2 - x^2)}}$

by means of the substitution * $x = \phi(u) = au$, $u = \psi(x) = x/a$, $dx = a\,du$, by which, using No. 13 of the table on p. 206, we obtain

$$\int \frac{dx}{\sqrt{(a^2 - x^2)}} = \int \frac{a\,du}{a\sqrt{(1-u^2)}} = \text{arc sin } u = \text{arc sin } \frac{x}{a}, \text{for } |x| < |a|.$$

By the same substitution we likewise obtain

$$\int \frac{dx}{a^2 + x^2} = \int \frac{a\,du}{a^2(1 + u^2)} = \frac{1}{a} \text{ arc tan } u = \frac{1}{a} \text{ arc tan } \frac{x}{a},$$

$$\int \frac{dx}{\sqrt{(a^2 + x^2)}} = \text{ar sinh } \frac{x}{a},$$

$$\int \frac{dx}{\sqrt{(x^2 - a^2)}} = \text{ar cosh } \frac{x}{a}, \text{ for } |x| > |a|,$$

$$\int \frac{dx}{a^2 - x^2} = \begin{cases} \dfrac{1}{a} \text{ ar tanh } \dfrac{x}{a} \text{ for } |x| < |a| \\ \dfrac{1}{a} \text{ ar coth } \dfrac{x}{a} \text{ for } |x| > |a| \end{cases};$$

formulæ which occur very frequently and which can easily be verified by differentiating the right-hand side.

In conclusion, we again emphasize the following point. In our substitution process we have made the assumption that the substitution has a unique inverse $x = \phi(u)$, and indeed that $\psi'(x)$ is nowhere equal to zero in the interval under consideration. If our assumption is not fulfilled, application of the substitution formula may easily lead to wrong conclusions. If $\psi'(x) = 0$ at isolated points of the interval of integration only, we can avoid these difficulties by subdividing this interval in such a way that $\psi'(x)$ vanishes only at the ends of a sub-interval; we can then apply the substitution to each sub-interval separately.†

---

* For the sake of brevity we take the liberty of writing the symbols dx and du separately, i.e. $dx = \phi'(u)\,du$ instead of $dx/du = \phi'(u)$ (cf. pp. 106–107).

† An application of this method at once leads to the following result, which applies to many special cases: if the derivative $\psi'(x)$ vanishes at a finite number of points, but the function $\psi(x)$ remains monotonic, then the substitution formula remains valid.

3. Further Examples of the Substitution Method

In this section we bring together a number of examples which the reader may consider carefully by way of practice.

By the substitution $u = 1 \pm x^2$, $du = \pm 2x \, dx$, we deduce that

$$\int \frac{x \, dx}{\sqrt{(1 \pm x^2)}} = \pm \sqrt{(1 \pm x^2)},$$

$$\int \frac{x \, dx}{1 \pm x^2} = \pm \tfrac{1}{2} \log |1 \pm x^2|.$$

In these formulæ we must take either the sign $+$ in all three places or the sign $-$ in all three places.

By the substitution $u = ax + b$, $du = a \, dx$ $(a \neq 0)$, we obtain

$$\int \frac{dx}{ax + b} = \frac{1}{a} \log |ax + b|,$$

$$\int (ax + b)^\alpha \, dx = \frac{1}{a(\alpha + 1)} (ax + b)^{\alpha + 1} \qquad (\alpha \neq -1),$$

$$\int \sin(ax + b) \, dx = -\frac{1}{a} \cos(ax + b);$$

similarly, by means of the substitution $u = \cos x$, $du = -\sin x \, dx$, we obtain

$$\int \tan x \, dx = -\log |\cos x|,$$

and by means of the substitution $u = \sin x$, $du = \cos x \, dx$,

$$\int \cot x \, dx = \log |\sin x|$$

(cf. p. 208). Using the analogous substitutions $u = \cosh x$, $du = \sinh x \, dx$ and $u = \sinh x$, $du = \cosh x \, dx$, we obtain the formulæ

$$\int \tanh x \, dx = \log |\cosh x|,$$

$$\int \coth x \, dx = \log |\sinh x|.$$

In virtue of the substitution $u = \frac{a}{b} \tan x$, $du = \frac{a}{b} \sec^2 x \, dx$, we arrive at the two formulæ

$$\int \frac{dx}{a^2 \sin^2 x + b^2 \cos^2 x} = \frac{1}{b^2} \int \frac{1}{\frac{a^2}{b^2} \tan^2 x + 1} \cdot \frac{dx}{\cos^2 x}$$

$$= \frac{1}{ab} \arctan \left(\frac{a}{b} \tan x \right)$$

and

$$\int \frac{dx}{a^2 \sin^2 x - b^2 \cos^2 x} = \begin{cases} -\dfrac{1}{ab} \ \text{ar tanh} \left(\dfrac{a}{b} \tan x \right) \\ -\dfrac{1}{ab} \ \text{ar coth} \left(\dfrac{a}{b} \tan x \right). \end{cases}$$

We evaluate the integral

$$\int \frac{dx}{\sin x}$$

by writing $\sin x = 2 \sin \dfrac{x}{2} \cos \dfrac{x}{2} = 2 \tan \dfrac{x}{2} \cos^2 \dfrac{x}{2}$, and putting $u = \tan \dfrac{x}{2}$, so that $du = \frac{1}{2} \sec^2 \dfrac{x}{2} dx$; the integral then becomes

$$\int \frac{dx}{\sin x} = \int \frac{du}{u} = \log \left| \tan \frac{x}{2} \right|.$$

If we replace x by $x + \pi/2$, this formula becomes

$$\int \frac{dx}{\cos x} = \log \left| \tan \left(\frac{x}{2} + \frac{\pi}{4} \right) \right|.$$

The substitution $u = 2x$ yields, if we also apply the known trigonometrical formulæ $2 \cos^2 x = 1 + \cos 2x$ and $2 \sin^2 x = 1 - \cos 2x$, the frequently used formulæ

$$\int \cos^2 x \, dx = \frac{1}{2} (x + \sin x \cos x)$$

and

$$\int \sin^2 x \, dx = \frac{1}{2} (x - \sin x \cos x).$$

By the substitution $x = \cos u$, equivalent to $u = \text{arc} \cos x$, or, more generally, $x = a \cos u \ (a \neq 0)$, we can reduce

$$\int \sqrt{(1 - x^2)} \, dx \quad \text{and} \quad \int \sqrt{(a^2 - x^2)} \, dx$$

respectively to these formulæ. We thus obtain

$$\int \sqrt{(a^2 - x^2)} \, dx = -\frac{a^2}{2} \ \text{arc} \cos \frac{x}{a} + \frac{x}{2} \sqrt{(a^2 - x^2)}.$$

Similarly, by the substitution $x = a \cosh u$ we obtain the formula

$$\int \sqrt{(x^2 - a^2)} \, dx = -\frac{a^2}{2} \ \text{ar} \cosh \frac{x}{a} + \frac{x}{2} \sqrt{(x^2 - a^2)}$$

and by the substitution $x = a \sinh u$

$$\int \sqrt{(a^2 + x^2)} \, dx = \frac{a^2}{2} \ \text{ar} \sinh \frac{x}{a} + \frac{x}{2} \sqrt{(a^2 + x^2)}.$$

The substitution $u = \dfrac{a}{x}$, $dx = -\dfrac{a}{u^2}\,du$ leads to the formulæ

$$\int \frac{dx}{x\sqrt{(x^2 - a^2)}} = -\frac{1}{a}\,\text{arc sin}\,\frac{a}{x},$$

$$\int \frac{dx}{x\sqrt{(x^2 + a^2)}} = -\frac{1}{a}\,\text{ar sinh}\,\frac{a}{x},$$

$$\int \frac{dx}{x\sqrt{(a^2 - x^2)}} = -\frac{1}{a}\,\text{ar cosh}\,\frac{a}{x}.$$

Finally, we consider the three integrals

$$\int \sin mx \sin nx\, dx, \quad \int \sin mx \cos nx\, dx, \quad \int \cos mx \cos nx\, dx,$$

where m and n are positive integers. By well-known trigonometrical formulæ we can divide each of these integrals into two parts, writing

$$\sin mx \sin nx = \frac{1}{2}\{\cos(m - n)x - \cos(m + n)x\},$$

$$\sin mx \cos nx = \frac{1}{2}\{\sin(m + n)x + \sin(m - n)x\},$$

$$\cos mx \cos nx = \frac{1}{2}\{\cos(m + n)x + \cos(m - n)x\}.$$

If we now make use of the substitutions $u = (m + n)x$ and $u = (m - n)x$ respectively, we directly obtain the following system of formulæ:

$$\int \sin mx \sin nx\, dx = \begin{cases} \dfrac{1}{2}\left\{\dfrac{\sin(m - n)x}{m - n} - \dfrac{\sin(m + n)x}{m + n}\right\} \text{ if } m \neq n, \\[2ex] \dfrac{1}{2}\left(x - \dfrac{\sin 2mx}{2m}\right) \text{ if } m = n; \end{cases}$$

$$\int \sin mx \cos nx\, dx = \begin{cases} -\dfrac{1}{2}\left\{\dfrac{\cos(m + n)x}{m + n} + \dfrac{\cos(m - n)x}{m - n}\right\} \text{ if } m \neq n, \\[2ex] -\dfrac{1}{2}\left(\dfrac{\cos 2mx}{2m}\right) \text{ if } m = n; \end{cases}$$

$$\int \cos mx \cos nx\, dx = \begin{cases} \dfrac{1}{2}\left\{ \dfrac{\sin(m+n)x}{m+n} + \dfrac{\sin(m-n)x}{m-n} \right\} \text{ if } m \neq n, \\ \dfrac{1}{2}\left(\dfrac{\sin 2mx}{2m} + x \right) \text{ if } m = n. \end{cases}$$

If in particular we now integrate from $-\pi$ to $+\pi$, we obtain from these formulæ the extremely important relations

$$\int_{-\pi}^{+\pi} \sin mx \sin nx\, dx = \begin{cases} 0 \text{ if } m \neq n, \\ \pi \text{ if } m = n, \end{cases}$$

$$\int_{-\pi}^{+\pi} \sin mx \cos nx\, dx = 0,$$

$$\int_{-\pi}^{+\pi} \cos mx \cos nx\, dx = \begin{cases} 0 \text{ if } m \neq n, \\ \pi \text{ if } m = n. \end{cases}$$

These are the "orthogonality relations" of the trigonometric functions, which we shall meet with again in Chap. IX (p. 438).

EXAMPLES

Evaluate the following integrals and verify the results by differentiation:

1. $\int xe^{x^2}\, dx.$

2. $\int x^3 e^{-x^4}\, dx.$

3. $\int x^2 \sqrt{1+x^3}\, dx.$

4. $\int \dfrac{\log x}{x}\, dx.$

5. $\int \dfrac{dx}{x(\log x)^n}.$

6. $\int \dfrac{3\, dx}{9x^2 - 6x + 2}.$

7. $\int \dfrac{dx}{\sqrt{(x^2 - 2x + 5)}}.$

8. $\int \dfrac{6x}{2+3x}\, dx.$

9. $\int \dfrac{x+1}{\sqrt{(1-x^2)}}\, dx.$

10. $\int \dfrac{dx}{\sqrt{(5 + 2x + x^2)}}.$

11. $\int \dfrac{dx}{\sqrt{(3 - 2x - x^2)}}.$

12. $\int \dfrac{x\, dx}{x^2 - x + 1}.$

13. $\int \dfrac{x\, dx}{\sqrt{(x^2 - 4x + 1)}}.$

14. $\int \dfrac{(x+1)\, dx}{\sqrt{(2 + 2x - 3x^2)}}.$

15. $\int \dfrac{dx}{x^2 + x + 1}.$

16. $\int \dfrac{dx}{x^2 - x + 1}.$

8●

(E 798)

17. $\int \dfrac{dx}{x^2 + 2ax + b}.$

18. $\int \dfrac{x^4}{1-x} \, dx.$

19. $\int \sin^3 x \, \cos^4 x \, dx.$

20. $\int \sin^2 x \, \cos^5 x \, dx.$

21. $\int x^3 (\sqrt{1-x^2})^5 \, dx.$

22. $\int \dfrac{x^2}{\sqrt{(1-x^2)}} \, dx.$

23. $\int_0^1 \dfrac{\arctan x}{1+x^2} \, dx.$

24. $\int_0^{\pi} \cos^n x \, \sin x \, dx.$

25. $\int_0^4 \dfrac{x \, dx}{\sqrt{(1+3x^2)}}.$

26. $\int_a^b \dfrac{x}{(1+x^2)^2} \, dx.$

27. $\int_a^b \dfrac{x^3}{1-x} \, dx \; (1 < a < b).$

28. $\int_0^{\pi/2} x \, \sin 2x^2 \, dx.$

29. Evaluate $\int_0^1 (1-x)^n \, dx$ (where n is a positive integer) by substitution.

4. Integration by Parts

The second useful method for dealing with integration problems is given by the formula for differentiating a product:

$$(fg)' = f'g + fg'.$$

1. General Remarks.

If we write this formula as an integral formula, we obtain (cf. p. 141)

$$f(x)g(x) = \int g(x)f'(x) \, dx + \int f(x)g'(x) \, dx$$

or

$$\int f(x)g'(x) \, dx = f(x)g(x) - \int g(x)f'(x) \, dx.$$

This formula will be referred to as the formula for *integration by parts*. The calculation of one integral is thereby reduced to the calculation of another integral. For if we split up the integrand of an integral $\int \omega(x) \, dx$ into a product $\omega(x) = f(x)\phi(x)$, and if we can find the indefinite integral

$$g(x) = \int \phi(x) \, dx$$

of the one factor $\phi(x)$, so that $\phi(x) = g'(x)$, then by our formula the integral $\int \omega(x) \, dx = \int f(x)\phi(x) \, dx = \int f(x)g'(x) \, dx$ is reduced to $\int g(x)f'(x) \, dx$, which in some cases can be found more readily than the original form. Since a given function $\omega(x)$ which occurs as an integrand can be regarded as a product $f(x)\,\phi(x) = f(x)g'(x)$ in a great many different ways, this formula provides us with a very effective tool for the transformation of integrals.

Written as a formula for *definite integration,* the formula for integration by parts is

$$\int_a^b f(x)g'(x)\,dx = f(x)g(x)\Big|_a^b - \int_a^b g(x)f'(x)\,dx$$
$$= f(b)g(b) - f(a)g(a) - \int_a^b g(x)f'(x)\,dx.$$

For in order to obtain the formula for definite integration from the formula for indefinite integration (Chap. II, § 4, p. 117) we have only to replace the variable appearing on both sides in the formula for the indefinite integral (1) by the value $x = b$, (2) by the value $x = a$, and write down the difference of these two expressions.

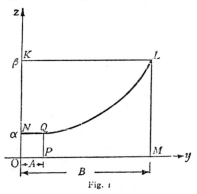

Fig. 1

A simple interpretation of this formula, at least with suitable restrictions on the functions involved, can be given. Let us suppose that $y = f(x)$ and $z = g(x)$ are monotonic, and that $f(a) = A$, $f(b) = B$, $g(a) = \alpha$, $g(b) = \beta$; we can then form the inverse of the first function and substitute in the equation, thus obtaining z as a function of y. We assume that this function is monotonic increasing. Since $dy = f'(x)\,dx$ and $dz = g'(x)\,dx$ the formula for integration by parts can be written

$$\int_A^B z\,dy + \int_\alpha^\beta y\,dz = B\beta - A\alpha,$$

in agreement with the relation made clear by fig. 1,

area $NQLK$ + area $PMLQ$ = area $OMLK$ — area $OPQN$.

The following example may serve as a first illustration:

$$\int \log x\,dx = \int \log x \cdot 1 \cdot dx.$$

We write the integrand in this way in order to indicate that we intend to put $f(x) = \log x$ and $g'(x) = 1$, so that we have $f'(x) = 1/x$ and $g(x) = x$. Our formula then becomes

$$\int \log x \, dx = x \log x - \int \frac{x}{x} \, dx = x \log x - x.$$

This last expression is therefore the integral of the logarithm, as may be verified at once by differentiation.

2. Examples.

The following further examples may help the reader to grasp this method.

If we put $f(x) = x$, $g'(x) = e^x$, we have $f'(x) = 1$, $g(x) = e^x$, and

$$\int x e^x \, dx = e^x (x - 1).$$

In a similar way we obtain

$$\int x \sin x \, dx = -x \cos x + \sin x$$

and

$$\int x \cos x \, dx = x \sin x + \cos x.$$

For $f(x) = \log x$, $g'(x) = x^a$, we have the relation

$$\int x^a \log x \, dx = \frac{x^{a+1}}{a+1} \left(\log x - \frac{1}{a+1} \right).$$

Here we must assume $a \neq -1$. For $a = -1$ we obtain (cf. p. 208)

$$\int \frac{1}{x} \log x \, dx = (\log x)^2 - \int \log x \cdot \frac{dx}{x};$$

transferring the integral on the right-hand side over to the left, we have

$$\int \frac{1}{x} \log x \, dx = \frac{1}{2} (\log x)^2.$$

We calculate the integral $\int \text{arc} \sin x \, dx$ by taking $f(x) = \text{arc} \sin x$, $g'(x) = 1$. From this we obtain

$$\int \text{arc} \sin x \, dx = x \, \text{arc} \sin x - \int \frac{x \, dx}{\sqrt{(1 - x^2)}}.$$

The integration on the right-hand side can be performed as in § 3 (p. 214); we thus find that

$$\int \text{arc} \sin x \, dx = x \, \text{arc} \sin x + \sqrt{(1 - x^2)}.$$

In the same way we calculate the integral

$$\int \text{arc} \tan x \, dx = x \, \text{arc} \tan x - \frac{1}{2} \log (1 + x^2)$$

and many others of a similar type.

The following examples are of a somewhat different nature; here a double application of the method of integration by parts brings us back to the original integral, for which we thus obtain an equation.

Integrating by parts twice, we obtain

$$\int e^{ax} \sin bx \, dx = - \frac{1}{b} e^{ax} \cos bx + \frac{a}{b} \int e^{ax} \cos bx \, dx$$

$$= - \frac{1}{b} e^{ax} \cos bx + \frac{a}{b^2} e^{ax} \sin bx - \frac{a^2}{b^2} \int e^{ax} \sin bx \, dx,$$

and, solving this equation for the integral $\int e^{ax} \sin bx \, dx$,

$$\int e^{ax} \sin bx \, dx = \frac{1}{a^2 + b^2} e^{ax} (a \sin bx - b \cos bx).$$

In a similar way it follows that

$$\int e^{ax} \cos bx \, dx = \frac{1}{a^2 + b^2} e^{ax} (a \cos bx + b \sin bx).$$

3. Recurrence Formulæ.

In many cases the integrand is a function not only of the independent variable, but also of an integral index n, and on integrating by parts we obtain, instead of the value of the integral, another similar expression in which the index n has a smaller value. We thus arrive after a number of steps at an integral which we can deal with by means of our table of integrals. Such a process is called a *recurrence process*. The following examples illustrate this: by repeated integration by parts we can calculate the trigonometrical integrals

$$\int \cos^n x \, dx, \quad \int \sin^n x \, dx, \quad \int \sin^m x \cos^n x \, dx,$$

provided that m and n are integers. For we find that

$$\int \cos^n x \, dx = \cos^{n-1} x \sin x + (n - 1) \int \cos^{n-2} x \sin^2 x \, dx;$$

we can write the right-hand side in the form

$$\cos^{n-1} x \sin x + (n - 1) \int \cos^{n-2} x \, dx - (n - 1) \int \cos^n x \, dx,$$

thus obtaining the recurrence relation

$$\int \cos^n x \, dx = \frac{1}{n} \cos^{n-1} x \sin x + \frac{n-1}{n} \int \cos^{n-2} x \, dx.$$

This formula enables us to keep on diminishing the index in the integrand until we finally arrive at the integral

$$\int \cos x \, dx = \sin x \quad \text{or} \quad \int dx = x,$$

according as n is odd or even. In a similar way we obtain the analogous recurrence formulæ

$$\int \sin^n x \, dx = -\frac{1}{n} \sin^{n-1} x \cos x + \frac{n-1}{n} \int \sin^{n-2} x \, dx$$

and

$$\int \sin^m x \cos^n x \, dx = \frac{\sin^{m+1} x \cos^{n-1} x}{m+n} + \frac{n-1}{m+n} \int \sin^m x \cos^{n-2} x \, dx.$$

In particular, these formulæ enable us to calculate the integrals

$$\int \sin^2 x \, dx = \frac{1}{2} (x - \sin x \cos x)$$

and

$$\int \cos^2 x \, dx = \frac{1}{2} (x + \sin x \cos x),$$

as we have already done by the method of substitution (p. 215).

It need hardly be mentioned that the corresponding integrals for the hyperbolic functions can be calculated in exactly the same way.

Further recurrence formulæ are given by the following transformations:

$$\int (\log x)^m \, dx = x (\log x)^m - m \int (\log x)^{m-1} \, dx,$$

$$\int x^m e^x \, dx = x^m e^x - m \int x^{m-1} e^x \, dx,$$

$$\int x^m \sin x \, dx = -x^m \cos x + m \int x^{m-1} \cos x \, dx,$$

$$\int x^m \cos x \, dx = x^m \sin x - m \int x^{m-1} \sin x \, dx,$$

$$\int x^a (\log x)^m \, dx = \frac{x^{a+1} (\log x)^m}{a+1} - \frac{m}{a+1} \int x^a (\log x)^{m-1} \, dx \quad (a \neq -1).$$

4. Wallis's Product.

The recurrence formula for the integral $\int \sin^n x \, dx$ leads in an elementary way to a most remarkable expression for the number π as an infinite product. We suppose that $n > 1$, and in the formula

$$\int \sin^n x \, dx = -\frac{1}{n} \sin^{n-1} x \cos x + \frac{n-1}{n} \int \sin^{n-2} x \, dx$$

we insert the limits 0 and $\pi/2$, thus obtaining

$$\int_0^{\pi/2} \sin^n x \, dx = \frac{n-1}{n} \int_0^{\pi/2} \sin^{n-2} x \, dx \quad \text{for } n > 1.$$

If we again apply the recurrence formula to the right-hand side and continue the process, we obtain, distinguishing between the cases $n = 2m$ and $n = 2m + 1$,

$$\int_0^{\pi/2} \sin^{2m} x \, dx = \frac{2m-1}{2m} \cdot \frac{2m-3}{2m-2} \cdots \frac{1}{2} \cdot \int_0^{\pi/2} dx,$$

$$\int_0^{\pi/2} \sin^{2m+1} x \, dx = \frac{2m}{2m+1} \cdot \frac{2m-2}{2m-1} \cdots \frac{2}{3} \cdot \int_0^{\pi/2} \sin x \, dx,$$

whence

$$\int_0^{\pi/2} \sin^{2m} x \, dx = \frac{2m-1}{2m} \cdot \frac{2m-3}{2m-2} \cdots \frac{1}{2} \cdot \frac{\pi}{2},$$

$$\int_0^{\pi/2} \sin^{2m+1} x \, dx = \frac{2m}{2m+1} \cdot \frac{2m-2}{2m-1} \cdots \frac{2}{3}.$$

By division this yields

$$\frac{\pi}{2} = \frac{2.2}{1.3} \cdot \frac{4.4}{3.5} \cdot \frac{6.6}{5.7} \cdots \frac{2m . 2m}{(2m-1) . (2m+1)} \frac{\int_0^{\pi/2} \sin^{2m} x \, dx}{\int_0^{\pi/2} \sin^{2m+1} x \, dx}.$$

The quotient of the two integrals on the right-hand side converges to 1 as m increases, as we recognize from the following considerations. In the interval $0 < x < \pi/2$ we have

$$0 < \sin^{2m+1} x \leqq \sin^{2m} x \leqq \sin^{2m-1} x;$$

consequently

$$0 < \int_0^{\pi/2} \sin^{2m+1} x \, dx \leqq \int_0^{\pi/2} \sin^{2m} x \, dx \leqq \int_0^{\pi/2} \sin^{2m-1} x \, dx.$$

If we here divide each term by $\int_0^{\pi/2} \sin^{2m+1} x \, dx$ and notice that by the first formula proved above

$$\frac{\displaystyle\int_0^{\pi/2} \sin^{2m-1} x \, dx}{\displaystyle\int_0^{\pi/2} \sin^{2m+1} x \, dx} = \frac{2m+1}{2m} = 1 + \frac{1}{2m},$$

we have
$$1 \leqq \frac{\displaystyle\int_0^{\pi/2} \sin^{2m} x \, dx}{\displaystyle\int_0^{\pi/2} \sin^{2m+1} x \, dx} \leqq 1 + \frac{1}{2m},$$

from which the above statement follows.

The relation

$$\frac{\pi}{2} = \lim_{m\to\infty} \frac{2}{1} \frac{2}{3} \frac{4}{3} \frac{4}{5} \frac{6}{5} \frac{6}{7} \cdots \frac{2m}{2m-1} \frac{2m}{2m+1}$$

consequently holds.

This product formula (due to Wallis), with its simple law of formation, gives a remarkable relation between the number π and the integers. If we observe that

$$\lim_{m\to\infty} \frac{2m}{2m+1} = 1, \text{ we can write}$$

$$\lim_{m\to\infty} \frac{2^2 . 4^2 \ldots (2m-2)^2}{3^2 . 5^2 \ldots (2m-1)^2} \, 2m = \frac{\pi}{2},$$

and if we take the square root and then multiply mumerator and denominator by $2 . 4 \ldots (2m-2)$ we find that

$$\sqrt{\frac{\pi}{2}} = \lim_{m\to\infty} \frac{2 . 4 \ldots (2m-2)}{3 . 5 \ldots (2m-1)} \sqrt{2m} = \lim_{m\to\infty} \frac{2^2 . 4^2 \ldots (2m-2)^2}{(2m-1)!} \sqrt{2m}$$

$$= \lim_{m\to\infty} \frac{2^2 . 4^2 \ldots (2m)^2}{(2m)!} \frac{\sqrt{2m}}{2m}.$$

From this we finally obtain

$$\lim_{m \to \infty} \frac{(m!)^2 \, 2^{2m}}{(2m)! \, \sqrt{m}} = \sqrt{\pi},$$

a form of Wallis's product which will be of use to us later (cf. Chap. VII, Appendix, p. 363).

EXAMPLES

Evaluate the integrals in Ex. 1–14:

1. $\int \dfrac{x \cos x}{\sin^2 x} \, dx.$ 　　　 2. $\int \dfrac{x^7}{(1-x^4)^2} \, dx.$ 　　　 3. $\int x^2 \cos x \, dx.$

4. $\int x^3 e^{-x^2} \, dx.$ 　　　 5. $\displaystyle\int_{-\pi}^{\pi} x^2 \cos nx \, dx$ 　 (n a positive integer).

6. $\displaystyle\int_{-\pi}^{\pi} x^2 \sin nx \, dx$ 　 (n a positive integer). 　　　 7. $\int x^3 \cos x^2 \, dx.$

8. $\int \sin^4 x \, dx.$ 　　　 9. $\int \cos^6 x \, dx.$ 　　　 10. $\int x^4 \sqrt{1-x^2} \, dx.$

11. $\int x^2 e^x \, dx.$ 　　　 12. $\int \dfrac{\log x}{x^n} \, dx$ 　 ($n \neq 1$).

13. $\int x^m \log x \, dx$ 　 ($m \neq 1$). 　　　 14. $\int x^2 (\log x)^2 \, dx.$

15. Prove the formula
$$\int e^x p(x) \, dx = e^x \{p(x) - p'(x) + p''(x) - + \ldots\},$$
where $p(x)$ is any polynomial.

16. Show that for all odd positive values of n the integral $\int e^{-x^2} x^n \, dx$ can be evaluated in terms of elementary functions.

17. Show that if n is even the integral $\int e^{-x^2} x^n \, dx$ can be evaluated in terms of elementary functions and the integral $\int e^{-x^2} dx$ (for which tables have been constructed).

18. Prove that
$$\int_0^x \left(\int_0^u f(t) \, dt \right) du = \int_0^x f(u)(x-u) \, du.$$

19.* Ex. 18 gives a formula for the second iterated integral. Prove that the n-th iterated integral of $f(x)$ is given by
$$\frac{1}{(n-1)!} \int_0^x f(u)(x-u)^{n-1} \, du.$$

5. Integration of Rational Functions

The most important general class of functions integrable in terms of elementary functions consists of the rational functions

$$R(x) = \frac{f(x)}{g(x)},$$

where $f(x)$ and $g(x)$ are polynomials:

$$f(x) = a_m x^m + a_{m-1} x^{m-1} + \ldots + a_0,$$
$$g(x) = b_n x^n + b_{n-1} x^{n-1} + \ldots + b_0 \ (b_n \neq 0).$$

We recall that every polynomial can be integrated at once and that the integral is itself a polynomial. We therefore need only consider those rational functions for which the denominator is not a constant. Moreover, we can always assume that the degree of the numerator is less than the degree (n) of the denominator. For otherwise we can divide the polynomial $f(x)$ by the polynomial $g(x)$ and obtain a remainder of degree less than n; in other words, we can write $f(x) = q(x)g(x) + r(x)$, where $q(x)$ and $r(x)$ are also polynomials and $r(x)$ is of lower degree than n. The integration of $\frac{f(x)}{g(x)}$ is then reduced to the integration of the polynomial $q(x)$ and of the " proper " fraction $\frac{r(x)}{g(x)}$. We further notice that the function $\frac{f(x)}{g(x)}$ can be represented as the sum of the functions $\frac{a_\nu x^\nu}{g(x)}$, so that we need only consider integrands of the form $\frac{x^\nu}{g(x)}$.

1. The Fundamental Types.

We shall not at once proceed to the integration of the most general rational function of the above type, but shall instead study only those functions in which the denominator $g(x)$ is of a particularly simple type, namely,

$$g(x) = x, \quad g(x) = 1 + x^2,$$

or, more generally,

$$g(x) = x^n, \quad g(x) = (1 + x^2)^n$$

where n is any positive integer.

To this case we can reduce the somewhat more general case in which $g(x) = (ax + \beta)^n$, a power of a linear expression $ax + \beta$ $(a \neq 0)$, or $g(x) = (ax^2 + 2bx + c)^n$, a power of a definite * quadratic expression. In the first case we introduce a new variable $\xi = ax + \beta$. Then $d\xi/dx = a$, and $x = (\xi - \beta)/a$ is also a linear function of ξ. Each numerator $f(x)$ becomes a polynomial $\phi(\xi)$ of the same degree, and consequently

$$\int \frac{f(x)}{(ax + \beta)^n}\, dx = \frac{1}{a} \int \frac{\phi(\xi)}{\xi^n}\, d\xi.$$

In the second case, we write

$$ax^2 + 2bx + c = \frac{1}{a}(ax + b)^2 + \frac{d^2}{a} \qquad (d^2 = ac - b^2, d > 0),$$

observing that, since we have assumed our expression to be definite, $ac - b^2$ must be positive and $a \neq 0$. By introducing the new variable

$$\xi = \frac{ax + b}{d}$$

we arrive at an integral with the denominator $\left[\dfrac{d^2}{a}(1 + \xi^2)\right]^n$.

Hence in order to integrate rational functions whose denominators are powers of a linear expression or of a definite quadratic expression it is sufficient to be able to integrate the following types of functions:

$$\frac{1}{x^n}, \quad \frac{x^{2\nu}}{(x^2 + 1)^n}, \quad \frac{x^{2\nu+1}}{(x^2 + 1)^n}.$$

We shall, in fact, see that even these types need not be treated in general, for we can reduce the integration of every rational function to the integration of the very special forms of these three functions obtained by taking $\nu = 0$. Accordingly we now consider the integration of the three expressions

$$\frac{1}{x^n}, \quad \frac{1}{(x^2 + 1)^n}, \quad \frac{x}{(x^2 + 1)^n}.$$

* A quadratic expression $Q(x) = ax^2 + 2bx + c$ is said to be *definite* if for all real values of x it takes values having one and the same sign, i.e. if the equation $Q(x) = 0$ has no real roots. For this it is necessary and sufficient that $ac - b^2$ should be positive.

2. Integration of the Fundamental Types.

Integration of the first type of function, $\frac{1}{x^n}$, immediately yields the expression $\log |x|$ if $n = 1$, and the expression $-\frac{1}{(n-1)x^{n-1}}$ if $n > 1$, so that in both cases the integral is again an elementary function. Functions of the third type can be integrated immediately by introducing the new variable $\xi = x^2 + 1$, whence we obtain $2x\,dx = d\xi$ and

$$\int \frac{x}{(x^2+1)^n}\,dx = \frac{1}{2}\int \frac{d\xi}{\xi^n} = \begin{cases} \frac{1}{2}\log(x^2+1) & \text{if } n=1, \\ -\dfrac{1}{2(n-1)(x^2+1)^{n-1}} & \text{if } n > 1. \end{cases}$$

Finally, in order to calculate the integral

$$I_n = \int \frac{dx}{(x^2+1)^n},$$

where n has any value exceeding 1, we make use of a recurrence method. For if we put

$$\frac{1}{(x^2+1)^n} = \frac{1}{(x^2+1)^{n-1}} - \frac{x^2}{(x^2+1)^n},$$

so that $$\int \frac{dx}{(x^2+1)^n} = \int \frac{dx}{(x^2+1)^{n-1}} - \int \frac{x^2\,dx}{(x^2+1)^n},$$

we can transform the right-hand side by integrating by parts, using the formula on p. 218 with

$$f(x) = x, \quad g'(x) = \frac{x}{(x^2+1)^n}.$$

Then, as we have just found above,

$$g(x) = -\frac{1}{2}\frac{1}{(n-1)(x^2+1)^{n-1}},$$

and consequently we obtain

$$I_n = \int \frac{dx}{(x^2+1)^n} = \frac{x}{2(n-1)(x^2+1)^{n-1}} + \frac{2n-3}{2(n-1)}\int \frac{dx}{(x^2+1)^{n-1}}.$$

The calculation of the integral I_n is thus reduced to that of the integral I_{n-1}. If $n-1 > 1$ we apply the same process to the latter integral, and continue the process until we finally arrive at the expression

$$\int \frac{dx}{(x^2+1)} = \text{arc tan } x.$$

We thus see that the integral * I_n can be explicitly expressed in terms of rational functions and the function arc tan x.

Incidentally, we could also have integrated the function $\dfrac{1}{(x^2 + 1)^n}$ directly using the substitution $x = \tan t$; we should then have obtained $dx = \sec^2 t\, dt$ and $1/(1 + x^2) = \cos^2 t$, so that

$$\int \frac{dx}{(x^2 + 1)^n} = \int \cos^{2n-2} t\, dt,$$

and we have already learned (p. 222) how to evaluate this integral.

3. Partial Fractions.

We are now in a position to integrate the most general rational functions, in virtue of the fact that every such function can be represented as the sum of so-called *partial fractions*, i.e. as the sum of a polynomial and a finite number of rational functions, each one of which has either a power of a linear expression for its denominator and a constant for its numerator, or else a power of a definite quadratic expression for its denominator and a linear function for its numerator.   If the degree of the numerator $f(x)$ is less than that of the denominator $g(x)$ the polynomial does not occur.   We are now in a position to integrate each partial fraction.   For according to p. 226 the denominator can be reduced to one of the special forms x^n and $(x^2 + 1)^n$, and the fraction is then a combination of the fundamental types integrated on p. 228.

We shall not give the general proof of the possibility of this resolution into partial fractions.   On the contrary, we shall confine ourselves to making the statement of the theorem intelligible to the reader and to showing by examples how the resolution into partial fractions can be carried out in typical cases.   In actual practice only comparatively simple functions are dealt with, for otherwise the computations become far too complicated.

---

* The integral of the function $\dfrac{1}{(x^2 - 1)^n}$ can be calculated in the same way; by the corresponding recurrence method we reduce it to the integral

$$\int \frac{dx}{1 - x^2} = \text{ar tanh } x \text{ (or ar coth } x).$$

As we know from elementary algebra, every polynomial $g(x)$ can be written in the form

$$g(x) = a(x - a_1)^{l_1}(x - a_2)^{l_2}\ldots(x^2 + 2b_1x + c_1)^{r_1}(x^2 + 2b_2x + c_2)^{r_2}\ldots$$

Here the numbers a_1, a_2, ... are the real and distinct roots of the equation $g(x) = 0$, and the positive integers l_1, l_2 ... indicate the numbers of times they are repeated; the factors $x^2 + 2b_\nu x + c_\nu$ indicate definite quadratic expressions, of which no two are the same, with conjugate complex roots, and the positive integers r_1, r_2, ... give the numbers of times that these roots are repeated.

We assume that the denominator is either given to us in this form or else that we have brought it to this form by calculating the real and imaginary roots. Let us further suppose that the numerator $f(x)$ is of lower degree than the denominator (cf. p. 226). Then the theorem on resolution into partial fractions can be stated as follows. For each factor $(x - a)^l$, where a is any one of the real roots and l is the number of times it is repeated, we can determine an expression of the form

$$\frac{A_1}{(x - a)} + \frac{A_2}{(x - a)^2} + \ldots + \frac{A_l}{(x - a)^l},$$

and for each quadratic factor $Q(x) = x^2 + 2bx + c$ in our product which is raised to the power r we can determine an expression of the form

$$\frac{B_1 + C_1x}{Q} + \frac{B_2 + C_2x}{Q^2} + \ldots + \frac{B_r + C_rx}{Q^r},$$

in such a way that the function $\dfrac{f(x)}{g(x)}$ is the sum of all these expressions. In other words, the quotient $\dfrac{f(x)}{g(x)}$ can be represented as a sum of fractions each of which belongs to one or other of the types integrated on p. 228.*

---

* Here we give a brief sketch of the method by which the possibility of this decomposition into partial fractions is proved. If $g(x) = (x - a)^k h(x)$ and $h(a) \neq 0$, then on the right-hand side of the equation

$$\frac{f(x)}{g(x)} - \frac{f(a)}{h(a)(x - a)^k} = \frac{1}{h(a)}\frac{f(x)h(a) - f(a)h(x)}{(x - a)^k h(x)}$$

the numerator obviously vanishes for $x = a$; it is therefore of the form

In particular cases the splitting up into partial fractions can be done easily by inspection. If, for example, $g(x) = x^2 - 1$, we see at once that

$$\frac{1}{x^2 - 1} = \frac{1}{2}\frac{1}{x - 1} - \frac{1}{2}\frac{1}{x + 1},$$

so that

$$\int \frac{dx}{x^2 - 1} = \frac{1}{2} \log \left| \frac{x - 1}{x + 1} \right|.$$

More generally, if $g(x) = (x - \alpha)(x - \beta)$, that is, if $g(x)$ is a non-definite quadratic expression with two real zeros α and β, we have

$$\frac{1}{(x - \alpha)(x - \beta)} = \frac{1}{\alpha - \beta}\frac{1}{x - \alpha} - \frac{1}{\alpha - \beta}\frac{1}{x - \beta}$$

so that

$$\int \frac{dx}{(x - \alpha)(x - \beta)} = \frac{1}{\alpha - \beta} \log \left| \frac{x - \alpha}{x - \beta} \right|.$$

4. Example. The Bimolecular Reaction.

A simple example of the application of this easy reduction to partial fractions is given by the so-called bimolecular reaction. Let us suppose that we have two reagents whose original concentrations in mols per unit volume are a and b, where we assume that $a < b$, and let us suppose that in time t there is formed in the unit volume a quantity x (mols) of the product of reaction. Then, according to the law of mass action (cf. p. 182), in the simplest case—reaction between one molecule of each of the reagents—the rate of increase of the quantity x is given by the equation $\frac{dx}{dt} = k(a - x)(b - x)$. The problem is to determine the function $x(t)$. If, inversely, we think of the time t as a function of x, we have

$$\frac{dt}{dx} = \frac{1}{k(a - x)(b - x)} = \frac{1}{k(b - a)}\left(\frac{1}{a - x} - \frac{1}{b - x}\right);$$

hence by integration

$$kt = \frac{1}{a - b} \log \frac{a - x}{b - x} + c, \quad \text{for } x < a < b.$$

$h(a)(x - a)^m f_1(x)$, where $f_1(x)$ is also a polynomial, the integer $m \geq 1$, and $f_1(a) \neq 0$. Writing $\frac{f(a)}{h(a)} = \beta$, this gives us

$$\frac{f(x)}{g(x)} - \frac{\beta}{(x - a)^k} = \frac{f_1(x)}{(x - a)^{k-m} h(x)}.$$

Continuing the process, we can keep on diminishing the degree of the power of $(x - a)$ occurring in the denominator until finally no such factor is left. On the remaining fraction we repeat the process for some other root of $g(x)$, and do this as many times as $g(x)$ has distinct factors. This being done not only for the real but also for the complex roots, we eventually arrive at the complete analysis into partial fractions.

We determine the constant of integration c by the condition that at time $t = 0$ no product of reaction has yet been formed, so that

$$\frac{1}{a-b} \log \frac{a}{b} + c = 0.$$

We thus obtain finally

$$kt = \frac{1}{a-b} \log \frac{1 - \dfrac{x}{a}}{1 - \dfrac{x}{b}},$$

and if we solve for x this gives the required function $x(t)$:

$$x = \frac{ab(1 - e^{(a-b)kt})}{b - ae^{(a-b)kt}}.$$

5. Further Examples of Resolution into Partial Fractions. The Method of Undetermined Coefficients.

If $g(x) = (x - a_1)(x - a_2) \dots (x - a_n)$, where $a_i \neq a_k$ if $i \neq k$, i.e. if the equation $g(x) = 0$ has only single real roots, the expression in terms of partial fractions has the simple form

$$\frac{1}{g(x)} = \frac{a_1}{x - a_1} + \frac{a_2}{x - a_2} + \dots + \frac{a_n}{x - a_n}.$$

We obtain explicit expressions for the coefficients $a_1, a_2, \dots$ if we multiply both sides of this equation by $(x - a_1)$, cancel the common factor $(x - a_1)$ in the numerator and denominator on the left and in the first term on the right, and then put $x = a_1$. This gives *

$$a_1 = \frac{1}{(a_1 - a_2)(a_1 - a_3) \dots (a_1 - a_n)}.$$

As a typical example of a denominator $g(x)$ with multiple roots, we consider the function $\dfrac{1}{x^2(x - 1)}$. The preliminary statement

$$\frac{1}{x^2(x - 1)} = \frac{a}{x - 1} + \frac{b}{x} + \frac{c}{x^2}$$

in accordance with p. 230 leads us to the required result. If we multiply both sides of this equation by $x^2(x - 1)$ we obtain the equation

$$1 = (a + b)x^2 - (b - c)x - c,$$

---

* The reader will observe that the denominator on the right is $g'(a_1)$, i.e. the derivative of the function $g(x)$ at the point $x = a_1$.

true for all values of x, from which we have to determine the coefficients a, b, c. This condition cannot hold unless all the coefficients of the polynomial $(a + b)x^2 - (b - c)x - c - 1$ are zero, i.e. we must have $a + b = b - c = c + 1 = 0$ or $c = -1, b = -1, a = 1$. We thus obtain the resolution

$$\frac{1}{x^2(x - 1)} = \frac{1}{x - 1} - \frac{1}{x} - \frac{1}{x^2},$$

and consequently

$$\int \frac{dx}{x^2(x - 1)} = \log|x - 1| - \log|x| + \frac{1}{x}.$$

We shall now split up the function $\dfrac{1}{x(x^2 + 1)}$ (which is an example of the case where the zeros of the denominator are complex) in accordance with the equation

$$\frac{1}{x(x^2 + 1)} = \frac{a}{x} + \frac{bx + c}{x^2 + 1}.$$

For the coefficients we obtain $a + b = c = a - 1 = 0$, so that

$$\frac{1}{x(x^2 + 1)} = \frac{1}{x} - \frac{x}{x^2 + 1},$$

and consequently

$$\int \frac{dx}{x(x^2 + 1)} = \log|x| - \frac{1}{2}\log(x^2 + 1).$$

As a third example we consider the function $\dfrac{1}{x^4 + 1}$. Even Leibnitz found this a troublesome integration. We can represent the denominator as the product of two quadratic factors:

$$x^4 + 1 = (x^2 + 1)^2 - 2x^2 = (x^2 + 1 + \sqrt{2}\,x)(x^2 + 1 - \sqrt{2}\,x).$$

We know, therefore, that the resolution into partial fractions will have the form

$$\frac{1}{x^4 + 1} = \frac{ax + b}{x^2 + \sqrt{2}\,x + 1} + \frac{cx + d}{x^2 - \sqrt{2}\,x + 1}.$$

To determine the coefficients a, b, c, d, we have the equation

$$(a + c)x^3 + (b + d - a\sqrt{2} + c\sqrt{2})x^2 + (a + c - b\sqrt{2} + d\sqrt{2})x + (b + d - 1) = 0,$$

which is satisfied by the values

$$a = \frac{1}{2\sqrt{2}}, \quad b = \frac{1}{2}, \quad c = -\frac{1}{2\sqrt{2}}, \quad d = \frac{1}{2}.$$

We therefore have

$$\frac{1}{x^4 + 1} = \frac{1}{2\sqrt{2}} \cdot \frac{x + \sqrt{2}}{x^2 + \sqrt{2}\,x + 1} - \frac{1}{2\sqrt{2}} \cdot \frac{x - \sqrt{2}}{x^2 - \sqrt{2}\,x + 1},$$

and, applying the method given on p. 227, we obtain

$$\int \frac{dx}{x^4 + 1} = \frac{1}{4\sqrt{2}} \log |x^2 + \sqrt{2}x + 1| - \frac{1}{4\sqrt{2}} \log |x^2 - \sqrt{2}x + 1|$$

$$+ \frac{1}{2\sqrt{2}} \text{arc tan} (\sqrt{2}x + 1) + \frac{1}{2\sqrt{2}} \text{arc tan} (\sqrt{2}x - 1),$$

which may easily be verified by differentiation.

EXAMPLES

Integrate:

1. $\int \dfrac{dx}{2x - 3x^2}.$

2. $\int \dfrac{dx}{x^2 - x}.$

3. $\int \dfrac{3\,dx}{x(x+1)^3}.$

4. $\int \dfrac{x^2 + x + 1}{3x^2 - 2x - 5}\,dx.$

5. $\int \dfrac{dx}{(x-1)^2(x^2+1)}.$

6. $\int \dfrac{x^2\,dx}{(x-1)^2(x^2+1)}.$

7. $\int \dfrac{dx}{1 - x^3}.$

8. $\int \dfrac{dx}{1 + x^3}.$

9. $\int \dfrac{(x-4)}{(x^2+1)(x-2)}\,dx.$

10. $\int \dfrac{x+4}{(x^2-1)(x+2)}\,dx.$

11. $\int \dfrac{x^6}{1 - x^4}\,dx.$

12.* $\int \dfrac{dx}{x^6 + 1}.$

13. $\int \dfrac{x^2}{x^4 + x^2 - 2}\,dx.$

14. $\int \dfrac{dx}{x^2(x^2+1)^2}.$

6. Integration of Some Other Classes of Functions

1. Preliminary Remarks on the Rational Representation of the Trigonometric and Hyperbolic Functions.

The integration of some other general classes of functions can be reduced to the integration of rational functions. We shall be better able to understand this reduction if we begin by stating certain elementary facts about the trigonometric and hyperbolic functions.   If we put $t = \tan \dfrac{x}{2}$, elementary trigonometry gives us the simple formulæ

$$\sin x = \frac{2t}{1 + t^2}, \quad \cos x = \frac{1 - t^2}{1 + t^2};$$

for
$$\frac{1}{1 + t^2} = \cos^2 \frac{x}{2} \quad \text{and} \quad \frac{t^2}{1 + t^2} = \sin^2 \frac{x}{2},$$

and hence from the elementary formulæ

$$\sin x = 2 \cos^2 \frac{x}{2} \tan \frac{x}{2} \quad \text{and} \quad \cos x = \cos^2 \frac{x}{2} - \sin^2 \frac{x}{2}$$

we obtain the above equations. These equations show that $\sin x$ *and* $\cos x$ *can be expressed rationally in terms of the quantity* $t = \tan \dfrac{x}{2}$. From $t = \tan \dfrac{x}{2}$ we have by differentiation

$$\frac{dt}{dx} = \frac{1}{2 \cos^2 x/2} = \frac{1 + t^2}{2}, \text{ so that } \frac{dx}{dt} = \frac{2}{1 + t^2};$$

hence the derivative $\dfrac{dx}{dt}$ is also a rational expression in t.

The geometrical representation of our formulæ and their geometrical meaning are given in fig. 2. Here the circle $u^2 + v^2 = 1$ in a uv-plane is shown. If x denotes the angle POT in the figure, then $u = \cos x$ and $v = \sin x$. The angle OSP with its vertex at the point $u = -1$, $v = 0$ is equal to $x/2$, by a theorem in elementary geometry, and we can read off the geometrical meaning of the parameter t from the figure; $t = \tan \frac{1}{2}x = OR$. If the point P starts from S and runs once round the circle in the positive direction, i.e. if x runs through the interval from $-\pi$ to $+\pi$, the quantity t will run through the whole range of values from $-\infty$ to $+\infty$ exactly once.

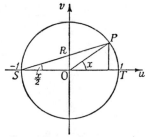

Fig. 2.—Parametric representation of the trigonometric functions

We may correspondingly express the hyperbolic functions $\cosh x = \frac{1}{2}(e^x + e^{-x})$ and $\sinh x = \frac{1}{2}(e^x - e^{-x})$ as rational functions of a third quantity. The most obvious way is to put $e^x = \tau$, so that we have

$$\cosh x = \frac{1}{2}\left(\tau + \frac{1}{\tau}\right), \quad \sinh x = \frac{1}{2}\left(\tau - \frac{1}{\tau}\right),$$

which are rational expressions for $\sinh x$ and $\cosh x$. Here again $dx/dt = 1/\tau$ is rational in τ. But we obtain a closer analogy with the trigonometric functions by introducing the quantity $t = \tanh\dfrac{x}{2}$; we then arrive at the formulæ

$$\sinh x = \frac{2t}{1 - t^2}, \quad \cosh x = \frac{1 + t^2}{1 - t^2}.$$

By differentiating $t = \tanh\dfrac{x}{2}$ we obtain, as on p. 235, the rational expression

$$\frac{dx}{dt} = \frac{2}{1 - t^2}$$

for the derivative dx/dt. Here again the quantity t has a geometrical meaning similar to that which it has in the case of the trigonometric functions, as we see at once from fig. 3.

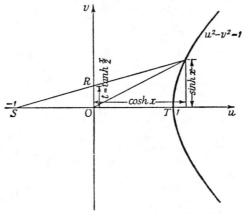

Fig. 3.—Parametric representation of the hyperbolic functions

But whereas in the case of the trigonometric functions t must run through the whole range of values from $-\infty$ to $+\infty$ in order to give all pairs of values of $\cos x$ and $\sin x$, in the case of the hyperbolic functions t is limited to the interval $-1 < t < 1$.

Having made these preliminary remarks, we proceed to our integration problem.

2. Integration of $R(\cos x, \sin x)$.

Let $R(\cos x, \sin x)$ denote an expression which is rational in the two functions $\sin x$ and $\cos x$, i.e. an expression which is formed rationally from these two functions and constants, such as

$$\frac{3 \sin^2 x + \cos x}{3 \cos^2 x + \sin x}.$$

If we apply the substitution $t = \tan \dfrac{x}{2}$ the integral

$$\int R(\cos x,\ \sin x)\, dx$$

is transformed into the integral

$$\int R\left(\frac{1 - t^2}{1 + t^2},\ \frac{2t}{1 + t^2}\right)\ \frac{2}{1 + t^2}\, dt,$$

and under the integral sign we now have a rational function of t. Thus we have in theory obtained the integral of our expression, since we can now perform the integration by the methods of the preceding section.

3. Integration of $R(\cosh x,\ \sinh x)$.

In the same way, if $R(\cosh x,\ \sinh x)$ is an expression which is rational in terms of the hyperbolic functions $\cosh x$ and $\sinh x$, we can effect its integration by means of the substitution $t = \tanh \dfrac{x}{2}$. Recalling that

$$\frac{dx}{dt} = \frac{2}{1 - t^2}$$

we have

$$\int R(\cosh x,\ \sinh x)\, dx = \int R\left(\frac{1 + t^2}{1 - t^2},\ \frac{2t}{1 - t^2}\right)\ \frac{2}{1 - t^2}\, dt.$$

(According to a previous remark we could also have introduced $\tau = e^x$ as a new variable and expressed $\cosh x$ and $\sinh x$ in terms of τ.) The integration is once again reduced to that of a rational function.

4. Integration of $R\{x,\ \sqrt{(1 - x^2)}\}$.

The integral $\int R\{x,\ \sqrt{(1 - x^2)}\}\, dx$ can be reduced to the type treated in No. 2 by using the substitution

$$x = \cos u,\quad \sqrt{(1 - x^2)} = \sin u,\quad dx = -\sin u\, du;$$

from this stage the transformation $t = \tan \dfrac{x}{2}$ brings us to the integration of a rational function. Incidentally, we could have carried out the reduction in one step instead of two by using the substitution

$$t = \sqrt{\left(\frac{1-x}{1+x}\right)}; \quad x = \frac{1-t^2}{1+t^2}; \quad \sqrt{(1-x^2)} = \frac{2t}{1+t^2};$$

$$\frac{dx}{dt} = \frac{-4t}{(1+t^2)^2};$$

that is, we could have introduced $t = \tan \dfrac{u}{2}$ directly as the new variable and thereby obtained a rational function to integrate.

5. Integration of $R\{x, \sqrt{(x^2-1)}\}$.

The integral $\int R\{x, \sqrt{(x^2-1)}\}\,dx$ is transformed by the substitution $x = \cosh u$ into the type treated in No. 3 (p. 237). Here again we can arrive at our goal directly by introducing

$$t = \sqrt{\left(\frac{x-1}{x+1}\right)} = \tanh \frac{u}{2}.$$

6. Integration of $R\{x, \sqrt{(x^2+1)}\}$.

The integral $\int R\{x, \sqrt{(x^2+1)}\}\,dx$ is reduced by the transformation $x = \sinh u$ to the type considered in No. 3 (p. 237), and can therefore be integrated in terms of elementary functions. Instead of the further reduction to the integral of a rational function by the substitution $e^u = \tau$ or $\tanh \dfrac{u}{2} = t$, we could have reached the integral of a rational function at a single stroke by either of the substitutions

$$\tau = x + \sqrt{(x^2+1)}, \quad t = \frac{-1 + \sqrt{(x^2+1)}}{x}.$$

7. Integration of $R\{x, \sqrt{(ax^2 + 2bx + c)}\}$.

The integral $\int R\{x, \sqrt{(ax^2 + 2bx + c)}\}\,dx$ of an expression which is rational in terms of x and the square root of an arbitrary polynomial of the second degree in x can immediately be reduced to one of the types just treated. We write (cf. p. 227)

$$ax^2 + 2bx + c = \frac{1}{a}(ax+b)^2 + \frac{ac-b^2}{a}.$$

If $ac - b^2 > 0$ we introduce a new variable ξ by means of the transformation $\xi = \dfrac{ax+b}{\sqrt{(ac-b^2)}}$, whereupon the surd takes the form $\sqrt{\left\{ \dfrac{ac-b^2}{a}(\xi^2+1) \right\}}$. Hence our integral when expressed in terms of ξ is of the type of No. 6. The constant a must here be positive in order that the square root may have real values. If $ac - b^2 = 0$, $a > 0$, then by way of the formula

$$\sqrt{(ax^2 + 2bx + c)} = \sqrt{a}\left(x + \frac{b}{a} \right)$$

we see that the integrand was rational in x to begin with.

If, finally, $ac - b^2 < 0$, we put $\xi = \dfrac{ax+b}{\sqrt{(b^2-ac)}}$ and obtain for the surd the expression $\sqrt{\left\{ \dfrac{b^2-ac}{a}(\xi^2-1) \right\}}$. If a is positive, our integral is thus reduced to the type of No. 5 (p. 238); if, on the other hand, a is negative, we write the surd in the form $\sqrt{\left(\dfrac{b^2-ac}{-a} \right)} \sqrt{(1-\xi^2)}$ and see that the integral is thus reduced to the type of No. 4 (p. 237).

8. Further Examples of Reduction to Integrals of Rational Functions.

Of other types of functions which can be integrated by reduction to rational functions we shall briefly mention two: (1) rational expressions involving two different surds of linear expressions, $R\{x, \sqrt{(ax+b)}, \sqrt{(\alpha x + \beta)}\}$; (2) expressions of the form $R\left\{ x, \sqrt[n]{\left(\dfrac{ax+b}{\alpha x + \beta} \right)} \right\}$, where a, b, α, β are constants. In the first case we introduce the new variable $\xi = \sqrt{(\alpha x + \beta)}$, so that $\alpha x + \beta = \xi^2$, and consequently

$$x = \frac{\xi^2 - \beta}{\alpha} \quad \text{and} \quad \frac{dx}{d\xi} = \frac{2\xi}{\alpha};$$

then $\int R\{x, \sqrt{(ax+b)}, \sqrt{(ax+\beta)}\}\,dx$

$$= \int R\left\{\frac{\xi^2-\beta}{a}, \sqrt{\frac{1}{a}\{a\xi^2-(a\beta-ba)\}}, \xi\right\}\frac{2\xi}{a}\,d\xi,$$

which is of the type discussed in No. 7.

If in the second case we introduce the new variable

$$\xi = \sqrt[n]{\left(\frac{ax+b}{ax+\beta}\right)},$$

we have

$$\xi^n = \frac{ax+b}{ax+\beta}, \quad x = \frac{-\beta\xi^n+b}{a\xi^n-a}, \quad \frac{dx}{d\xi} = \frac{a\beta-ba}{(a\xi^n-a)^2}\cdot n\xi^{n-1},$$

and we immediately arrive at the formula

$$\int R\left(x, \sqrt[n]{\left(\frac{ax+b}{ax+\beta}\right)}\right)dx$$

$$= \int R\left(\frac{-\beta\xi^n+b}{a\xi^n-a}, \xi\right)\frac{a\beta-ba}{(a\xi^n-a)^2}n\xi^{n-1}d\xi,$$

which is the integral of a rational function.

9. Remarks on the Examples.

The preceding discussions are chiefly of theoretical interest. In the case of complicated expressions the actual calculations would be far too involved. It is therefore expedient to make use, when possible, of the special form of the integrand to simplify the work. For example, in order to integrate the expression $\dfrac{1}{a^2\sin^2 x + b^2\cos^2 x}$ it is better to use the substitution $t = \tan x$ instead of that given on p. 237, for $\sin^2 x$ and $\cos^2 x$ can be expressed rationally in terms of $\tan x$, and it is therefore unnecessary to go back to $t = \tan\dfrac{x}{2}$. The same is true for every expression formed rationally from* $\sin^2 x$, $\cos^2 x$, and $\sin x \cos x$. Moreover, for the calculation of many integrals a trigonometrical form is to be preferred to a rational one, provided that the trigonometrical form can be evaluated by some simple recurrence method.

---

* For $\sin x \cos x = \tan x \cos^2 x$ can of course be expressed rationally in terms of $\tan x$.

For example, although the integrand in $\int x^n \{\sqrt{(1-x^2)}\}^m \, dx$ can be reduced to a rational form, it is better to write $x = \sin u$ and bring it to the form $\int \sin^n u \cos^{m+1} u \, du$, since this can easily be treated by the recurrence method on p. 222 (or by using the addition theorems to reduce the powers of the sine and cosine to sines and cosines of multiple angles).

For the evaluation of the integral

$$\int \frac{dx}{a \cos x + b \sin x} \quad (a^2 + b^2 > 0),$$

instead of referring to the general theory we determine a number A and an angle θ in such a way that

$$a = A \sin \theta, \quad b = A \cos \theta;$$

that is, we write

$$A = \sqrt{a^2 + b^2}, \quad \sin \theta = \frac{a}{A}, \quad \cos \theta = \frac{b}{A}.$$

The integral then takes the form

$$\frac{1}{A} \int \frac{dx}{\sin(x + \theta)},$$

and on introducing the new variable $x + \theta$ we find (cf. p. 215) that the value of the integral is

$$\frac{1}{A} \log \left| \tan \frac{x + \theta}{2} \right|.$$

EXAMPLES

Integrate:

1. $\int \dfrac{dx}{1 + \sin x}.$

2. $\int \dfrac{dx}{1 + \cos x}.$

3. $\int \dfrac{dx}{2 + \sin x}.$

4. $\int \dfrac{dx}{\sin^3 x}.$

5. $\int \dfrac{dx}{\cos x}.$

6. $\int_0^{\pi/2} \dfrac{dx}{3 + \cos x}.$

7. $\int \dfrac{dx}{1 + \cos^2 x}.$

8. $\int \dfrac{dx}{3 + \sin^2 x}.$

9. $\int \tan^3 x \, dx.$

10. $\int \dfrac{dx}{\sin x + \cos x}.$

11. $\int \dfrac{\sin^2 x + \cos^3 x}{3 \cos^2 x + \sin^4 x} \sin x \, dx.$

12. $\int \sqrt{(x^2 - 4)} \, dx.$

9

(E 798)

13. $\int \sqrt{(4 + 9x^2)}\, dx.$

16. $\int \dfrac{dx}{\sqrt{x} + \sqrt{(1 - x)}}.$

14. $\int \dfrac{dx}{(x - 2)\sqrt{(x^2 - 4x + 3)}}.$

17. $\int \dfrac{\sqrt{(1 + x)} + \sqrt{(1 - x)}}{\sqrt{(1 + x)} - \sqrt{(1 - x)}}\, dx.$

15. $\int x\sqrt{(x^2 + 4x)}\, dx.$

18. $\int \dfrac{\sqrt{(x - a)}}{1 + \sqrt{(x - a + 1)}}\, dx.$

19. $\int \dfrac{dx}{\sqrt{(x - a)} + \sqrt{(x - b)}}.$

7. Remarks on Functions which are not Integrable in Terms of Elementary Functions

1. Definition of Functions by means of Integrals. Elliptic Integrals.

With the above examples of types of functions which can be integrated by reduction to rational functions, we have practically exhausted the list of functions which are integrable in terms of elementary functions. Attempts to express general integrals such as

$$\int \frac{dx}{\sqrt{(a_0 + a_1 x + \ldots + a_n x^n)}}, \quad \int \sqrt{(a_0 + a_1 x + \ldots + a_n x^n)}\, dx$$

or $\int \dfrac{e^x}{x}\, dx$ in terms of elementary functions have always ended in failure; and in the nineteenth century it was finally proved that it is actually impossible to carry out these integrations in terms of elementary functions.

If, therefore, the object of the integral calculus were to integrate functions in terms of elementary functions, we should have come to a definite halt. But such a restricted object has no intrinsic justification; indeed, it is of a somewhat artificial nature. We know that the integral of every continuous function exists and is itself a continuous function of the upper limit, and this fact has nothing to do with the question whether the integral can be expressed in terms of elementary functions or not. The distinguishing features of the elementary functions are based on the fact that their properties are easily recognized, that their application to numerical problems is often facilitated by convenient tables or, as in the case of the rational functions, that they can easily be calculated with as great a degree of accuracy as we please.

Where the integral of a function cannot be expressed by means of functions with which we are already acquainted, there is nothing to hinder us from introducing this integral as a new " higher " function in analysis, which really means no more than giving it a name. Whether the introduction of such a new function is convenient or not depends on the properties which it possesses, the frequency with which it occurs, and the ease with which it can be manipulated in theory and in practice. In this sense the process of integration therefore forms a basis for the generation of new functions.

After all, we are already acquainted with this principle from our dealings with the elementary functions. Thus we found ourselves obliged (p. 167) to introduce the previously unknown integral of $1/x$ as a new function, which we called the logarithm and whose properties we could easily determine. We could have introduced the trigonometric functions in a similar way, making use only of the rational functions, the process of integration, and the process of inversion. For this purpose we need only take one or other of the equations

$$\text{arc tan} \, x = \int_0^x \frac{dt}{1 + t^2} \quad \text{or} \quad \text{arc sin} \, x = \int_0^x \frac{dt}{\sqrt{(1 - t^2)}}$$

as the *definition* of the function arc tan x or arc sin x respectively, in order to arrive at the trigonometric functions by inversion. By this process the definition of these functions is separated from geometry, but we are naturally left with the task of developing their properties, also independently of geometry.*

The first and most important example which leads us beyond the region of elementary functions is given by the *elliptic integrals*. These are integrals in which the integrand is formed in a rational way from the variable of integration and the square root of an expression of the third or fourth degree. Among these integrals the function

$$u(s) = \int_0^s \frac{dx}{\sqrt{(1 - x^2)(1 - k^2 x^2)}}$$

turns out to be of particular importance. Its inverse function $s(u)$ plays a correspondingly important part. In particular, for

---

* We shall not go into the development of these ideas here. The essential step is to prove the addition theorems for the inverse functions, i.e. for the sine and the tangent.

$k = 0$ we obtain $u(s) = \text{arc sin}\, x$ and $s(u) = \sin u$ respectively. The function $s(u)$ has been as thoroughly examined and tabulated as the elementary functions. This, however, leads us away from the line of the present discussion and into the realm of the so-called elliptic functions, which occupy a central position in the theory of functions of a complex variable.

Here we shall merely remark that the name "elliptic integral" arises from the fact that such integrals enter into the problem of determining the length of an arc of an ellipse. (Cf. Chap. V, p. 289.)

We may further point out that integrals which at first sight have quite a different appearance turn out after a simple substitution to be elliptic integrals. As an example, the integral

$$\int \frac{dx}{\sqrt{(\cos \alpha - \cos x)}}$$

is transformed by means of the substitution $u = \cos \dfrac{x}{2}$ into the integral

$$-k\sqrt{2} \int \frac{du}{\sqrt{(1 - u^2)(1 - k^2 u^2)}}, \quad k = \frac{1}{\cos \alpha/2};$$

the integral
$$\int \frac{dx}{\sqrt{(\cos 2x)}}$$

by means of the substitution $u = \sin x$ becomes

$$\int \frac{du}{\sqrt{(1 - u^2)(1 - 2u^2)}};$$

and finally the integral $\quad \displaystyle\int \frac{dx}{\sqrt{(1 - k^2 \sin^2 x)}}$

is transformed by the substitution $u = \sin x$ into

$$\int \frac{du}{\sqrt{(1 - u^2)(1 - k^2 u^2)}}.$$

2. On Differentiation and Integration.

Another remark on the relation between differentiation and integration may be inserted here. Differentiation may be considered a more elementary process than integration, since it does not lead us away from the domain of known functions. On the other hand, we must remember that the differentiability of an arbitrary continuous function is by no means a foregone conclusion, but a very stringent additional assumption. We have, in fact, seen that there are continuous functions which are non-differentiable at isolated points, and we may mention without proof that since Weierstrass' time many examples have been constructed of continuous functions which do not possess a derivative anywhere at all.* (There is therefore much less in the mathematical definition of continuity than simple intuition would lead us to suppose.) In contrast to this, even though integration in terms of elementary functions is not always possible, in all circumstances we are certain at least that the integral of a continuous function exists.

Taken all in all, we see that integration and differentiation cannot be simply classified as more elementary and less elementary, but that from some points of view the one and from other points of view the other should be thought of as the more elementary.

In so far as the concept of integral is concerned, we shall see in the next section that it is not closely bound up with the assumption that the integrand is continuous, but that it may be extended to wide classes of functions with discontinuities.

8. Extension of the Concept of Integral. Improper Integrals

1. Functions with Jump Discontinuities.

In the first instance we see that there is no difficulty in extending the concept of integral to the case where the function to be integrated has jump discontinuities at one or more points in the interval of integration. For we need only take the integral of the function as the sum of the integrals over the separate sub-intervals in which the function is continuous.† The integral then retains its intuitive meaning as an area (cf. fig. 4).

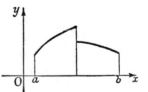

Fig. 4.—The integral of a discontinuous function

* Cf. Titchmarsh, *The Theory of Functions* (Oxford, 1932), §§ 11·21–11·23 (pp. 350–354).

† We should really observe that in our previous definition of integral we took the interval as closed and the function as continuous in the closed interval. This gives us no trouble, since in each closed sub-interval we can extend the function so that it is continuous by taking for the value of the function at the end-point the limit of the function as x approaches the end-point from the interior of the interval.

2. Functions with Infinite Discontinuities.

It is quite a different matter when the function has an infinite discontinuity in the interior of the interval or at one of its ends. In order even to formulate the notion of integral in this case we must introduce a further limiting process. Before stating the general definition we shall illustrate some of the possibilities by means of examples.

We begin with the integral $\int \dfrac{dx}{x^a}$,

where α is a positive number. The integrand $1/x^a$ becomes infinite as $x \to 0$, and we therefore cannot extend the integral to the lower limit 0. We can, however, try to find what happens when we take the integral from the positive limit ε to the limit 1, say, and finally let ε tend to 0. According to the elementary rules of integration, we obtain, provided $\alpha \neq 1$,

$$\int_\varepsilon^1 \frac{dx}{x^a} = \frac{1}{1-\alpha} (1 - \varepsilon^{1-\alpha}).$$

We immediately recognize that the following possibilities occur: (1) α is greater than 1; then as $\varepsilon \to 0$ the right-hand side tends to ∞: (2) α is less than 1; then the right-hand side tends to the limit $1/(1 - \alpha)$. In the second case, therefore, we shall simply take this limiting value as the integral between the limits 0 and 1. In the first case we shall say that the integral from 0 to 1 does not exist. (3) In the third case, where $\alpha = 1$, the integral will be equal to $-\log \varepsilon$ and therefore as $\varepsilon \to 0$ it approaches no limit, but tends to ∞; that is, the integral from 0 to 1 does not exist.

Another example of the extension of the integral of a function up to an infinite discontinuity is given by the integrand $\dfrac{1}{\sqrt{(1-x^2)}}$. We find that

$$\int_0^{1-\varepsilon} \frac{dx}{\sqrt{(1-x^2)}} = \arcsin (1 - \varepsilon).$$

If we let ε tend to 0, the right-hand side converges to a definite limit, $\pi/2$; we therefore call this the value of the integral $\int_0^1 \dfrac{dx}{\sqrt{(1-x^2)}}$, even though the integrand becomes infinite at the point $x = 1$.

In order to extract a perfectly general concept from these examples, we notice in the first place that it clearly makes no essential difference whether the discontinuity of the integrand lies at the upper end or the lower end of the interval of integration. We now make the following statement:

If in an interval a $\leqq$ x $\leqq$ b *the function* f(x) *is continuous with the single exception of the end-point* b, *we define* $\int_a^b$ f(x) dx *as the limit*

$$\lim_{\epsilon \to 0} \int_a^{b-\epsilon} f(x)\,dx$$

—*where the point* b — ϵ *approaches the end-point* b *from the interior of the interval—provided that such a limit exists.*

In this case we say that the *improper integral* $\int_a^b$ f(x) dx *converges*. If, however, no such limit exists, we say that the integral $\int_a^b f(x)\,dx$ does not exist or does not converge or that it *diverges*.

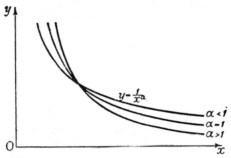

Fig. 5.—To illustrate the convergence or divergence of improper integrals

An analogous definition holds for the case where the lower limit of the interval of integration, and not the upper, is the exceptional point.

Even improper integrals can be interpreted as areas. In the first instance, of course, there is no sense in speaking of the area of a region which extends to infinity; yet one may attempt to define such an area by means of a passage to the limit from a bounded region with a finite area. For example, the above results for the function $1/x^a$ imply that the area bounded by the x-axis, the line $x = 1$, the line $x = \epsilon$, and the curve $y = 1/x^a$ tends to a finite limit as $\epsilon \to 0$, provided that $a < 1$, and that it tends to infinity if $a \geqq 1$. This fact may be simply expressed as follows: the area between the x-axis, the y-axis, the curve, and the line $x = 1$ is finite or infinite according as $a < 1$ or $a \geqq 1$.

Intuition can, of course, give us no precise information about the finiteness or infiniteness of the area of a region stretching to infinity. Of such a region we can only say that the more closely its sides approach one another the more likely it is to have a finite area. In this sense fig. 5 illustrates the fact that for $\alpha < 1$ the area under our curve remains finite, while for $\alpha \geq 1$ it is infinite.

In order to find out whether a function $f(x)$ which has an infinite discontinuity at the point $x = b$ can be integrated up to b, we can often save ourselves a special investigation by using the following criterion:

Let the function $f(x)$ be positive* in the interval $a \leq x \leq b$, and let $\lim\limits_{x \to b} f(x) = \infty$. Then the integral $\int_a^b f(x)\,dx$ *converges* if there exist both a positive number μ less than 1, and a fixed number M independent of x, such that everywhere in the interval $a \leq x < b$ the inequality $f(x) \leq \dfrac{M}{(b - x)^\mu}$ is true; in other words, *if at the point* x = b *the function* f(x) *becomes infinite of a lower order than the first.* On the other hand, the integral *diverges*, if there exist both a number $\nu \geq 1$ and a fixed number N, such that everywhere in the interval $a \leq x < b$ the inequality $f(x) \geq \dfrac{N}{(b - x)^\nu}$ is true; in other words, *if at the point* x = b *the function* f(x) *becomes infinite of the first order at least.*

The proof follows almost immediately by comparison with the very simple special case discussed above. In order to prove the first part of the theorem we observe that for $0 < \epsilon < b - a$ we have

$$0 \leq \int_a^{b-\epsilon} f(x)\,dx \leq \int_a^{b-\epsilon} \frac{M}{(b - x)^\mu}\,dx.$$

As $\epsilon \to 0$ the integral on the right, which is obtained from the integral $\int \dfrac{dx}{x^\alpha}$ (p. 128) by a simple change of notation, has a limit, and therefore remains bounded. Moreover, the values of $\int_a^{b-\epsilon} f(x)\,dx$ increase monotonically as $\epsilon \to 0$; since they are also bounded, they must possess a limit, and the integral $\int_a^b f(x)\,dx$ therefore converges.

The parallel proof of the second part of the theorem is left as an exercise for the reader.

---

* In the Appendix to Chap. VIII (p. 418) we shall see that this restriction of sign can easily be removed.

We likewise see at once that exactly analogous theorems hold where the *lower* limit of the integral is a point of infinite discontinuity. If a point of infinite discontinuity lies in the interior of the interval of integration, we merely use this point to divide the interval into two sub-intervals and then apply the above considerations to each of these separately.

As a further example we consider the elliptic integral

$$\int_0^1 \frac{dx}{\sqrt{(1 - x^2)(1 - k^2x^2)}} \quad (k^2 < 1).$$

From the identity $1 - x^2 = (1 - x)(1 + x)$ we see at once that as $x \to 1$ the integrand becomes infinite only of order $\frac{1}{2}$, whence it follows that the improper integral exists.

3. Infinite Interval of Integration.

Another important extension of the concept of integral consists in taking one of the limits of integration as infinite. In order to make this extension precise, we introduce the following notation: if the integral

$$\int_a^A f(x)\,dx,$$

where a is fixed, tends to a definite limit when A increases positively beyond all bounds, we denote the limit by

$$\int_a^\infty f(x)\,dx,$$

and call it the integral from a to ∞ of the function $f(x)$. Of course such an integral does not necessarily exist or, as we often say, *converge*.

Simple examples of the various possibilities are again yielded by the functions $f(x) = 1/x^a$,

$$\int_1^A \frac{dx}{x^a} = \frac{1}{1 - \alpha}(A^{1-\alpha} - 1).$$

Here we see that, if we again exclude the case $\alpha = 1$, the integral to infinity exists for the case $\alpha > 1$, and in fact

$$\int_1^\infty \frac{dx}{x^a} = \frac{1}{\alpha - 1};$$

on the contrary, when $\alpha < 1$ the integral no longer exists. For the case $\alpha = 1$ the integral again clearly fails to exist, since $\log x$ tends to infinity as x does. We see, therefore, that with regard to integration over an

9 *

infinite interval the functions $1/x^a$ do not behave in the same way as for integration up to the origin. This statement also is made plausible by a glance at fig. 5. For we see that the larger α is, the more closely do the curves draw in towards the x-axis when x is large, so that we can readily suppose that the area under consideration tends to a definite limit for sufficiently large values of α.

The following criterion for the existence of an integral with an infinite limit is often useful. We again assume that for sufficiently large values of x, say for $x \geqq a$, the integrand has always the same sign, which without loss of generality we can choose to be positive.* Then we have the following statement:

The integral $\int_a^\infty \mathbf{f}(\mathbf{x})\,d\mathbf{x}$ *converges if the function* $\mathbf{f}(\mathbf{x})$ *vanishes at infinity to a higher order than the first,* that is, if there is a number $\nu > 1$ such that for all values of x, no matter how large, the relation $0 < f(x) \leqq \dfrac{M}{x^\nu}$ is true, where M is a fixed number independent of x. Again, *the integral diverges if the function remains positive and vanishes at infinity to an order not higher than the first,* that is, if there is a fixed number $N > 0$ such that $xf(x) \geqq N$.

The proof of these criteria, which runs exactly parallel to the previous argument, can be left to the reader.

A very simple example is the integral $\int_a^\infty \dfrac{1}{x^2}\,dx\,(a > 0)$. The integrand vanishes at infinity to the second order. As a matter of fact, we see at once that the integral does converge, for $\int_a^A \dfrac{1}{x^2}\,dx = \dfrac{1}{a} - \dfrac{1}{A}$, and therefore

$$\int_a^\infty \frac{1}{x^2}\,dx = \frac{1}{a}.$$

Another equally simple example is

$$\int_0^\infty \frac{1}{1 + x^2}\,dx = \lim_{A \to \infty} (\text{arc tan } A - \text{arc tan } 0) = \frac{\pi}{2}.$$

4. The Gamma Function.

A further example of particular importance in analysis is offered by the so-called gamma function

$$\Gamma(n) = \int_0^\infty e^{-x} x^{n-1}\,dx \qquad (n > 0).$$

---

* As we shall see in the Appendix to Chap. VIII (p. 418), this restriction of sign can easily be removed.

Here also the criterion of convergence is satisfied; e.g. if we choose $\nu = 2$, we have $\lim\limits_{x \to \infty} x^\nu \cdot e^{-x} x^{n-1} = 0$, since the exponential function e^{-x} tends to zero to a higher order than any power $1/x^m$ $(m > 0)$. This gamma function, which we can think of as a function of the number n (not necessarily an integer), satisfies a remarkable relation, which we can arrive at in the following way by integration by parts. To begin with, we have

$$\int e^{-x} x^{n-1} dx = - e^{-x} x^{n-1} + (n-1) \int e^{-x} x^{n-2} dx.$$

If we take this formula between the limits 0 and A and then let A increase beyond all bounds, we immediately obtain

$$\Gamma(n) = (n-1) \int_0^\infty e^{-x} x^{n-2} dx = (n-1) \Gamma(n-1),$$

and by this recurrence formula, provided μ is an integer and $0 < \mu < n$,

$$\Gamma(n) = (n-1)(n-2) \ldots (n-\mu) \int_0^\infty e^{-x} x^{n-\mu-1} dx.$$

In particular, if n is a positive integer, we have

$$\Gamma(n) = (n-1)(n-2) \ldots 3 . 2 . 1 \int_0^\infty e^{-x} dx,$$

and since

$$\int_0^\infty e^{-x} dx = 1,$$

it follows finally that

$$\Gamma(n) = (n-1)(n-2) \ldots 2 . 1 = (n-1)!$$

This expression of a factorial by an integral is of importance in many applications.

The integrals $\qquad \int_0^\infty e^{-x^2} dx, \quad \int_0^\infty x^n e^{-x^2} dx$

also converge, as we may easily convince ourselves by means of our criterion.

5. The Dirichlet Integral.

A convergent integral, important in many applications, whose convergence does not follow directly from our criterion, and which is a simple case of a type investigated by Dirichlet, is

$$I = \int_0^\infty \frac{\sin x}{x} dx.$$

This integral is easily seen to be convergent if the upper limit is finite, for $\frac{\sin x}{x} \to 1$ as $x \to 0$. Its convergence in the infinite interval is due to the periodic change of sign of the integrand, which causes the contributions to the integral from neighbouring intervals of length π almost to cancel one another. In order to make use of this fact we write the expression

$$D_{AB} = \int_A^B \frac{\sin x}{x}\, dx$$

in the form

$$D_{AB} = \int_A^{A+\pi} \frac{\sin x}{x}\, dx - \int_B^{B+\pi} \frac{\sin x}{x}\, dx + \int_{A+\pi}^{B+\pi} \frac{\sin t}{t}\, dt,$$

introduce in the last of the three integrals on the right the new variable $x = t - \pi$, whence $\sin t = -\sin x$, and obtain

$$D_{AB} = \int_A^{A+\pi} \frac{\sin x}{x}\, dx - \int_B^{B+\pi} \frac{\sin x}{x}\, dx - \int_A^B \frac{\sin x}{x + \pi}\, dx.$$

Addition of this to the original expression for D_{AB} gives us

$$2D_{AB} = \int_A^{A+\pi} \frac{\sin x}{x}\, dx - \int_B^{B+\pi} \frac{\sin x}{x}\, dx + \pi \int_A^B \frac{\sin x}{x(x + \pi)}\, dx.$$

From this it follows, if we assume that $B > A > 0$, that

$$|\, 2D_{AB}\,| < \frac{2\pi}{A} + \pi \int_A^B \frac{dx}{x^2};$$

for we may use the method of p. 127, observing that

$$-\frac{1}{x} \leqq \frac{\sin x}{x} \leqq \frac{1}{x}$$

and
$$-\frac{1}{x^2} \leqq \frac{\sin x}{x(x + \pi)} \leqq \frac{1}{x^2}$$

for positive values of x. The integral on the right is convergent, by our criterion, and our formula shows that $|\, D_{AB}\,| \to 0$ as A and B both tend to infinity. Now

$$|\, D_{0B} - D_{0A}\,| = |\, D_{AB}\,|,$$

and it follows from Cauchy's convergence test that D_{0B} tends to a definite limit as $B \to \infty$. In other words, the integral I

exists. Another proof of this is given in the Appendix to Chap. VIII (p. 418), and on p. 450 we shall further show that I has the value $\pi/2$.

6. Substitution.

It is obvious that all rules for the substitution of new variables, &c., remain valid for convergent improper integrals. As an example, in order to calculate $\int_0^\infty x e^{-x^2}\, dx$ we introduce the new variable $u = x^2$ and obtain

$$\int_0^\infty x e^{-x^2}\, dx = \frac{1}{2}\int_0^\infty e^{-u}\, du = \lim_{A \to \infty} \frac{1}{2}(1 - e^{-A}) = \frac{1}{2}.$$

Another example of the use of substitution in the investigation of improper integrals is given by the Fresnel integrals, which occur in the theory of diffraction of light:

$$F_1 = \int_0^\infty \sin(x^2)\, dx, \quad F_2 = \int_0^\infty \cos(x^2)\, dx.$$

The substitution $x^2 = u$ yields

$$F_1 = \frac{1}{2}\int_0^\infty \frac{\sin u}{\sqrt{u}}\, du, \quad F_2 = \frac{1}{2}\int_0^\infty \frac{\cos u}{\sqrt{u}}\, du.$$

Integrating by parts, we have

$$\int_A^B \frac{\sin u}{\sqrt{u}}\, du = \frac{\cos A}{\sqrt{A}} - \frac{\cos B}{\sqrt{B}} - \frac{1}{2}\int_A^B \frac{\cos u}{u^{3/2}}\, du.$$

As A and B tend to ∞ the first two terms on the right tend to 0, and by the criterion of p. 250 the integral also tends to 0. Hence by the same argument as for the Dirichlet integral we see that the integral F_1 converges. The convergence of the integral F_2 is proved in exactly the same way.

These Fresnel integrals show that an improper integral may exist even although the integrand does not tend to zero as $x \to \infty$. In fact, an improper integral can exist even when the integrand is unbounded, as is shown by the example

$$\int_0^\infty 2u \cos(u^4)\, du.$$

When $u^4 = n\pi$, i.e. when $u = \sqrt[4]{n\pi}$, $n = 0, 1, 2, \ldots$, the integrand becomes $2\sqrt[4]{n\pi}\cos n\pi = \pm 2\sqrt[4]{n\pi}$, so that the integrand is unbounded. By the substitution $u^2 = x$, however, the integral is reduced to

$$\int_0^\infty \cos(x^2)\,dx,$$

which we have just shown to be convergent.

By means of a substitution an improper integral may often be transformed into a proper one. For example, the transformation $x = \sin u$ gives

$$\int_0^1 \frac{dx}{\sqrt{(1-x^2)}} = \int_0^{\pi/2} du = \frac{\pi}{2}.$$

On the other hand, integrals of continuous functions may be transformed into improper integrals; this occurs if the transformation $u = \phi(x)$ is such that at the end of the interval of integration the derivative $\phi'(x)$ vanishes, so that dx/du is infinite.

EXAMPLES

Test the convergence of the improper integrals in Ex. 1–11:

1. $\int_{-3}^3 \frac{dx}{x^2}$. 2. $\int_{-1}^1 \frac{dx}{\sqrt[3]{x}}$. 3. $\int_{-\infty}^\infty \frac{dx}{1+x^6}$.

4. $\int_0^\infty \frac{dx}{(1+x)\sqrt{x}}$. 5. $\int_0^\pi \frac{dx}{1-\cos x}$.

6. $\int_A^B \frac{dx}{\sqrt{(x-a_1)(x-a_2)(x-a_3)(x-a_4)}}$, where a_1, a_2, a_3, a_4 are all different and lie between A and B.

7. $\int_0^\infty \frac{\arctan x}{1+x^2}\,dx$. 8. $\int_0^\infty \frac{\arctan x}{1-x^3}\,dx$.

9. $\int_1^\infty \frac{x}{1-e^x}\,dx$. 10. $\int_0^\infty \frac{x}{e^x-1}\,dx$. 11. $\int_0^{\pi/2} \log\tan x\,dx$.

12.* Prove that $\int_0^\infty \sin^2\left[\pi\left(x+\frac{1}{x}\right)\right]dx$ does not exist.

13.* Prove that $\lim\limits_{k\to\infty} \int_0^\infty \frac{dx}{1+kx^{10}} = 0$.

14. For what values of s is (a) $\int_0^\infty \frac{x^{s-1}}{1+x}\,dx$, (b) $\int_0^\infty \frac{\sin x}{x^s}\,dx$ convergent?

15.* Does $\int_0^\infty \frac{\sin t}{1+t}\,dt$ converge?

16.* (*a*) If *a* is a fixed positive number, prove that

$$\lim_{h \to 0} \int_{-a}^{a} \frac{h}{h^2 + x^2}\, dx = \pi.$$

(*b*) If $f(x)$ is continuous in the interval $-1 \leq x \leq 1$, prove that

$$\lim_{h \to 0} \int_{-1}^{1} \frac{h}{h^2 + x^2}\, f(x)\, dx = \pi f(0).$$

MISCELLANEOUS EXAMPLES

Evaluate the integrals in Ex. 1–7:

1. $\int e^{\arcsin x}\, dx.$

2. $\int \sin^3 x \cos^6 x\, dx.$ (By a shorter method than that of the text, using trigonometrical identities.)

3. $\int (\log x)^2\, dx.$　　　4. $\int \dfrac{\sin x\, dx}{3 + \sin^2 x}.$　　　5. $\int \sqrt{1 - e^{-2x}}\, dx.$

6. $\int_{-1}^{+1} x e^{-x^2 \tan^2 x}\, dx.$　　　　7. $\int_{\frac{1}{2}}^{2} \dfrac{1}{x} \sin\left(x - \dfrac{1}{x}\right) dx.$

8.* Prove that $\lim\limits_{x \to \infty} e^{-x^2} \int_0^x e^{t^2}\, dt = 0.$

9. Assuming that $|\alpha| \neq |\beta|$, prove that

$$\lim_{T \to \infty} \frac{1}{T} \int_0^T \sin \alpha x \sin \beta x\, dx = 0.$$

10. Evaluate $\int_{-1}^{1} x^3 e^{-x^4} \cos 2x\, dx.$

11.* Prove that the substitution $x = \dfrac{\alpha t + \beta}{\gamma t + \delta}$, where $\alpha\delta - \gamma\beta \neq 0$, transforms the integral

$$\int \frac{dx}{\sqrt{ax^4 + bx^3 + cx^2 + dx + e}}$$

into an integral of similar type; and that if the biquadratic

$$ax^4 + bx^3 + cx^2 + dx + e$$

has no repeated factors, neither has the new biquadratic in t which takes its place.

Prove that the same statements are true for

$$\int R(x, \sqrt{ax^4 + bx^3 + cx^2 + dx + e})\, dx,$$

where R is a rational function.

12. Find the limit as $n \to \infty$ of $a_n = \dfrac{1}{n+1} + \dfrac{1}{n+2} + \ldots + \dfrac{1}{2n}$.

13.* Find the limit of

$$b_n = \frac{1}{\sqrt{n^2 - 0}} + \frac{1}{\sqrt{n^2 - 1}} + \frac{1}{\sqrt{n^2 - 4}} + \ldots + \frac{1}{\sqrt{n^3 - (n-1)^2}}.$$

14.* Prove that $\displaystyle\lim_{n \to \infty} \sqrt[n]{\dfrac{n!}{n^n}} = \dfrac{1}{e}$.

15.* If α is any real number greater than -1, evaluate

$$\lim_{n \to \infty} \frac{1^\alpha + 2^\alpha + 3^\alpha + \ldots + n^\alpha}{n^{\alpha+1}}.$$

Appendix to Chapter IV

The Second Mean Value Theorem of the Integral Calculus

The method of integration by parts affords us an easy method for proving an important theorem on the estimation of integrals, usually called the second mean value theorem of the integral calculus.

Let us suppose that the function $\phi(x)$ is monotonic and continuous in the interval $a \leq x \leq b$, and that the derivative $\phi'(x)$ is continuous; and let us further suppose that $f(x)$ is an arbitrary function continuous in the same interval. Then the second mean value theorem of the integral calculus is expressed as follows. There exists a number ξ, such that $a \leq \xi \leq b$, for which

$$\int_a^b f(x)\,\phi(x)\,dx = \phi(a) \int_a^\xi f(x)\,dx + \phi(b) \int_\xi^b f(x)\,dx.$$

To prove this we notice first that we can assume that $\phi(b) = 0$; for replacing $\phi(x)$ by $\phi(x) - \phi(b)$ changes both sides of the equation by the same amount, and gives us a function which vanishes at $x = b$. Moreover, we can assume that $\phi(a) > 0$; for if $\phi(a) < 0$ we need only replace $\phi(x)$ by $-\phi(x)$, which changes the sign of both sides of the equation. (The case $\phi(a) = 0$ is trivial; for if both $\phi(a)$ and $\phi(b)$ vanish, $\phi(x)$ must be identically zero, and our equation becomes $0 = 0$.) We therefore need only prove that if $\phi(x)$ is continuous and monotonic decreasing, and $\phi(b) = 0$, then

$$\int_a^b f(x)\,\phi(x)\,dx = \phi(a)\int_a^\xi f(x)\,dx.$$

We now put $F(x) = \int_a^x f(x)\,dx$ and apply the formula for integration by parts to the left-hand side of the last equation; we then have

$$\int_a^b f(x)\,\phi(x)\,dx = F(x)\,\phi(x)\,\Big|_a^b + \int_a^b F(x)\,\{-\phi'(x)\}\,dx.$$

The integrated part vanishes, since $F(a)$ and $\phi(b)$ are zero. The expression $-\phi'(x)$ is everywhere positive, so that we can apply the first mean value theorem of the integral calculus. We thus find that the integral on the right has the value

$$F(\xi)\int_a^b \{-\phi'(x)\}\,dx, \quad a \leqq \xi \leqq b.$$

But

$$F(\xi) = \int_a^\xi f(x)\,dx \quad \text{and} \quad \int_a^b \{-\phi'(x)\}\,dx = \phi(a) - \phi(b) = \phi(a)$$

and our theorem is established.

This theorem can be extended (although we shall not carry out the proof) to more general classes of functions. For the theorem remains true for all continuous monotonic functions $\phi(x)$, whether they have derivatives or not. In fact, it is true for any discontinuous monotonic function for which we are in a position to integrate $f(x)\,\phi(x)$.

CHAPTER V

Applications

In this chapter, after disposing of a few preliminaries, we shall illustrate how what we have now learned may be applied in a great variety of ways in geometry and physics.

1. REPRESENTATION OF CURVES

1. Parametric Representation.

As we saw in Chap. I (p. 17), in representing a curve by means of an equation $y = f(x)$ we must always restrict ourselves to a single-valued branch. Hence it is often more convenient—when we are dealing with a closed curve, in particular—to introduce other analytical methods of representation. The most general and at the same time the most useful representation of a curve is *parametric representation*. Instead of considering one of the rectangular co-ordinates as a function of the other, we think of *both* the co-ordinates x and y as functions of a *third* independent variable, a so-called *parameter*; the point with the co-ordinates x and y then describes the curve as t traverses a definite interval. Such parametric representations have already been encountered. For example, for the circle $x^2 + y^2 = a^2$ we obtain a parametric representation in the form $x = a \cos t$, $y = a \sin t$. Here, as we already know, t has the geometrical meaning of an angle at the centre of the circle. For the ellipse $x^2/a^2 + y^2/b^2 = 1$ we likewise have the parametric representation $x = a \cos t$, $y = b \sin t$, where t is the so-called eccentric angle, that is, the angle at the centre corresponding to the point of the circumscribed circle lying vertically above or below the point $P(a \cos t, b \sin t)$ of the ellipse (fig. 1). In both these cases the point with the co-ordinates x, y describes the complete circle or ellipse as the parameter t traverses the interval from 0 to 2π.

In general, we can seek to represent a curve parametrically by taking

$$x = \phi(t) = x(t), \quad y = \psi(t) = y(t),$$

that is, by considering two functions of a parameter t; the shorter notation $x(t)$ and $y(t)$ will henceforth be used where there is no danger of confusion.   For a given curve these two functions $\phi(t)$ and $\psi(t)$ must be determined in such a way that the totality of pairs of functional values $x(t)$ and $y(t)$ corresponding to a given interval of values of t gives all the points on the curve and no points that are not on the curve.   If a curve is in the first instance given in the form $y = f(x)$, we can arrive at a representation of this kind by first writing $x = \phi(t)$, where $\phi(t)$ is any continuous mono-tonic function which in a definite interval passes exactly once through each of the values of x in question; it then follows that $y = f\{\phi(t)\}$, that is, the second function $\psi(t)$ is deter-mined by compounding f and ϕ. We thus see that owing to the

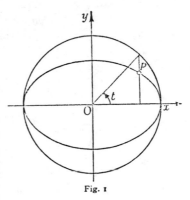

Fig. 1

arbitrariness in the choice of the function ϕ we have a great deal of freedom in representing a given curve parametrically; in particular, we may actually take $t = x$ and may thus think of the original representation $y = f(x)$ as a parametric repre-sentation with the parameter $t = x$.

The advantage of the parametric representation is that this arbitrari-ness may be utilized for purposes of simplification. For example, we repre-sent the curve $y = \sqrt[3]{x^2}$ by taking $x = t^3$, $y = t^2$, so that $\varphi(t) = t^3$, $\psi(t) = t^2$. The point with the co-ordinates x, y will then describe the whole curve (semicubical parabola) as t varies from $-\infty$ to $+\infty$.

If, on the other hand, a curve is originally given by a para-metric representation $x = \phi(t)$, $y = \psi(t)$, and we wish to obtain the equation of the curve in non-parametric form, that is, in the form $y = f(x)$, we have only to eliminate the parameter t from the two equations.   In the case of the parametric representations of the circle and ellipse given above we can do this at once by squaring and using the equation $\sin^2 t + \cos^2 t = 1$. (For a further example see below.) In general, we should have to find an expression for t from the equation $x = \phi(t)$ by means of the inverse function $t = \Phi(x)$ and substitute this in $y = \psi(t)$, in order to obtain the representation * $y = \psi\{\Phi(x)\} = f(x)$. In such an elimination, of course, we must ordinarily restrict ourselves to a portion of the curve; in fact, to a portion which is not cut twice by any line parallel to the y-axis.

The parametric representation has associated with it a definite *sense* in which the curve is described, corresponding to the direction in which the values of the parameter increase; this direction we shall call the *positive sense*. If, for example, the point $x = x(t)$, $y = y(t)$ describes a curve C as t traverses an interval $t_0 \leqq t \leqq t_1$ and the end-points P_0 and P_1 of the curve correspond respectively to t_0 and t_1, then the curve is traversed positively in the direction from P_0 to P_1. If we introduce $\tau = -t$ as a new parameter, the curve C will correspond to the values $-t_1 \leqq \tau \leqq -t_0$ of the variable τ, and the points P_0 and P_1 will correspond to $\tau = -t_0$ and $\tau = -t_1$ respectively. If we now traverse the curve from P_0 to P_1 we proceed in the direction in which the values of the parameter τ decrease, that is, in the negative sense. In general, a change of parameter $t = t(\tau)$ preserves the sense in which the curve is described if the function $t(\tau)$ is monotonic increasing, but reverses it if the function $t(\tau)$ is monotonic decreasing.

2. Interpretation of the Parameter. Change of Parameter.

In many cases we can give an immediate physical interpretation to the parameter t, namely, time. Any motion of a point in the plane may be expressed mathematically by the fact that the co-ordinates x and y appear as functions of the time. These two functions therefore determine the *motion along a path or trajectory* in parametric form.

As an example of this we have the cycloids which arise when a circle rolls along a straight line or another circle. Here we limit ourselves to the simplest case, in which a circle of radius a rolls along the x-axis, and

---

* It may happen, however, that the equation $y = f(x)$ obtained in this way represents *more* than the original parametric representation. Thus for example the equations $x = a \sin t$, $y = b \sin t$ represent only the finite portion of the line $y = bx/a$ lying between the points $x = -a$, $y = -b$ and $x = a$, $y = b$, whereas the equation $y = bx/a$ represents the whole of the line.

we consider a point on its circumference. This point then describes a
" common " cycloid. If we choose the origin of the co-ordinate system
and the initial time in such a way that for time $t = 0$ the corresponding
point of the curve coincides with the origin, we obtain (cf. fig. 2) the
parametric representation

$$x = a(t - \sin t), \quad y = a(1 - \cos t)$$

for the cycloid; here t denotes the angle through which the circle has
turned from its original position; in the case where the velocity of
rolling is uniform it is proportional to the time.

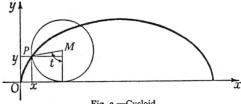

Fig. 2.—Cycloid

By eliminating the parameter t we can obtain the equation of the curve
in non-parametric form, at the cost, however, of neatness of expression.
We have

$$\cos t = \frac{a - y}{a}, \quad t = \text{arc cos} \frac{a - y}{a}, \quad \sin t = \pm \sqrt{\left\{1 - \frac{(a - y)^2}{a^2}\right\}},$$

and hence

$$x = a \text{ arc cos} \frac{a - y}{a} \mp \sqrt{\{(2a - y)y\}},$$

thus obtaining x as a function of y.

In the parametric representation of a given curve we have a
great deal of freedom in the choice of parameter (p. 259). For
example, instead of the time t we could take the quantity
$\tau = t^2$ as parameter, or indeed any arbitrary quantity τ which
is related to the original parameter t by an arbitrary equation
of the form $\tau = \omega(t)$, where we assume that for the whole interval
of values of t considered this function has a unique inverse $t = \kappa(\tau)$.
If increasing values of τ correspond to increasing values of t,
the positive sense of description remains the same; otherwise
it is reversed.

Parametric representation is, of course, not limited to rectangular co-ordinates, e.g. it can just as well be used with the
polar co-ordinates r and θ, which are connected with the rectangular co-ordinates by the well-known equations $x = r \cos \theta$,

$y = r \sin \theta$, or $r = \sqrt{(x^2 + y^2)}$, $\sin \theta = y/r$, $\cos \theta = x/r$; the equations of the curve would then be $r = r(t)$, $\theta = \theta(t)$.

As an example, the straight line may be represented parametrically (see fig. 3) in the form

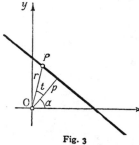

$$r = \frac{p}{\cos t}, \quad \theta = \alpha + t$$

(p and α being constants), from which we immediately obtain the equation of the line in polar co-ordinates,

$$r = \frac{p}{\cos(\theta - \alpha)},$$

Fig. 3

by eliminating the parameter t.

3. The Derivatives for a Curve Represented Parametrically.

If on the one hand a curve is given by an equation $y = f(x)$, and on the other hand it is given parametrically by $x = x(t)$, $y = y(t)$, then we must have $y(t) = f\{x(t)\}$. By the chain rule for differentiation it follows that

$$\frac{dy}{dt} = \frac{dy}{dx} \frac{dx}{dt}$$

or

$$y' = \frac{dy}{dx} = \frac{\dot{y}}{\dot{x}},$$

where as an abbreviation for differentiation with respect to the parameter t we use a dot over the variable (Newton's notation), instead of the dash $'$; the latter we shall reserve for differentiation with respect to x.

For the cycloid, for example, we have

$$\dot{x} = a(1 - \cos t) = 2a \sin^2 \frac{t}{2},$$

$$\dot{y} = a \sin t = 2a \sin \frac{t}{2} \cos \frac{t}{2}.$$

These formulæ show that the cycloid has a cusp with a vertical tangent at the points $t = 0$, $\pm 2\pi$, $\pm 4\pi$, ... at which it meets the x-axis, for on approaching these points the derivative $y' = \dot{y}/\dot{x} = \cot(t/2)$ becomes infinite. At these points y is equal to 0; everywhere else $y > 0$.

The equation of the tangent to the curve is

$$(\xi - x)\dot{y} - (\eta - y)\dot{x} = 0,$$

where ξ and η are the "current" co-ordinates, that is, the variable co-ordinates corresponding to an arbitrary point on the tangent. For the equation of the normal, i.e. the straight line through a point of the curve perpendicular to the tangent at that point, we likewise obtain

$$(\xi - x)\dot{x} + (\eta - y)\dot{y} = 0.$$

The *direction cosines of the tangent*, that is, the cosines of the angles α, β which the tangent makes with the x and y axes respectively, are given by the expressions

$$\cos\alpha = \frac{\dot{x}}{\pm\sqrt{(\dot{x}^2 + \dot{y}^2)}}, \quad \cos\beta = \frac{\dot{y}}{\pm\sqrt{(\dot{x}^2 + \dot{y}^2)}},$$

as we may verify by elementary methods. The corresponding *direction cosines of the normal* are given by

$$\cos\alpha' = \frac{-\dot{y}}{\pm\sqrt{(\dot{x}^2 + \dot{y}^2)}}, \quad \cos\beta' = \frac{\dot{x}}{\pm\sqrt{(\dot{x}^2 + \dot{y}^2)}}.$$

(See fig. 4.)

These formulæ show us that at every point at which $\dot{x}$ and $\dot{y}$ are continuous and $\dot{x}^2 + \dot{y}^2 \neq 0$ the direction of the tangent varies continuously with t. This is the most important case for us; it is interesting, however, to illustrate by examples the various possibilities that arise when our assumptions are not fulfilled and we cannot state directly that the tangent keeps on turning continuously. At a point at which $\dot{x} = \dot{y} = 0$ the tangent may or may not turn continuously. As one example we have the curve $x = t^3$, $y = t^2$ discussed on pp. 99, 259, which has a cusp at the origin

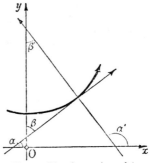

Fig. 4.—Direction cosines of the tangent and the normal

even though $\dot{x}$ and $\dot{y}$ are continuous everywhere. As another example we consider the curve $x = t^3$, $y = t^3$, which is the straight line $y = x$. This curve has the same tangent direction everywhere; the latter is therefore continuous, although the derivatives $\dot{x}$ and $\dot{y}$ both vanish for $t = 0$. Moreover, at a point at which $\dot{x}$ and $\dot{y}$ are discontinuous the direction of the tangent may or may not be continuous. For let $\phi(t)$ be any continuous monotonic increasing function, defined for $t_1 \leqq t \leqq t_2$, which has a sharp corner at $t = t_3$, $t_1 < t_3 < t_2$. Then the curve $x = t$, $y = \phi(t)$, which is the same curve as $y = \phi(x)$, has a sharp corner at $x = t_3$; while the curve $x = \phi(t)$, $y = \phi(t)$, which is a segment of the straight line $y = x$, has a constant tangent direction, even though the derivatives $\dot{x}$ and $\dot{y}$ do not exist at $t = t_3$. This indicates that if we wish to investigate the behaviour of the tangent at a point where our theorem does not apply, we should first use the formulæ to find $\cos \alpha$ or $\cos \beta$ as functions of t and then investigate these direction cosines themselves.

From a well-known formula in trigonometry or analytical geometry we find that the angle between the two curves represented parametrically by $x = x_1(t)$, $y = y_1(t)$ and $x = x_2(t)$, $y = y_2(t)$ respectively (that is, the angle between their tangents or normals) is given by the expression

$$\cos \delta = \frac{\dot{x}_1 \dot{x}_2 + \dot{y}_1 \dot{y}_2}{\pm \sqrt{(\dot{x}_1{}^2 + \dot{y}_1{}^2)} \sqrt{(\dot{x}_2{}^2 + \dot{y}_2{}^2)}}.$$

The indeterminacy of the signs of the square roots in the last few formulæ suggests that the angles are not completely determined, since we can still specify either sense of direction on the tangent or normal as " positive ". Taking the square root as positive, as is usually done, corresponds to choosing for the positive direction on the tangent the direction in which the parameter increases, and for the positive direction on the normal the direction obtained by rotating the tangent through an angle $\pi/2$ in the positive * sense.

The second derivative $y'' = \dfrac{d^2 y}{dx^2}$ is obtained in the following way by means of the chain rule and the rule for differentiating a quotient:

$$y'' = \frac{dy'}{dx} = \frac{dy'}{dt} \frac{dt}{dx} = \frac{d}{dt}\left(\frac{\dot{y}}{\dot{x}}\right) \frac{1}{\dot{x}} = \frac{\dot{x}\ddot{y} - \dot{y}\ddot{x}}{\dot{x}^2} \frac{1}{\dot{x}},$$

---

* I.e. in the counter-clockwise sense.

whence

$$y'' = \frac{d^2y}{dx^2} = \frac{\dot{x}\ddot{y} - \dot{y}\ddot{x}}{\dot{x}^3}.$$

4. Change of Axes for Curves Represented Parametrically.

If we rotate the axes through an angle a in the positive direction, the new rectangular co-ordinates ξ, η and the old ones x, y are related by the equations

$$x = \xi \cos a - \eta \sin a, \quad \xi = x \cos a + y \sin a,$$
$$y = \xi \sin a + \eta \cos a, \quad \eta = -x \sin a + y \cos a.$$

Thus the new co-ordinates ξ and η are specified along with x and y as functions of the parameter t. By differentiation we at once obtain

$$\dot{x} = \dot{\xi} \cos a - \dot{\eta} \sin a, \quad \dot{\xi} = \dot{x} \cos a + \dot{y} \sin a,$$
$$\dot{y} = \dot{\xi} \sin a + \dot{\eta} \cos a, \quad \dot{\eta} = -\dot{x} \sin a + \dot{y} \cos a.$$

Let us suppose that the curve is given in polar co-ordinates and that both polar co-ordinates and rectangular co-ordinates are given as functions of a parameter t. Then by differentiation with respect to t we obtain from the equations $x = r \cos\theta$, $y = r \sin\theta$ the formulæ

$$\left. \begin{array}{l} \dot{x} = \dot{r} \cos\theta - r \sin\theta . \dot{\theta}, \\ \dot{y} = \dot{r} \sin\theta + r \cos\theta . \dot{\theta}, \end{array} \right\} \quad \ldots \quad (a)$$

which are frequently used in passing from rectangular co-ordinates to polar. As an example we consider the polar equation of a curve, $r = f(\theta)$, which might, for example, arise from a parametric representation $r = r(t)$, $\theta = \theta(t)$ by elimination of the parameter t. The angle ψ between the radius vector to a point on the curve and the tangent to the curve at that point is then given by

$$\tan\psi = \frac{f(\theta)}{f'(\theta)}.$$

We can convince ourselves of this in the following way. If we think of the curve as given by an equation $y = F(x)$ and use θ as parameter, so that $\dot{\theta} = 1$ and $\dot{r} = f'(\theta)$, we have

$$\tan a = y' = \frac{\dot{y}}{\dot{x}} = \frac{\dot{r}\tan\theta + r}{\dot{r} - r\tan\theta}$$

(cf. fig. 5 and equations (a) above). In addition, $\psi = a - \theta$, and hence

$$\tan \psi = \frac{y' - \tan\theta}{1 + y' \tan\theta} = \frac{r + r\tan^2\theta}{\dot{r} + \dot{r}\tan^2\theta} = \frac{r}{\dot{r}}.$$

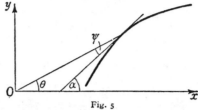

Fig. 5

This formula can also be established by geometrical methods.

5. General Remarks.

In discussing given curves we sometimes consider properties which do not assert anything about the form of the curve itself, but merely something about the position of the curve with respect to the co-ordinate system; for example, the occurrence of a horizontal tangent, expressed by the equation $\dot{y} = 0$, or the occurrence of a vertical tangent, expressed by $\dot{x} = 0$. Such properties do not persist when the axes are rotated.

In contrast to this, a point of inflection will still be a point of inflection after the axes have been rotated. According to the formula on p. 265 the condition for a point of inflection is

$$\dot{x}\ddot{y} - \ddot{x}\dot{y} = 0.$$

If on the left we replace the expressions $\dot{x}$, $\dot{y}$, $\ddot{x}$, $\ddot{y}$ by their values in terms of the new co-ordinates ξ, η, we readily obtain

$$\dot{x}\ddot{y} - \ddot{x}\dot{y} = \dot{\xi}\ddot{\eta} - \ddot{\xi}\dot{\eta}.$$

Hence from the equation $\dot{x}\ddot{y} - \ddot{x}\dot{y} = 0$ it follows that $\dot{\xi}\ddot{\eta} - \ddot{\xi}\dot{\eta} = 0$, so that our equation expresses a property of the point of the curve which is independent of the co-ordinate system.

We shall often see later that properties which are truly geometrical are expressed by formulæ which are unaltered in form by rotation of the axes.

1. Find the equation, in non-parametric form, of the curve

$$x = a \cos 2\theta \cos \theta$$
$$y = a \cos 2\theta \sin \theta.$$

2. A circle c, of radius r, rolls on the outside of a fixed circle C of radius R. The point P on the circumference of c moves with c, and describes a curve called the *epicycloid*. Find the parametric representation of the epicycloid (consider c to rotate with constant velocity, and measure time so that at $t = 0$ the point P is in contact with the circle C).

3. Sketch the epicycloid for the special case $r = R$, and find its parametric equations. (This particular epicycloid is called the *cardioid*.)

4. If in Ex. 2 the radius r is less than R and c rolls *inside* C, the point P describes a *hypocycloid*. Find its parametric equations.

5. Sketch the hypocycloid (1) for $R = 2r$, (2) for $R = 3r$.

6. Sketch the hypocycloid for $R = 4r$ (the *astroid*) and find its non-parametric equation.

7. Find the parametric equations for the curve $x^3 + y^3 = 3axy$ (the *folium of Descartes*), choosing as parameter t the tangent of the angle between the x-axis and the radius vector from the origin to the point (x, y).

8. Find the formula for the angle α between two curves $r = f(\theta)$ and $r = g(\theta)$ in polar co-ordinates.

9. Find the equation of the curves which everywhere intersect the straight lines through the origin at the same angle α.

10. Let C be a fixed curve and P a fixed point with co-ordinates x_0, y_0. The *pedal curve* of C with respect to P is defined to be the locus of the foot of the perpendicular from P on the tangent to C. Find the parametric representation of the pedal of C if C is itself given parametrically by $x = f(t), y = g(t)$.

11. Find the pedal curve of the circle C, (a) with respect to its centre M, (b) with respect to a point P on its circumference.

12. Find the pedal curve of the ellipse $x = a \cos \theta$, $y = b \sin \theta$ with respect to the origin.

2. Applications to the Theory of Plane Curves

We shall consider two different kinds of geometrical properties or quantities associated with curves. The first type consists of properties or quantities which depend only on the *behaviour of the curve in the small*, i.e. in the immediate neighbourhood of a point, and which can be expressed analytically by means of the derivative at the point. Properties of the second type depend on the whole course of the curve or of a portion of the curve, and are expressed analytically by means of the concept of integral. We shall begin by considering properties of the second type.

1. Orientation of Areas.

The idea of area was our starting-point for the definition of the integral; but the connexion between definite integral and area is still somewhat incomplete. The areas with which we are concerned in geometry are bounded by given *closed* curves; on the other hand, the area measured by the integral $\int_{x_1}^{x_2} f(x)\, dx$ is bounded only in part by the given curve $y = f(x)$, the rest of the boundary consisting of lines which depend on the choice of the co-ordinate system. If we wished to determine the area interior to a closed curve, such as a circle or ellipse, by means of integrals of this type, we should have to use some such device as breaking up the area into several parts, each of which is bounded by a single-valued branch of the curve and also by the x-axis and the corresponding ordinates.

For the discussion of this general case it is convenient first to make some remarks on the determination of the sign of the area considered. For any surface bounded by an arbitrary closed curve which does not intersect itself, we can relate the sign of the area to the purely geometrical idea of the sense in which the curve is described, according to the following convention. We say that the boundary of a region is described in the *positive* sense if we go round the boundary in such a direction that the interior of the region is on the left;* the opposite sense we call negative. If then we consider a region whose boundary is traversed in an assigned sense, a so-called *oriented* region, we reckon the area as positive if this sense is positive, and negative if this sense is negative (cf. fig. 6).

Suppose, in particular, that in the interval $a \leqq x \leqq b$ the function $f(x)$ is everywhere positive. We consider the closed curve obtained by starting at the point $x = b = x_1$, $y = 0$, traversing the x-axis back to the point $x = a = x_0$, $y = 0$, then

---

* If we wish to avoid the words "right" and "left" in such a context, we say that the triangle, whose vertices in order are the origin, the point $x = 1$, $y = 0$, and the point $x = 0$, $y = 1$, is described in the positive sense if the vertices are passed in the order mentioned. For every other region, we say that the boundary is positively described if it is described in the same sense as this triangle; otherwise it is negatively described.

proceeding along the ordinate to the curve $y = f(x)$, then along this curve to the ordinate $x = b$, and finally along this ordinate to the x-axis (cf. fig. 7). The absolute value of the area interior to this curve—the number of square units contained in it—is, as we know, $\int_a^b f(x)\,dx$. Hence, denoting by A_{01} the area with its sign as determined above, the integral gives us the value A_{01} except for sign. To determine the sign we need only observe

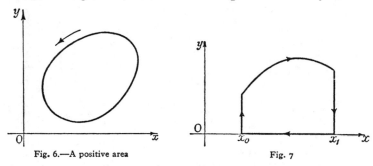

Fig. 6.—A positive area   Fig. 7

that the boundary of the region is traversed in the negative sense, so that A_{01} is negative; hence we have

$$A_{01} = -\int_a^b f(x)\,dx.$$

Similarly, if $a > b$, we find that according to our convention A_{01} is positive, while the integral $\int_a^b f(x)\,dx$ is negative; hence in either case A_{01} is given by the above equation.

2. The General Formula for the Area as an Integral.

After these preliminaries, the difficulties mentioned at the beginning can now be avoided in a simple way by representing our curve parametrically. If we introduce t formally as a new independent variable in the above integral, writing $x = x(t)$, $y = y(t) = f\{x(t)\}$, we have

$$A_{01} = -\int_{t_0}^{t_1} y(t)\,\dot{x}(t)\,dt,$$

where t_0 and t_1 are the values of the parameter corresponding to the abscissæ $x_0 = a$ and $x_1 = b$ respectively. Here we suppose that the branch in question of the curve $y = f(x)$ is related to an interval $t_0 \leqq t \leqq t_1$ by a $(1, 1)$ correspondence,* that $f(x)$ is everywhere positive, and that $\dot{x}(t)$ never vanishes in this interval. As we have seen, our expression then gives us the area of the region bounded by the curve, the lines $x = a$ and $x = b$, and the x-axis. It is, of course, still subject to the disadvantages mentioned above. We shall now show that if the curve $x = x(t)$, $y = y(t)$, $t_0 \leqq t \leqq t_1$, is a closed curve bounding a region of area A_{01}, the area A_{01} is given by an integral which in form is exactly the same as the preceding.

Let us then consider a closed curve which is represented parametrically by the equations $x = x(t)$, $y = y(t)$, the curve being

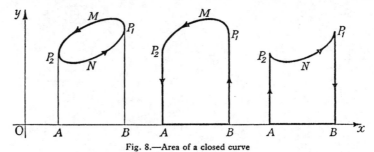

Fig. 8.—Area of a closed curve

described just once as t describes the interval $t_0 \leqq t \leqq t_1$. In order that the curve may be closed it is essential that $x(t_0) = x(t_1)$ and $y(t_0) = y(t_1)$. We shall assume that the derivatives are continuous except for a finite number of jump-discontinuities at most, and that $\dot{x}^2 + \dot{y}^2$ is different from zero except perhaps at a finite number of points which may be corners † of the curve.

We shall first consider a closed curve which has no corners and is convex and of such a type that no straight line intersects it in more than two points. We denote by P_1 and P_2 the points at which the curve possesses a vertical tangent; these tangents are said to be " lines of support " at P_1 and P_2 respectively, because the points of the curve in the neighbourhood of P_1 and P_2 lie entirely on one side of the line. We can then (cf. fig. 8)

* I.e. is such that every point of it corresponds to a single value of t in the interval $t_0 \leqq t \leqq t_1$, and conversely.

† A continuous curve $x = x(t)$, $y = y(t)$ is said to have a corner at $t = t_0$ if the positive direction of the tangent approaches a limit as $(t - t_0) \to 0$ through positive values, and approaches a limit as $(t - t_0) \to 0$ through negative values, but the two limits are not the same.

regard the area bounded by the curve as the sum of the area A_{12} bounded by the closed curve $P_1MP_2ABP_1$, formed as in the preceding section, and the area A_{21} bounded by the closed curve $P_2NP_1BAP_2$. Here we assume that the curve is described in the positive sense, as in the figure; by our sign convention A_{12} is then positive and A_{21} negative. We suppose that the point $x(t)$, $y(t)$ describes the upper part of the curve from P_1 to P_2 as t goes from t_0 to τ, and the lower part from P_2 to P_1 as t goes from τ to t_1. We then immediately obtain

$$A_{12} = -\int_{t_0}^{\tau} y(t)\dot{x}(t)\,dt$$

and
$$A_{21} = -\int_{\tau}^{t_1} y(t)\dot{x}(t)\,dt;$$

hence, for the total area bounded by the convex curve, we have

$$A = -\int_{t_0}^{t_1} y(t)\dot{x}(t)\,dt.$$

If we denote by "absolute area" of a region the number of square units contained in it—which is, of course, never negative—then the above expression always gives us the absolute area bounded by the curve, except perhaps for sign. In order to see what happens when we reverse the sense in which the curve is described, we simply take the same integral from t_1 to t_0 instead of from t_0 to t_1; our integral becomes

$$-\int_{t_1}^{t_0} y\dot{x}\,d\tau,$$

which is equal to $-A$. We thus recognize the truth of the following statement:

*The area represented by our formula is positive or negative, according as the sense in which the boundary is described is positive or negative.**

* In drawing the figure we have assumed that $y > 0$ for all points of the curve. This really does not restrict the generality of the result. For if we displace the curve through a distance a parallel to the y-axis, without rotating it, in other words, replace y by $y + a$, the area is unchanged; the value of the integral is likewise unaltered, for the above integral is replaced by

$$-\int_{t_0}^{t_1} (y + a)\dot{x}(t)\,dt,$$

and since the curve is closed

$$\int_{t_0}^{t_1} a\dot{x}\,dt = a\{x(t_1) - x(t_0)\} = 0.$$

Two simple observations enable us to extend our results. Firstly, our formula remains valid for closed curves which do not intersect themselves, even when they are not convex, but have a more general form as illustrated in fig. 9. Secondly, the derivatives may have jump discontinuities or may both vanish at a finite number of points, which may represent corners; according to Chap. IV, § 8, p. 245, the function $y\dot{x}$ remains integrable. (The ordinate to a corner-point is considered to be a line of support if the curve in the neighbourhood of the point lies entirely to one side of the ordinate). We assume that the curve has only a finite number of lines of support, corresponding to the points P_1, P_2, ..., P_n, and we subdivide the

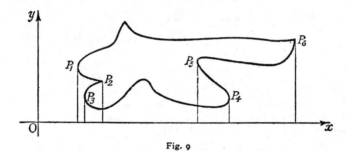

Fig. 9

curve into the single-valued branches P_1P_2, ..., $P_{n-1}P_n$, P_nP_1. Then as in fig. 9 we obtain the area bounded by the curve in the form $A = A_{12} + A_{23} + \ldots + A_{n-1,\,n} + A_{n1}$. (See fig. 9, which illustrates this for the case $n = 6$.) If we express each of these portions of area parametrically and combine the expressions into a single integral, we find that the area bounded by the curve is given by the expression

$$-\int_{t_0}^{t_1} y\dot{x}\,dt,$$

which as before has the same sign as the sense in which the boundary curve is traversed.

Our formula even gives us the area, in a certain sense, in the case where the curve intersects itself. But we shall not enter into such a discussion here; the reader may if he wishes turn to § 2 of the appendix to this chapter (p. 311).

We can express our formula for the area in a more elegant symmetrical form if we first transform the integral by integration by parts:

$$\int_{t_0}^{t_1} y\dot{x}\, dt = -\int_{t_0}^{t_1} x\dot{y}\, dt + xy \Big|_{t_0}^{t_1}.$$

Since the curve is closed,

$$x(t_0) = x(t_1), \quad y(t_0) = y(t_1),$$

and therefore *

$$A = -\int_{t_0}^{t_1} y\dot{x}\, dt = \int_{t_0}^{t_1} x\dot{y}\, dt.$$

If we form the arithmetic mean of the two expressions we obtain the *symmetrical form*

$$A = -\frac{1}{2}\int_{t_0}^{t_1} (y\dot{x} - x\dot{y})\, dt.$$

3. Remarks and an Example.

In connexion with these expressions we must make a remark of a fundamental nature. Both the proof and the statement of the formulæ depend on a particular system of rectangular co-ordinates. But the value of the area, a purely geometrical quantity, cannot depend on the particular co-ordinate system chosen. It is therefore important to show that our integrals are unaltered in value by a change of co-ordinates.

If the axes are merely displaced without rotation the integrals are clearly unaltered (see the footnote on p. 271). Let us then suppose that the axes are rotated through an angle a; instead of x and y we now have new variables ξ and η, defined by the equations $x = \xi \cos a - \eta \sin a$, $y = \xi \sin a + \eta \cos a$, the new variables being also functions of the parameter t. If we recall that $\dot{x} = \dot{\xi} \cos a - \dot{\eta} \sin a$ and $\dot{y} = \dot{\xi} \sin a + \dot{\eta} \cos a$, a short calculation gives us $y\dot{x} - x\dot{y} = \eta\dot{\xi} - \xi\dot{\eta}$, so that

$$A = -\frac{1}{2}\int_{t_0}^{t_1} (y\dot{x} - x\dot{y})\, dt = -\frac{1}{2}\int_{t_0}^{t_1} (\eta\dot{\xi} - \xi\dot{\eta})\, dt.$$

---

* Instead of finding the second expression for the area by integration by parts, we could have derived it by using the fact that as regards the definition of area the x-axis and the y-axis are interchangeable, except that the sense of rotation which brings the x-axis into the y-axis in the shortest way is opposite to the sense which brings the y-axis into the x-axis in the shortest way.

This equation expresses the fact that the area is independent of the co-ordinate system.

Our integral expression for the area is also independent of the choice of parameter. For suppose that we introduce a new parameter τ by the equation $\tau = \tau(t)$; we have

$$\frac{dx}{dt} = \frac{dx}{d\tau}\frac{d\tau}{dt}, \quad \frac{dy}{dt} = \frac{dy}{d\tau}\frac{d\tau}{dt},$$

so that

$$-\int_{t_0}^{t_1}\left(y\frac{dx}{dt} - x\frac{dy}{dt}\right)dt = -\int_{t_0}^{t_1}\left(y\frac{dx}{d\tau} - x\frac{dy}{d\tau}\right)\frac{d\tau}{dt}\,dt$$

$$= -\int_{\tau_0}^{\tau_1}\left(y\frac{dx}{d\tau} - x\frac{dy}{d\tau}\right)d\tau,$$

where τ_0 and τ_1 are the initial and final values of the new parameter, corresponding to the parametric values t_0 and t_1 respectively.*

As an example of the application of our formula for the area we consider the ellipse $y = \frac{b}{a}\sqrt{(a^2 - x^2)}$. In order to find its area we take the upper and lower halves of the ellipse separately and in this way express the area by the integral

$$2\frac{b}{a}\int_{-a}^{+a}\sqrt{(a^2 - x^2)}\,dx.$$

If, however, we use the parametric representation $x = a\cos t$, $y = b\sin t$, we find immediately that the area is given by the expression

$$ab\int_{0}^{2\pi}\sin^2 t\,dt.$$

This can be integrated as on p. 215; it has the value $ab\pi$.

---

* In this section we have based the definition of the area on the concept of integral and have shown that this analytical definition has a truly geometrical character, since it yields a quantity independent of the co-ordinate system. It is, however, easy to give a direct geometrical definition of the area bounded by a closed curve which does not intersect itself, as follows: the area is the upper bound of the areas of all polygons lying interior to the curve. The proof that the two definitions are equivalent is quite simple, but will not be given here.

4. Areas in Polar Co-ordinates.

For many purposes it is important to be able to calculate areas using polar co-ordinates. Let $r = f(\theta)$ be the equation of a curve in polar co-ordinates. Let $A(\theta)$ be the area of the region which is bounded by the x-axis (that is, the line $\theta = 0$), the line through the origin making an angle θ with the x-axis, and the portion of the curve between these two lines.   Then

$$A'(\theta) = \tfrac{1}{2} r^2.$$

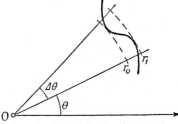

Fig. 10.—Element of area in polar co-ordinates

For if we consider the radius vector corresponding to the angle θ and that corresponding to the angle $\theta + \Delta\theta$, and denote the smallest radius vector in this angular interval (cf. fig. 10) by r_0 and the greatest by r_1, the sector lying between the radius vector θ and the radius vector $\theta + \Delta\theta$ will have an area ΔA which lies between the bounds $\tfrac{1}{2} r_0^2 \Delta\theta$ and $\tfrac{1}{2} r_1^2 \Delta\theta$.   Consequently

$$\frac{1}{2} r_0^2 \leqq \frac{\Delta A}{\Delta\theta} \leqq \frac{1}{2} r_1^2,$$

and on passing to the limit as $\Delta\theta \to 0$, we obtain the relation given above.   By the fundamental theorem of the integral calculus, the area of the sector between the polar angles α and β is then given by the expression

$$\frac{1}{2} \int_\alpha^\beta r^2 \, d\theta.$$

If $\beta > \alpha$, this expression cannot be less than zero.   Since we readily see that as θ increases the point with co-ordinates (r, θ) describes the boundary of the region in the positive sense, this is in agreement with our previous convention for sign.

As an example, let us consider the area bounded by one loop of a lemniscate.   The equation of the lemniscate (cf. p. 73) is $r^2 = 2a^2 \cos 2\theta$, and we obtain one loop by letting θ vary from $-\dfrac{\pi}{4}$ to $+\dfrac{\pi}{4}$.   This gives us the expression

$$a^2 \int_{-\pi/4}^{\pi/4} \cos 2\theta \, d\theta$$

for the area. This can be integrated at once by introducing the new variable $u = 2\theta$; we find the value of the integral to be a^2.

5. Length of a Curve.

Another important geometrical concept associated with a curve leads to an integration. This is the *length of arc*.

We shall first explain geometrically how we are led to a definition of the length of an arbitrary curve. The elementary process of measuring a length consists of comparing the length to be measured with rectilinear standards of length. The simplest method is to apply our standard length to the curve, with its ends on the curve, and count the number of times that we have to repeat the process in order to pass from the beginning to the end of the curve; we can refine the method as required by using smaller and smaller standards of length. By analogy with this elementary intuitive idea, we set up the definition of the length of a curve in the following manner. We suppose that our curve is given by the equations $x = x(t)$, $y = y(t)$, $a \leqq t \leqq \beta$. (This includes curves in the form $y = f(x)$, since these can be written $y = f(t)$, $x = t$.) In the interval between a and β we choose points $t_0 = a, t_1, t_2, \ldots, t_n = \beta$, in that order. The points on the curve corresponding to these values t_ν we join in order by line segments, thus obtaining part of a polygon inscribed in the curve; we now measure the perimeter of this polygon. This length will depend on the way in which the points t_ν, or, as we may also say, the vertices of the polygon, are chosen. We now let the number of the points t_ν increase beyond all bounds, in such a way that the length of the longest sub-interval in the interval $a \leqq t \leqq \beta$ at the same time tends to 0; this makes the number of sides of our polygon increase without limit, while the length of the longest side tends to 0. The length of the curve is then defined to be the limit of the perimeters of these inscribed polygons, *provided* that such a limit does exist and is independent of the particular way in which the polygons are chosen. It is only when this assumption that the limit exists (assumption of *rectifiability*) is fulfilled that we can speak of the length of the curve. We shall soon see that very wide classes of curves can be proved to be rectifiable.

To express the length analytically by an integral, in fact, we think of the curve as represented in the first instance by a function

$y = f(x)$ with a continuous derivative y'. By the points $a = x_1$, $x_2, \ldots, x_n = b$ we divide up the interval $a \leqq x \leqq b$ of the x-axis, over which our curve lies, into $(n-1)$ sub-intervals of lengths $\Delta x_1, \ldots, \Delta x_{n-1}$. In the curve we inscribe a polygon whose vertices lie vertically above these points. The total length of this inscribed polygon is given according to Pythagoras' theorem (cf. fig. 11) by the expression

$$\sum_{\nu=1}^{n-1} \sqrt{(\Delta x_\nu{}^2 + \Delta y_\nu{}^2)} = \sum_{\nu=1}^{n-1} \sqrt{\left\{1 + \left(\frac{\Delta y_\nu}{\Delta x_\nu}\right)^2\right\}} \, \Delta x_\nu.$$

But by the mean value theorem of the differential calculus the difference quotient $\Delta y_\nu / \Delta x_\nu$ is equal to $f'(\xi_\nu)$, where ξ_ν is an

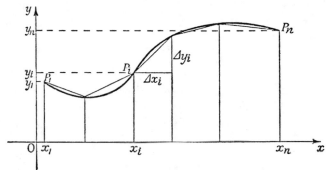

Fig. 11.—Rectification of curves

intermediate value in the interval Δx_ν. If we now let n increase beyond all bounds and at the same time let the length of the longest sub-interval Δx_ν tend to zero, then by the definition of integral our expression will tend to the limit

$$\int_a^b \sqrt{(1 + y'^2)} \, dx.$$

Since this passage to the limit always leads us to the same result, namely, the integral, no matter how the subdivision of the interval is made, we have established the following theorem:

Every curve $y = f(x)$ *for which the derivative* $f'(x)$ *is continuous is a rectifiable curve, and its length between* $x = a$ *and* $x = b$ $(b \geqq a)$ *is given by the formula*

$$s(a, b) = \int_a^b \sqrt{(1 + y'^2)} \, dx.$$

If by s we denote the length of arc measured from an arbitrary fixed point to the point with abscissa x, the above equation gives us the following expression for the derivative of the length of arc with respect to x:

$$\frac{ds}{dx} = \sqrt{(1 + y'^2)}.$$

Our expression for the length of arc is still subject to the special and artificial assumption that the curve consists of one single-valued branch above the x-axis. Parametric representation frees us from this restriction. If a curve of the kind which we have been considering is given in parametric form by the equations $x = x(t)$, $y = y(t)$, then by introducing the parameter t in the above expression we obtain the parametric form of the length of arc

$$s(\alpha, \beta) = \int_\alpha^\beta \sqrt{(\dot{x}^2 + \dot{y}^2)}\, dt,$$

where α and β are the values of t which correspond respectively to the points of the curve $x = a$ and $x = b$.

This parametric expression for the length of a curve has a considerable advantage over the previous form in that it is not restricted to single-valued branches of curves, represented by the equation $y = f(x)$, but instead holds for any arbitrary arcs of curves, including closed curves, provided that the derivatives $\dot{x}$ and $\dot{y}$ are continuous along the arcs.

We recognize this most easily if we go back again to the formula for the length of the inscribed polygon. We suppose that along the arc $\dot{x}$ and $\dot{y}$ are continuous. As in the definition, we subdivide the interval $a \leqq t \leqq \beta$ by points $t_0 = a$, $t_1, \ldots, t_n = \beta$, with the differences Δt_ν and use the corresponding points on the curve as vertices of an inscribed polygon; in the passage to the limit $n \to \infty$ we assume that the greatest difference Δt_ν tends to 0. If we now write the length of the polygon in the form

$$\sum_{\nu=1}^{n} \sqrt{(\Delta x_\nu^2 + \Delta y_\nu^2)} = \sum_{\nu=1}^{n} \sqrt{\left\{ \left(\frac{\Delta x_\nu}{\Delta t_\nu}\right)^2 + \left(\frac{\Delta y_\nu}{\Delta t_\nu}\right)^2 \right\}}\, \Delta t_\nu,$$

we see at once that this sum tends to the integral $\int_\alpha^\beta \sqrt{(\dot{x}^2 + \dot{y}^2)}\, dt$; we need only recall the generalized method of formation of an integral (cf. p. 133). If the curve is composed of several arcs of this type, which may join one another at corners, the expression for the length of the curve is simply the sum of the corresponding integrals. Collecting the results, we have the following statement:

If in the interval $a \leqq t \leqq \beta$ *the functions* x(t) *and* y(t) *are continuous and their derivatives* $\dot{x}(t)$, $\dot{y}(t)$ *are also continuous, except perhaps for a finite number of jump discontinuities, the arc of* x = x(t), y = y(t) *has a length given by the expression*

$$\int_a^\beta \sqrt{(\dot{x}^2 + \dot{y}^2)}\,dt,$$

where this integral, if necessary, is to be taken as an improper integral in the sense of Chap. IV (p. 245). In virtue of this formula, in which a must be less than β, there is a meaning in ascribing a negative length, given by the same formula, to an arc of a curve traversed in the direction in which the value of the parameter t decreases. The sign of the length of arc therefore depends on the choice of the parameter. If we introduce a new parametric expression for the same curve which does not reverse the sense of description, that is, if we introduce a new parameter by the equation $\tau = \tau(t)$, where $d\tau/dt > 0$, we see *a priori* that our integral formula should give the same value no matter whether t or τ is used as parameter; for the two integrals give the length of the same curve and must therefore be equal. This, however, may also be verified directly, for

$$\int \sqrt{(\dot{x}^2 + \dot{y}^2)}\,dt = \int \sqrt{\left\{\left(\frac{dx}{d\tau}\right)^2 \left(\frac{d\tau}{dt}\right)^2 + \left(\frac{dy}{d\tau}\right)^2 \left(\frac{d\tau}{dt}\right)^2\right\}}\,dt$$

$$= \int \sqrt{\left\{\left(\frac{dx}{d\tau}\right)^2 + \left(\frac{dy}{d\tau}\right)^2\right\}}\,d\tau.$$

We now give the expression for the length of arc when the curve is expressed in *polar co-ordinates*. In the last expression we have only to substitute for $\dot{x}$ and $\dot{y}$ their values as given in formula (a) on p. 265 in order to obtain

$$\dot{x}^2 + \dot{y}^2 = \dot{r}^2 + r^2\dot{\theta}^2,$$

whence

$$s(a, \beta) = \int_a^\beta \sqrt{(\dot{r}^2 + r^2\dot{\theta}^2)}\,dt.$$

If we now change over from the parametric expression to the equation in the form $r = f(\theta)$, by introducing as parameter $t = \theta$ itself, so that $\dot\theta = 1$, we have the expression

$$s(\theta_0,\, \theta_1) = \int_{\theta_0}^{\theta_1} \sqrt{(\dot r^2 + r^2)}\, d\theta$$

for the length of arc.

A simple example of the explicit calculation of the length of an arc is given by the parabola $y = \frac{1}{2} x^2$; for its length of arc we immediately obtain the integral $\int_a^b \sqrt{(1 + x^2)}\, dx$, which with the substitution $x = \sinh u$ becomes

$$\int_{\text{ar sinh}\, a}^{\text{ar sinh}\, b} \cosh^2 u\, du = \frac{1}{2} \int_{\text{ar sinh}\, a}^{\text{ar sinh}\, b} (1 + \cosh 2u)\, du = \frac{1}{2} (u + \sinh u\ \cosh u) \Big|_{\text{ar sinh}\, a}^{\text{ar sinh}\, b},$$

so that the length of arc of the parabola between the abscissæ $x = a$ and $x = b$ is given by the expression

$$s(a,\, b) = \frac{1}{2} \{\text{ar sinh}\, b + b \sqrt{(1 + b^2)} - \text{ar sinh}\, a - a\sqrt{(1 + a^2)}\}.$$

For the catenary $y = \cosh x$ we find that

$$s(a,\, b) = \int_a^b \sqrt{(1 + \sinh^2 x)}\, dx = \int_a^b \cosh x\, dx, \quad \text{or } s(a,\, b) = \sinh b - \sinh a.$$

Finally, let it be noted that in many cases it is convenient to introduce as parameter the length of arc reckoned from some fixed point P_0 on the curve, that is, to take $x = x(s)$ and $y = y(s)$. Points of the curve on opposite sides of P_0 will correspond to values of s with opposite signs. In this case we have

$$\dot x^2 + \dot y^2 = \left(\frac{ds}{dt}\right)^2 = 1,$$

whence by differentiation

$$\dot x \ddot x + \dot y \ddot y = 0;$$

these two relations find frequent application.

6. Curvature of a Curve.

The area and the length of arc of a curve depend on the complete course of the curve. We now insert a discussion of a concept which has reference only to the behaviour of a curve in the neighbourhood of a point, namely, the *curvature*.

If we think of the curve as described uniformly in the positive sense, in such a way that equal lengths of arc are passed over in equal periods of time, the direction of the curve will vary at a definite rate, which we take as a measure of the curvature of the curve. If, therefore, we denote the angle between the positive direction of the tangent (p. 264) and the positive x-axis by a, and if we think of a as a function of the length of arc s, we shall define the curvature k at the point corresponding to the length of arc s by the equation $k = da/ds$. We know that $a = \arc\tan y'$, and hence by the chain rule

$$\frac{da}{ds} = \frac{da}{dx} \div \frac{ds}{dx} = \frac{y''}{1 + y'^2} \cdot \frac{1}{\sqrt{(1 + y'^2)}}$$

(where the positive sign of the square root means that increasing values of x correspond to increasing values of s). The curvature is consequently given by the expression

$$k = \frac{y''}{(1 + y'^2)^{3/2}}.$$

Using the parametric formulæ for y' and y'' we obtain the following simple expression for the curvature of a curve represented parametrically:

$$k = \frac{\dot{x}\ddot{y} - \dot{y}\ddot{x}}{(\dot{x}^2 + \dot{y}^2)^{3/2}},$$

which, of course, can also be found directly from the equation

$$a = \arc\tan\frac{\dot{y}}{\dot{x}} = \arc\cot\frac{\dot{x}}{\dot{y}}.$$

In contrast with the previous expression, which is dependent on the equation $y = f(x)$ and consequently involves a special assumption about the position of the arc with respect to the x-axis, the parametric expression for the curvature holds for all arcs along which $\dot{x}$, $\dot{y}$, $\ddot{x}$, and $\ddot{y}$ are continuous functions of t and $\dot{x}^2 + \dot{y}^2 \neq 0$. In particular, it holds for points where $\dot{x} = 0$, i.e. where dy/dx becomes infinite.

If we introduce the length of arc s as parameter and recall that $\dot{x}^2 + \dot{y}^2 = 1$ and $\dot{x}\ddot{x} + \dot{y}\ddot{y} = 0$, we have

$$k = \dot{x}\ddot{y} - \dot{y}\ddot{x} = \ddot{y}\left(\dot{x} + \dot{y}\frac{\dot{y}}{\dot{x}}\right) = \frac{\ddot{y}}{\dot{x}} = -\frac{\ddot{x}}{\dot{y}}.$$

(E 798)

We thus obtain a particularly simple expression for the curvature.

The *sign* of the curvature is changed if we reverse the sense of description of the curve, that is, if we replace the parameter t or s by the new parameter $\tau = -t$ or $\sigma = -s$. For then $\dot{x}$ and $\dot{y}$ change sign, but not $\ddot{x}$, $\ddot{y}$, $\dot{x}^2$ or $\dot{y}^2$, as the following simple calculation shows:

$$\frac{d}{d\tau} x\{t(\tau)\} = \frac{dx}{dt} \frac{dt}{d\tau} = (\dot{x})(-1);$$

$$\frac{d^2}{d\tau^2} x\{t(\tau)\} = \frac{d}{d\tau}\left[-\dot{x}\{t(\tau)\}\right] = -\frac{d\dot{x}}{dt}\frac{dt}{d\tau} = (-\ddot{x})(-1).$$

(A similar calculation can be made for y.) In the case of the expression $k = \dfrac{y''}{(1+y'^2)^{\frac{3}{2}}}$ first found, this fact is concealed, since it is natural and customary to think of the curve as described from left to right, in which case the square root can only be positive.

As an example we consider the curvature of a positively described circle with radius a. If we start from the parametric representation $x = a\cos t$, $y = a\sin t$, we immediately obtain

$$k = \frac{1}{a}.$$

The curvature of a positively described circle is therefore the reciprocal of its radius. This result assures us that our definition of curvature is really a suitable one; for in the case of a circle we naturally think of the reciprocal of the radius as a measure of the curvature.

Let us put $\rho = \dfrac{1}{k}$. The quantity $|\rho| = \dfrac{1}{|k|}$ is generally called the *radius of curvature* of the curve at the point in question. For a given point on the curve, that circle which touches the curve at the point and there has the same sense of description and the same curvature as the curve, and, moreover, has its centre on the positive or negative side of the normal according as k is positive or negative, is called the *circle of curvature* corresponding to the point. Let us think of the equation of the circle (or an arc of the circle containing the point in question) as written in the form $y = g(x)$. Then at the point in question we have not only $f(x) = g(x)$ and $f'(x) = g'(x)$, as follows from the fact that the circle and curve touch, but in virtue of the relation

$$\frac{f''(x)}{\sqrt{\{1 + f'(x)^2\}^3}} = k = \frac{g''(x)}{\sqrt{\{1 + g'(x)^2\}^3}}$$

we also have

$$f''(x) = g''(x).$$

The centre of the circle of curvature is called the *centre of curvature* corresponding to the given point. Its co-ordinates are expressed parametrically by

$$\xi = x - \frac{\rho \dot{y}}{\sqrt{(\dot{x}^2 + \dot{y}^2)}}, \quad \eta = y + \frac{\rho \dot{x}}{\sqrt{(\dot{x}^2 + \dot{y}^2)}}.$$

To prove this we need only make use of the formulæ for the direction cosines of the normal, on which the centre of curvature lies at a distance $1/|k| = |\rho|$ from the tangent. These formulæ give us an expression for the centre of curvature in terms of the parameter t. As t describes its range the centre of curvature describes a curve, the so-called *evolute* of the given curve; and since, with x and y, we have to regard $\dot{x}$, $\dot{y}$, and ρ as known functions of t, the formulæ above give parametric equations for this evolute.

For special examples the reader may be referred to § 3 (p. 287 *et seq.*) and to the appendix (p. 307 *et seq.*).

7. Centre of Mass and Moment of a Curve.

We now come to some applications which bring us into the realm of mechanics. We consider a system of n particles lying in a plane. Let $m_1, m_2, \ldots, m_n$ be the masses of these particles, and let $y_1, y_2, \ldots, y_n$ be their respective ordinates. We then call

$$T = \sum_{\nu=1}^{n} m_\nu y_\nu = m_1 y_1 + m_2 y_2 + \ldots + m_n y_n$$

the *moment of the system of particles with respect to the x-axis*. The expression $\eta = T/M$, where M denotes the total mass $m_1 + m_2 + \ldots + m_n$ of the system, gives us the *height of the centre of mass* of the system of particles above the x-axis. We define the moment with respect to the y-axis and the abscissa of the centre of mass in a corresponding way.

We shall now see that this idea can easily be extended to give us a definition of the moment of a curve along which a mass is uniformly distributed, and of the co-ordinates ξ and η of the centre of mass of such a curve. Merely for the sake of brevity we assume that the density has a constant value, say μ, along the curve; any continuous distribution could equally well be discussed.

To arrive at this extension we go back to the consideration of a system of a finite number of particles and then pass to the limit. For this purpose we suppose that the length of arc s is introduced as a parameter on the curve, and that the curve is subdivided by $(n-1)$ points of division into arcs of lengths $\Delta s_1, \Delta s_2, \ldots, \Delta s_n$. The mass $\mu \Delta s_i$ of each arc Δs_i we represent as concentrated at an arbitrary point of the arc, say that with the ordinate y_i.

By definition the moment of this system of particles with respect to the x-axis has the value

$$T = \mu \sum y_i \Delta s_i.$$

If now the greatest of the quantities Δs_i tends to 0, this sum tends to a definite limit given by the expression

$$T = \mu \int_{s_0}^{s_1} y \, ds = \mu \int_{x_0}^{x_1} y \sqrt{(1 + y'^2)} \, dx,$$

which we shall therefore naturally accept as the definition of the moment of the curve with respect to the x-axis. Since the total mass of the curve is equal to its length multiplied by μ,

$$\mu \int_{s_0}^{s_1} ds = \mu (s_1 - s_0),$$

we are immediately led to the following expressions for the co-ordinates of the centre of mass of the curve:

$$\eta = \frac{\int_{s_0}^{s_1} y \, ds}{s_1 - s_0}, \quad \xi = \frac{\int_{s_0}^{s_1} x \, ds}{s_1 - s_0}.$$

These statements are actually *definitions* of the moment and centre of mass of a curve; but they are such straightforward extensions of the simpler case of a number of particles that we naturally expect that—as is actually the case—any statement in mechanics which involves the centre of mass or the moment of a system of particles will be valid for curves also. In particular, the position of the centre of mass with respect to the curve is independent of the system of co-ordinates.

8. Area and Volume of a Surface of Revolution.

If we rotate the curve $y = f(x)$, for which $f(x) \geqq 0$, about the x-axis, it describes a so-called *surface of revolution*. The area of this surface, whose abscissæ we suppose to lie between the bounds x_0 and $x_1 > x_0$, can be obtained by a discussion analogous to the preceding. For if we replace the curve by an inscribed polygon, instead of the curved surface we shall have a figure composed of a number of thin truncated cones. Following the suggestions of intuition, we define the area of the surface of revolution as the limit of the areas of these conical surfaces when the length of the longest side of the inscribed polygon tends to zero. We know from elementary geometry that the area of each truncated cone is equal to its slant height multiplied by the circumference of the circular section of mean radius. If we add these expressions and then carry out the passage to the limit, we obtain the expression

$$A = 2\pi \int_{x_0}^{x_1} y \sqrt{(1 + y'^2)}\, dx = 2\pi \int_{s_0}^{s_1} y\, ds$$

for the area. Expressed in words, this result states that the area of a surface of revolution is equal to the length of the curve generating it multiplied by the distance traversed by the centre of mass (Guldin's rule).

In the same way we find that the volume interior to the surface of revolution and bounded at the ends by the planes $x = x_0$ and $x = x_1 > x_0$ is given by the expression

$$V = \pi \int_{x_0}^{x_1} y^2\, dx.$$

This formula is obtained by following the suggestion of intuition that the volume in question is the limit of the volumes of the above-mentioned figures consisting of truncated cones. The rest of the proof is left to the reader.

9. Moment of Inertia.

In the study of rotatory motion in mechanics an important part is played by certain quantities called moments of inertia. These expressions will be briefly mentioned here.

We suppose that a particle m at a distance y from the x-axis rotates uniformly about that axis with angular velocity ω (that is, in unit time it rotates through an angle ω). The *kinetic energy* of the particle, expressed by half the product of the mass and the square of the velocity, is obviously

$$\frac{m}{2}(y\omega)^2.$$

We call the coefficient of $\frac{1}{2}\omega^2$, that is, the quantity my^2, the *moment of inertia of the particle about the x-axis*.

Similarly, if we have n particles with masses $m_1, m_2, \ldots, m_n$ and ordinates $y_1, y_2, \ldots, y_n$ we call the expression

$$T = \sum_i m_i y_i^2$$

the moment of inertia of the system of masses about the x-axis. The moment of inertia is a quantity which belongs to the system of masses itself, without reference to its state of motion. Its importance lies in the fact that if the whole system is set in rigid rotation about an axis, without change of the distances between pairs of particles, the kinetic energy is obtained by multiplying the moment of inertia about that axis by half the square of the angular velocity. Thus the moment of inertia about an axis plays the same part in rotation about an axis as is played by the mass in rectilinear motion.

Suppose now that we have an arbitrary curve $y = f(x)$ lying between the abscissæ x_0 and x_1 ($> x_0$), along which a mass is uniformly distributed with unit density. In order to define the moment of inertia of this curve we proceed just as we did in the sub-section 7 (p. 284); as before, we arrive at an expression for the moment of inertia about the x-axis, namely,

$$T_x = \int_{s_0}^{s_1} y^2\,ds = \int_{x_0}^{x_1} y^2 \sqrt{(1 + y'^2)}\,dx.$$

For the moment of inertia about the y-axis we have the corresponding expression

$$T_y = \int_{s_0}^{s_1} x^2\,ds = \int_{x_0}^{x_1} x^2 \sqrt{(1 + y'^2)}\,dx.$$

3. Examples

The theory of plane curves with its great variety of special forms and properties offers us a rich store of examples of these abstract concepts. But to avoid being lost in a mass of detail we must limit ourselves to a few typical applications.

1. The Common Cycloid.

From the equations (cf. p. 261) $x = a(t - \sin t)$, $y = a(1 - \cos t)$, we at once obtain $\dot{x} = a(1 - \cos t)$, $\dot{y} = a \sin t$, whence the length of arc is

$$s = \int_0^a \sqrt{(\dot{x}^2 + \dot{y}^2)}\, dt = \int_0^a \sqrt{\{2a^2(1 - \cos t)\}}\, dt.$$

But since $1 - \cos t = 2 \sin^2 \dfrac{t}{2}$ the integrand is equal to $2a \sin \dfrac{t}{2}$, and hence for $0 \leqq \alpha \leqq 2\pi$

$$s = 2a \int_0^a \sin \frac{t}{2}\, dt = -4a \cos \frac{t}{2} \bigg|_0^a = 4a\left(1 - \cos \frac{\alpha}{2}\right) = 8a \sin^2 \frac{\alpha}{4}.$$

If, in particular, we consider the length of arc between two successive cusps we must put $\alpha = 2\pi$, since the interval $0 \leqq t \leqq 2\pi$ of values of the parameter corresponds to one revolution of the rolling circle. We thus obtain the value $8a$; that is, the length of arc of the cycloid between successive cusps is equal to four times the diameter of the rolling circle.

Similarly, we calculate the area bounded by one arch of the cycloid and the x-axis:

$$I = \int_0^{2\pi} y\dot{x}\, dt = a^2 \int_0^{2\pi} (1 - \cos t)^2\, dt$$

$$= a^2 \int_0^{2\pi} (1 - 2\cos t + \cos^2 t)\, dt$$

$$= a^2 \left(t - 2\sin t + \frac{t}{2} + \frac{\sin 2t}{4}\right)\bigg|_0^{2\pi} = 3a^2\pi.$$

This area is therefore three times the area of the rolling circle.

For the radius of curvature $\rho = 1/k$ we have

$$\rho = \frac{(\dot{x}^2 + \dot{y}^2)^{3/2}}{\dot{x}\ddot{y} - \dot{y}\ddot{x}} = -2a\sqrt{\{2(1 - \cos t)\}} = -4a\left|\sin \frac{t}{2}\right|;$$

at the points $t = 0$, $t = \pm 2\pi$, ... this expression has the value zero. These are actually the cusps, where the cycloid meets the x-axis at right angles.

The area of the surface of revolution formed by rotating an arch of the cycloid about the x-axis is given according to our formula (p. 285) by

$$A = 2\pi \int_0^{8a} y\,ds = 2\pi \int_0^{2\pi} a(1 - \cos t)\,.\,2a \sin \frac{t}{2}\,dt$$

$$= 8a^2\pi \int_0^{2\pi} \sin^3 \frac{t}{2}\,dt = 16a^2\pi \int_0^{\pi} \sin^3 u\,du$$

$$= 16a^2\pi \int_0^{\pi} (1 - \cos^2 u)\,\sin u\,du.$$

The last integral can be evaluated by means of the substitution $\cos u = v$; we find that

$$A = 16a^2\pi\left(-\cos u + \frac{1}{3}\cos^3 u\right)\Big|_0^{\pi} = \frac{64a^2\pi}{3}.$$

As an exercise the reader may calculate for himself the height η of the centre of mass of the cycloid above the x-axis, and also the moment of inertia T_x. The results are

$$\eta = \frac{4}{3}a = \frac{A}{2\pi s} \quad \text{and} \quad T_x = \frac{256}{15}a^3.$$

2. The Catenary.

The length of arc of the catenary has already been calculated as an example in the preceding section (p. 280), and we found its value to be

$$s = \int_a^b \cosh x\,dx = \sinh b - \sinh a.$$

For the area of the surface of revolution obtained by rotating the catenary about the x-axis, the so-called catenoid, we find

$$A = 2\pi \int_a^b \cosh^2 x\,dx = 2\pi \int_a^b \frac{1 + \cosh 2x}{2}\,dx$$

$$= \pi\left(b - a + \frac{1}{2}\sinh 2b - \frac{1}{2}\sinh 2a\right).$$

From this we further obtain the height of the centre of mass of the arc from a to b:

$$\eta = \frac{A}{2\pi s} = \frac{b - a + \frac{1}{2}\sinh 2b - \frac{1}{2}\sinh 2a}{2\,(\sinh b - \sinh a)}.$$

Finally, for the curvature we have

$$k = \frac{y''}{(1 + y'^2)^{3/2}} = \frac{\cosh x}{\cosh^3 x} = \frac{1}{\cosh^2 x}.$$

3. The Ellipse and the Lemniscate.

The lengths of arc of these two curves cannot be reduced to elementary functions, but belong to the class of " elliptic integrals " mentioned on p. 243.

For the ellipse $y = \dfrac{b}{a} \sqrt{(a^2 - x^2)}$ we obtain

$$ s = \frac{1}{a} \int \sqrt{\left\{ \frac{a^4 - (a^2 - b^2)x^2}{a^2 - x^2} \right\}} \, dx = a \int \frac{1 - \varkappa^2 \xi^2}{\sqrt{(1 - \xi^2)(1 - \varkappa^2 \xi^2)}} \, d\xi, $$

where we have put $x/a = \xi$, $1 - b^2/a^2 = \varkappa^2$. By the substitution $\xi = \sin \varphi$ this integral can be expressed in the form

$$ s = \int \sqrt{\{a^2 - (a^2 - b^2)\sin^2 \varphi\}} \, d\varphi = a \int \sqrt{(1 - \varkappa^2 \sin^2 \varphi)} \, d\varphi. $$

Here, to obtain the semi-perimeter of the ellipse, we must let x traverse the interval from $-a$ to $+a$, which corresponds to the interval

$$ -1 \leqq \xi \leqq +1 \quad \text{or} \quad -\pi/2 \leqq \varphi \leqq +\pi/2. $$

For the lemniscate, whose equation in polar co-ordinates is $r^2 = 2a^2 \cos 2t$, we similarly obtain

$$ s = \int \sqrt{(r^2 + \dot{r}^2)} \, dt = \int \sqrt{\left(2a^2 \cos 2t + 2a^2 \frac{\sin^2 2t}{\cos 2t} \right)} \, dt $$

$$ = a \sqrt{2} \int \frac{dt}{\sqrt{(\cos 2t)}} = a \sqrt{2} \int \frac{dt}{\sqrt{(1 - 2 \sin^2 t)}}. $$

If we introduce $u = \tan t$ as independent variable in the last integral, we have

$$ \sin^2 t = \frac{u^2}{1 + u^2}, \quad dt = \frac{du}{1 + u^2}, $$

and consequently

$$ s = a \sqrt{2} \int \frac{du}{\sqrt{(1 - u^4)}}. $$

In a complete loop of the lemniscate u runs from -1 to $+1$, and the length of arc is therefore equal to

$$ a \sqrt{2} \int_{-1}^{+1} \frac{du}{\sqrt{(1 - u^4)}}, $$

a special elliptic integral which played a great part in the researches of Gauss.

EXAMPLES

1. Calculate the area bounded by the semicubical parabola $y = x^{3/2}$, the x-axis, and the lines $x = a$ and $x = b$.

2. Calculate the area of the region bounded by the line $y = x$ and the lower half of the loop of the folium of Descartes. (Use the parametric representation found in Ex. 7, p. 267.)

3. Calculate the area of a sector of the Archimedean spiral $r = a\theta$ $(a > 0)$.

4. Calculate the area of the cardioid (Ex. 3, p. 267), using polar coordinates.

5. Calculate the area of the astroid (Ex. 6, p. 267).

6. Calculate the area of the pedal curve of the circle $x^2 + y^2 = 1$ with respect to a point $P(x_0, 0)$ on the x-axis. Show that this area is least when P is at the origin.

7. Do the same for the ellipse $\dfrac{x^2}{a^2} + \dfrac{y^2}{b^2} = 1$.

8. Find the parametric representation of the cardioid when the length of arc is used as parameter.

9. Do the same for the cycloid.

10. Calculate the length of arc of the semicubical parabola $y = x^{3/2}$.

11. Calculate the length of the astroid.

12. Calculate the length of arc of:
   (a) The Archimedean spiral $r = a\theta$ $(a > 0)$.
   (b) The logarithmic spiral $r = e^{m\theta}$.
   (c) The cardioid (Ex. 3, p. 267).
   (d) The curve $r = a(\theta^2 - 1)$.

13. Find the radius of curvature of (a) the parabola $y = x^2$; (b) the ellipse $x = a\cos\varphi$, $y = b\sin\varphi$, as a function of x and of φ respectively. Find the maxima and minima of the radius of curvature and the points at which these maxima and minima occur.

14. Sketch the curve

$$x = \int_0^t \frac{\cos u}{\sqrt{u}}\, du, \quad y = \int_0^t \frac{\sin u}{\sqrt{u}}\, du$$

and determine its radius of curvature (ρ).

15. Show that the expression for the curvature of a curve $x = x(t)$, $y = y(t)$ is unaltered by rotation of axes and also by change of parameter given by $t = \varphi(\tau)$, where $\varphi'(\tau) > 0$.

16. Let $r = f(\theta)$ be the equation of a curve in polar co-ordinates. Prove that the curvature is given by the formula

$$k = \frac{2r'^2 - rr'' + r^2}{(r'^2 + r^2)^{3/2}},$$

where $$r' = \frac{df}{d\theta}, \quad r'' = \frac{d^2f}{d\theta^2}.$$

17. Find the volume and surface area of a zone of a sphere of radius r, i.e. of the portion of the sphere cut off by two parallel planes distant h_2, h_1 respectively from the centre.

18. Find the volume and surface area of the *torus* or *anchor ring* obtained by rotating a circle about a line which does not intersect it.

19. Find the area of the *catenoid*, the surface obtained by rotating an arc of the catenary $y = \cosh x$ about the x-axis.

20. Sketch the curve defined by the equations

$$x = \int_0^t \cos\left(\tfrac{1}{2}\pi t^2\right) dt, \quad y = \int_0^t \sin\left(\tfrac{1}{2}\pi t^2\right) dt.$$

What is the behaviour of the curve as t runs from $-\infty$ to $+\infty$? Calculate the curvature k as a function of the length of arc.

21. The curve for which the length of the tangent intercepted between the point of contact and the y-axis is always equal to 1 is called the *tractrix*. Find its equation. Show that the radius of curvature at each point of the curve is inversely proportional to the length of the normal intercepted between the point on the curve and the y-axis. Calculate the length of arc of the tractrix and find the parametric equations in terms of the length of arc.

22. Let $x = x(t)$, $y = y(t)$ be a closed curve. A constant length p is measured off along the normal to the curve. The extremity of this segment describes a curve which is called a *parallel curve* to the original curve. Find the area, the length of arc, and the radius of curvature of the parallel curve.

23. Find the centre of mass of an arbitrary arc (*a*) of a circle of radius r, (*b*) of a catenary.

24. Calculate the moment of inertia about the x-axis of the boundary of the rectangle $a \leqq x \leqq b$, $\alpha \leqq y \leqq \beta$.

25. Calculate the moment of inertia of an arc of the catenary $y = \cosh x$ (*a*) about the x-axis, (*b*) about the y-axis.

26. The equation $y = f(x) + \alpha$, $a \leqq x \leqq b$, represents a family of curves, one for each value of the parameter α. Prove that in this family the curve with the least moment of inertia about the x-axis is that which has its centre of mass on the x-axis.

4. SOME VERY SIMPLE PROBLEMS IN THE MECHANICS OF A PARTICLE

Next to geometry the differential and integral calculus are especially indebted to the science of mechanics for their early development. Mechanics rests upon certain basic principles which were first laid down by Newton; the statement of these principles involves the concept of the derivative, and their application requires the theory of integration. Without analysing these basic principles in detail, we shall illustrate by some simple examples how the integral and differential calculus are applied in mechanics.

1. The Fundamental Hypotheses of Mechanics.

Here we shall restrict ourselves to the consideration of a single particle, that is, of a point at which a mass m is imagined to be concentrated. We shall further assume that motion can only take place along a certain fixed curve, on which the position of the particle is specified by the length of arc s measured from a fixed point on the curve; in particular, the curve may be a straight line, in which case we use the abscissa x as the co-ordinate of the point instead of s. The motion of the point is determined by expressing the co-ordinate $s = \phi(t)$ as a function of the time. By the *velocity* of motion we shall mean the derivative $\phi'(t)$, or, as we shall also write,

$$\frac{ds}{dt} = \phi'(t) = \dot{s}.$$

The second derivative,

$$\frac{d^2s}{dt^2} = \phi''(t) = \ddot{s},$$

we call the *acceleration*.

In mechanics we start from the assumption that the motion of a point can be explained by means of *forces* of definite direction and magnitude. Newton's second fundamental law of mechanics may, in the case of motion on our given curve, be expressed as follows:

The mass multiplied by the acceleration is equal to the force acting on the particle in the direction of the curve; in symbols

$$m\ddot{s} = F.$$

Thus the direction of the force is always the same as that of the acceleration; its direction is that of increasing values of s if the velocity in that direction is increasing, otherwise it is opposed to the direction of increasing values of s.

The law of Newton is in the first instance nothing more than a definition of the concept of force. The left-hand side of our equation is a quantity which can be determined by observation of the motion, by means of which we measure the force. But this equation has a far deeper meaning. As a matter of fact, it turns out that in many cases we can determine the acting force from other physical assumptions, without any consideration of the corresponding motion. The above fundamental law of Newton is then no longer a *definition of force*, but is instead a relation from which we can draw important conclusions about the motion.

The most important example of a known force is given us by *gravity*. From direct measurements we know that the force of gravity acting on a mass m is directed vertically downwards and is of magnitude mg, where the constant g, the so-called gravitational acceleration, is approximately equal to 981 if the time

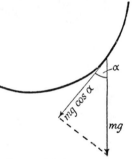

Fig. 12.—Motion on a given curve under gravity

is measured in seconds and the lengths in centimetres. If a mass moves along a given curve, we learn by experiment that the force of gravity in the direction of this curve is equal to $mg \cos \alpha$, where α denotes the angle between the vertical and the tangent to the curve at the point under consideration (cf. fig. 12).

In the case of motion on our given curve the basic problem of mechanics is as follows: if we know the force acting on the particle (e.g. the force of gravity), we have to determine the position of the point, that is, its co-ordinate s or x, as a function of the time.

If we restrict ourselves to the simplest case, in which this force * $mf(s)$ is known at the outset as a function of the length of arc—so that the force is independent of the time—we shall

---

* The separation of the factor m in the expression for the given force is not essential, but makes the formulæ simpler.

show how the course of the motion along the curve can be found from the equation

$$\ddot{s} = \frac{1}{m} F = f(s).$$

Here we have to deal with a *differential equation*, that is, an equation from which an unknown function—here $s(t)$—is to be determined and in which the derivative of this function occurs as well as the function itself (cf. Chap. III, § 7, p. 178).

2. Body Falling Freely. Resistance of the Air.

In the case of the free fall of a particle along the vertical x-axis. Newton's law gives us the differential equation

$$\ddot{x} = g.$$

From this follows $\dot{x}(t) = gt + v_0$, where v_0 is a constant of integration. Its meaning is easily found by putting $t = 0$. We then find $\dot{x}(0) = v_0$; that is, v_0 is the velocity of the particle at the instant from which the time is reckoned, the *initial velocity*. By another integration, we obtain

$$x(t) = \tfrac{1}{2}gt^2 + v_0 t + x_0,$$

where x_0 is also a constant of integration, whose value is again found by putting $t = 0$; we thus find that x_0 is the *initial position*, that is, the coordinate of the point at the beginning of the motion.

Conversely, we can choose the initial position x_0 and the initial velocity v_0 arbitrarily, and then obtain the complete representation of the motion from the equation $x = \tfrac{1}{2}gt^2 + v_0 t + x_0$.

If we wish to take account of the effect of the *friction* or *air resistance* acting on the particle, we have to consider this as a force whose direction is opposite to the direction of motion and concerning which we must make definite physical assumptions.* We shall work out the results of different physical assumptions: (a) the resistance is proportional to the velocity, being given by an expression of the form $-r\dot{x}$, where r is a positive constant; (b) the resistance is proportional to the square of the velocity, being of the form $-r\dot{x}^2$. In accordance with Newton's law we obtain for the equations of motion

$$(a)\ \ m\ddot{x} = mg - r\dot{x}, \quad (b)\ \ m\ddot{x} = mg - r\dot{x}^2.$$

If we at first consider $\dot{x} = u(t)$ as the function sought, we have $\ddot{x}(t) = \dot{u}(t)$, so that

$$(a)\ \ m\dot{u} = mg - ru, \quad (b)\ \ m\dot{u} = mg - ru^2.$$

---

* These assumptions must be chosen to suit the particular system under consideration; for example, the law of resistance for low speeds is not the same as that for high (e.g. bullet velocities).

Instead of determining u as a function of t by these equations, we determine t as a function of u, writing our differential equations in the form

$$(a)\quad \frac{dt}{du} = \frac{1}{g - ru/m}, \quad (b)\quad \frac{dt}{du} = \frac{1}{g - ru^2/m}.$$

With the help of the methods given in the preceding chapter we can immediately carry out the integrations and obtain

$$(a)\quad t(u) = -\frac{m}{r} \log\left(1 - \frac{r}{mg} u\right) + t_0,$$

$$(b)\quad t(u) = -\frac{1}{2} k \log \frac{kg - u}{kg + u} + t_0,$$

where we have put $\sqrt{(m/rg)} = k$ and where t_0 is a constant of integration. Solving these equations for u, we have

$$(a)\quad u(t) = -\frac{mg}{r} (e^{-r(t-t_0)/m} - 1),$$

$$(b)\quad u(t) = -gk \frac{e^{-2(t-t_0)/k} - 1}{e^{-2(t-t_0)/k} + 1}.$$

These equations at once reveal an important property of the motion. The velocity does not increase with time beyond all bounds, but tends to a definite limit depending on the mass m. For

$$(a)\quad \lim_{t \to \infty} u(t) = \frac{mg}{r}, \quad (b)\quad \lim_{t \to \infty} u(t) = \sqrt{\frac{mg}{r}}.$$

A second integration, performed on our expressions for $u(t) = \dot{x}$ with the help of the methods of the preceding chapter, gives the results (which may be verified by differentiation)

$$(a)\quad x(t) = \frac{m^2}{r^2} g e^{-r(t-t_0)/m} + \frac{mg}{r} t + c,$$

$$(b)\quad x(t) = \frac{m}{r} \log \cosh \sqrt{\frac{rg}{m}} (t - t_0) + c,$$

where c is a new constant of integration. The two constants of integration t_0 and c are readily determined if we know the initial position $x(0) = x_0$ and the initial velocity $\dot{x}(0) = u(0) = v_0$ of the falling particle.

3. The Simplest Type of Elastic Vibration.

As a second example we consider the motion of a particle which moves along the x-axis and is pulled back towards the origin by an elastic force. As regards the elastic force we assume that it is always directed towards the origin and that its magnitude is proportional to the distance from the origin. In other words, we take the force as equal to $-kx$, where the coefficient k is a measure of the stiffness of the elastic connexion. Since

k is assumed positive, the force is negative when x is positive and positive when x is negative. Newton's law now tells us that

$$m\ddot{x} = -kx.$$

We cannot expect that this differential equation will determine the motion completely, but it is plausible to suppose that for a given instant of time, say $t = 0$, we can arbitrarily assign the initial position $x(0) = x_0$ and the initial velocity $\dot{x}(0) = v_0$; that is, in physical language, that we can start off the particle from an arbitrary position with an arbitrary velocity and that thereafter the motion is determined by the differential equation. Mathematically this is expressed by the fact that the general solution of our differential equation contains two constants of integration, at first undetermined, whose values we find by means of the initial conditions. This fact we shall prove immediately.

We can easily state such a solution directly. If we put $\omega = \sqrt{(k/m)}$, we may at once verify by differentiation that our differential equation is satisfied by all the functions

$$x(t) = c_1 \cos \omega t + c_2 \sin \omega t,$$

where c_1 and c_2 denote constants chosen arbitrarily. On p. 297 we shall see that there are no other solutions of our differential equation and hence that every such motion under the influence of an elastic force is given by the above expression. This expression can easily be put in the form

$$x(t) = a \sin \omega (t - \delta) = -a \sin \omega \delta \cos \omega t + a \cos \omega \delta \sin \omega t;$$

we need only write $-a \sin \omega \delta = c_1$ and $a \cos \omega \delta = c_2$, thus introducing instead of c_1 and c_2 the new constants a and δ. Motions of this type are said to be *sinusoidal* or *simple harmonic*. They are periodic; any state (i.e. position $x(t)$ and velocity $\dot{x}(t)$) is repeated after the time $T = 2\pi/\omega$, which is called the *period*, since the functions $\sin \omega t$ and $\cos \omega t$ have the period T. The number a is called the *maximum displacement* or *amplitude* of the oscillation. The number $1/T = \omega/2\pi$ is called the *frequency* of the oscillation; it measures the number of oscillations per unit time. We shall return to the theory of oscillations in Chap. XI (p. 501).

4. Motion on a Given Curve.

Finally, we shall discuss the most general form of the problem stated above, namely, the problem of motion along a given curve under an arbitrary pre-assigned force $mf(s)$.

The point in question here is the determination of the function $s(t)$ as a function of t by means of the differential equation

$$\ddot{s} = f(s),$$

where $f(s)$ is a given function. This differential equation in s can be solved completely by the following device.

We begin by considering any primitive function $F(s)$ of $f(s)$, so that

$F'(s) = f(s)$, and multiply both sides of the equation $\ddot{s} = f(s) = F'(s)$ by $\dot{s}$. We can then write the left-hand side in the form $\dfrac{d}{dt}\left(\dfrac{1}{2}\dot{s}^2\right)$, as we see at once by differentiating the expression $\dot{s}^2$; the right side $F'(s)\dot{s}$, however, is the derivative of $F(s)$ with respect to the time t, if in $F(s)$ we regard the quantity s as a function of t. Hence we immediately have

$$\frac{d}{dt}\left(\frac{1}{2}\dot{s}^2\right) = \frac{d}{dt}F(s),$$

or by integration

$$\frac{1}{2}\dot{s}^2 = F(s) + c,$$

where c denotes a constant yet to be determined.

Let us write this equation in the form $\dfrac{ds}{dt} = \sqrt{2(F(s) + c)}$. We see that from this we cannot immediately find s as a function of t by integration. But we arrive at a solution of the problem if we at first content ourselves with finding the inverse function $t(s)$, that is, the time taken by the particle to reach a definite position s. For this we have the equation

$$\frac{dt}{ds} = \frac{1}{\sqrt{2\{F(s) + c\}}};$$

thus the derivative of the function $t(s)$ is known, and we have

$$t = \int \frac{ds}{\sqrt{2\{F(s) + c\}}} + c_1,$$

where c_1 is another constant of integration. As soon as we have performed this last integration we have solved the problem, for while we have not determined the position s as a function of t, we have inversely found the time t as a function of the position s. The fact that the two constants of integration c and c_1 are still available enables us to make the general solution fit special initial conditions.

In the above example of elastic motion we have to identify x with s; we have $f(s) = -\omega^2 s$ and correspondingly, say, $F(s) = -\frac{1}{2}\omega^2 s^2$. We therefore obtain

$$\frac{dt}{ds} = \frac{1}{\sqrt{(2c - \omega^2 s^2)}},$$

and further

$$t = \int \frac{ds}{\sqrt{(2c - \omega^2 s^2)}} + c_1.$$

This integral, however, can easily be evaluated by introducing $\omega s/\sqrt{2c}$ as a new variable; we thus obtain

$$t = \frac{1}{\omega}\arcsin\frac{\omega s}{\sqrt{2c}} + c_1,$$

or, forming the inverse function,

$$s = \frac{\sqrt{2c}}{\omega} \sin\omega(t - c_1).$$

We are thus led to exactly the same statement of the solution as before.

From this example we also see what the constants of integration mean and how they are to be determined. If, for example, we require that at the time $t = 0$ the particle shall be at the point $s = 0$ and at that instant shall have the velocity $\dot{s}(0) = 1$, we obtain the two equations

$$0 = \frac{\sqrt{2c}}{\omega} \sin\omega c_1, \quad 1 = \sqrt{2c} \cos\omega c_1,$$

from which we find that the constants have the values $c_1 = 0$, $c = \frac{1}{2}$. The constants of integration c and c_1 can be determined in exactly the same way when the initial position s_0 and the initial velocity $\dot{s}_0$ (at time $t = 0$) are prescribed arbitrarily.

EXAMPLES

1. A point A moves with constant velocity 1 on a circle with radius r and centre the origin. The point A is connected to a point B by a line of constant length $l(>r)$; B is constrained to move on the x-axis (cf. the crank, connecting-rod, and piston of a steam engine). Calculate the velocity and acceleration of B as functions of the time.

2. A particle starts from the origin with velocity 4, and under the influence of gravity slides down a straight wire until it reaches the vertical line $x = 2$. What must the slope of the path be in order that the point may reach the vertical line in the shortest time?

3. A particle moves in a straight line subject to a resistance producing the retardation ku^3, where u is the velocity and k a constant. Find expressions for the velocity (u) and the time (t) in terms of s, the distance from the initial position, and v_0, the initial velocity.

4. A particle of unit mass moves along the x-axis and is acted upon by a force $f(x) = -\sin x$.

(a) Determine the motion of the point if at time $t = 0$ it is at the point $x = 0$ and has velocity $v_0 = 2$. Show that as $t \to \infty$ the particle approaches a limiting position, and find this limiting position.

(b) If the conditions are the same, except that v_0 may have any value, show that if $v_0 > 2$ the point moves to an infinite distance as $t \to \infty$, and that if $v_0 < 2$ the point oscillates about the origin.

5. Choose axes with their origin at the centre of the earth, whose radius we shall denote by R. According to Newton's law of gravitation, a particle of unit mass lying on the y-axis is attracted by the earth with a force $-\frac{\mu M}{y^2}$, where μ is the "gravitational constant" and M is the mass of the earth.

(a) Calculate the motion of the particle after it is released at the point $y_0 \, (> R)$; that is, if at time $t = 0$ it is at the point $y = y_0$ and has the velocity $v_0 = 0$.

(b) Find the velocity with which the particle in (a) strikes the earth.

(c) Using the result of (b), calculate the velocity of a particle falling to the earth from infinity.†

6.* A particle of mass m moves along the ellipse $r = k/(1 - e \cos \theta)$. The force on the particle is cm/r^2 directed towards the origin. Describe the motion of the particle, find its period, and show that the radius vector to the particle sweeps out equal areas in equal times.

5. Further Applications: Particle Sliding Down a Curve

1. General Remarks.

The case of a particle sliding along a frictionless curve under the influence of gravity can be treated very simply by the method just described. We shall first discuss this motion in general, and then with special reference to the cases of the ordinary pendulum and the cycloidal pendulum. We choose axes in such a way that the y-axis points vertically upwards, that is, opposite to the direction of the force of gravity, and consider the curve as given in terms of a parameter θ by the parametric equations $x = \varphi(\theta) = x(\theta)$, $y = \psi(\theta) = y(\theta)$. A portion of the curve, for which the motion will be studied, is shown in fig. 13.

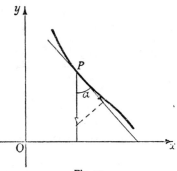

Fig. 13

At every point of the curve the force of gravity acts downwards (that is, in the direction of decreasing y) on the particle with magnitude mg. If we denote the angle between the negative y-axis and the tangent to the curve by α, according to the hypothesis stated on p. 293 the force acting along the direction of the curve is

$$mg \cos \alpha = -mg \frac{y'}{\sqrt{(x'^2 + y'^2)}},$$

where

$$x' = \frac{d\varphi}{d\theta} = \varphi'(\theta), \quad y' = \frac{d\psi}{d\theta} = \psi'(\theta).$$

(Note that here the dash denotes the derivative with respect to θ, and not with respect to x.) If in particular we introduce the length of arc s

---

† This is the same as the least velocity with which a projectile would have to be fired in order that it should leave the earth and never return.

as parameter in place of θ, we obtain the expression $-mg\dfrac{dy}{ds}$ for the force along the curve. By Newton's law, therefore, the function $s(t)$ satisfies the differential equation

$$\ddot{s} = -g\frac{dy}{ds}.$$

The right-hand side of this equation is a known function of s, since we know the curve and must therefore regard the quantities x and y as known functions of s.

As in the last section, we multiply both sides of this equation by $\dot{s}$. The left-hand side then becomes the derivative of $\dfrac{1}{2}\dot{s}^2$ with respect to t. If in the function $y(s)$ we regard s as a function of t, the right-hand side of our equation is the derivative of $-gy$ with respect to t. On integrating. we therefore have

$$\frac{1}{2}\dot{s}^2 = -gy + c,$$

where c is a constant of integration. In order to fix the meaning of this constant, we suppose that at the time $t = 0$ our particle is at the point of the curve for which the value of the parameter is θ_0 and the co-ordinates are $x_0 = \varphi(\theta_0)$, $y_0 = \psi(\theta_0)$, and that at this instant its velocity is zero, that is, $\dot{s}(0) = 0$. Then putting $t = 0$ we immediately have $-gy_0 + c = 0$, so that

$$\frac{1}{2}\dot{s}^2 = -g(y - y_0).$$

Now instead of regarding s as a function of t we shall consider the inverse function $t(s)$. For this we at once obtain

$$\frac{dt}{ds} = \pm \frac{1}{\sqrt{\{2g(y_0 - y)\}}},$$

which is equivalent to

$$t = c_1 \pm \int \frac{ds}{\sqrt{\{2g(y_0 - y)\}}},$$

where c_1 is a new constant of integration. As regards the sign of the square root, which is the same as the sign of $\dot{s}$, we notice that if the particle moves along an arc which is lower than y_c everywhere except at the ends, the sign cannot change. For the sign of $\dot{s}$ can change only where $\dot{s} = 0$, that is, where $y - y_0 = 0$. The integrand on the right is known in terms of the parameter θ, since the curve is known. Introducing θ as independent variable, we obtain

$$t = c_1 \pm \int \frac{ds}{d\theta}\,\frac{d\theta}{\sqrt{\{2g(y_0 - y)\}}} = c_1 \pm \int \sqrt{\left(\frac{x'^2 + y'^2}{2g(y_0 - y)}\right)}\,d\theta,$$

where the functions $x' = \varphi'(\theta)$, $y' = \psi'(\theta)$, $y = \psi(\theta)$ are known. In order to determine the constant of integration c_1 we note that for $t = 0$ the value of the parameter must be θ_0. This immediately gives us our solution in the form

$$t = \pm \int_{\theta_0}^{\theta} \sqrt{\left(\frac{x'^2 + y'^2}{2g(y_0 - y)}\right)}\, d\theta.$$

When integrated this equation represents the time taken by the particle to move from the parameter value θ_0 to the parameter value θ. The inverse function $\theta(t)$ of this function $t(\theta)$ enables us to describe the motion completely; for at each instant t we can determine the point $x = \varphi\{\theta(t)\}$, $y = \psi\{\theta(t)\}$ which the particle is then passing.

2. Discussion of the Motion.

From the equations just found, without an explicit expression for the result of the integration we can deduce the general nature of the motion by simple intuitive reasoning. We suppose that our curve is of the type shown in fig. 14, that is, that it consists of an arc convex downwards; we take s as increasing from left to right. If we initially release the particle at the point A with co-ordinates $x = x_0$, $y = y_0$, corresponding to $\theta = \theta_0$, the velocity increases, for the acceleration $\ddot{s}$ is positive. The

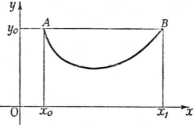

Fig. 14

particle travels from A to the lowest point with ever-increasing velocity. After the lowest point is passed, however, the acceleration is negative, since the right-hand side $-g\dfrac{dy}{ds}$ of the equation of motion is negative. The velocity therefore decreases. From the equation $\dot{s}^2 = -2g(y - y_0)$ we see at once that the velocity reaches the value 0 when the particle reaches the point B whose height is the same as that of the initial position A. Since the acceleration is still negative, the motion of the particle must be reversed at this point, so that the particle will swing back to the point A; this action will repeat itself indefinitely. (The reader will recall that friction has been disregarded.) In this oscillatory motion the time which the point takes to return from B to A must clearly be the same as the time taken to move from A to B. If we denote the time required for a complete journey from A to B and back again by T, the motion will obviously be periodic with period T. If θ_0 and θ_1 are the values of the parameter corresponding to the points A and B respectively, the half-period is given by the expression

$$\frac{T}{2} = \frac{1}{\sqrt{2g}}\left|\int_{\theta_0}^{\theta_1} \sqrt{\left(\frac{x'^2 + y'^2}{y_0 - y}\right)}\, d\theta\right|$$

$$= \frac{1}{\sqrt{2g}}\left|\int_{\theta_0}^{\theta_1} \sqrt{\left(\frac{\varphi'^2(\theta) + \psi'^2(\theta)}{\psi(\theta_0) - \psi(\theta)}\right)}\, d\theta\right|.$$

If θ_2 is the value of the parameter corresponding to the lowest point of the curve, the time which the particle takes to fall from A to this lowest point is

$$\frac{1}{\sqrt{2g}} \left| \int_{\theta_0}^{\theta_2} \sqrt{\left(\frac{x'^2 + y'^2}{y_0 - y}\right)} \, d\theta \right|.$$

3. The Ordinary Pendulum.

The simplest example is given by the so-called simple pendulum. Here the curve under consideration is a circle of radius l:

$$x = l \sin \theta, \quad y = -l \cos \theta,$$

where the angle θ is measured in the positive sense from the position of rest. From the general expression above we at once obtain

$$T = \sqrt{\frac{2l}{g}} \int_{-\alpha}^{\alpha} \frac{d\theta}{\sqrt{(\cos\theta - \cos\alpha)}} = \sqrt{\frac{l}{g}} \int_{-\alpha}^{\alpha} \frac{d\theta}{\sqrt{\left(\sin^2\frac{\alpha}{2} - \sin^2\frac{\theta}{2}\right)}},$$

where $\alpha \, (0 < \alpha < \pi)$ denotes the amplitude of oscillation of the pendulum, that is, the angular position from which the particle is released at time $t = 0$ with velocity 0. By the substitution

$$u = \frac{\sin(\theta/2)}{\sin(\alpha/2)}, \quad \frac{du}{d\theta} = \frac{\cos(\theta/2)}{2\sin(\alpha/2)}$$

our expression for the period of oscillation of the pendulum becomes

$$T = 2\sqrt{\frac{l}{g}} \int_{-1}^{1} \frac{du}{\sqrt{(1 - u^2)(1 - u^2 \sin^2(\alpha/2))}}.$$

We have therefore expressed the period of oscillation of the pendulum by an elliptic integral.

If we assume that the amplitude of the oscillation is small, so that we may with sufficient accuracy replace the second factor under the square root sign by 1, we obtain the expression

$$2\sqrt{\frac{l}{g}} \int_{-1}^{1} \frac{du}{\sqrt{(1 - u^2)}}$$

as an approximation for the period of oscillation. We can evaluate this last integral by formula 13 in our table of integrals (p. 206), and obtain the expression $2\pi \sqrt{\dfrac{l}{g}}$ as an approximate value for T.

4. The Cycloidal Pendulum.

The fact that the period of oscillation of the ordinary pendulum is not strictly independent of the amplitude of oscillation caused Christian Huygens, in his prolonged efforts to construct accurate clocks, to seek for a curve such that the period of oscillation is strictly independent of

the particular position on the curve at which the oscillating particle begins
its motion.* Huygens recognized that the cycloid is such a curve.

In order that a particle may actually be able to oscillate on a cycloid
the cusps of the cycloid must point in the direction opposite to that of
the force of gravity; that is, we must rotate the cycloid considered previously (p. 261) about the x-axis (cf. fig. 15). We therefore write the equations of the cycloid in the form

$$x = a(\theta - \sin\theta),$$
$$y = a(1 + \cos\theta),$$

which also involves a translation of the curve through a distance $2a$ in
the positive y-direction. The time which the particle takes to travel from
a point at the height

$$y_0 = a(1 + \cos\alpha) \quad (0 < \alpha < \pi)$$

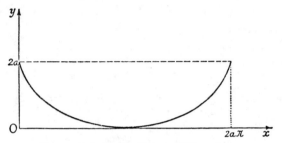

Fig. 15.—Path described by a cycloidal pendulum

down to the lowest point, by the formula worked out on p. 301, is

$$\frac{T}{4} = \sqrt{\frac{1}{2g}} \int_{a}^{\pi} \sqrt{\left(\frac{x'^2 + y'^2}{y_0 - y}\right)} \, d\theta = \sqrt{\frac{a}{g}} \int_{a}^{\pi} \sqrt{\left(\frac{1 - \cos\theta}{\cos\alpha - \cos\theta}\right)} \, d\theta.$$

We now use the equation

$$\cos\alpha - \cos\theta = 2\left(\cos^2\frac{\alpha}{2} - \cos^2\frac{\theta}{2}\right);$$

this gives

$$\frac{T}{4} = \sqrt{\frac{a}{g}} \int_{a}^{\pi} \frac{\sin\frac{\theta}{2}}{\sqrt{\left(\cos^2\frac{\alpha}{2} - \cos^2\frac{\theta}{2}\right)}} \, d\theta.$$

We then work out the definite integral, making use of the substitution

$$\cos\frac{\theta}{2} = u\cos\frac{\alpha}{2}, \quad \sin\frac{\theta}{2}\,d\theta = -2\cos\frac{\alpha}{2}\,du.$$

---

* The oscillations are then said to be *isochronous.*

This gives

$$\int \frac{\sin\frac{\theta}{2}}{\sqrt{\left(\cos^2\frac{\alpha}{2} - \cos^2\frac{\theta}{2}\right)}}\, d\theta = -2\int \frac{du}{\sqrt{(1-u^2)}} = -2 \arcsin u,$$

and we therefore obtain

$$T = -8\sqrt{\frac{a}{g}} \arcsin \left. \frac{\cos\frac{\theta}{2}}{\cos\frac{\alpha}{2}} \right|_a^\pi = 4\pi\sqrt{\frac{a}{g}}.$$

The period of oscillation T, therefore, is actually independent of the amplitude α.

6. Work

1. General Remarks.

The concept of *work* throws new light on the considerations of the last section and on many other questions of mechanics and physics.

Let us again think of the particle as moving on a curve under the influence of a force acting along the curve, and let us suppose that its position is specified by the length of arc measured from any fixed initial point. The force itself will then, as a rule, be a function of s. We assume that it is a continuous function $f(s)$ of the length of arc. This function will have positive values where the direction of the force is the same as the direction of increasing values of s, and negative values where the direction of the force is opposite to that of increasing values of s.

If the magnitude of the force is constant along the path, by the *work done by the force* we mean the product of the force by the distance $(s_1 - s_0)$ traversed, where s_1 denotes the final point and s_0 the initial point of the motion. If the force is not constant we define the work by means of a limiting process. We subdivide the interval from s_0 to s_1 into n equal or unequal sub-intervals and notice that if the sub-intervals are small, the force in each one is nearly constant; if σ_ν is a point chosen arbitrarily in the ν-th sub-interval, then throughout this sub-interval the force will be approximately $f(\sigma_\nu)$. If the force throughout the ν-th sub-interval were exactly $f(\sigma_\nu)$, the work done by our force would be exactly

$$\sum_{\nu=1}^{n} f(\sigma_\nu)\Delta s_\nu,$$

where Δs_ν as usual denotes the length of the ν-th sub-interval. If we now pass to the limit, letting n increase beyond all bounds while the length of the longest sub-interval tends to zero, then by the definition of an integral our sum will tend to

$$W = \int_{s_0}^{s_1} f(s)\, ds,$$

which we naturally call the work done by the force.

If the direction of the force and that of the motion are the same, the work done by the force is positive; we then say that *the force does work*. On the other hand, if the direction of the force and that of the motion are opposed, the work done by the force is negative; we then say that *work is done against the force*.*

If we regard the co-ordinate of position s as a function of the time t, so that the force $f(s) = p$ is also a function of t, then in a plane with rectangular co-ordinates s and p we can plot the point with co-ordinates $s = s(t)$, $p = p(t)$ as a function of the time. This point will describe a curve, which may be called the work diagram of the motion. If we are dealing with a periodic motion, as in the case of any machine, then after a certain time T (one period) the moving point $s(t)$, $p(t)$ will return to the same point; that is, the work diagram will be a closed curve. In this case the curve may consist simply of one and the same arc, traversed first forwards and then backwards; this happens, for instance, in elastic oscillations. But it is also possible for the curve to be a more general closed curve, enclosing an area; this is the case e.g. with machines in which the pressure on a piston is not the same during the forward stroke as during the backward stroke. The work done in one cycle, that is, in time T, will then be given simply by the negative area of the work diagram, or in other words, by the integral

$$\int_{t_0}^{t_0 + T} p(t) \frac{ds}{dt} dt,$$

where the interval of time from t_0 to $t_0 + T$ represents exactly one period of the motion. If the boundary of the area is positively traversed the work done is negative, if negatively traversed the work done is positive. If the curve consists of several loops, some traversed positively and some traversed negatively, the work done is given by the sum of the areas of the loops, each with its sign changed.

These considerations are illustrated in practice by the *indicator diagram* of a steam engine. By a suitably-designed mechanical device a pencil is made to move over a sheet of paper; the horizontal motion of the pencil relative to the paper is proportional to the distance s of the piston from its extreme position, while the vertical motion is proportional to the steam pressure, and hence proportional to the total force p of the steam on the piston. The piston therefore describes the work diagram for the engine on a known scale. The area of this diagram is measured (usually by means of a planimeter) and the work done by the steam on the piston is thus found. Here we also see that our convention for the sign of an area, as discussed in § 2, No. 1 of this chapter (p. 271), is not of exclusively theoretical interest. For it sometimes happens when an engine is running light, that the highly expanded steam at the end of the stroke has a pres-

---

* Note that here we must carefully distinguish the force of which we are speaking. For example, in lifting a weight the work done by the force of gravity is negative: work is done against gravity. But from the point of view of the person doing the lifting the work done is positive, for the person must exert a force opposed to gravity.

sure lower than that required to expel it on the return stroke; on the diagram this is shown by a positively traversed loop; the engine itself is drawing energy from the flywheel instead of furnishing energy.

2. The Mutual Attraction of Two Masses.

Let us suppose that a particle attracts another particle according to Newton's law of attraction; as a first example we shall consider the work done by the force of attraction as the second particle moves along the line joining the two particles. According to Newton's law of gravitation, the attracting force is inversely proportional to the square of the distance. If we imagine the first particle at rest at the origin and the second particle at the distance r from the origin, the attracting force is given by the expression

$$f(r) = -\mu \frac{1}{r^2},$$

where μ is a positive constant. The work done by this force when the particle moves from the distance r to the distance $r_1(<r)$ is therefore positive and equal to the integral

$$-\mu \int_r^{r_1} \frac{ds}{s^2} = \mu \left(\frac{1}{r_1} - \frac{1}{r} \right).$$

If, by means of an opposing force, the particle is moved farther away from the origin, going from the distance r to a distance $r_1 > r$, the work done by the force of attraction will, of course, still be given by this integral (now negative). The work done by the opposing force has the same numerical value, but the opposite sign; it is therefore equal to $\mu \left(\frac{1}{r} - \frac{1}{r_1} \right)$. If we think of the final position as being chosen farther and farther away, this approaches the limiting value $\frac{\mu}{r}$, which we may call the work which must be done against the force of attraction in order to move the particle from the distance r to "infinity". This important expression is called the *mutual potential* of the two particles. Here, therefore, the potential is defined as the work required to separate two attracting masses completely; for example, the work required in order to tear an electron completely away from an atom (ionization potential).

3. The Stretching of a Spring.

As a second example we consider the work done in stretching a spring. As is usual in the theory of elasticity, we assume (cf. p. 295 also) that the force needed to stretch the spring is proportional to x, the increase in the length of the spring, that is, $p = kx$, where k is a constant. The work which must be done in order to stretch the spring from the unstressed position $x = 0$ to the final position $x = x_1$ is therefore given by the integral

$$\int_0^{x_1} kx \, dx = \tfrac{1}{2} k x_1^2.$$

4. The Charging of a Condenser.

The concept of work in other branches of physics can be treated in a similar way. For example, we may consider the charging of a condenser. If we denote the quantity of electricity in the condenser by Q, its capacity by C, and the difference of potential (voltage) across the condenser by V, then we know from physics that $Q = CV$. Moreover, the work done in moving a charge Q through a difference of potential V is equal to QV. Since in the charging of the condenser the difference of potential V is not constant but increases with Q, we perform a passage to the limit exactly analogous to that on p. 304 and as the expression for the work done in charging the condenser we obtain

$$\int_0^{Q_1} V \, dQ = \frac{1}{C} \int_0^{Q_1} Q \, dQ = \frac{1}{2} \frac{Q_1^2}{C} = \frac{1}{2} Q_1 V_1,$$

where Q_1 is the total quantity of electricity passed into the condenser and V_1 is the difference of potential across the condenser at the end of the charging process.

Appendix to Chapter V

1. Properties of the Evolute

The parametric equations

$$\xi = x - \rho \frac{\dot{y}}{\sqrt{(\dot{x}^2 + \dot{y}^2)}}, \qquad \eta = y + \rho \frac{\dot{x}}{\sqrt{(\dot{x}^2 + \dot{y}^2)}},$$

for the evolute of a given curve $x = x(t)$, $y = y(t)$ (cf. p. 283) enable us to deduce some interesting geometrical relations between it and the given curve. For convenience we use the length of arc s as parameter, so that

$$\dot{x}^2 + \dot{y}^2 = 1 \quad \text{and} \quad \dot{x}\ddot{x} + \dot{y}\ddot{y} = 0,$$

$$\frac{1}{\rho} = k = \frac{\ddot{y}}{\dot{x}} = -\frac{\ddot{x}}{\dot{y}},$$

or $\qquad \rho\ddot{y} = \dot{x} \quad \text{and} \quad \rho\ddot{x} = -\dot{y}.$

We thus have $\qquad \xi = x - \rho\dot{y}, \quad \eta = y + \rho\dot{x};$

on differentiation these give

$$\dot{\xi} = \dot{x} - \rho\ddot{y} - \dot{\rho}\dot{y} = -\dot{\rho}\dot{y}, \quad \dot{\eta} = \dot{y} + \rho\ddot{x} + \dot{\rho}\dot{x} = \dot{\rho}\dot{x},$$

and therefore $\qquad \dot{\xi}\dot{x} + \dot{\eta}\dot{y} = 0.$

Since the direction cosines of the normal to the curve are given by $-\dot{y}$ and $\dot{x}$, it follows that *the normal to the curve is tangent to the evolute at the centre of curvature*; or, the tangents to the evolute are the normals of the given curve; or, *the evolute is the " envelope " of the normals* (cf. fig. 16).

If further we denote the length of arc of the evolute, measured from an arbitrary fixed point, by σ, we have

$$\left(\frac{d\sigma}{ds}\right)^2 = \dot{\sigma}^2 = \dot{\xi}^2 + \dot{\eta}^2.$$

From the above formulæ, since $\dot{x}^2 + \dot{y}^2 = 1$, we obtain

$$\dot{\sigma}^2 = \dot{\rho}^2,$$

so that if we choose the direction in which σ is measured in a suitable way,

$$\dot{\sigma} = \dot{\rho},$$

provided that $\sigma \neq 0$,

or on integration

$$\sigma_1 - \sigma_0 = \rho_1 - \rho_0.$$

That is, *the length of arc of the evolute between two points is equal to the difference of the corresponding radii of curvature, provided that $\dot{\rho}$ remains different from zero for the arc under consideration.*

This last condition is not superfluous. For if $\dot{\rho}$ changes sign, then from the formula $\dot{\sigma} = \dot{\rho}$ we see that on passing the corresponding point of the evolute the length of arc σ has a maximum or minimum; that is, on passing this point we do not simply continue to reckon σ onward, but must reverse the sense in which σ is measured. If we wish to avoid this, on passing such a point we must change the sign in the above formula, i.e. put $\dot{\sigma} = -\dot{\rho}$.

It may also be noted that the centres of curvature which correspond to maxima or minima of the radius of curvature are *cusps of the evolute*. (The proof will not be given here.)

The geometrical relationship which we have just found can be expressed in yet another way. If we imagine a flexible inextensible thread laid along an arc of the evolute and stretched so that a part of it extends away from the curve tangentially to it, and if in addition the end-point Q of this thread lies on the original curve C, then as we unwind the thread the point Q will describe the curve C. This accounts for the name evolute

(*evolvere*, to unwind).   The curve C is called an *involute* of the
evolute E.   On the other hand we may start with an arbitrary
curve E and construct its involute C by this unwinding process.
We then see that E conversely is the evolute of C.

To prove this we consider the curve E, which is now the
given curve, as given in the form $\xi = \xi(\sigma)$, $\eta = \eta(\sigma)$, where the
current rectangular co-ordinates are denoted by ξ and η and
the parameter σ is the length of arc.   The winding is done as
indicated in fig. 17; when the thread is completely wound on to
the evolute E, its end Q coincides with the point A of E corre-

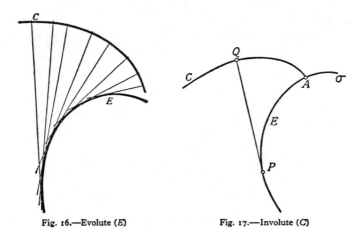

Fig. 16.—Evolute (*E*)   Fig. 17.—Involute (*C*)

sponding to the length of arc a.   If the thread is now unwound
until it is tangent to the evolute at the point P, corresponding
to the length of arc $\sigma \leqq a$, the length of the segment PQ will be
$(a - \sigma)$ and its direction cosines will be $\dot{\xi}$ and $\dot{\eta}$, where the dot
denotes differentiation with respect to σ.   Thus for the co-
ordinates x, y of the point Q we obtain the expressions

$$x = \xi + (a - \sigma)\dot{\xi}, \quad y = \eta + (a - \sigma)\dot{\eta},$$

which give the equations for the involute described by the point
Q in terms of the parameter σ.   By differentiation with respect
to σ it follows that

$$\dot{x} = \dot{\xi} - \dot{\xi} + (a - \sigma)\ddot{\xi} = (a - \sigma)\ddot{\xi},$$
$$\dot{y} = \dot{\eta} - \dot{\eta} + (a - \sigma)\ddot{\eta} = (a - \sigma)\ddot{\eta}.$$

And since $\dot{\xi}\ddot{\xi} + \dot{\eta}\ddot{\eta} = 0$, we at once find that

$$\dot{\xi}\dot{x} + \dot{\eta}\dot{y} = 0,$$

which shows that the line PQ is normal to the involute C. We can therefore state that the normals to the curve C are tangent to the curve E. But this is the characteristic property of E, the evolute of C. Hence *every curve is the evolute of all its involutes*.

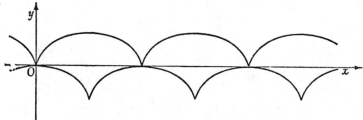

Fig. 18.—The cycloid as evolute and involute

As a particular case we consider the evolute of the cycloid $x = t - \sin t$, $y = 1 - \cos t$. By pp. 281, 283,

$$\xi = x - \dot{y}\,\frac{\dot{x}^2 + \dot{y}^2}{\dot{x}\ddot{y} - \dot{y}\ddot{x}}, \quad \eta = y + \dot{x}\,\frac{\dot{x}^2 + \dot{y}^2}{\dot{x}\ddot{y} - \dot{y}\ddot{x}};$$

we therefore obtain the evolute in the form $\xi = t + \sin t$, $\eta = -1 + \cos t$. If we put $t = \tau + \pi$, then $\xi - \pi = \tau - \sin \tau$ and $\eta + 2 = 1 - \cos \tau$, and these equations show that the evolute is itself a cycloid which is similar to the original curve, and can be obtained from it by translation, as indicated in fig. 18.

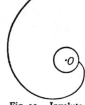

Fig. 19.—Involute of the circle

As a further example we shall work out the equation for the involute of the circle. We begin with the circle $\xi = \cos t$, $\eta = \sin t$ and unwind the tangent, as indicated in fig. 19. The involute of the circle is then given in the form

$$x = \cos t + t \sin t, \quad y = -\sin t + t \cos t.$$

Finally, we shall determine the evolute of the ellipse $x = a \cos t$, $y = b \sin t$. We at once have

$$\xi = x - \dot{y}\,\frac{\dot{x}^2 + \dot{y}^2}{\dot{x}\ddot{y} - \dot{y}\ddot{x}} = \frac{a^2 - b^2}{a}\cos^3 t$$

and

$$\eta = y + \dot{x}\,\frac{\dot{x}^2 + \dot{y}^2}{\dot{x}\ddot{y} - \dot{y}\ddot{x}} = -\frac{a^2 - b^2}{b}\sin^3 t,$$

which is a parametric representation of the evolute. If from these equations we eliminate t in the usual way, we obtain the equation of the evolute in non-parametric form:

$$(a\xi)^{2/3} + (b\eta)^{2/3} = (a^2 - b^2)^{2/3}.$$

This curve is called an *astroid*. Its graph is given in fig. 20. By means of the parametric equations we may readily convince ourselves that the centres of curvature corresponding to the vertices of the ellipse are actually the cusps of the astroid.

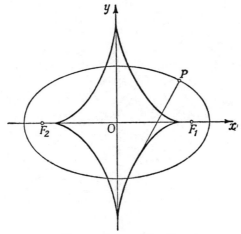

Fig. 20.—Evolute of the ellipse

1. Show that the evolute of an epicycloid (Ex. 2, p. 267) is another epicycloid similar to the first, which can be obtained from the first by rotation and contraction.

2. Show that the evolute of a hypocycloid (Ex. 4, p. 267) is another hypocycloid, which can be obtained from the first by rotation and expansion.

2. Areas Bounded by Closed Curves

We saw in § 2 (p. 271) that the area bounded by a closed curve $x = x(t)$, $y = y(t)$, $t_0 \leqq t \leqq t_1$, which nowhere intersects itself (a so-called simple closed curve) is given by the integral

$$-\int_{t_0}^{t_1} y(t) \dot{x}(t) \, dt,$$

where the value obtained is positive or negative according as the sense in which the boundary is described is positive or negative. We shall now extend this result to more general curves. Suppose that the curve C, given by the equation $x = x(t)$, $y = y(t)$, intersects itself in a finite number of points, thus dividing the plane into a finite number of portions $R_1, R_2, \ldots$. Suppose further that the derivatives are continuous, except perhaps for a finite number of jump discontinuities, and that $\dot{x}^2 + \dot{y}^2 \neq 0$, except perhaps at a finite number of values of t which may correspond to corners. Finally, it is assumed that the curve has a finite number of lines of support (p. 270).

To each region R_i we then assign an *index* μ_i defined in the following way: we choose an arbitrary point Q in R_i, not lying on any line of support, and erect the line extending from Q

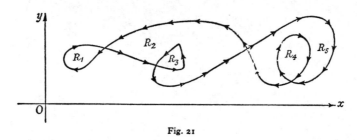

Fig. 21

upwards, in the direction of the positive y-axis. We count the number of times the curve C crosses the half-line from right to left, and subtract the number of times the curve C crosses from left to right; the difference is the index μ_i. For example, the interior of the curve illustrated in fig. 6 (p. 269) has the index $\mu = +1$; and in fig. 21 the regions $R_1, \ldots, R_5$ have the indices $\mu_1 = -1$, $\mu_2 = +1$, $\mu_3 = +2$, $\mu_4 = -2$, $\mu_5 = -1$. This number μ_i actually does depend on the region R_i and not on the particular point Q chosen in R_i, as we readily see in the following manner. We choose any other point Q' in R_i, not on a line of support, and join Q to Q' by a broken line lying entirely in the region R_i. As we proceed along this broken line from Q to Q' the number of right-to-left crossings minus the number of left-to-right crossings is constant; for between lines of support the number of crossings of either type is unchanged, while on crossing a line of support either the number of crossings of both types increases by one, or else both the numbers decrease by one; in either case, the difference is unaltered. In the case where the line of support meets the curve at several different points, say $A, B, \ldots, H$, we consider it as several different lines of support, $FA, FB, \ldots, FH$, where F is the point of the x-axis vertically below all the points $A, B, \ldots, H$. Our argument then applies to each of these lines. Hence the number μ_i has the same value whether we use Q or Q' in determining it.

In particular, if our curve does not intersect itself, the interior of the curve consists of a single region R whose index is $+1$ or -1 according as the sense in which the boundary is described is positive or negative. To see this we draw any vertical line (not a line of support) intersecting the curve; on this line we find the highest point of intersection (P) with the curve, and in R we choose a point Q below P and so near it that no point of intersection lies between P and Q. Then above Q there lies one crossing of the curve, which if the curve is traversed positively must be a right-to-left crossing, so that $\mu = +1$; otherwise $\mu = -1$. As we have just seen, this same value of μ holds for every other point of R. For such a curve, and in fact for all closed curves, one of the regions, the "outside" of the curve, extends unboundedly in all directions; we see immediately that this region has index 0, and henceforth neglect it.

Our theorem about the area is now as follows: the value of the integral $-\int_{t_0}^{t_1} y\dot{x}\,dt$ is equal to the sum of the absolute areas of the regions R_i, each area R_i being counted μ_i times; in symbols

$$-\int_{t_0}^{t_1} y\dot{x}\,dt = \Sigma\mu_i \,|\,\text{area } R_i\,|.$$

The proof is simple. We assume, as we are entitled to do, that the whole of the curve lies above the x-axis (cf. footnote, p. 271). The lines of support cut R_i into a finite number of portions; let r be one of these portions. Then on taking the integral $-\int y\dot{x}\,dt$ for each single-valued branch of the curve, we find that the absolute area of r is counted $+1$ times for each right-to-left branch over r and -1 times for each left-to-right branch over r, in all μ_i times. The same is true for every other portion of R_i; hence R_i is counted μ_i times. Thus the integral round the complete curve has the value $\Sigma\mu_i \,|\,\text{area } R_i\,|$, as stated. This formula agrees with what we have found for simple closed curves,

11*   (E 793)

as we recognize from the discussion of the values of μ for such curves.

The definition given for the index μ_i has the disadvantage of being stated in terms of a particular co-ordinate system. As a matter of fact, however, it can be shown that the value of μ_i is independent of the co-ordinate system and depends solely on the curve; but we shall not prove this here.

CHAPTER VI

Taylor's Theorem and the Approximate Expression of Functions by Polynomials

In many respects the rational functions are the simplest functions of analysis. They are formed by a finite number of applications of the rational operations of calculation, while in the last resort the formation of every other function involves a more or less concealed passage to the limit from rational functions. The questions whether and how a given function can be expressed approximately by rational functions, in particular by polynomials, are therefore of great importance both in theory and in practice.

1. THE LOGARITHM AND THE INVERSE TANGENT

1. The Logarithm.

We begin by considering some special cases in which the integration of the geometrical progression leads almost at once to the desired approximations. We first remind the reader of the following fact. For $q \neq 1$ and for positive integers n, we have

$$\frac{1}{1-q} = 1 + q + q^2 + \ldots + q^{n-1} + r_n,$$

where

$$r_n = \frac{q^n}{1-q}.$$

If $|q| < 1$ the remainder r_n tends to 0 as n increases, and we then obtain (pp. 34–35) the *infinite geometric series*

$$1 + q + q^2 + \ldots \text{ with the sum } \frac{1}{1-q}.$$

As our starting-point we take the formula

$$\log(1+x) = \int_0^x \frac{dt}{1+t}$$

and expand the integrand in accordance with the above formula, putting $q = -t$. Then by integration we at once obtain

$$\log(1+x) = x - \frac{x^2}{2} + \frac{x^3}{3} - \frac{x^4}{4} + - \ldots + (-1)^{n-1}\frac{x^n}{n} + R_n,$$

where
$$R_n = \int_0^x r_n \, dt = (-1)^n \int_0^x \frac{t^n \, dt}{1+t}.$$

Hence for any positive integer n we have expressed the function $\log(1+x)$ approximately by a polynomial of the n-th degree, namely,

$$x - \frac{x^2}{2} + \frac{x^3}{3} - + \ldots + (-1)^{n-1}\frac{x^n}{n};$$

at the same time the quantity R_n, the *remainder*, specifies the amount of the *error* made in this approximation.

In order to estimate the accuracy of this approximation, we need only have an estimate for the remainder R_n; and such an estimate is given us immediately by the integral estimates on p. 126. If at first we suppose that $x \geqq 0$, then in the whole interval of integration the integrand is nowhere negative and nowhere exceeds t^n. Consequently

$$|R_n| \leqq \int_0^x t^n \, dt = \frac{x^{n+1}}{n+1},$$

and we therefore see that for every value of x in the interval $0 \leqq x \leqq 1$ this remainder can be made as small as we like by choosing n large enough (cf. p. 32). If, on the other hand, the quantity x is in the interval $-1 < x \leqq 0$, the integrand will not change sign and its absolute value will not exceed $|t|^n/(1+x)$, and we thus obtain the estimate for the remainder

$$|R_n| \leqq \frac{1}{1+x} \int_0^{|x|} t^n \, dt = \frac{|x|^{n+1}}{(1+x)(n+1)}.$$

We see, therefore, that here again the remainder is arbitrarily small when n is sufficiently large. Of course, our estimate has no meaning when we put $x = -1$.

Summing up, we can say that

$$\log(1+x) = x - \frac{x^2}{2} + \frac{x^3}{3} - + \ldots + (-1)^{n-1}\frac{x^n}{n} + R_n,$$

where the remainder R_n tends to zero as n increases, provided that x lies in the interval * $-1 < x \leqq 1$. From the above inequalities we can, in fact, find an estimate for the remainder, independent of x, which is valid for all values of x in the interval $-1+h \leqq x \leqq 1$, where h is a number such that $0 < h \leqq 1$. For then

$$|R_n| \leqq \frac{1}{h}\frac{1}{n+1},$$

and this formula shows us that in the whole interval the function $\log(1+x)$ is expressed approximately by our polynomial of the n-th degree, the error being nowhere greater than $\frac{1}{h}\frac{1}{n+1}$. We leave it to the reader to convince himself that for all values of x for which $|x| > 1$ the remainder not only fails to approach zero, but, in fact, increases numerically beyond all bounds as n increases, so that for such values of x our polynomial does not give us an approximation to the logarithm.

The fact that in the above interval the remainder R_n tends to zero may be expressed by saying that in this interval we have the *infinite series* †

$$\log(1+x) = x - \frac{x^2}{2} + \frac{x^3}{3} - \frac{x^4}{4} + - \ldots$$

for the logarithm. If in this series we insert the particular value $x = 1$, we obtain the remarkable formula

$$\log 2 = 1 - \frac{1}{2} + \frac{1}{3} - \frac{1}{4} + - \ldots$$

This is one of the relations whose discovery made a deep impression on the minds of the first pioneers of the differential and integral calculus.

---

* It is to be noted that this interval is open on the left and closed on the right.

† Infinite series will be considered in detail in Chap. VIII (p. 365).

The above approximation for the logarithm leads us to another formula which is useful for many purposes, particularly in numerical calculations. Provided that $-1 < x < 1$, we have only to write $-x$ in place of x in the above formula to obtain

$$\log(1-x) = -x - \frac{x^2}{2} - \frac{x^3}{3} - \frac{x^4}{4} - \ldots - S_n.$$

Taking n as even and subtracting, we have

$$\frac{1}{2} \log \frac{1+x}{1-x} = x + \frac{x^3}{3} + \frac{x^5}{5} + \ldots + \frac{x^{n-1}}{n-1} + R_n,$$

where $\bar{R}_n$ is given by the expression

$$\bar{R}_n = \frac{1}{2}(R_n + S_n) = \frac{1}{2} \int_0^x t^n \left(\frac{1}{1+t} + \frac{1}{1-t} \right) dt$$

$$= \int_0^x \frac{t^n}{1-t^2} dt.$$

On account of the relation

$$|\bar{R}_n| \leqq \frac{|x^{n+1}|}{n+1} \frac{1}{1-x^2},$$

the remainder tends to zero as n increases, a fact which we again express by writing the expansion as an infinite series:

$$\frac{1}{2} \log \frac{1+x}{1-x} = \operatorname{ar\,tanh} x = x + \frac{x^3}{3} + \frac{x^5}{5} + \frac{x^7}{7} + \ldots,$$

for all values of x such that $|x| < 1$.

An advantage of this formula is that as x traverses the interval from -1 to 1, the expression $\dfrac{1+x}{1-x}$ ranges over all positive numbers. Hence if the value of x is suitably chosen this series enables us to calculate the value of the logarithm of any positive number, with an error not exceeding the above estimate for R_n.

2. The Inverse Tangent.

We can treat the inverse tangent in a similar way if we begin with the formula, true for every positive integer n,

$$\frac{1}{1+t^2} = 1 - t^2 + t^4 - + \ldots + (-1)^{n-1} t^{2n-2} + r_n,$$

where
$$r_n = (-1)^n \frac{t^{2n}}{1+t^2}.$$

By integration, we obtain

$$\text{arc tan} x = x - \frac{x^3}{3} + \frac{x^5}{5} - + \ldots + (-1)^{n-1} \frac{x^{2n-1}}{2n-1} + R_n,$$

$$R_n = (-1)^n \int_0^x \frac{t^{2n}}{1+t^2}\, dt,$$

and we see at once that in the interval $-1 \leq x \leq 1$ the remainder R_n tends to zero as n increases, for by the mean value theorem of the integral calculus

$$|R_n| \leq \int_0^{|x|} t^{2n}\, dt = \frac{|x|^{2n+1}}{2n+1}.$$

From the formula for the remainder we can also show fairly easily that for $|x| > 1$ the absolute value of the remainder increases beyond all bounds as n increases. We have accordingly deduced the infinite series

$$\text{arc tan} x = x - \frac{x^3}{3} + \frac{x^5}{5} - + \ldots ,$$

valid for $|x| \leq 1$. For $x = 1$, since $\text{arc tan} 1 = \frac{\pi}{4}$, we have

$$\frac{\pi}{4} = 1 - \frac{1}{3} + \frac{1}{5} - + \ldots ,$$

as remarkable a formula as that previously found for $\log 2$.

EXAMPLES

1. Prove that $x - \frac{x^2}{2} + \frac{x^3}{3(1+x)} < \log(1+x) < x - \frac{x^2}{2} + \frac{x^3}{3}$ $(x > 0)$.
Hence find $\log \frac{4}{3}$ to 2 places.

2. Calculate $\log \frac{6}{5}$ to 3 places, using the series

$$\log(1+x) = x - \frac{x^2}{2} + \frac{x^3}{3} - \ldots .$$

Prove that the result is accurate to 3 places.

3. How many terms of the series for $\log(1+x)$ must be used in order to obtain $\log(1+x)$ to within 10 per cent if $30 \leq x \leq 31$?

2. TAYLOR'S THEOREM

An approximate representation by rational functions, as in the special cases above, can also be obtained in the case of an arbitrary function $f(x)$, about which we assume only that for all values of the independent variable in an assigned closed interval the function possesses continuous derivatives up to the $(n + 1)$-th order at least. In most of the cases which actually occur the existence and continuity of *all* the derivatives of the function is known to begin with, so that for n we can choose any arbitrary integer.

The approximation formula which we shall now derive was discovered in the early days of the differential and integral calculus by Taylor, a student of Newton's, and is known as Taylor's theorem *.

1. Taylor's Theorem for Polynomials.

In order to get a clear idea of the problem, we shall begin by considering the case where $f(x) = a_0 + a_1 x + a_2 x^2 + \ldots + a_n x^n$ is itself a polynomial of the n-th degree. We can then easily express the coefficients of this polynomial in terms of the derivatives of $f(x)$ at the point $x = 0$. For if we differentiate both sides of the equation once, twice, &c., with respect to x and then put $x = 0$, we at once find that the coefficients are

$$a_0 = f(0), \quad a_1 = f'(0), \quad a_2 = \frac{1}{2!} f''(0), \ldots, \quad a_n = \frac{1}{n!} f^{(n)}(0).$$

Any polynomial $f(x)$ of the n-th degree can therefore be written in the form

$$f(x) = f(0) + x f'(0) + \frac{x^2}{2!} f''(0) + \frac{x^3}{3!} f'''(0) + \ldots + \frac{x^n}{n!} f^{(n)}(0).$$

This formula merely states that the coefficients a_ν can be expressed in terms of the derivatives at $x = 0$ and gives the expressions for them.

We can generalize this " Taylor series " for the polynomial slightly if we replace x by $\xi = x + h$ and consider the function $f(\xi) = f(x + h) = g(h)$ as a function of h, for

---

* A special case of this theorem is often referred to, without historical justification, as *Maclaurin's theorem*. We shall not follow this usage.

the moment thinking of x as fixed and h as the independent variable. It then follows that

$$g'(h) = f'(\xi), \quad \ldots, \quad g^{(n)}(h) = f^{(n)}(\xi),$$

and hence, if we put $h = 0$,

$$g'(0) = f'(x), \quad \ldots, \quad g^{(n)}(0) = f^{(n)}(x).$$

If we apply the previous formula to the function $f(x + h) = g(h)$, which is itself a polynomial of the n-th degree in h, we immediately obtain the Taylor series

$$f(\xi) = f(x + h) = f(x) + hf'(x) + \frac{h^2}{2!} f''(x) + \frac{h^3}{3!} f'''(x) + \cdots$$

$$+ \frac{h^n}{n!} f^{(n)}(x).$$

2. Taylor's Theorem for an Arbitrary Function.

These formulæ suggest that we should seek a similar formula in the case of an arbitrary function $f(x)$, not necessarily a polynomial; in this case, however, the formula can lead only to an approximation to the function by a polynomial.

We wish to compare the values of the function f at the point x and at the point $\xi = x + h$, so that $h = \xi - x$. If now n is any positive integer whatever, the expression

$$f(x) + (\xi - x)f'(x) + \ldots + \frac{(\xi - x)^n}{n!} f^{(n)}(x)$$

will not, as a rule, be an exact expression for the functional value $f(\xi)$. We must therefore put

$$f(\xi) = f(x) + (\xi - x)f'(x) + \frac{(\xi - x)^2}{2!} f''(x) + \cdots$$

$$+ \frac{(\xi - x)^n}{n!} f^{(n)}(x) + R_n,$$

where the expression R_n denotes the *remainder* when $f(\xi)$ is replaced by the expression $f(x) + f'(x)(\xi - x) + \ldots$. In the first instance this equation is nothing but a formal definition of the expression R_n. Its significance lies in the fact that we can easily find a neat and useful expression for this remainder R_n. For this purpose we think of the quantity ξ as fixed and the quantity x as the independent variable. The remainder is then a function $R_n(x)$. By the above equation this function vanishes for $x = \xi$:

$$R_n(\xi) = 0.$$

Further, by differentiation, we obtain

$$R_n'(x) = -\frac{(\xi - x)^n}{n!} f^{(n+1)}(x).$$

For if we differentiate the equation defining the remainder with respect to x, we obtain 0 on the left, since $f(\xi)$ does not depend on x and is therefore to be regarded as a constant. On the right we differentiate each term by the rule for products, and find that all the terms cancel out except the last one, which is written above with a minus sign.

Now by the fundamental theorem of the integral calculus

$$R_n(x) = R_n(x) - R_n(\xi) = \int_\xi^x R_n'(t)\,dt = -\int_x^\xi R_n'(t)\,dt,$$

so that we obtain the formula

$$R_n(x) = \int_x^{x+h} \frac{(x + h - t)^n}{n!} f^{(n+1)}(t)\,dt.$$

If we introduce a new variable of integration τ by means of the equation $\tau = t - x$, this becomes

$$R_n = \frac{1}{n!} \int_0^h (h - \tau)^n f^{(n+1)}(x + \tau)\,d\tau.$$

Collecting these results, we have the following statement:

If the function f(x) *has continuous derivatives up to the* (n + 1)-th *order in the interval under consideration, then*

$$f(x + h) = f(x) + hf'(x) + \frac{h^2}{2!}f''(x) + \frac{h^3}{3!}f'''(x) + \cdots$$

$$+ \frac{h^n}{n!}f^{(n)}(x) + R_n,$$

or (the equivalent expression for h = ξ − x*)*

$$f(\xi) = f(x) + (\xi - x)f'(x) + \frac{(\xi - x)^2}{2!}f''(x) + \cdots$$

$$+ \frac{(\xi - x)^n}{n!}f^{(n)}(x) + R_n,$$

where the remainder R_n is given by the formula

$$R_n = \frac{1}{n!} \int_0^h (h - \tau)^n f^{(n+1)}(x + \tau)\, d\tau.$$

If in particular we put $x = 0$ and then replace h by x, we obtain the formula

$$f(x) = f(0) + \frac{x}{1!} f'(0) + \frac{x^2}{2!} f''(0) + \ldots + \frac{x^n}{n!} f^{(n)}(0) + R_n$$

with the remainder

$$R_n = \frac{1}{n!} \int_0^x (x - \tau)^n f^{(n+1)}(\tau)\, d\tau.$$

These formulæ are known as Taylor's theorem. They give expressions for the functions $f(x + h)$ and $f(x)$ respectively in terms of polynomials of degree n in h and in x respectively (the so-called polynomial of approximation), and a remainder. The polynomial of approximation is characterized by the fact that when $h = 0$ (or $x = 0$, as the case may be) its value and that of its first n derivatives are the same as those of the given function and its first n derivatives. In contrast with the Taylor series for polynomials * the remainder and the expression for it are *essential* here. The significance of the formula lies in the fact that the remainder, even though it has a more complicated form than the other terms of the formula, nevertheless affords us a useful means for estimating the accuracy with which the sum of the first $n + 1$ terms,

$$f(0) + \frac{x}{1!} f'(0) + \frac{x^2}{2!} f''(0) + \ldots + \frac{x^n}{n!} f^{(n)}(0),$$

represents the function $f(x)$.

3. Estimation of the Remainder.

Whether the first $n + 1$ terms of Taylor's series actually give a sufficiently good approximation to the function naturally depends on whether the remainder is sufficiently small. We therefore now turn our attention to the estimation of this remainder. Such an estimate can most easily be made by means of the mean value theorem of the integral calculus (Chap. II, § 7, p. 127).

---

* Whose representation requires no remainder.

We use this theorem in the form

$$\int_0^h p(\tau)\,\phi(\tau)\,d\tau = \phi(\theta h)\int_0^h p(\tau)\,d\tau,$$

where $p(\tau)$ is a continuous function which is nowhere negative - in the interval of integration, and $\phi(\tau)$ is merely a continuous function there, while θ is a number in the interval * $0 \leqq \theta \leqq 1$. If in the formula for the remainder we take $(h - \tau)^n$ to be $p(\tau)$ we obtain

$$R_n = \frac{h^{n+1}}{(n+1)!}\,f^{(n+1)}(x + \theta h);$$

while if instead we put $p(\tau) = 1$ we obtain the expression

$$R_n = \frac{h^{n+1}}{n!}\,(1 - \theta)^n f^{(n+1)}(x + \theta h),$$

which is less important for us and is stated here only for the sake of completeness. In these formulæ θ denotes a certain number in the interval $0 \leqq \theta \leqq 1$, whose value we cannot in general specify more accurately; as a rule, of course, this value is different in the two formulæ for the remainder, and in addition depends on n, x, and h. The first form of the remainder was given by Lagrange, the second by Cauchy, and they are correspondingly named.†

Our interest will be directed chiefly towards finding out whether the remainder R_n tends to zero as n increases; if this is the case, the larger we choose n the more accurately is $f(x + h)$ represented by the corresponding polynomial in h. In this case

* We may in fact assume that $0 < \theta < 1$, but this is of no importance here.

† These expressions for the remainder, as well as others, can be derived from the mean value theorem of the differential calculus and from the generalized mean value theorem (p. 203) respectively. We apply these theorems to the function $R_n(x) = R_n(x) - R_n(\xi)$ and to the pair of functions $R_n(x)$ and $(x - \xi)^{n+1}$ respectively, where we consider ξ as fixed and make use of the formula

$$R_n'(x) = -\frac{(\xi - x)^n}{n!}\,f^{(n+1)}(x).$$

These methods of deriving the formulæ for the remainder throw more stress on the fact that Taylor's theorem is a generalization of the mean value theorem; they also offer the advantage, which for many theoretical purposes is important, that we need only assume the *existence* and not the *continuity* of the $(n + 1)$-th derivative. On the other hand, however, we lose the advantage of having an exact expression for the remainder in the form of an integral.

we say that we have *expanded the function in an infinite Taylor series*

$$f(x+h) = f(x) + \frac{h}{1!}f'(x) + \frac{h^2}{2!}f''(x) + \frac{h^3}{3!}f'''(x) + \cdots,$$

or, in particular, if we first put $x = 0$ and then write x in place of h,

$$f(x) = f(0) + \frac{x}{1!}f'(0) + \frac{x^2}{2!}f''(0) + \frac{x^3}{3!}f'''(0) + \cdots.$$

We shall meet with examples of this in the next section.

First, however, we wish to point out the second important point of view arising from consideration of Taylor's series. If in the first formula we think of the quantity h as becoming smaller and tending to zero, then in the terminology of Chap. III, § 9 (p. 195), the various terms of the series will tend to zero with different orders of magnitude; we accordingly call the expression $f(x)$ the term of zero order in Taylor's series, the expression $hf'(x)$ the term of first order, the expression $\frac{h^2}{2!}f''(x)$ the term of second order, and so on. From the form of our remainder we see the following fact:

In expanding a function as far as the term of n-th *order we make an error which tends to zero with order* (n + 1) *as* h → 0.

On this fact many important applications depend. It shows us that the nearer the point $x + h$ lies to the point x, the better is the representation of the function $f(x + h)$ by the polynomial of approximation, and that in a given case the approximation in the immediate neighbourhood of the point x can be improved by increasing the value of n.

EXAMPLES

1. Let $f(x)$ have a continuous derivative in the interval $a \leq x \leq b$, and let $f''(x) \geq 0$ for every value of x. Then if ξ is any point in the interval, the curve nowhere falls below its tangent at the point $x = \xi$, $y = f(\xi)$. (Use the Taylor expansion to three terms.)

2. Find the value of θ in Lagrange's form of the remainder R_n for the expansions of $\dfrac{1}{1-x}$ and $\dfrac{1}{1+x}$ in powers of x.

3. Applications. Expansions of the Elementary Functions

We shall now use the general results of the preceding section in order to express the elementary functions approximately by means of polynomials and to expand them in Taylor series. We shall, however, restrict ourselves to those functions for which the coefficients of the expansion in series are given by simple laws of formation. The series for certain other functions will be discussed in Chap. VIII (p. 405 *et seq.*).

1. The Exponential Function.

The simplest example is offered by the exponential function, $f(x) = e^x$. Here all the derivatives are identical with $f(x)$ and therefore have the value 1 for $x = 0$. Hence by using Lagrange's form for the remainder we at once obtain the formula

$$e^x = 1 + \frac{x}{1!} + \frac{x^2}{2!} + \frac{x^3}{3!} + \ldots + \frac{x^n}{n!} + \frac{x^{n+1}}{(n+1)!} e^{\theta x}$$

in accordance with § 2 (p. 320 *et seq.*). If we now let n increase beyond all bounds the remainder will tend to zero, no matter what fixed value of x we have chosen. For $|e^{\theta x}| \leq e^{|x|}$ to begin with. We now choose a fixed integer m greater than $2|x|$. Then for $n \geq m$ we have

$$\frac{|x|}{n} < \frac{1}{2}; \quad \left| \frac{x^{n+1}}{(n+1)!} \right| = \frac{|x^m|}{m!} \cdot \frac{|x|}{m+1} \cdots \frac{|x|}{n+1}$$

$$\leq \frac{|x^m|}{m!} \frac{1}{2^{n+1-m}} \leq \frac{|2x|^m}{m!} \frac{1}{2^n},$$

so that
$$|R_n| \leq \frac{|2x|^m}{m!} e^{|x|} \frac{1}{2^n}.$$

Since the first two factors on the right are independent of n, while the number $1/2^n$ tends to zero as n increases, our statement is proved. If we think of the number x not as fixed, but as free to vary in the interval $-a \leq x \leq a$, where a is a fixed positive number, it follows from the above that if we choose $m > 2a$ the estimate

$$|R_n| \leq \frac{|2a|^m}{m!} e^a \frac{1}{2^n}$$

is valid provided $n \geqq m$. For the remainder we have therefore specified a bound which holds for all values of x in the interval $-a \leqq x \leqq a$, and which tends to zero as $n \to \infty$. For the function e^x we can therefore write the expansion as an infinite series

$$e^x = 1 + \frac{x}{1!} + \frac{x^2}{2!} + \frac{x^3}{3!} + \ldots = \sum_{\nu=0}^{\infty} \frac{x^\nu}{\nu!},$$

the last expression being merely an abbreviated expression for the series. This expansion is valid for all values of x. Thus we have again proved that the number e considered in Chap. I (cf. p. 43) is the same as the base of the natural logarithms (cf. Chap. III, § 6). For numerical calculations we must, of course, make use of the finite form of Taylor's theorem with the remainder; for $x = 1$, for example, this gives

$$e = 1 + 1 + \frac{1}{2!} + \frac{1}{3!} + \ldots + \frac{1}{n!} + \frac{e^\theta}{(n+1)!}.$$

If we wish to calculate e with an error of at most $1/10,000$, we need only choose n so large that the remainder is certainly less than $1/10,000$; and since this remainder is certainly less * than $3/(n+1)!$, it is sufficient to choose $n = 7$, since $8! > 30,000$. We thus obtain the approximate value

$$e = 2 \cdot 71822$$

with an error less than $0 \cdot 0001$. Here we do not take any account of the error due to neglecting the figures in the sixth decimal place.

2. $\operatorname{Sin} x$, $\cos x$, $\sinh x$, $\cosh x$.

For the functions $\sin x$, $\cos x$, $\sinh x$, $\cosh x$ we find the following formulæ: †

$f(x)$	$=$	$\sin x$	$\cos x$	$\sinh x$	$\cosh x,$
$f'(x)$	$=$	$\cos x$	$-\sin x$	$\cosh x$	$\sinh x,$
$f''(x)$	$=$	$-\sin x$	$-\cos x$	$\sinh x$	$\cosh x,$
$f'''(x)$	$=$	$-\cos x$	$\sin x$	$\cosh x$	$\sinh x,$
$f^{iv}(x)$	$=$	$\sin x$	$\cos x$	$\sinh x$	$\cosh x.$

. . . .

---

* Here we have made use of the fact that $e < 3$. This follows immediately (cf. p. 43 also) from our series for e; for it is always true that $\frac{1}{n!} \leqq \frac{1}{2^{n-1}}$, and therefore

$$e < 1 + 1 + \tfrac{1}{2} + \tfrac{1}{4} + \ldots = 1 + 1/(1 - \tfrac{1}{2}) = 3.$$

† If $f(x) = \sin x$ or $f(x) = \cos x$, the n-th derivative can always be represented by the expression

$$f^{(n)}(x) = f(x + \tfrac{1}{2}n\pi).$$

Hence in the polynomials of approximation for $\sin x$ and $\sinh x$ the coefficients of the even powers of x vanish, while in the polynomials of approximation for $\cos x$ and $\cosh x$ the coefficients of the odd powers vanish. Thus in the first case the $(2n + 1)$-th and the $(2n + 2)$-th polynomials are identical, while in the second case the $2n$-th and the $(2n + 1)$-th are identical. If in each case we use the higher of these polynomials we at once obtain, using Lagrange's form of the remainder,

$$\sin x = x - \frac{x^3}{3!} + \frac{x^5}{5!} - + \ldots + (-1)^n \frac{x^{2n+1}}{(2n+1)!}$$
$$+ (-1)^{n+1} \frac{x^{2n+3}}{(2n+3)!} \cos(\theta x),$$

$$\cos x = 1 - \frac{x^2}{2!} + \frac{x^4}{4!} - + \ldots + (-1)^n \frac{x^{2n}}{(2n)!}$$
$$+ (-1)^{n+1} \frac{x^{2n+2}}{(2n+2)!} \cos(\theta x),$$

$$\sinh x = x + \frac{x^3}{3!} + \frac{x^5}{5!} + \ldots + \frac{x^{2n+1}}{(2n+1)!}$$
$$+ \frac{x^{2n+3}}{(2n+3)!} \cosh(\theta x),$$

$$\cosh x = 1 + \frac{x^2}{2!} + \frac{x^4}{4!} + \ldots + \frac{x^{2n}}{(2n)!}$$
$$+ \frac{x^{2n+2}}{(2n+2)!} \cosh(\theta x),$$

where in each of the four formulæ, of course, θ denotes a different number in the interval $0 \leqq \theta \leqq 1$, a number which in addition depends on n and on x. In these formulæ we can also make the approximation as exact as we please for each value of x, since the remainder tends to 0 as n increases. We thus obtain the four series

$$\sin x = x - \frac{x^3}{3!} + \frac{x^5}{5!} - + \ldots = \sum_{\nu=0}^{\infty} (-1)^\nu \frac{x^{2\nu+1}}{(2\nu+1)!},$$

$$\cos x = 1 - \frac{x^2}{2!} + \frac{x^4}{4!} - + \ldots = \sum_{\nu=0}^{\infty} (-1)^\nu \frac{x^{2\nu}}{(2\nu)!},$$

$$\sinh x = x + \frac{x^3}{3!} + \frac{x^5}{5!} + \quad \ldots = \sum_{\nu=0}^{\infty} \frac{x^{2\nu+1}}{(2\nu+1)!},$$

$$\cosh x = 1 + \frac{x^2}{2!} + \frac{x^4}{4!} + \quad \ldots = \sum_{\nu=0}^{\infty} \frac{x^{2\nu}}{(2\nu)!}.$$

The last two can also be obtained formally from the series for e^x in accordance with the definitions of the hyperbolic functions.

3. The Binomial Series.

We may pass over the Taylor series for the functions $\log(1 + x)$ and arc $\tan x$, which have already been treated directly in § 1 (p. 315). We must, however, take up the generalization of the binomial theorem for arbitrary indices, which is one of the most fruitful of Newton's mathematical discoveries, and which represents one of the most important cases of expansion in Taylor series. Our object is the expansion of the function

$$f(x) = (1 + x)^a$$

in a Taylor series, where $x > -1$ and a is an arbitrary number, positive or negative, rational or irrational. We have chosen the function $(1 + x)^a$ instead of x^a because at the point $x = 0$ it is not true that all the derivatives of x^a are continuous, except in the trivial case of non-negative integral values of a. We first calculate the derivatives of $f(x)$, obtaining

$$f'(x) = a(1 + x)^{a-1}, \quad f''(x) = a(a - 1)(1 + x)^{a-2}, \ldots,$$
$$f^{(\nu)}(x) = a(a - 1) \ldots (a - \nu + 1)(1 + x)^{a-\nu}.$$

In particular, for $x = 0$ we have

$$f'(0) = a, \, f''(0) = a(a - 1), \, \ldots, \, f^{(\nu)}(0) = a(a - 1)(a - \nu + 1).$$

Taylor's theorem then gives

$$(1 + x)^a = 1 + ax + \frac{a(a - 1)}{2!} x^2 + \ldots$$

$$+ \frac{a(a - 1)(a - 2) \ldots (a - n + 1)}{n!} x^n + R_n.$$

We have still to discuss the remainder. This problem is not very difficult, but nevertheless is not quite so simple as in the cases treated previously. Here we shall pass over the estimation of the remainder, since the general binomial theorem will be proved completely in a somewhat different and simpler way in Chap. VIII (p. 406 *et seq.*; cf. also p. 336). The result, which we mention here in advance, is that whenever $|x| < 1$ the remainder tends to 0 and therefore the expression $(1 + x)^a$ can be expanded in the infinite *binomial series*

$$(1 + x)^a = 1 + \frac{a}{1!}x + \frac{a(a-1)}{2!}x^2 + \ldots = \sum_{\nu=0}^{\infty} \binom{a}{\nu} x^\nu,$$

where for brevity we have introduced the general binomial coefficients $\binom{a}{\nu} = \dfrac{a(a-1)\ldots(a-\nu+1)}{\nu!}$ (for $\nu > 0$), $\binom{a}{0} = 1$.

EXAMPLES

1. Expand $(1 + x)^{\frac{1}{2}}$ to two terms plus remainder. Estimate the remainder.

2. Use the expansion of Ex. 1 (discarding the remainder) to calculate $\sqrt{2}$. What is the degree of accuracy of the approximation?

3. What linear function best approximates to $\sqrt[3]{(1 + x)}$ in the neighbourhood of $x = 0$? Between what values of x is the error of the approximation less than $\cdot 01$?

4. What quadratic function best approximates to $\sqrt[3]{(1 + x)}$ in the neighbourhood of $x = 0$? What is the greatest error in the interval $-0\cdot 1 \leqq x \leqq 0\cdot 1$?

5. (*a*) What linear function, (*b*) what quadratic function, best approximates to $\sqrt[n]{(1 + x)}$ in the neighbourhood of $x = 0$? What are the greatest errors when $-\cdot 1 \leqq x \leqq \cdot 1$?

6. Calculate $\sin(\cdot 01)$ to 4 places.

7. Do the same for (*a*) $\cos(\cdot 01)$, (*b*) $\sqrt[3]{126}$, (*c*) $\sqrt{97}$.

8. Expand $\sin(x + h)$ in a Taylor series in h. Use this to find $\sin 31°$ ($= \sin(30° + 1°)$) to 3 places.

Expand the functions in Ex. 9–18 in the neighbourhood of $x = 0$ to three terms plus remainder (writing the remainder in Lagrange's form).

9. $\sin^2 x$.

10. $\cos^3 x$.

11. $\log \cos x$.

12. $\tan x$.

13. $\log \dfrac{1}{\cos x}$.

14. e^{-x^2}.

15. $\dfrac{1}{\cos x}$.

16. $\cot x - \dfrac{1}{x}$.

17. $\dfrac{1}{\sin x} - \dfrac{1}{x}$.

18. $\dfrac{\log(1 + x)}{1 + x}$.

19. (a) Expand $e^{\sin x}$ to five terms plus remainder; (b) in the power series for e^z substitute for z the power series for $\sin x$, taking enough terms to secure that the coefficient of x^4 is correct. Compare with (a).

20. Find the polynomial of fourth degree which best approximates to $\tan x$ in the neighbourhood of $x = 0$. In what interval does this polynomial represent $\tan x$ to within 5 per cent?

21. Find the first 6 terms of the Taylor series for y in powers of x for the functions defined by

(a) $x^2 + y^2 = y$, $y(0) = 0$; (b) $x^2 + y^2 = y$, $y(0) = 1$;
(c) $x^3 + y^3 = y$, $y(0) = 0$.

4. Geometrical Applications

The behaviour of a function $f(x)$ in the neighbourhood of a point $x = a$, or the behaviour of a given curve in the neighbourhood of a point, can be studied with increased accuracy by means of Taylor's theorem, for this theorem resolves the increment of the function on passing to a neighbouring point $x = a + h$ into a sum of quantities of the first order, second order, &c.

1. Contact of Curves.

We shall now make use of this method in order to investigate the concept of *contact* of two curves.

If at a point, say the point $x = a$, two curves $y = f(x)$ and $y = g(x)$ not only intersect, but also have a common tangent, we shall say that at this point the curves touch one another or have *contact of the first order*. The Taylor expansions of the functions $f(a + h)$ and $g(a + h)$ then have the same terms of zero order and of first order in h. If at the point $x = a$ the second derivatives of $f(x)$ and $g(x)$ are also equal to one another, we say that the curves have *contact of the second order*. In the Taylor expansions the terms of second order are then also the same, and if we assume that both functions have continuous derivatives of the third order at least, the difference $D(x) = f(x) - g(x)$ can be expressed in the form

$$D(a + h) = f(a + h) - g(a + h) = \frac{h^3}{3!} D'''(a + \theta h) = \frac{h^3}{3!} F(h),$$

where the expression $F(h)$ tends to $f'''(a) - g'''(a)$ as h tends to zero. The difference $D(a + h)$ therefore vanishes to at least the third order with h.

We can proceed in this way and consider the general case, where the Taylor series for $f(x)$ and $g(x)$ are the same up to terms of the n-th order, that is,

$$f(a) = g(a), f'(a) = g'(a), f''(a) = g''(a), \ldots, f^{(n)}(a) = g^{(n)}(a).$$

We here assume that the $(n+1)$-th derivatives are also continuous. Under these conditions we say that at this point the curves have *contact of the* n-th *order*. The difference of the two functions will then be of the form

$$f(a+h) - g(a+h) = \frac{h^{n+1}}{(n+1)!} F(h),$$

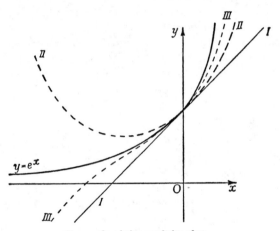

Fig. 1.—Osculating parabolas of e^x

where since $0 \leq \theta \leq 1$ the quantity $F(h) = D^{(n+1)}(a + \theta h)$ tends to $f^{(n+1)}(a) - g^{(n+1)}(a)$ as h tends to 0. We recognize from this formula that at the point of contact the difference $f(x) - g(x)$ vanishes to the $(n+1)$-th order at least.

The Taylor polynomials are simply defined geometrically by the fact that they are those parabolas of the n-th order which at the given point have contact of the highest possible order with the graph of the given function. Hence they are sometimes called *osculating parabolas*. For the case $y = e^x$ fig. 1 gives us the first three osculating parabolas at the point $x = 0$.

If two curves $y = f(x)$ and $y = g(x)$ have contact of the n-th order, the definition does not exclude the possibility that the contact may be of still higher order, i.e. that the equation $f^{(n+1)}(a) = g^{(n+1)}(a)$ is also true. If this is not the case, i.e. if $f^{(n+1)}(a) \neq g^{(n+1)}(a)$, we speak of a contact of exactly the n-th order or say that the order of the contact * is *exactly n*.

From our formulæ as well as from our figures we can at once state a remarkable fact which is often unnoticed by beginners. If the contact of two curves is exactly of an even order, that is, if an even number n of derivatives of the two functions have the same value at the point in question, while the $(n + 1)$-th derivatives differ, then in conformity with the above formulæ the difference $f(a + h) - g(a + h)$ will have different signs for small positive values of h and for numerically small negative values of h. The two curves will then cross at the point of contact. This case occurs e.g. in contact of the second order if the third derivatives have different values. If, however, we consider the case of contact exactly of an odd order, e.g. the case of an ordinary contact of the first order, the difference $f(a+h) - g(a+h)$ will have the same sign for all numerically small values of h, whether positive or negative; the two curves therefore will not cross in the neighbourhood of the point of contact. The simplest example of this is the contact of a curve with its tangent. The tangent can cross the curve only at points where the contact is of the second order at least; it will actually cross the curve at points where the order of contact is even, e.g. at an ordinary point of inflection, where $f''(x) = 0$ but $f'''(x) \neq 0$. At points where the order of contact is odd it will not cross the curve; as examples we may take an ordinary point of the curve where the second derivative is not zero, or the curve $y = x^4$ at the origin.

2. The Circle of Curvature as Osculating Circle.

When looked at from this point of view the concept of the curvature of a curve $y = f(x)$ gains a new intuitive significance. Through a definite point of the curve with the co-ordinates $x = a$, $y = b$ there pass an infinite number of circles which touch the curve at the point. The centres of these circles lie on the normal to the curve, and to each point of this normal

---

* That the order of contact of two curves is a genuine geometrical relation which is unaffected by change of axes is a fact which can easily be established by means of the formulæ for change of axes.

there corresponds just one such tangent circle. We may expect that by proper choice of the centre of the circle we can bring about a contact of the *second* order between the curve and the circle.

As a matter of fact, we know from Chap. **V** (p. 283) that for the circle of curvature at the point $x = a$, whose equation is, say, $y = g(x)$, we not only have $g(a) = f(a)$ and $g'(a) = f'(a)$, but also $g''(a) = f''(a)$. Hence the circle of curvature is at the same time the osculating circle at the point of the curve under discussion; that is, it is the circle which at that point has

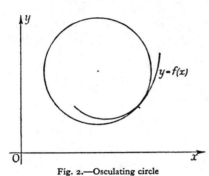

Fig. 2.—Osculating circle

contact of the second order with the curve. In the limiting case of a point of inflection, or in general of a point at which the curvature is zero and the radius of curvature is infinite, the circle of curvature degenerates into the tangent. In ordinary cases, that is, when the contact at the point in question does not happen to be of an order higher than the second, the circle of curvature will not merely touch the curve, but will also cross it (cf. fig. 2).

3. On the Theory of Maxima and Minima.

As we have already seen in Chap. III (p. 161), a point $x = a$ at which $f'(a) = 0$ gives a maximum of the function $f(x)$ if $f''(a)$ is negative, a minimum if $f''(a)$ is positive. These last conditions, therefore, are *sufficient* conditions for the occurrence of a maximum or minimum. They are by no means *necessary*; for in the case where $f''(a) = 0$ there are three possibilities open; at the point in question the function may have a maximum or a minimum or neither. Examples of the three possibilities are given by the functions $y = -x^4$, $y = x^4$, and $y = x^3$ at the point $x = 0$. Taylor's theorem at once enables us to make a general statement of sufficient conditions for a maximum or a minimum. We need only expand the function $f(a + h)$ in powers of h; the essential point is then to find whether the first non-vanishing term contains an even power of h or an odd power. In the first case we have a maximum or a minimum according as the co-efficient of h is negative or positive; in the second case we have a horizontal inflectional tangent and neither maximum nor minimum. The reader may complete the argument for himself by using the formula for the remainder.*

EXAMPLES

1. What is the order of contact of the curves $y = e^x$ and $y = 1 + x + \frac{1}{2}\sin^2 x$ at $x = 0$?

2. What is the order of contact of $y = \sin^4 x$ and $y = \tan^4 x$ at $x = 0$?

3. Determine the constants a, b, c, d in such a way that the curves $y = e^{2x}$ and $y = a\cos x + b\sin x + c\cos 2x + d\sin 2x$ have contact of order 3 at $x = 0$.

4. What is the order of contact of the curves
$$x^3 + y^3 = xy, \quad x^2 + y^2 = x$$
at their points of intersection? Plot the curves.

5. What is the order of contact of the curves
$$x^2 + y^2 = y \quad x^3 = y$$
at their points of intersection?

6. The curve $y = f(x)$ passes through the origin O and touches the x-axis at O. Show that the radius of curvature of the curve at O is given by $\rho = \lim\limits_{x \to 0} \dfrac{x^2}{2y}$.

7.* K is a circle which touches a given curve at a point P and passes through a neighbouring point Q of the curve. Show that the limit of the circle K as $Q \to P$ is the circle of curvature of the curve at P.

8.* R is the point of intersection of the two normals to a given curve at the neighbouring points P, Q of the curve. Show that, as $Q \to P$, R tends to the centre of curvature of the curve for the point P. (The centre of curvature is the intersection of neighbouring normals.)

9.* Show that the order of contact of a curve and its osculating circle is at least three at points where the radius of curvature is a maximum or minimum.

10. Determine the maxima and minima of the function $y = e^{-1/x^2}$. (See p. 336.)

---

* The necessary and sufficient condition given previously (p. 161), however, is more general and more convenient in applications, namely: provided the first derivative $f'(x)$ vanishes at only a finite number of points, a necessary and sufficient condition for the occurrence of a maximum or minimum at one of these points is that the first derivative $f'(x)$ changes sign as it passes through the point.

Appendix to Chapter VI

1. Example of a Function which cannot be expanded in a Taylor Series

The possibility of expressing a function by means of a Taylor series with a remainder of the $(n + 1)$-th order depends essentially on the differentiability of the function at the point in question. For this reason the function $\log x$ cannot be represented by a Taylor series in powers of x, and the same is true of the function $\sqrt[3]{x}$, whose derivative is infinite at $x = 0$.

In order that a function may be capable of being expanded in an *infinite* Taylor series, *all* its derivatives must necessarily exist at the point in question; this condition, however, is by no means sufficient. A function for which all the derivatives exist and are continuous throughout an interval still need not necessarily be capable of expansion in a Taylor series; that is, the remainder R_n in Taylor's theorem may fail to tend to zero as n increases, no matter how small the interval in which we wish to expand the function.

The simplest example of this phenomenon is offered by the function $y = f(x) = e^{-1/x^2}$ for $x \neq 0$, $f(0) = 0$, which we have already considered in the appendix to Chap. III (p. 196). This function, with all its derivatives, is continuous in every interval, even at $x = 0$, and we have seen that at this point all the derivatives vanish, i.e. that $f^{(n)}(0) = 0$ for every value of n. Hence in Taylor's theorem all the coefficients of the polynomials of approximation vanish, no matter what value we choose for n. In other words, the remainder is and remains equal to the function itself, and therefore, except when $x = 0$, does not approach 0 as n increases, since the function is positive for every non-zero value of x.

2. Proof that e is Irrational

From the formula $e = 2 + \dfrac{1}{2!} + \ldots + \dfrac{1}{n!} + \dfrac{1}{(n+1)!} e^\theta$, we immediately deduce that the number e is irrational. For if the contrary is true, that is, if $e = p/q$, where p and q are integers, we can certainly choose n larger than q. Then $n!e = n! \dfrac{p}{q}$ must be an integer. On the other hand, $n!e = 2n! + \dfrac{n!}{2!} + \ldots + \dfrac{n!}{n!} + \dfrac{1}{n+1} e^\theta$, and since $e^\theta < e < 3$, we must have $0 < \dfrac{e^\eta}{n+1} < 1$. Hence the integer $n!e =$ the integer $2n! + \dfrac{n!}{2} +$ $\ldots + 1$ plus a non-vanishing proper fraction, which is impossible.

3. Proof that the Binomial Series Converges

In § 3 (p. 329) we postponed the estimation of the remainder R_n in the expansion of $f(x) = (1 + x)^a$ for $|x| < 1$. This estimation we shall carry out here. It is most convenient to separate the cases $x > 0$ and $x < 0$.

For $f^{(n+1)}(x)$ we have the expression

$$f^{(n+1)}(x) = a(a-1)\ldots(a-n)\frac{(1+x)^a}{(1+x)^{n+1}}.$$

If $x > 0$, we write the remainder in Lagrange's form,

$$R_n(x) = \frac{x^{n+1}}{(n+1)!}\, a(a-1)\ldots(a-n)\frac{(1+\theta x)^a}{(1+\theta x)^{n+1}}$$

so that

$$|R_n(x)| \leqq \left|\frac{a(a-1)\ldots(a-n)}{(n+1)!}\right| \frac{x^{n+1}(1+x)^a}{1^{n+1}}.$$

Writing $b = [|a|] + 1$, where $[|a|]$ means the greatest integer which does not exceed $|a|$, we have

$$|R_n(x)| \leqq 2^b \frac{b(b+1)\ldots(b+n)}{(n+1)!}\, x^{n+1}$$

$$\leqq \frac{2^b}{(b-1)!}\frac{1\,.\,2\ldots(n+1)\,(n+2)\ldots(n+b)}{(n+1)!}\, x^{n+1}$$

$$\leqq \frac{2^b}{(b-1)!}\,(n+b)^{b-1}x^{n+1},$$

and since b is fixed, if $0 < x < 1$, this approaches 0 as n increases.

For the case $-1 < x < 0$ we write the remainder in Cauchy's form,

$$R_n(x) = \frac{x^{n+1}}{n!}\,(1-\theta)^n a(a-1)\ldots(a-n)\frac{(1+\theta x)^{a-1}}{(1+\theta x)^n},$$

so that

$$|R_n(x)| \leqq \frac{(1-\theta)^n}{(1-\theta|x|)^n}\,|x|^{n+1}\left|\frac{a(a-1)\ldots(a-n)}{n!}\right| |(1+\theta x)^{a-1}|.$$

(E 798)

Since $|x| < 1$, the last factor cannot exceed a constant K, independent of n. Also $(1 - \theta) / (1 - \theta |x|) < 1$. As before, writing $b = [|a|] + 1$, we have

$$|R_n(x)| \leqq K |x|^{n+1} \frac{1}{(b-1)!} (n+2)(n+3)\ldots(n+b)$$

$$\leqq \frac{K}{(b-1)!} (n+b)^{b-1} |x|^{n+1},$$

which approaches 0 as n increases.

Thus in either case when $|x| < 1$ the remainder tends to zero as n increases, justifying the expansion in § 3 (p. 330).

4. Zeros and Infinities of Functions and So-called Indeterminate Expressions

The Taylor series for a function in the neighbourhood of a point $x = a$ enables us to characterize the behaviour of the function in the neighbourhood of this point in the following way. We say that a function $f(x)$ at $x = a$ has an *exactly* n-*tuple zero* or it vanishes there *exactly to the order* n, if $f(a) = 0$, $f'(a) = 0$, $f''(a) = 0, \ldots, f^{(n-1)}(a) = 0$, and $f^{(n)}(a) \neq 0$. Here we assume that in the neighbourhood of the point the function possesses continuous derivatives to the n-th order at least. By our definition we seek to indicate that the Taylor series for the function in the neighbourhood of the point can be written in the form

$$f(a+h) = \frac{h^n}{n!} F(h),$$

in which as $h \to 0$ the factor $F(h)$ tends to a limit different from 0, namely, the value $f^n(a)$.

If a function $\phi(x)$ is defined at all points in the neighbourhood of a point $x = a$, except perhaps at $x = a$ itself, and if

$$\phi(x) = \frac{f(x)}{g(x)},$$

where at the point $x = a$ the numerator does not vanish, but the denominator possesses a ν-tuple zero, we say that the function $\phi(x)$ becomes *infinite of the ν-th order* at the point $x = a$. If at the point $x = a$ the numerator also possesses a μ-tuple zero and if $\mu > \nu$, we say that the function has a $(\mu - \nu)$-tuple zero there; while if $\mu < \nu$ we say that the function has a $(\nu - \mu)$-tuple infinity.

All these definitions are in agreement with the conventions already laid down (cf. Chap. III, § 9, p. 194) regarding the behaviour of a function. In order to make these relations precise, we expand numerator and denominator by Taylor's theorem, using Lagrange's form of the remainder; the function then has the form

$$\phi(a + h) = \frac{f(a + h)}{g(a + h)} = \frac{\nu!}{\mu!} \frac{h^{\mu} f^{(\mu)}(a + \theta h)}{h^{\nu} g^{(\nu)}(a + \theta_1 h)},$$

in which θ and θ_1 are two numbers between 0 and 1 and the factors by which $\dfrac{h^{\mu}}{\mu!}$ and $\dfrac{h^{\nu}}{\nu!}$ are multiplied do not tend to zero as h does, since they approach the limits $f^{(\mu)}(a)$ and $g^{(\nu)}(a)$ respectively, which differ from zero. If $\mu > \nu$, we then have

$$\lim_{h \to 0} \phi(a + h) = \lim_{h \to 0} \frac{\nu!}{\mu!} h^{\mu - \nu} \frac{f^{(\mu)}(a)}{g^{(\nu)}(a)} = 0.$$

The expression $\phi(x)$ accordingly vanishes to the order $\mu - \nu$. If $\nu > \mu$, we see that the expression $\phi(a + h)$ becomes infinite of the order $\nu - \mu$ as $h \to 0$. If $\mu = \nu$, we obtain the equation

$$\lim_{h \to 0} \phi(a + h) = \frac{f^{(\mu)}(a)}{g^{(\nu)}(a)}.$$

We can express the content of the last equations in the following way: if the numerator and denominator of a function $\phi(x) = \dfrac{f(x)}{g(x)}$ both vanish at $x = a$, we can determine the limiting value as $x \to a$ by differentiating the numerator and denominator an equal number of times until one at least of the derivatives is other than zero. If this happens for numerator and denominator simultaneously, the limit which we are seeking is equal to the quotient of these two derivatives. If we encounter a non-vanishing derivative in the denominator earlier than in the numerator, the fraction tends to zero. If we encounter a non-vanishing derivative in the numerator earlier than in the denominator, the absolute value of the fraction increases beyond all bounds.

We thus have a rule for evaluating the so-called indeterminate expression 0/0—a subject that is discussed at exaggerated length in many textbooks on the differential and integral calculus. In reality the point in question is merely the very simple determination of the *limiting value* of a quotient in which the numerator and the denominator tend to zero. The name "indeterminate expression" usually found in the literature is misleading and vague.

We can arrive at our results in a somewhat different way by basing the proof on the generalized mean value theorem * instead of on Taylor's theorem (cf. p. 135). According to this, if $g'(x) \neq 0$, we have

$$\frac{f(a + h) - f(a)}{g(a + h) - g(a)} = \frac{f'(a + \theta h)}{g'(a + \theta h)},$$

where θ is the same in both numerator and denominator. Hence, in particular, when $f(a) = 0 = g(a)$,

$$\frac{f(a + h)}{g(a + h)} = \frac{f'(a + \theta h)}{g'(a + \theta h)}.$$

Here θ is a value in the interval $0 < \theta < 1$, and if we put $k = \theta h$, we obtain

$$\lim_{h \to 0} \frac{f(a + h)}{g(a + h)} = \lim_{k \to 0} \frac{f'(a + k)}{g'(a + k)},$$

it being assumed that the limit on the right exists. If

$$f'(a) = 0 = g'(a),$$

we can proceed in the same manner until we come to the first index for which it is no longer true that $f^{(\mu)}(a) = 0 = g^{(\mu)}(a)$. Then

$$\lim_{h \to 0} \frac{f(a + h)}{g(a + h)} = \lim_{l \to 0} \frac{f^{(\mu)}(a + l)}{g^{(\mu)}(a + l)},$$

in which we also include the case in which both sides have the limit infinity.

---

* This method of deriving our rule has the advantage that in it no use is made of the *existence* of the derivative at the point $x = a$ itself: further, it includes the case in which $\phi(x)$ is defined for $x \geqq a$ only, so that the passage to the limit $x \to a$ or $h \to 0$ is made from one side only.

As examples we consider

$$\frac{\sin x}{x}, \quad \frac{1 - \cos x}{x}, \quad \frac{e^{2x} - 1}{\log(1 + x)}, \quad \frac{x^2 \tan x}{\sqrt{(1 - x^2)} - 1}$$

as $x \to 0$. We have

$$\lim_{x \to 0} \frac{\sin x}{x} = \frac{\cos 0}{1} = 1; \quad \lim_{x \to 0} \frac{1 - \cos x}{x} = \frac{\sin 0}{1} = 0;$$

$$\lim_{x \to 0} \frac{e^{2x} - 1}{\log(1 + x)} = \lim_{x \to 0} \frac{2e^{2x}}{1/(1 + x)} = 2;$$

$$\lim_{x \to 0} \frac{x^2 \tan x}{\sqrt{(1 - x^2)} - 1} = \lim_{x \to 0} \frac{2x \tan x + x^2/\cos^2 x}{-x/\sqrt{(1 - x^2)}}$$

$$= -\lim_{x \to 0} \left(2 \tan x + \frac{x}{\cos^2 x} \right) \sqrt{(1 - x^2)} = 0.$$

We further note that other so-called indeterminate forms can also be reduced to the case we have considered; for example, the limit of $\dfrac{1}{\sin x} - \dfrac{1}{x}$ as $x \to 0$, being the limit of the difference of two expressions which both become infinite, is an " indeterminate form " $\infty - \infty$. By the transformation

$$\frac{1}{\sin x} - \frac{1}{x} = \frac{x - \sin x}{x \sin x}$$

we at once arrive at an expression whose limit as $x \to 0$ is determined by our rule to be

$$\lim_{x \to 0} \frac{1 - \cos x}{x \cos x + \sin x} = \lim_{x \to 0} \frac{\sin x}{2 \cos x - x \sin x} = 0.$$

EXAMPLES

Evaluate the limits in Ex. 1–12:

1. $\lim\limits_{x \to a} \dfrac{x^n - a^n}{x - a}$.

2. $\lim\limits_{x \to 0} \dfrac{x - \sin x}{x^3}$.

3. $\lim\limits_{x \to 0} \dfrac{24 - 12x^2 + x^4 - 24 \cos x}{(\sin x)^6}$.

4. $\lim\limits_{x \to 0} \dfrac{e^x - e^{-x}}{\sin x}$.

5. $\lim\limits_{x \to 0} \dfrac{\text{arc} \sin x}{x}$.

6. $\lim\limits_{x \to \pi/2} \dfrac{\tan 5x}{\tan x}$.

7. $\lim\limits_{x \to 1} \left(\dfrac{2}{x^2 - 1} - \dfrac{1}{x - 1} \right)$.

8. $\lim\limits_{x \to 0} \left(\dfrac{1}{\sin^2 x} - \dfrac{1}{x^2} \right)$.

9. $\lim\limits_{x \to 0} x^{\sin x}$.

10. $\lim\limits_{x \to 0} (1 + x)^{1/x}$.

11. $\lim\limits_{x \to 0} \dfrac{e^{2x} - 1}{\log(1 + x)}$.

12. $\lim\limits_{x \to 0} \dfrac{x \tan x}{\sqrt{(1 - x^2)} - 1}$.

13. Prove that the function $y = (x^2)^x$, $y(0) = 1$ is continuous at $x = 0$.

CHAPTER VII

Numerical Methods

Preliminary Remarks

Anyone who has to use analysis as an instrument for investigating physical or technical phenomena is faced by the question whether and how the theory can be adapted to yield useful practical methods for actual numerical calculations. Yet even from the point of view of the theorist, who desires only to recognize the connexions between natural phenomena, not to conquer them, these questions are of no trifling interest. For a systematic treatment of numerical methods we refer the reader to special textbooks on the subject.* Here we can only discuss some particularly important points which are more or less closely connected with the preceding ideas. We wish to direct special attention to the fundamental fact that the meaning of an approximate calculation is not precise unless it is supplemented by an estimate of the errors occurring, i.e. unless it is accompanied by definite knowledge of the degree of accuracy attained.

1. Numerical Integration

We have seen that even relatively simple functions cannot be integrated in terms of elementary functions and that it is futile to make this unattainable goal the aim of the integral calculus. On the other hand, the definite integral of a continuous function does *exist*, and this fact raises the problem of finding methods for calculating it numerically. Here we shall discuss the simplest and most obvious of these methods with

---

* Cf. e.g. Whittaker and Robinson, *The Calculus of Observations* (Blackie & Son, Ltd., 1929).

the aid of geometrical intuition, and we shall then consider the estimation of errors.

Our object is to calculate the integral $I = \int_a^b f(x)\,dx$, where a is less than b. We imagine the interval of integration divided into n equal parts of length $h = (b - a)/n$, and we denote the points of subdivision by $x_0 = a$, $x_1 = a + h, \ldots, x_n = b$, the values of the function at the points of division by $f_0, f_1, \ldots, f_n$, and similarly the values of the function at the midpoints of the intervals by $f_{1/2}, f_{3/2}, \ldots, f_{(2n-1)/2}$. We interpret our integral as an area and cut up the region under the curve into strips of breadth h in the usual manner. We must now obtain an approximation for each such strip of surface, that is, for the integral

$$I_\nu = \int_{x_\nu}^{x_\nu + h} f(x)\,dx.$$

1. The Rectangle Rule.

The crudest and most obvious method of approximating to I is directly connected with the definition of integral; we replace the area of the strip I_ν by the rectangle of area $f_\nu h$ and obtain for the integral I the approximate expression *

$$I \approx h(f_0 + f_1 + \ldots + f_{n-1}).$$

2. The Trapezoid and Tangent Formulæ.

We obtain a closer approximation with no greater trouble if we replace the area of the strip I_ν, not by the above rectangular area, but by the trapezoid of area $\frac{1}{2}(f_\nu + f_{\nu+1})h$ shown in fig. 1. For the whole integral this gives the approximate expression

$$I \approx h(f_1 + f_2 + \ldots + f_{n-1}) + \frac{h}{2}(f_0 + f_n)$$

(the *trapezoid formula*), since, when the areas of the trapezoids are added, each value of the function except the first and the last occurs twice.

---

* Here and hereafter the symbol $\approx$ means " is approximately equal to ".

As a rule the approximation becomes even better if instead of choosing the trapezoid under the chord AB as an approximation to the area of I we choose the trapezoid under the tangent to the curve at the point with the abscissa $x = x_\nu + h/2$. The area of this trapezoid is simply $h f_{\nu + \frac{1}{2}}$, and the approximation for the entire integral is

$$I \approx h(f_{1/2} + f_{3/2} + \ldots + f_{(2n-1)/2}),$$

which is called the *tangent formula*.

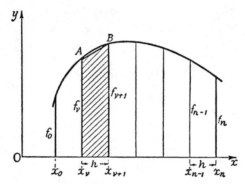

Fig. 1.—The trapezoid formula

3. Simpson's Rule.

By means of Simpson's rule we arrive, with very little more trouble, at a numerical result which is generally much more exact. This rule depends on estimating the area $I_\nu + I_{\nu+1}$ of the double strip between the abscissæ $x = x_\nu$ and $x = x_\nu + 2h = x_{\nu+2}$ by considering the upper boundary to be no longer a straight line but a parabola; to be specific, that parabola which passes through the three points of the curve with abscissæ x_ν, $x_{\nu+1} = x_\nu + h$, and $x_{\nu+2} = x_\nu + 2h$ (cf. fig. 2). The equation of this parabola is

$$y = f_\nu + (x - x_\nu) \frac{f_{\nu+1} - f_\nu}{h}$$
$$+ \frac{(x - x_\nu)(x - x_\nu - h)}{2} \frac{f_{\nu+2} - 2f_{\nu+1} + f_\nu}{h^2}.$$

(The student may verify by direct substitution that for the three values of x in question this equation gives the proper values of y, namely, f_ν, $f_{\nu+1}$, and $f_{\nu+2}$ respectively.) If we integrate this polynomial of the second degree between the limits x_ν and $x_\nu + 2h$, we obtain, after a brief calculation, the following expression for the area under the parabola:

$$\int_{x_\nu}^{x_\nu+2h} y\, dx = 2hf_\nu + 2h(f_{\nu+1}-f_\nu) + \frac{1}{2}\left(\frac{8}{3}h - 2h\right)(f_{\nu+2}-2f_{\nu+1}+f_\nu)$$

$$= \frac{h}{3}(f_\nu + 4f_{\nu+1} + f_{\nu+2}).$$

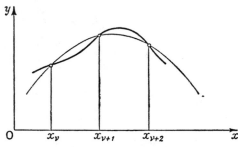

Fig. 2.—Simpson's rule

This represents the required approximation to the area of our strip, $I_\nu + I_{\nu+1}$.

If we now assume that $n = 2m$, i.e. that n is an even number, by the addition of the areas of such strips we obtain Simpson's rule:

$$I \approx \frac{4h}{3}(f_1 + f_3 + + \ldots f_{2m-1})$$

$$+ \frac{2h}{3}(f_2 + f_4 + \ldots + f_{2m-2}) + \frac{h}{3}(f_0 + f_{2m}).$$

4. Examples.

We now apply these methods to the calculation of $\log_e 2 = \int_1^2 \dfrac{dx}{x}$. If we divide the integral from 1 to 2 into ten equal parts, h will be equal to $\frac{1}{10}$, and by the trapezoid formula we obtain

$$
\begin{array}{ll}
x_1 = 1 \cdot 1 & f_1 = 0 \cdot 90909 \\
x_2 = 1 \cdot 2 & f_2 = 0 \cdot 83333 \\
x_3 = 1 \cdot 3 & f_3 = 0 \cdot 76923 \\
x_4 = 1 \cdot 4 & f_4 = 0 \cdot 71429 \\
x_5 = 1 \cdot 5 & f_5 = 0 \cdot 66667 \\
x_6 = 1 \cdot 6 & f_6 = 0 \cdot 62500 \\
x_7 = 1 \cdot 7 & f_7 = 0 \cdot 58824 \\
x_8 = 1 \cdot 8 & f_8 = 0 \cdot 55556 \\
x_9 = 1 \cdot 9 & f_9 = 0 \cdot 52632 \\
\end{array}
$$

$$\text{Sum} \quad 6 \cdot 18773$$

$$
\begin{array}{ll}
x_0 = 1 \cdot 0 & \tfrac{1}{2} f_0 = 0 \cdot 5 \\
x_{10} = 2 \cdot 0 & \tfrac{1}{2} f_{10} = 0 \cdot 25 \\
\end{array}
$$

$$6 \cdot 93773 \times \tfrac{1}{10}$$

$$\log_e 2 \approx 0 \cdot 69377$$

This value, as was to be expected, is too large, since the curve has its convex side turned towards the x-axis.

By the tangent rule, we have

$$
\begin{array}{ll}
x_0 + \tfrac{1}{2} h = 1 \cdot 05 & f_{1/2} = 0 \cdot 95238 \\
x_1 + \tfrac{1}{2} h = 1 \cdot 15 & f_{3/2} = 0 \cdot 86957 \\
x_2 + \tfrac{1}{2} h = 1 \cdot 25 & f_{5/2} = 0 \cdot 80000 \\
x_3 + \tfrac{1}{2} h = 1 \cdot 35 & f_{7/2} = 0 \cdot 74074 \\
x_4 + \tfrac{1}{2} h = 1 \cdot 45 & f_{9/2} = 0 \cdot 68966 \\
x_5 + \tfrac{1}{2} h = 1 \cdot 55 & f_{11/2} = 0 \cdot 64516 \\
x_6 + \tfrac{1}{2} h = 1 \cdot 65 & f_{13/2} = 0 \cdot 60606 \\
x_7 + \tfrac{1}{2} h = 1 \cdot 75 & f_{15/2} = 0 \cdot 57143 \\
x_8 + \tfrac{1}{2} h = 1 \cdot 85 & f_{17/2} = 0 \cdot 54054 \\
x_9 + \tfrac{1}{2} h = 1 \cdot 95 & f_{19/2} = 0 \cdot 51282 \\
\end{array}
$$

$$6 \cdot 92836 \times \tfrac{1}{10}$$

$$\log_e 2 \approx 0 \cdot 69284$$

Owing to the convexity of the curve, this value is too small.

For the same set of subdivisions we obtain the most exact result by means of Simpson's rule. We have

$x_1 = 1\cdot1$	$f_1 = 0\cdot90909$	$x_2 = 1\cdot2$	$f_2 = 0\cdot83333$
$x_3 = 1\cdot3$	$f_3 = 0\cdot76923$	$x_4 = 1\cdot4$	$f_4 = 0\cdot71429$
$x_5 = 1\cdot5$	$f_5 = 0\cdot66667$	$x_6 = 1\cdot6$	$f_6 = 0\cdot62500$
$x_7 = 1\cdot7$	$f_7 = 0\cdot58824$	$x_8 = 1\cdot8$	$f_8 = 0\cdot55556$
$x_9 = 1\cdot9$	$f_9 = 0\cdot52632$		

Sum $2\cdot72818 \times 2$

Sum $3\cdot45955 \times 4$

$5\cdot45636$

$13\cdot83820$ $\qquad 13\cdot83820$

$x_0 = 1\cdot0 \qquad f_0 = 1\cdot0$
$x_{10} = 2\cdot0 \qquad f_{10} = 0\cdot5$

$20\cdot79456 \times \frac{1}{30}$

$\log_e 2 \approx 0\cdot69315$

As a matter of fact

$$\log_e 2 = 0\cdot693147 \ldots$$

5. Estimation of the Error.

It is easy to give an estimate of the error in each of our methods of integration if the derivatives of the function $f(x)$ are known throughout the interval of integration. We take M_1, M_2, ... as the upper bounds of the absolute values of the first, second, ... derivatives, respectively; that is, we assume that throughout the interval $|f^{(\nu)}(x)| < M_\nu$. Then the estimation formulæ are as follows:

For the rectangle rule

$$|I_\nu - hf_\nu| < \frac{1}{2} M_1 h^2 \text{ or } \left| I - h \sum_{\nu=0}^{n-1} f_\nu \right| < \frac{1}{2} M_1 n h^2 = \frac{1}{2} M_1 (b-a) h.$$

For the tangent rule

$$|I_\nu - hf_{\nu+\frac{1}{2}}| < \frac{M_2}{24} h^3 \text{ or } \left| I - h \sum_{\nu=0}^{n-1} f_{\nu+\frac{1}{2}} \right| < \frac{M_2}{24} (b-a) h^2.$$

For the trapezoid rule

$$\left| I_\nu - \frac{h}{2}(f_\nu + f_{\nu+1}) \right| < \frac{M_2}{12} h^3.$$

For Simpson's rule

$$\left| I_\nu + I_{\nu+1} - \frac{h}{3}(f_\nu + 4f_{\nu+1} + f_{\nu+2}) \right| < \frac{M_4}{90} h^5.$$

From the last two estimates there also follow estimates for the entire integral I. We see that Simpson's rule has an error of much higher order in the small quantity h than the other rules, so that where M_4 is not too large it is very advantageous for practical calculations. To avoid wearying the reader with the details of the proofs of these estimates, which are fundamentally quite simple, we shall content ourselves with the proof for the tangent formula. For this purpose we expand the function $f(x)$ in the $(\nu + 1)$-th strip by Taylor's theorem:

$$f(x) = f_{\nu+\frac{1}{2}} + \left(x - x_\nu - \frac{h}{2}\right)f'\left(x_\nu + \frac{h}{2}\right) + \frac{1}{2}\left(x - x_\nu - \frac{h}{2}\right)^2 f''(\xi),$$

where ξ is a certain intermediate value in the strip. If we integrate the right-hand side over the interval $x_\nu \leqq x \leqq x_\nu + h$, the integral of the middle term is zero. Since

$$\frac{1}{2}\int_{x_\nu}^{x_\nu+h}\left(x - x_\nu - \frac{h}{2}\right)^2 dx = \frac{h^3}{24},$$

as is easily verified, it follows immediately that

$$\left|\int_{x_\nu}^{x_\nu+h} f(x)\,dx - hf_{\nu+\frac{1}{2}}\right| < M_2\frac{h^3}{24},$$

which proves our assertion.

EXAMPLES

1. From the formula $\dfrac{\pi}{4} = \displaystyle\int_0^1 \frac{1}{1+x^2}\,dx$, calculate π

    (a) using the trapezoid formula with $h = 0.1$;

    (b) using Simpson's rule with $h = 0.1$.

2. Calculate $\displaystyle\int_0^\infty e^{-x^2}\,dx$ numerically to within $\frac{1}{100}$ (cf. p. 496).

3. Calculate $\displaystyle\int_0^1 \frac{1}{\sqrt{1+x^4}}$ numerically with an error less than 0.1.

2. Applications of the Mean Value Theorem and of Taylor's Theorem. The Calculus of Errors

1. The " Calculus of Errors ".

We now come to quite a different type of numerical calculations. These are applications of the mean value theorem, or more generally of Taylor's theorem with remainder, or finally of the infinite Taylor series. As an application which, though simple, is quite important in practice we shall consider the calculus of errors. This rests upon the idea—which lies at the root of the whole of the differential calculus—that a function $f(x)$ which is differentiable a sufficient number of times can be represented in the neighbourhood of a point by a linear function with an error of order less than the first, by a quadratic function with an error of order less than the second, and so on. Let us consider the linear approximation to a function $y = f(x)$. If $y + \Delta y = f(x + \Delta x) = f(x + h)$, by Taylor's theorem we have

$$\Delta y = h f'(x) + \frac{h^2}{2} f''(\xi),$$

where $\xi = x + \theta h \, (0 < \theta < 1)$ is an intermediate value, which need not be more precisely known. If $h = \Delta x$ is small we obtain as a practical approximation

$$\Delta y \approx h f'(x).$$

In other words, we replace the difference quotient by the derivative to which it is approximately equal and the increment of y by the approximately equal linear expression in h.

We use this fairly obvious fact for practical purposes in the following way. Suppose two physical quantities x and y are connected by the relation $y = f(x)$. The question then arises, what effect an inaccuracy in the measurement of x has on the determination of y. If instead of the " true " value x we happen to use the inaccurate value $x + h$, then the corresponding value of y differs from the true value $y = f(x)$ by the amount $\Delta y = f(x + h) - f(x)$. The error is therefore given approximately by the above relation.

We shall understand the use of these relations better if we consider a few examples.

Ex. 1. *The tangent galvanometer.* In determining current by a tangent galvanometer we use the formula $y = c \tan \alpha$, where α is the angle of deflection of the magnetic needle, c is the constant of the apparatus, and $y = I$ is the intensity of the current. Then

$$\frac{dy}{d\alpha} = \frac{c}{\cos^2 \alpha}$$

and therefore $\Delta y \approx \dfrac{c}{\cos^2 \alpha} \Delta \alpha$. The percentage error in the measurement is given by

$$\frac{100 \Delta y}{y} \approx \frac{100 c \Delta \alpha}{c \cos^2 \alpha \tan \alpha} = \frac{200}{\sin 2\alpha} \Delta \alpha.$$

From this we see that the accuracy reaches the greatest possible value, i.e. to a given error in the measurement of the angle there corresponds

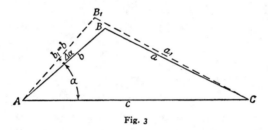

Fig. 3

the least possible error in the determination of the current, when the angle α is equal to $\pi/4$ or $45°$.

In particular, let us suppose that it is possible to read the tangent galvanometer to within half a degree; then $|\Delta \alpha|$ in radians $< \frac{1}{2} \times 0 \cdot 01745 \ldots$, and the percentage error will be $\dfrac{1 \cdot 745}{\sin 2\alpha}$. If the reading on the galvanometer is $30°$, $\sin 2\alpha = \frac{1}{2} \sqrt{3} = \frac{1}{2} \times 1 \cdot 73205 \ldots$, and the percentage error is less than $2 \times \dfrac{1 \cdot 745}{1 \cdot 732}$, which is about 2 per cent.

Ex. 2. In a triangle ABC (cf. fig. 3) we suppose that the sides b and c are measured accurately, while the angle $\alpha = x$ can only be measured with an error $|\Delta x| < \delta$. Within what limit of error does the value $y = a = \sqrt{(b^2 + c^2 - 2bc \cos \alpha)}$ vary?

We have
$$\Delta a \approx \frac{1}{a} bc \sin \alpha \, \Delta \alpha;$$

the percentage error is therefore $\dfrac{100 \Delta a}{a} \approx \dfrac{100 bc}{a^2} \sin \alpha \, \Delta \alpha$. If we take the

special case where $b = 400$ metres, $c = 500$ metres, and $\alpha = 60°$, then
by the cosine formula $y = a = 458\cdot2576$ metres, and

$$\Delta a \approx \frac{200000}{458\cdot2576} \times \frac{1}{2} \sqrt{3}\,\Delta\alpha.$$

If $\Delta\alpha$ can be measured to within ten seconds of arc, that is, if

$$\Delta\alpha = 10'' = 4848 \times 10^{-8} \text{ radians,}$$

we find that at worst

$$\Delta a \approx 1\cdot83 \text{ cm.;}$$

that is. the error is at most about 0·004 per cent.

Ex. 3. The following illustrates a type of application of the above
methods by which we can often save ourselves considerable trouble in
physical problems.

It is known experimentally that if an iron rod has the length l_0 at
temperature 0, then at temperature t its length will be $l = l_0(1 + \alpha t)$,
where α depends only on the material of which the rod is made. If now
a pendulum clock keeps correct time at temperature t_1, how many
seconds will it lose per day if the temperature rises to t_2?

For the period of oscillation we have the formula

$$T(l) = 2\pi \sqrt{\frac{l}{g}}, \quad \text{whence} \quad \frac{dT}{dl} = \frac{\pi}{\sqrt{lg}}.$$

Hence if the change of length is Δl the corresponding change in the
period of oscillation is

$$\Delta T \approx \frac{\pi\,\Delta l}{\sqrt{l_1 g}},$$

where $l_1 = l_0(1 + \alpha t_1)$ and $\Delta l = \alpha l_0(t_2 - t_1)$. This is the time lost per
oscillation. The time lost per second is $\Delta T/T \approx \Delta l/2l_1$; hence in one day
the clock loses $43200\,\Delta l/l_1$ seconds.

Here the application of our methods has saved us a number of multi-
plications and two extractions of the square root. In the longer direct
process, moreover, we should finally have to subtract $T(l_1)$ from the almost
equal value $T(l_2)$, and a very small error in calculation would cause a relatively large percentage error in the result.*

In this case, and in most cases where the function under consideration
has several factors or fractional indices, we can reduce the calculation
even more by taking the logarithms of both sides before differentiating.
In the present example we have

$$\log T = \log 2\pi - \frac{1}{2}\log g + \frac{1}{2}\log l;$$

by differentiating, we have

$$\frac{dT}{dl}\bigg/ T = \frac{1}{2l}.$$

---

* It is a point of this nature that makes the calculations of applied optics
so extremely laborious.

If we replace $\dfrac{dT}{dl}$ by $\dfrac{\Delta T}{\Delta l}$ this gives

$$\frac{\Delta T}{T} = \frac{\Delta l}{2l},$$

in agreement with the preceding result.

2. Evaluation of π.

Gregory's series,* $\dfrac{\pi}{4} = 1 - \dfrac{1}{3} + \dfrac{1}{5} - \dfrac{1}{7} + - \ldots$, which we obtained in Chap. VI, § 1 (p. 319) using the series for the inverse tangent, is not suitable for the calculation of π, on account of the slowness of its convergence. We may, however, calculate π with comparative ease by the following artifice. From the addition theorem for the tangent,

$$\tan(a + \beta) = \frac{\tan a + \tan \beta}{1 - \tan a \, \tan \beta},$$

if we change to the inverse functions $a = \text{arc} \tan u$, $\beta = \text{arc} \tan v$, we obtain the formula

$$\text{arc} \tan u + \text{arc} \tan v = \text{arc} \tan \left(\frac{u + v}{1 - uv} \right).$$

If we now choose u and v in such a way that $\dfrac{u + v}{1 - uv} = 1$, we obtain the value $\dfrac{\pi}{4}$ on the right-hand side, and if u and v are small numbers we can easily calculate the left-hand side by means of known series. If, for example, we put $u = \dfrac{1}{2}$, $v = \dfrac{1}{3}$, as Euler did, we obtain

$$\frac{\pi}{4} = \text{arc} \tan \frac{1}{2} + \text{arc} \tan \frac{1}{3}.$$

If we further notice that $\left(\dfrac{1}{3} + \dfrac{1}{7} \right) \div \left(1 - \dfrac{1}{21} \right) = \dfrac{1}{2}$, we have $\text{arc} \tan \dfrac{1}{2} = \text{arc} \tan \dfrac{1}{3} + \text{arc} \tan \dfrac{1}{7}$, so that

$$\frac{\pi}{4} = 2 \, \text{arc} \tan \frac{1}{3} + \text{arc} \tan \frac{1}{7}.$$

Using this formula Vega calculated the number π to 140 places.

* Sometimes called Leibnitz's series.

By means of the equation $\left(\dfrac{1}{5} + \dfrac{1}{8}\right) \div \left(1 - \dfrac{1}{40}\right) = \dfrac{1}{3}$, we further obtain

$$\text{arc tan} \frac{1}{3} = \text{arc tan} \frac{1}{5} + \text{arc tan} \frac{1}{8}$$

or

$$\frac{\pi}{4} = 2\,\text{arc tan} \frac{1}{5} + \text{arc tan} \frac{1}{7} + 2\,\text{arc tan} \frac{1}{8}.$$

This expansion is extremely useful for the calculation of π by means of the series $\text{arc tan}\, x = x - \dfrac{x^3}{3} + \dfrac{x^5}{5} - + \dots\,;$ for if we substitute for x the value $\dfrac{1}{5}, \dfrac{1}{7},$ or $\dfrac{1}{8}$ we obtain with but few terms a high degree of accuracy, since the terms diminish rapidly. We can, however, perform the calculation even more conveniently if we base it on the formula

$$\frac{\pi}{4} = \text{arc tan} \frac{120}{119} - \text{arc tan} \frac{1}{239} = 4\,\text{arc tan} \frac{1}{5} - \text{arc tan} \frac{1}{239},$$

obtained by considerations similar to those above.

3. Calculation of Logarithms.

For the numerical calculation of logarithms we transform the logarithmic series $\dfrac{1}{2} \log \dfrac{1+x}{1-x} = x + \dfrac{x^3}{3} + \dfrac{x^5}{5} + \dots (|x| < 1)$ where $0 < x < 1$ by the substitution

$$\frac{1+x}{1-x} = \frac{p^2}{p^2 - 1}, \quad x = \frac{1}{2p^2 - 1}$$

into the series

$$\log p = \frac{1}{2} \log (p - 1) + \frac{1}{2} \log (p + 1) + \frac{1}{2p^2 - 1}$$
$$+ \frac{1}{3(2p^2 - 1)^3} + \dots ,$$

where $2p^2 - 1 > 1$, that is, $p^2 > 1$. If p is an integer and $p + 1$ can be resolved into smaller integral factors, this last series expresses the logarithm of p by the logarithms of smaller integers plus a series whose terms diminish very rapidly and whose sum can therefore be calculated accurately enough by use of only a few terms. From this series we can therefore calculate successively the logarithms of any prime number, and hence of any number, provided we have already calculated the value of $\log 2$.

The accuracy of this determination of $\log p$ can be estimated more easily by means of the geometric series than from the general formula for the remainder. For the remainder R_n of the series, i.e. the sum of all the terms following the term $\dfrac{1}{n(2p^2-1)^n}$, we have

$$R_n < \frac{1}{(n+2)(2p^2-1)^{n+2}}\left(1+\frac{1}{(2p^2-1)^2}+\frac{1}{(2p^2-1)^4}+\cdots\right)$$
$$= \frac{1}{(n+2)(2p^2-1)^n}\cdot\frac{1}{(2p^2-1)^2-1},$$

and this formula immediately gives the required estimate of the error.

Let us for example calculate $\log_e 7$, using the first four terms of the series. We have

$$p = 7, \quad 2p^2 - 1 = 97,$$

$$\log 7 = 2\log 2 + \frac{1}{2}\log 3 + \frac{1}{97} + \frac{1}{3.97^3} + \cdots;$$

$$\frac{1}{97} \approx 0\cdot01030928, \quad \frac{1}{3.97^3} \approx 0\cdot00000037,$$

$$2\log 2 \approx 1\cdot38629436, \quad \frac{1}{2}\log 3 \approx 0\cdot54930614;$$

hence

$$\log_e 7 \approx 1\cdot94591015.$$

Estimation of the error gives

$$R_n < \frac{1}{5.97^3} \times \frac{1}{97^2-1} < \frac{1}{36\times10^9}.$$

We must, however, note that each of the four numbers which we have added is only given to within an error of 5×10^{-9}, so that the last place in the value of $\log 7$ given above might be wrong by 2. As a matter of fact, however, the last place is right also.

1. To measure the height of a hill, a tower 100 metres high on top of the hill is observed from the plain. The angle of elevation of the base of the tower is 42° and the tower itself subtends an angle of 6°. What are the limits of error in the determination of the height if the angle 42° is subject to an error of 1°?

2. Calculate $\log_e 2$ to three decimal places by means of an expansion in series.

3. Calculate $\log_e 5$ to six decimal places, using the values of $\log_e 2$ and $\log_e 3$ given in the text.

4. Calculate π to five decimal places, using any one of the formulæ in sub-section 2 (pp. 352–3).

3. NUMERICAL SOLUTION OF EQUATIONS

In conclusion we shall add some remarks about the numerical solution of the equation $f(x) = 0$, where $f(x)$ need not necessarily be a polynomial.* Every such numerical method is based on the plan of starting with some known approximation x_0 of one of the roots and then improving this approximation. Whence this first approximation for the desired root of the equation is found, and how good the approximation is, do not particularly matter. We may perhaps take a rough guess as a first approximation, or better, obtain it from the graph of the function $y = f(x)$, whose intersection with the x-axis gives the required root (of course with an error depending on the scale and the accuracy of the drawing).

1. Newton's Method.

The following procedure which comes down to us from Newton is based on the fundamental principle of the differential calculus—the replacing of a curve by a straight line, the tangent, in the immediate neighbourhood of the point of contact. If we have an approximate value x_0 for a root of the equation $f(x) = 0$, we consider the point on the graph of the function $y = f(x)$ whose co-ordinates are $x = x_0$, $y = f(x_0)$. What we wish to find is the intersection of the curve with the x-axis; as an approximation to this we find the point where the tangent at the point

---

* Here, of course, we are only concerned with the determination of *real* roots of $f(x) = 0$.

$x = x_0$, $y = f(x_0)$ intersects the x-axis. The abscissa x_1 of this intersection of the tangent with the x-axis will then represent a new and, under certain circumstances, a better approximation than x_0 to the required root of the equation.

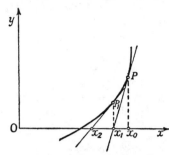

Fig. 4.—Newton's method of approximation

In virtue of the geometrical meaning of the derivative, fig. 4 at once gives

$$\frac{f(x_0)}{x_0 - x_1} = f'(x_0).$$

From this we obtain the formula for the calculation of the new approximation x_1:

$$x_1 = x_0 - \frac{f(x_0)}{f'(x_0)}.$$

If by this procedure we have found an approximation better than x_0, then we repeat the process to find x_2, and so on, and if the curve is of the form shown in fig. 3 these approximations will approach more and more nearly to the required solution.

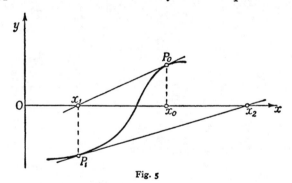

Fig. 5

The usefulness of this process depends essentially on the nature of the curve $y = f(x)$. In fig. 4 we see that the successive estimates converge with greater and greater accuracy to the required root. This is due to the fact that the curve has its convex side turned towards the x-axis. But in fig. 5 we see that if we choose the original value x_0 badly, our construction does not lead to the required root at all. From this we see that in using Newton's method we must examine each individual case to determine with what degree of accuracy we have really solved the equation. We shall return to this subject on p. 359.

2. The Rule of False Position.

Newton's method, in which the tangent to the curve plays a decisive part, is only the limiting case of an older method, known as the rule of false position, in which the secant appears in place of the tangent. Let us assume that we know two points (x_0, y_0) and (x_1, y_1) in the neighbourhood of the required intersection with the x-axis. If we replace the curve by the secant joining these two points the intersection of this secant with the x-axis will in certain circumstances be an improved approximation to the required root of the equation.

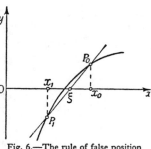

Fig. 6.—The rule of false position

If the abscissa of this point is denoted by ξ, we have (fig. 6) the equation

$$\frac{\xi - x_0}{f(x_0)} = \frac{\xi - x_1}{f(x_1)},$$

and from this we calculate ξ:

$$\xi = \frac{x_0 f(x_1) - x_1 f(x_0)}{f(x_1) - f(x_0)}$$

$$= \frac{x_0 f(x_1) - x_0 f(x_0) + x_0 f(x_0) - x_1 f(x_0)}{f(x_1) - f(x_0)}$$

or

$$\xi = x_0 - \frac{f(x_0)}{\{f(x_1) - f(x_0)\}/(x_1 - x_0)}.$$

This formula, which determines the further approximation ξ from x_0 and x_1, is called the *rule of false position*. We can use it with advantage if one value of the function is positive and the other negative, say as in fig. 6, where $y_0 > 0$ and $y_1 < 0$. Repetition of this process will always lead us to the required result if at each step we use a positive and a negative value of the function, between which the required root must necessarily lie.

The above formula of Newton results from the rule of false position as a limiting case if we let x_1 tend to x_0. For the denominator of the second term on the right-hand side of the statement of the rule of false position tends to $f'(x_0)$ as x_1 tends to x_0.

3. The Method of Iteration.

Another means of approximation to the roots of an equation $f(x) = 0$ is the method of *iteration*. Here we put $\phi(x) = f(x) + x$ and write our equation in the form $x = \phi(x)$. We then suppose that ξ is the true value of a solution of our equation, and x_0 a first approximation. We obtain a second approximation x_1 by putting $x_1 = \phi(x_0)$, a third approximation x_2 by putting $x_2 = \phi(x_1)$, &c. To investigate the convergence of these approximations, we apply the mean value theorem; recalling that $\xi = \phi(\xi)$, we have

$$\xi - x_1 = \phi(\xi) - \phi(x_0) = (\xi - x_0)\phi'(\bar{\xi})$$

where $\bar{\xi}$ lies between ξ and x_0. This shows that if for

$$|\xi - x| < |\xi - x_0|$$

the derivative $\phi'(x)$ is less in absolute value than $k < 1$, then the successive approximations converge, for

$$|\xi - x_1| < k|\xi - x_0|, \quad |\xi - x_2| < k^2|\xi - x_0|, \ldots,$$
$$|\xi - x_n| < k^n|\xi - x_0|,$$

and the errors therefore tend to zero. The smaller the absolute value of the derivative $\phi'(x)$ near ξ, the more rapid is the convergence.

If $\phi'(x) > 1$ in the neighbourhood of ξ the approximations no longer tend to ξ. We can then use the inverse function or else the following device. We choose a first approximation x_0, calculate $A = f'(x_0)$, and write

$$\phi(x) = -\frac{1}{A}f(x) + x.$$

Then the equation $f(x) = 0$ can be written $x = \phi(x)$, and here $\phi'(x) = -\frac{1}{A}f'(x) + 1$, which has the value 0 at $x = x_0$ and hence will usually be less in absolute value than a constant $k < 1$ if $|\xi - x| < |\xi - x_0|$.

Returning to Newton's method, we can now investigate its suitability for application at any given point. The equation $f(x) = 0$ is equivalent to $x = \phi(x) = x - \dfrac{f(x)}{f'(x)}$, provided that $f'(x) \neq 0$. Applying the method of iteration to this last equation, from a first approximation x_0 we obtain a second, $x_1 = x_0 - \dfrac{f(x_0)}{f'(x_0)}$; in other words, the same second approximation as Newton's method gives when applied to the equation $f(x) = 0$. We thus see that the smaller the value of

$$\phi'(x) = \frac{f(x)f''(x)}{(f'(x))^2},$$

the more rapidly do the successive approximations converge. In words, Newton's formula converges rapidly for large values of $f'(x_0)$ and small values of $f(x_0)$ and the curvature, as intuition would lead us to suspect.

We can also obtain an estimate of the accuracy of Newton's method, if we recall that, since $f(\xi) = 0$, the derivative $\phi'(\xi) = 0$. Applying Taylor's theorem, we have

$$\xi - x_1 = \phi(\xi) - \phi(x_0) = \frac{(\xi - x_0)^2}{2} \phi''(\bar{\xi}),$$

where $\bar{\xi}$ lies between ξ and x_0. Thus, if the error of the original estimate is small, the method converges much more rapidly than the method of iteration applied directly to $f(x) = 0$.

For example, if

$$\phi''(x) = \frac{\{f'(x)\}^2 f''(x) + f'(x) f(x) f'''(x) - 2f(x)\{f''(x)\}^2}{\{f'(x)\}^3} \qquad (a)$$

is everywhere less than 10, then a first approximation which is in error by less than $\cdot 001$ will yield a second approximation with an error of less than $(\cdot 001)^2 \times 10 \div 2 = \cdot 000005$.

4. Examples.

As an example we consider the equation

$$f(x) = x^3 - 2x - 5 = 0.$$

For $x_0 = 2$, we have $f(x_0) = -1$, while for $x_1 = 2 \cdot 1$ we have $f(x_1) = 0 \cdot 061$. By Newton's method,

$$x_2 = x_1 - \frac{f(x_1)}{f'(x_1)} = 2 \cdot 1 - \frac{0 \cdot 061}{3(2 \cdot 1)^2 - 2} = 2 \cdot 1 - \cdot 005431 = 2 \cdot 094569.$$

To estimate the error we find from the expression (*a*) above that $\varphi''(x)$ is about 1 and certainly less than 2 near $x = 2$. Moreover, the error of our first approximation is certainly less than 1/160, for the secant joining the points $x = 2$, $y = -1$ and $x = 2\cdot1$, $y = \cdot061$ cuts the x-axis at a distance less than 1/160 from $x = 2\cdot1$, and the curve, lying under the secant, cuts it even nearer $2\cdot1$. So the error * of our second approximation is less than

$$\frac{1}{2} \cdot \frac{2}{(160)^2} = \frac{1}{25,600} < \cdot00004.$$

If this degree of accuracy is not sufficient, we can repeat the process, calculating $f(x_2)$ and $f'(x_2)$ for $x_2 = 2\cdot094569$, and obtain a third approximation x_3 with an error less than $\dfrac{1}{(25,600)^2} < \cdot000000002$.

As a second example, let us solve the equation $f(x) = x \log_{10} x - 2 = 0$. We have $f(3) = -0\cdot6$ and $f(4) = +0\cdot4$, and therefore use $x_0 = 3\cdot5$ as a first approximation. Then by using ten-figure logarithmic tables we obtain the successive approximations

$$x_0 = 3\cdot5$$
$$x_1 = 3\cdot598$$
$$x_2 = 3\cdot5972849$$
$$x_3 = 3\cdot5972850235.$$

EXAMPLES

1. Using Newton's method, find the positive root of $x^6 + 6x - 8 = 0$ to four decimal places.

2. Find to four places the root of $x = \tan x$ between π and 2π. *Prove* that the result is accurate to four places.

3. Using Newton's method, find the value of x for which

$$\int_0^x \frac{u^2}{1 + u^2}\, du = \frac{1}{2}.$$

4. Find the roots of the equation $x = 2 \sin x$ to two places.

5. Determine the positive roots of the equation $x^5 - x - 0\cdot2 = 0$ by the method of iteration.

6. Determine the least positive root of $x^4 - 3x^3 + 10x - 10 = 0$ by the method of iteration.

7. Find the roots of $x^3 - 7x^2 + 6x + 20 = 0$ to four decimal places.

---

* Another way of estimating the error, without reference to the secant, is as follows: if we estimate that the error is less than 1/20, the error of our second approximation is less than $1/20^2 = \cdot0025$. Hence the root differs from $2\cdot1$ by less than $(2\cdot1 - 2\cdot0945) + \cdot0025 = \cdot008$. Therefore the error was not merely less than 1/20, but less than $\cdot008$, so that x_2 is in error by less than $(\cdot008)^2 = \cdot000064$.

Appendix to Chapter VII

STIRLING'S FORMULA

In very many applications, especially in statistics and in the theory of probability, we find it necessary to have a simple approximation to $n!$ as an elementary function of n. Such an expression is given by the following theorem, which bears the name of its discoverer Stirling:

As $n \to \infty$ $$\frac{n!}{\sqrt{2\pi}\, n^{n+\frac{1}{2}} e^{-n}} \to 1;$$

more exactly,

$$\sqrt{2\pi}\, n^{n+\frac{1}{2}} e^{-n} < n! < \sqrt{2\pi}\, n^{n+\frac{1}{2}} e^{-n}\left(1 + \frac{1}{4n}\right).$$

In other words, the expressions $n!$ and $\sqrt{2\pi}\, n^{n+\frac{1}{2}} e^{-n}$ differ only by a small percentage when the value of n is large—as we say, the two expressions are *asymptotically* equal—and at the same time the factor $1 + 1/4n$ gives us an estimate of the degree of accuracy of the approximation.

We are led to this remarkable formula if we attempt to evaluate the area under the curve $y = \log x$. By integration (p. 220) we find that A_n, the exact area under this curve between the ordinates $x = 1$ and $x = n$, is given by

$$\int_1^n \log x\, dx = x \log x - x \Big|_1^n = n \log n - n + 1.$$

If, however, we estimate the area by the trapezoid rule, erecting ordinates at $x = 1$, $x = 2$, ..., $x = n$ as in fig. 7, we obtain T_n, an approximate value for the area:

$$T_n = \log 2 + \log 3 + \ldots + \log(n-1) + \frac{1}{2}\log n$$

$$= \log n! - \frac{1}{2}\log n.$$

If we make the reasonable assumption that A_n and T_n are of the same order of magnitude, we find at once that $n!$ and $n^{n+\frac{1}{2}} e^{-n}$ are of the same order of magnitude, which is essentially what is stated in Stirling's formula.

To make this argument precise, we first show that the difference $a_n = A_n - T_n$ is bounded, from which it will immediately follow that $T_n = A_n\left(1 - \dfrac{a_n}{A_n}\right)$ is of the same order of magnitude as A_n.

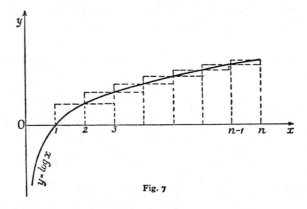

Fig. 7

The difference $a_{k+1} - a_k$ is the difference between the area under the curve and the area under the secant in the strip $k \leq x \leq k+1$.

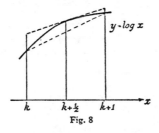

Fig. 8

Since the curve is concave downward and lies above the secant, $a_{k+1} - a_k$ is positive, and $a_n = (a_n - a_{n-1}) + (a_{n-1} - a_{n-2}) + \ldots + (a_2 - a_1) + a_1$ is monotonic increasing. Moreover, the difference $a_{k+1} - a_k$ is clearly less (cf. fig. 8) than the difference between the area under the tangent at $x = k + \dfrac{1}{2}$ and the area under the secant; hence we have the inequality

$$a_{k+1} - a_k < \log\left(k + \frac{1}{2}\right) - \frac{1}{2}\log k - \frac{1}{2}\log(k+1)$$

$$= \frac{1}{2}\log\left(1 + \frac{1}{2k}\right) - \frac{1}{2}\log\left(1 + \frac{1}{2\left(k + \frac{1}{2}\right)}\right)$$

$$< \frac{1}{2}\log\left(1 + \frac{1}{2k}\right) - \frac{1}{2}\log\left(1 + \frac{1}{2(k+1)}\right).$$

If we add these inequalities for $k = 1, 2, \ldots, n-1$, all the terms on the right except two will cancel out, and (since $a_1 = 0$), we have

$$a_n < \frac{1}{2} \log \frac{3}{2} - \frac{1}{2} \log \left(1 + \frac{1}{2n}\right) < \frac{1}{2} \log \frac{3}{2}.$$

Hence a_n is bounded, and being monotonic increasing it tends to a limit a as $n \to \infty$. Our inequality for $a_{k+1} - a_k$ now gives us

$$a - a_n = \sum_{k=n}^{\infty} (a_{k+1} - a_k) < \frac{1}{2} \log \left(1 + \frac{1}{2n}\right).$$

Since by definition $A_n - T_n = a_n$, we now have

$$\log n! = 1 - a_n + \left(n + \frac{1}{2}\right) \log n - n,$$

or, writing $\alpha_n = e^{1-a_n}$,

$$n! = \alpha_n n^{n+\frac{1}{2}} e^{-n}.$$

The sequence α_n is monotonic *decreasing* and tends to the limit $\alpha = e^{1-a}$; hence

$$1 < \frac{\alpha_n}{\alpha} = e^{a - a_n} < e^{\frac{1}{2} \log (1 + 1/2n)} = \sqrt{\left(1 + \frac{1}{2n}\right)} < 1 + \frac{1}{4n}.$$

Hence we have

$$\alpha n^{n+\frac{1}{2}} e^{-n} < n! < \alpha n^{n+\frac{1}{2}} e^{-n} \left(1 + \frac{1}{4n}\right).$$

It only remains for us to find the actual value of the limit α. Here we make use of the formula proved in Chap. IV, § 4 (p. 225):

$$\sqrt{\pi} = \lim_{n \to \infty} \frac{(n!)^2 2^{2n}}{(2n)! \sqrt{n}}.$$

Replacing $n!$ by $\alpha_n n^{n+\frac{1}{2}} e^{-n}$ and $(2n)!$ by $\alpha_{2n} 2^{2n+\frac{1}{2}} n^{2n+\frac{1}{2}} e^{-2n}$, we immediately obtain

$$\sqrt{\pi} = \lim_{n \to \infty} \frac{\alpha_n^2}{\alpha_{2n} \sqrt{2}} = \frac{\alpha^2}{\alpha \sqrt{2}},$$

whence $\alpha = \sqrt{2\pi}$. The proof of Stirling's formula is thus completed.

In addition to its theoretical interest, Stirling's formula is a very useful tool for the numerical calculation of $n!$ when n is large. Instead of multiplying together a large number of integers, we have merely to calculate Stirling's expression by means of logarithms, which involves far fewer operations. Thus for $n = 10$ we obtain the value 3598696 for Stirling's expression (using seven-figure tables), while the exact value of 10! is 3628800. The percentage error is barely $\frac{5}{6}$ per cent.

EXAMPLE

Prove that $\lim\limits_{n \to \infty} \dfrac{\sqrt[n]{n!}}{n} = \dfrac{1}{e}$.

CHAPTER VIII

Infinite Series and Other Limiting Processes

PRELIMINARY REMARKS

The geometric series, Taylor's series, and a number of special examples which we have already met in this book suggest that we may well study those limiting processes which are called the summation of *infinite series* from a rather more general point of view. From its nature any limiting value

$$S = \lim_{n \to \infty} s_n$$

can be written as an infinite series; for if n takes the values $1, 2, 3, \ldots$, we need only put $a_n = s_n - s_{n-1}$ for $n > 1$ and $a_1 = s_1$ to obtain

$$s_n = a_1 + a_2 + \ldots + a_n,$$

and the value S thus appears as the limit of s_n, the sum of n terms as n increases. We express this fact by saying that S is the " sum of the infinite series "

$$a_1 + a_2 + a_3 + \ldots$$

Thus an infinite series is simply a way of representing a limit where each successive approximation is found from the preceding by adding one more term. The expression of a number as a decimal is in principle merely the representation of a number a in the form of an infinite series $a = a_1 + a_2 + a_3 + \ldots$, where, if $0 \leqq a \leqq 1$, the term a_n is put equal to $a_n \times 10^{-n}$ and a_n is a whole number between 0 and 9 inclusive. Since every limiting value can be written in the form of an infinite series, it may seem that a special study of series is superfluous. But in many cases it happens that limiting values occur naturally in the form of infinite series, which often exhibit particularly simple laws of formation. Of course it is not true that *every* series has an easily recognizable law of formation. For example, the number π can certainly be represented as a decimal, yet we know no simple law enabling us to state the value of an arbitrary digit, say the 7000th, of this decimal. If, however, we set aside the representation of π by a decimal and consider Gregory's series instead, we have an expression with a perfectly clear general law of formation.

Analogous to infinite series, in which the approximations to the limit are found by repeated addition of new terms, are *infinite products*, in which the approximations to the limit arise from repeated multiplication by new factors. We shall not go deeply into the theory of infinite products, however; the principal subject of this chapter and of the following chapter will be infinite series.

1. The Concepts of Convergence and Divergence

1. The Fundamental Ideas.

We consider an infinite series whose "general term" we denote * by a_n; the series is then of the form

$$a_1 + a_2 + \ldots = \sum_{\nu=1}^{\infty} a_\nu.$$

The symbol on the right with the summation sign is merely an abbreviated way of writing the expression on the left.

If as n increases the *n-th partial sum*

$$s_n = a_1 + a_2 + \ldots + a_n = \sum_{\nu=1}^{n} a_\nu$$

approaches a limit

$$S = \lim_{n \to \infty} s_n$$

we say that the series is *convergent*, otherwise we say that it is *divergent*. In the first case we call S the *sum of the series*.

We have already met with many examples of convergent

---

* For formal reasons we include the possibility that certain of the numbers a_n may be zero. If *all* the a_n's from a number N onward (i.e. when $n > N$) vanish, we speak of a *terminating series*.

series; for instance, the geometric series $1 + q + q^2 + \ldots$, which converges to the sum $1/(1 - q)$ when $|q| < 1$, Gregory's series, the series for log 2, the series for e, and others. In the language of infinite series, Cauchy's convergence test (cf. Chap. I, § 6, p. 40) is expressed as follows:

A necessary and sufficient condition for the convergence of a series is that the number

$$|s_m - s_n| = |a_{n+1} + a_{n+2} + \ldots + a_m|$$

becomes arbitrarily small if m *and* n *are chosen sufficiently large* (m > n). *In other words:* A series converges if, and only if, the following condition is fulfilled: if a positive number ϵ is given, no matter how small, it is possible to choose an index N = N(ϵ), which in general increases beyond all bounds as $\epsilon \to 0$, in such a way that the above expression $|s_m - s_n|$ is less than ϵ, provided only that m > N and n > N.*

We can make the meaning of the convergence test clearer by considering the geometric series where $q = \frac{1}{2}$. If we choose $\epsilon = \frac{1}{10}$, we need only take $N = 4$. For

$$|s_m - s_n| = \frac{1}{2^n} + \ldots + \frac{1}{2^{m-1}}$$

$$= \frac{1}{2^{n-1}}\left(\frac{1}{2} + \frac{1}{2^2} + \ldots + \frac{1}{2^{m-n}}\right) < \frac{1}{2^{n-1}}$$

and

$$\frac{1}{2^{n-1}} < \frac{1}{10} \text{ if } n > 4.$$

If we choose ϵ equal to $\frac{1}{100}$, it is sufficient to take 7 as the corresponding value of N, as may easily be verified.

Obviously it is a *necessary* condition for the convergence of a series that

$$\lim_{n \to \infty} a_n = 0.$$

Otherwise the convergence criterion certainly cannot be fulfilled. But this necessary condition is by no means *sufficient* for convergence; on the contrary, it is easy to find infinite series whose general term a_n approaches 0 as n increases, but whose sum does not exist, as the partial sum s_n increases without limit as n increases.

An example of this is the series

$$1 + \frac{1}{\sqrt{2}} + \frac{1}{\sqrt{3}} + \ldots + \frac{1}{\sqrt{n}} + \ldots.$$

the general term of which is $\frac{1}{\sqrt{n}}$. We immediately see that

$$s_n > \frac{1}{\sqrt{n}} + \ldots + \frac{1}{\sqrt{n}} = \frac{n}{\sqrt{n}} = \sqrt{n}.$$

The n-th partial sum increases beyond all bounds as n increases, and therefore the series diverges.

The same is true for the classic example of the *harmonic series*

$$1 + \frac{1}{2} + \frac{1}{3} + \frac{1}{4} + \ldots.$$

Here $a_{n+1} + \ldots + a_{2n} = \frac{1}{n+1} + \ldots + \frac{1}{2n} > \frac{1}{2n} + \ldots + \frac{1}{2n} = \frac{1}{2}$. Since n and $m = 2n$ can be taken as large as we please, the series diverges, for Cauchy's test is not fulfilled; in fact, the n-th partial sum obviously tends to infinity, since all the terms are positive. On the other hand, the series formed from the same numbers with alternating signs,

$$1 - \frac{1}{2} + \frac{1}{3} - \frac{1}{4} + \frac{1}{5} - + \ldots + \frac{(-1)^{n-1}}{n} + \ldots,$$

converges (cf. Chap. VI, p. 317), and has the sum $\log 2$.

It is by no means true that in every divergent series s_n tends to $+\infty$ or $-\infty$. Thus, in the case of the series

$$1 - 1 + 1 - 1 + 1 + - \ldots,$$

we see that the partial sum s_n has the values 1 and 0 alternately, and on account of this oscillation backwards and forwards neither approaches a definite limit nor increases numerically beyond all bounds.

With regard to the convergence and divergence of an infinite series the following fact which, though self-evident, is very important, should be noted. *The convergence or divergence of a series is not changed by inserting a finite number of terms or by removing a finite number of terms.* So far as convergence or divergence is concerned, it does not matter in the least whether we begin the series at the term a_0, or a_1, or a_5, or any other term chosen arbitrarily.

2. Absolute Convergence and Conditional Convergence.

The series $1 + \dfrac{1}{2} + \dfrac{1}{3} + \dfrac{1}{4} \ldots$ diverges; but if we change

the sign of every second term the resulting series converges. On the other hand, the geometric series $1 - q + q^2 - q^3 + - \ldots$ converges and has the sum $1/(1 + q)$, provided that $0 \leqq q < 1$; and on making all the signs plus we obtain the series

$$1 + q + q^2 + q^3 + \ldots,$$

which is also convergent, having the sum $1/(1 - q)$.

Here there appears a distinction which we must examine a little more closely. With a series whose terms are all positive there are only two possible cases; either it converges or the partial sum increases beyond all bounds as n increases. For the partial sums, being a monotonic increasing sequence, must converge if they remain bounded. Convergence occurs if the terms approach zero rapidly enough as n increases; on the other hand, divergence occurs if the terms do not approach zero at all or if they approach zero too slowly. In series where some terms are positive and some negative, however, the changes of sign may bring about convergence, since a too great increase in the partial sums, due to the positive terms, is compensated by the negative terms, so that the final result is that a definite limit is approached.

In order to grasp this fact the better, with a series $\overset{\infty}{\underset{\nu=1}{\Sigma}} a_\nu$ having positive and negative terms we compare the series which has the same terms all with positive signs, that is,

$$|a_1| + |a_2| + \ldots = \overset{\infty}{\underset{\nu=1}{\Sigma}} |a_\nu|.$$

If this series converges, then for sufficiently large values of n and $m > n$, the expression

$$|a_{n+1}| + |a_{n+2}| + \ldots + |a_m|$$

will certainly be as small as we please; on account of the relation

$$|a_{n+1} + \ldots + a_m| \leqq |a_{n+1}| + \ldots + |a_m|$$

the expression on the left is also arbitrarily small, and so the original series $\overset{\infty}{\underset{\nu=1}{\Sigma}} a_\nu$ converges. In this case the original series is said to be *absolutely convergent*. Its convergence is due to the numerical smallness of its terms and does not depend on the change of the signs.

If, on the other hand, the series with all the terms taken positively diverges and the original series still converges we say that the original series is *conditionally convergent*. Conditional convergence results from the terms of opposite signs compensating one another.

For conditional convergence Leibnitz's convergence test is frequently useful:

If the terms of a series are of alternating sign and in addition their absolute values $|\,a_n\,|$ *tend monotonically to* 0 *(so that* $|\,a_{n+1}\,| < |\,a_n\,|$), *the series* $\overset{\infty}{\underset{\nu=1}{\Sigma}} a_\nu$ *converges.* (Example: Gregory's series (p. 352)).

In the proof we assume that $a_1 > 0$, which does not essentially limit the generality of the argument, and write our series in the form

$$b_1 - b_2 + b_3 - + \ldots,$$

where all the terms b_n are now positive, b_n tends to 0, and the condition $b_{n+1} < b_n$ is satisfied. If we bracket the terms together in the two ways

$$b_1 - (b_2 - b_3) - (b_4 - b_5) - \ldots$$

and $\quad\quad (b_1 - b_2) + (b_3 - b_4) + (b_5 - b_6) + \ldots$

we see at once that the two following relations are satisfied by the partial sums:

$$s_1 > s_3 > s_5 > \ldots > s_{2m+1} > \ldots$$
$$s_2 < s_4 < s_6 < \ldots < s_{2m} < \ldots :$$

On the other hand, $s_{2n} < s_{2n+1} < s_1$ and $s_{2n+1} > s_{2n} > s_2$. The odd partial sums $s_1, s_3, \ldots$ therefore form a monotonic decreasing sequence, which in no case falls below the value s_2; hence this sequence possesses a limit L (p. 61). The even partial sums $s_2, s_4, \ldots$ likewise form a monotonic increasing sequence whose

terms in no case exceed the fixed number s_1, and therefore this
sequence must have a limiting value L'. Since the numbers s_{2n}
and s_{2n+1} differ from one another only by the number b_{2n+1}
which approaches 0 as n increases, the limiting values L and L'
are equal to one another. That is, the even and the odd partial
sums approach the same limit, which we now denote by S (cf.
fig. 1). This, however, implies that our series is convergent, as
was asserted; its sum is S.

In conclusion, we make another general remark about the
fundamental difference between absolute convergence and con-
ditional convergence. We consider a convergent series $\sum\limits_{\nu=1}^{\infty} a_\nu$.
We denote the positive terms of the series by p_1, p_2, p_3, $\ldots$,
and the negative terms by $-q_1$, $-q_2$, $-q_3$, $\ldots$. If we form the
n-th partial sum $s_n = \sum\limits_{\nu=1}^{n} a_\nu$ of the given series, a certain number,
say n', of positive terms and a certain number, say n'', of nega-

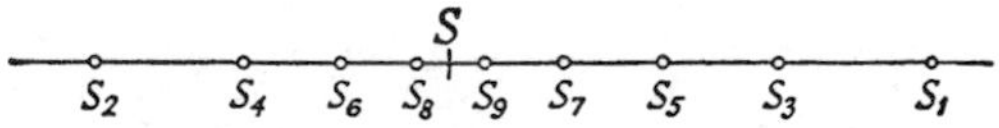

Fig. 1.—Convergence of an alternating series

tive terms must appear, where $n' + n'' = n$. Further, if the
number of positive terms as well as the number of negative terms
in the series is infinite, then the two numbers n' and n'' will
increase beyond all bounds as n does. We see immediately that
the partial sum s_n is simply equal to the partial sum $\sum\limits_{\nu=1}^{n'} p_\nu$ of the
positive terms of the series plus the partial sum $-\sum\limits_{\nu=1}^{n''} q_\nu$ of the
negative terms. If the given series converges absolutely, then
the series of positive terms $\sum\limits_{\nu=1}^{\infty} p_\nu$ and the series of absolute values
of the negative terms $\sum\limits_{\nu=1}^{\infty} q_\nu$ certainly both converge. For as m
increases the partial sums $\sum\limits_{\nu=1}^{m} p_\nu$ and $\sum\limits_{\nu=1}^{m} q_\nu$ are monotonic non-
decreasing sequences with the upper bound $\sum\limits_{\nu=1}^{\infty} |a_\nu|$.

*The sum of an absolutely convergent series is then simply equal
to the sum of the series consisting of the positive terms only plus the*

sum of the series consisting of the negative terms only, or, in other words, is equal to the difference of the two series with positive terms.

For $\sum\limits_{\nu=1}^{n} a_\nu = \sum\limits_{\nu=1}^{n'} p_\nu - \sum\limits_{\nu=1}^{n''} q_\nu$; as n increases n' and n'' must also increase beyond all bounds, and the limit of the left-hand side must therefore be equal to the difference of the two sums on the right. If the series contains only a finite number of terms of one particular sign the facts are correspondingly simplified. If, on the other hand, the series does not converge absolutely, but does converge conditionally, then the series $\sum\limits_{\nu=1}^{\infty} p_\nu$ and $\sum\limits_{\nu=1}^{\infty} q_\nu$ must both be divergent. For if both were convergent the series would converge absolutely, contrary to our hypothesis. If only one diverged, say $\sum\limits_{\nu=1}^{\infty} p_\nu$, and the other converged, then separation into positive and negative parts, $s_n = \sum\limits_{\nu=1}^{n'} p_\nu - \sum\limits_{\nu=1}^{n''} q_\nu$, shows that the series could not converge; for as n increases n' and $\sum\limits_{\nu=1}^{n'} p_\nu$ would increase beyond all bounds, while the term $\sum\limits_{\nu=1}^{n''} q_\nu$ would approach a definite limit, so that the partial sum s_n would increase beyond all bounds.

We see, therefore, that *a conditionally convergent series cannot be thought of as the difference of two convergent series, the one consisting of its positive terms and the other consisting of the absolute values of its negative terms.*

Closely connected with this fact there is another difference between absolutely and conditionally convergent series which we shall now briefly mention.

3. Rearrangement of Terms.

It is a property of finite sums that we can change the order of the terms or, as we say, rearrange the terms at will without changing the value of the sum. The question arises, what is the exact meaning of a change of the order of terms in an infinite series, and does such a rearrangement leave the value of the sum unchanged? While in the case of finite sums there is no difficulty, for example, in adding the terms in reverse order, in the case of infinite series such a possibility does not exist; there is no last

term with which to begin. Now a change of order in an infinite
series can only mean this: we say that a series $a_1 + a_2 + a_3 + \ldots$
is transformed by rearrangement into a series $b_1 + b_2 + b_3 + \ldots$,
provided that every term a_n of the first series occurs exactly
once in the second and conversely. For example, the amount by
which a_n is displaced may increase beyond all bounds as n does;
the only point is that it must appear *somewhere* in the new series.
If some of the terms are moved to later positions in the series,
other terms must, of course, be moved to earlier positions. For
example, the series

$$1 + q + q^2 + q^4 + q^3 + q^8 + q^7 + q^6 + q^5 + q^{16} + \ldots$$

is a rearrangement of the geometric series $1 + q + q^2 + \ldots$.

With regard to change of order there is a fundamental distinction between absolutely convergent series and conditionally
convergent series.

*In absolutely convergent series rearrangement of the terms does
not affect the convergence, and the value of the sum of the series is
unchanged, exactly as in the case of finite sums.*

*In conditionally convergent series, on the other hand, the value
of the sum of the series can be changed at will by suitable rearrangement of the series, and the series can even be made to diverge if
desired.*

The first of these facts, referring to absolutely convergent
series, is easily established. Let us assume to begin with that our
series has positive terms only, and let us consider the n-th partial
sum $s_n = \sum_{\nu=1}^{n} a_\nu$. All the terms of this partial sum occur in the
m-th partial sum $t_m = \sum_{\nu=1}^{m} b_\nu$ of the rearranged series, provided
only that m is chosen large enough. Hence $t_m \geqq s_n$. On the
other hand, we can determine an index n' so large that the
partial sum $s_{n'} = \sum_{\nu=1}^{n'} a_\nu$ of the first series contains all the terms
$b_1, b_2, \ldots, b_m$. It then follows that $t_m \leqq s_{n'} \leqq A$, where A is
the sum of the first series. Thus for all sufficiently large values
of m we have $s_n \leqq t_m \leqq A$; and since s_n can be made to differ
from A by an arbitrarily small amount, it follows that the
rearranged series also converges; and in fact to the same limit
A as the original series.

If the absolutely convergent series has both positive and negative terms, we may regard it as the difference of two series each of which has positive terms only. Since in the rearrangement of the original series each of these two series merely undergoes rearrangement and therefore converges to the same value as before, the same is true of the original series when rearranged. For by the case just considered the new series is absolutely convergent and is therefore the difference of the two rearranged series of positive terms.

To the beginner the fact just proved may seem a triviality. That it really does require proof, and that in this proof the absolute convergence is essential, can be shown by an example of the opposite behaviour of conditionally convergent series. We take the familiar series

$$1 - \frac{1}{2} + \frac{1}{3} - \frac{1}{4} + \frac{1}{5} - \frac{1}{6} + \frac{1}{7} - \frac{1}{8} + - \ldots = \log 2.$$

under it write the result of multiplication by the factor $\frac{1}{2}$,

$$\frac{1}{2} \quad - \frac{1}{4} \quad + \frac{1}{6} \quad - \frac{1}{8} + - \ldots = \frac{1}{2} \log 2,$$

and add, combining the terms placed in vertical columns.* We thus obtain

$$1 + \frac{1}{3} - \frac{1}{2} + \frac{1}{5} + \frac{1}{7} - \frac{1}{4} + \frac{1}{9} + \frac{1}{11} - \frac{1}{6} + + - \ldots = \frac{3}{2} \log 2.$$

This last series can obviously be obtained by rearranging the original series, and yet the value of the sum of the series has been multiplied by the factor 3/2. It is easy to imagine the effect that the discovery of this apparent paradox must have had on the mathematicians of the eighteenth century, who were accustomed to operate with infinite series without regard to their convergence.

We shall give the proof of the theorem stated above concerning the change in the sum of a conditionally convergent series which arises from change of order of the terms, although we shall have no occasion to make use of the result. Let p_1, p_2, ... be the positive terms and $-q_1$, $-q_2$, ... the negative terms of the series. Since the absolute value $|a_n|$ tends to 0 as n increases, the numbers p_n and q_n must also tend to 0 as n increases. As

---

* For the addition of series see No. 4, p. 376.

we have already seen, moreover, the sum* $\overset{\infty}{\underset{1}{\Sigma}} p_\nu$ must diverge, and the same is true of $\overset{\infty}{\underset{1}{\Sigma}} q_\nu$.

Now we can easily find a rearrangement of the original series which has an arbitrary number a as limit. Suppose, to be specific, that a is positive. We then add together the first n_1 positive terms, just enough to secure that the sum $\overset{n_1}{\underset{1}{\Sigma}} p_\nu$ is greater than a. Since the sum $\overset{n_1}{\underset{1}{\Sigma}} p_\nu$ increases with n_1 beyond all bounds, it is always possible by using enough terms to make the partial sum greater than a. The sum will then differ from the exact value a by p_{n_1} at most. We now add just enough negative terms $-\overset{m_1}{\underset{1}{\Sigma}} q_\nu$ to ensure that the sum $\overset{n_1}{\underset{1}{\Sigma}} p_\nu -\overset{m_1}{\underset{1}{\Sigma}} q_\nu$ is less than a; this is also possible, as follows from the divergence of the series $\overset{\infty}{\underset{1}{\Sigma}} q_\nu$. The difference between this sum and a is now q_{m_1} at most. We now add just enough other positive terms $\overset{n_2}{\underset{n_1+1}{\Sigma}} p_\nu$ to make the partial sum again greater than a, as is again possible, since the series of positive terms diverges. The difference between the partial sum and a is now p_{n_2} at most. We again add just enough negative terms $-\overset{m_2}{\underset{m_1+1}{\Sigma}} q_\nu$, beginning next after the last one previously used, to make the sum once more less than a, and continue in the same way. The values of the sums thus obtained will oscillate about the number a, and when the process is carried far enough the oscillation will only take place between arbitrarily narrow bounds; for since the terms p_ν and q_ν themselves tend to 0 when ν is sufficiently large, the length of the interval in which the oscillation takes place will also tend to 0. The theorem is thus proved.

In the same way we can rearrange the series in such a way as to make it diverge; we have only to choose such large numbers of the positive terms as compared with the negative that compensation no longer takes place.

---

* This abbreviated notation for $\overset{\infty}{\underset{\nu=1}{\Sigma}} p_\nu$, and analogous expressions for other series, will often be used in future.

4. Operations with Infinite Series.

It is clear that two convergent infinite series $a_1 + a_2 + \ldots = S$ and $b_1 + b_2 + \ldots = T$ can be added term by term, that is, that the series formed from the terms $c_n = a_n + b_n$ converges and has the value $S + T$ for its sum.* For

$$\sum_{\nu=1}^{n} c_\nu = \sum_{\nu=1}^{n} a_\nu + \sum_{\nu=1}^{n} b_\nu \to S + T.$$

It is also clear that if we multiply each term of a convergent infinite series by the same factor the series remains convergent, its sum being multiplied by the same factor.

In the cases just mentioned it is immaterial whether the convergence is absolute or conditional. On the other hand, further study, which is not necessary for us here, shows that if two infinite series are multiplied together by the method used in multiplying finite sums together, the product series will not usually converge or have the product of the two sums for its sum unless at least one of the two series is absolutely convergent (cf. appendix, p. 415).

EXAMPLES

1. Prove that $\displaystyle\sum_{\nu=1}^{\infty} \frac{1}{\nu(\nu+1)} = \frac{1}{1 . 2} + \frac{1}{2 . 3} + \frac{1}{3 . 4} + \ldots = 1.$

2. Prove that $\displaystyle\sum_{\nu=1}^{\infty} \frac{1}{\nu(\nu+1)(\nu+2)} = \frac{1}{4}.$

3. Prove that $\displaystyle\sum_{\nu=0}^{\infty} (-1)^\nu \frac{2\nu+3}{(\nu+1)(\nu+2)} = 1.$

4. For what values of α does the series $1 - \dfrac{1}{2^\alpha} + \dfrac{1}{3^\alpha} - \dfrac{1}{4^\alpha} + \ldots$ converge?

5.* Prove that if $\displaystyle\sum_{\nu=1}^{\infty} a_\nu$ converges, and $s_n = a_1 + a_2 + \ldots + a_n$, then the sequence

$$\frac{s_1 + s_2 + \ldots + s_N}{N}$$

also converges, and has $\displaystyle\sum_{\nu=1}^{\infty} a_\nu$ as its limit.

6. Is the series $\displaystyle\sum_{n=1}^{\infty} \left(\frac{2n}{2n+1} - \frac{2n-1}{2n} \right)$ convergent?

7. Is the series $\displaystyle\sum_{\nu=1}^{\infty} (-1)^\nu \frac{\nu}{\nu+1}$ convergent?

* This theorem is really nothing more than another statement of the fact (cf. Chap. I, § 6, p. 41) that the limit of the sum of two terms is the sum of their limits.

2. Tests for Convergence and Divergence

We have already met with a test of a general nature for the convergence of series, which applies to series with terms of alternating signs and decreasing absolute value and which asserts that such series are at least conditionally convergent. In the following pages we shall only consider criteria referring to *absolute* convergence.

1. The Comparison Test.

All such considerations of convergence depend on the comparison of the series in question with a second series; this second series is chosen in such a way that its convergence can readily be tested. The general *comparison test* may be stated as follows:

If the numbers b_1, b_2, ... *are all positive and the series* $\sum\limits_{\nu=1}^{\infty} b_\nu$ *converges, and if*

$$|a_n| \leq b_n$$

for all values of n, *then the series* $\sum\limits_{n=1}^{\infty} a_n$ *is absolutely convergent.*

If we apply Cauchy's test the proof becomes almost trivial. For if $m \geq n$, we have

$$|a_n + \ldots + a_m| \leq |a_n| + \ldots + |a_m| \leq b_n + \ldots + b_m.$$

Since the series $\sum\limits_{n=1}^{\infty} b_n$ converges, the right-hand side is arbitrarily small, provided that n and m are sufficiently large. It follows that for such values of n and m the left-hand side is also arbitrarily small, so that by Cauchy's test the given series converges. The convergence is absolute, since our argument applies equally well to the convergence of the series of absolute values $|a_n|$.

The analogous proof for the following fact can be left to the reader. *If*

$$|a_n| \geq b_n > 0,$$

and the series $\sum\limits_{n=1}^{\infty} b_n$ *diverges, then the series* $\sum\limits_{n=1}^{\infty} a_n$ *is certainly not absolutely convergent.*

13 *

2. Comparison with the Geometric Series.

In applications of the test the comparison series most frequently used is the geometric series. We at once obtain the following theorem:

The series $\sum\limits_{n=1}^{\infty} a_n$ is absolutely convergent if from a certain term onward a relation of the form

$$| a_n | < c\, q^n \tag{I}$$

holds, where c is a positive number independent of n *and* q *is any fixed positive number less than* 1.

This test is usually expressed in one of the following weaker forms: the series $\sum\limits_{n=1}^{\infty} a_n$ converges absolutely, if from a certain term onward a relation of the form

$$\left| \frac{a_{n+1}}{a_n} \right| < q \tag{IIa}$$

holds, where q is again a positive number less than 1 and independent of n, or: if from a certain term onward a relation of the form

$$\sqrt[n]{| a_n |} < q \tag{IIb}$$

holds, where q is a positive number less than 1. In particular, the conditions of these tests are satisfied if a relation of the form

$$\lim_{n \to \infty} \left| \frac{a_{n+1}}{a_n} \right| = k < 1 \tag{IIIa}$$

or

$$\lim_{n \to \infty} \sqrt[n]{| a_n |} = k < 1 \tag{IIIb}$$

is true. These statements are easily established in the following way.

Let us suppose that the criterion II*a*, the *ratio test*, is satisfied from the suffix n_0 onward, that is, when $n > n_0$. For brevity we put $a_{n_0+m+1} = b_m$ and find that

$$| b_1 | < q\, | b_0 |, \quad | b_2 | < q\, | b_1 | < q^2\, | b_0 |, \quad | b_3 | < q\, | b_2 | < q^3\, | b_0 |,$$

and so on; hence

$$| b_m | < q^m\, | b_0 |,$$

which establishes our statement. For the criterion IIb, the *root test*, we at once have $|a_n| < q^n$, and our statement follows immediately.

Finally, in order to prove criterion III, we consider an arbitrary number q such that $k < q < 1$. Then from a certain n_0 onward, that is, when $n > n_0$, it is certain that $\left|\dfrac{a_{n+1}}{a_n}\right| < q$ or $\sqrt[n]{|a_n|} < q$, as the case may be, since from a certain term onwards the values of $\left|\dfrac{a_{n+1}}{a_n}\right|$ or of $\sqrt[n]{|a_n|}$ differ from k by less than $(q - k)$. The statement is then established by a reference to the results already proved.

We stress the point that the four tests derived from the original criterion $|a_n| < cq^n$ are *not* equivalent to one another or to the original, that is, that they cannot be derived from one another in both directions. We shall soon see from examples that if a series satisfies one of the conditions, it need not by any means satisfy all the others.*

For completeness it may be pointed out that a series certainly diverges if from a certain term onward

$$|a_n| > c$$

for a properly chosen positive number c, or if from a certain term onward

$$\sqrt[n]{|a_n|} > 1,$$

or if $$\lim_{n \to \infty} \left|\frac{a_{n+1}}{a_n}\right| = k, \quad \text{or} \quad \lim_{n \to \infty} \sqrt[n]{|a_n|} = k,$$

where k is a number greater than 1. For, as we immediately recognize, in such a series the terms cannot tend to zero as n increases; the series must therefore diverge. (In these circumstances the series cannot even be conditionally convergent.)

Our tests furnish *sufficient* conditions for the absolute convergence of a series; that is, when they are satisfied we can conclude that the series converges absolutely. They are definitely not *necessary* conditions, however; that is, absolutely convergent series can be found which do not satisfy the conditions.

* More exactly: if IIIa is fulfilled, then IIa is fulfilled; if IIIb, then IIb; if IIIa, then IIIb; if IIa, then IIb; and if any of the four is satisfied, then so is I. None of these statements can be reversed.

For example, the knowledge that

$$\lim_{n \to \infty} \left| \frac{a_{n+1}}{a_n} \right| = 1 \quad \text{or} \quad \lim_{n \to \infty} \sqrt[n]{|a_n|} = 1$$

does not entitle us to make any statement about the convergence of the series. Such a series may converge or diverge. For example, the series

$$\sum_{n=1}^{\infty} \frac{1}{n},$$

for which $\lim_{n \to \infty} \sqrt[n]{|a_n|} = 1$ and $\lim_{n \to \infty} \left| \frac{a_{n+1}}{a_n} \right| = 1$, is divergent, as we saw on p. 368. On the other hand, we shall soon see that the series $\sum_{n=1}^{\infty} \frac{1}{n^2}$, which satisfies the same relations, is convergent.

As an example of the application of our tests we first consider the series

$$q + 2q^2 + 3q^3 + \ldots + nq^n + \ldots .$$

For this series

$$\lim_{n \to \infty} \sqrt[n]{|a_n|} = |q| \cdot \lim_{n \to \infty} \sqrt[n]{n} = |q|,$$

$$\lim_{n \to \infty} \left| \frac{a_{n+1}}{a_n} \right| = |q| \cdot \lim_{n \to \infty} \frac{n+1}{n} = |q|.$$

That the series converges if $|q| < 1$ follows from the ratio test and from the root test also, even in the weaker form III.

If, on the other hand, we consider the series

$$1 + 2q + q^2 + 2q^3 + \ldots + q^{2n} + 2q^{2n+1} + \ldots ,$$

we can no longer prove convergence by the ratio test when $\frac{1}{2} \leqq |q| < 1$; for then $\left| \frac{2q^{2n+1}}{q^{2n}} \right| = 2|q| \geqq 1$. But the root test immediately gives us $\lim_{n \to \infty} \sqrt[n]{|a_n|} = |q|$, and shows that the series converges provided that $|q| < 1$, which, of course, we could also have observed directly.

3. Comparison with an Integral.*

We now proceed to a discussion of convergence which is independent of the preceding. We shall carry it out for the particularly simple and important case of the series

$$\sum_{n=1}^{\infty} \frac{1}{n^\alpha} = 1 + \frac{1}{2^\alpha} + \frac{1}{3^\alpha} + \ldots ,$$

* In this connexion see also the appendix to Chap. VII (p. 361).

where the general term a_n is $1/n^\alpha$, α being a positive number. In order to investigate the convergence or divergence of this series, we consider the graph of the function $y = 1/x^\alpha$ and mark off on the x-axis the integral abscissæ $x = 1$, $x = 2$, ... We first construct the rectangle of height $1/n^\alpha$ over the interval $n - 1 \leqq x \leqq n$ of the x-axis $(n > 1)$, and compare it with the area of the region bounded by the same interval of the x-axis, the ordinates at the ends, and the curve $y = 1/x^\alpha$ (this region is shown shaded in fig. 2). Secondly, we construct the rectangle of height $1/n^\alpha$ lying above the interval $n \leqq x \leqq n + 1$, and similarly compare it with the area of the region lying above the same interval and below the curve (this region is cross-hatched in fig. 2). In the first case the area under the curve is obviously greater than the area of the rectangle; in the second case it is less than the area of the rectangle. In other words,

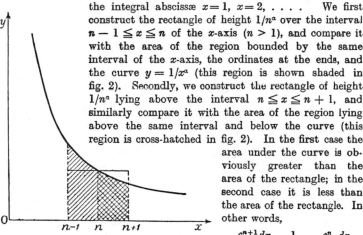

Fig. 2.—Comparison of a series with an integral

$$\int_n^{n+1} \frac{dx}{x^\alpha} < \frac{1}{n^\alpha} < \int_{n-1}^n \frac{dx}{x^\alpha},$$

as we may also prove directly from the integral itself (cf. Chap. II, § 7, p. 129). Writing down this inequality for $n = 2$, $n = 3$, . . . , $n = m$ and summing, we obtain the following estimate * for the m-th partial sum $s_m = \sum\limits_{n-1}^{m} \frac{1}{n^\alpha}$:

$$1 + \int_2^{m+1} \frac{dx}{x^\alpha} < s_m < 1 + \int_1^m \frac{dx}{x^\alpha}.$$

Now as m increases the integral $\int_1^m \frac{1}{x^\alpha} dx$ tends to a finite limit or increases without limit according as $\alpha > 1$ or $\alpha \leqq 1$. Consequently the monotonic sequence of numbers s_m is bounded or increases beyond all bounds according as $\alpha > 1$ or $\alpha \leqq 1$, and we thus have the following theorem:

---

* From this relation for $a = 1$ it follows at once that the sequence of numbers $C_n = 1 + \frac{1}{2} + \frac{1}{3} + \ldots + \frac{1}{n} - \log n$ is bounded below. Since from the inequality $\frac{1}{n+1} < \int_n^{n+1} \frac{dx}{x} = \log(n + 1) - \log n$ we see that the sequence is monotonic decreasing, it must approach a limit

$$\lim_{n \to \infty} C_n = \lim_{n \to \infty} (1 + \frac{1}{2} + \frac{1}{3} + \ldots + \frac{1}{n} - \log n) = C.$$

The number C, whose value is ·5772 . . . , is called *Euler's constant*. In contrast to the other important special numbers of analysis, such as π and e, no other expression with a simple law of formation has been found for Euler's constant.

The series

$$\sum_{n=1}^{\infty} \frac{1}{n^{\alpha}} = \frac{1}{1^{\alpha}} + \frac{1}{2^{\alpha}} + \frac{1}{3^{\alpha}} + \cdots$$

is convergent—and, of course, absolutely convergent—if, and only if, $\alpha > 1$.

The divergence of the harmonic series, which we previously proved in a different way, is an immediate consequence of this. In particular, we see that the series

$$\frac{1}{1^2} + \frac{1}{2^2} + \frac{1}{3^2} + \cdots,$$

$$\frac{1}{1^3} + \frac{1}{2^3} + \frac{1}{3^3} + \cdots,$$

$$\bullet \quad \bullet \quad \bullet \quad \bullet \quad \bullet \quad \bullet \quad \bullet \quad \bullet \quad,$$

converge.

The series $\sum_{\nu=1}^{\infty} \frac{1}{\nu^{\alpha}}$, whose convergence we have just studied, frequently serve as comparison series in investigations of convergence. For example, we see at once that for $\alpha > 1$ the series $\sum_{\nu=1}^{\infty} \frac{c_{\nu}}{\nu^{\alpha}}$ converges absolutely if the absolute values $|c_{\nu}|$ of the coefficients remain less than a fixed bound independent of ν.

EXAMPLES

Find whether the series in Ex. 1–6 are convergent or not:

1. $\sum_{\nu=1}^{\infty} \frac{1}{1 + \nu^2}$.

2. $\sum_{\nu=1}^{\infty} \frac{\nu!}{\nu^{\nu}}$.

3. $\sum_{\nu=1}^{\infty} \frac{1}{\sqrt{\nu(\nu+1)}}$.

4.* $\sum_{\nu=2}^{\infty} \frac{1}{(\log \nu)^{\alpha}}$, α fixed.

5. $\sum_{\nu=2}^{\infty} \frac{1}{(\log \nu)^{\log \nu}}$.

6. $\sum_{\nu=1}^{\infty} \frac{\nu}{2^{\nu}}$.

Estimate the error after n terms of the series in Ex. 7–10:

7. $\sum_{\nu=1}^{\infty} \frac{(-1)^{\nu+1}}{\nu^2}$.

8. $\sum_{\nu=1}^{\infty} \frac{1}{\nu!}$.

9. $\sum_{\nu=1}^{\infty} \frac{1}{\nu^{\nu}}$.

10. $\sum_{\nu=1}^{\infty} \frac{\nu}{2^{\nu}}$.

11. Prove that $\sum_{\nu=1}^{\infty} \sin^2\left[\pi\left(\nu + \frac{1}{\nu}\right)\right]$ converges.

12. Does $\sum_{\nu=-\infty}^{\infty} e^{-\nu^2}$ (that is, $1 + 2 \sum_{\nu=1}^{\infty} e^{-\nu^2}$) converge?

13.* Prove that $\sum_{\nu=2}^{\infty} \frac{1}{\nu(\log \nu)^{\alpha}}$ converges when $\alpha > 1$ and diverges when $\alpha \leqq 1$.

14.* Prove that $\sum\limits_{\nu=3}^{\infty} \dfrac{1}{\nu \log \nu \, (\log \log \nu)^{\alpha}}$ converges when $\alpha > 1$ and diverges when $\alpha \leqq 1$.

15. Prove that if $u_i \geqq 0$ $(i = 1, 2, 3, \ldots)$ and $\sum\limits_{i=1}^{\infty} u_i$ converges, then $\sum\limits_{i=1}^{\infty} u_i^2$ also converges.

16. Show that if $\sum\limits_{k=1}^{\infty} a_k^2$ and $\sum\limits_{k=1}^{\infty} b_k^2$ both converge, then $\sum\limits_{k=1}^{\infty} a_k b_k$ also converges.

17. Prove that

$$1 + \frac{1}{2} - \frac{2}{3} + \frac{1}{4} + \frac{1}{5} - \frac{2}{6} + \frac{1}{7} + \ldots + \frac{1}{3n+1}$$
$$+ \frac{1}{3n+2} - \frac{2}{3n+3} + \ldots = \log 3.$$

18.* Prove that if n is an arbitrary integer greater than 1

$$\sum_{\nu=1}^{\infty} \frac{a_\nu^n}{\nu} = \log n,$$

where a_ν^n is defined as follows:

$$a_\nu^n = \begin{cases} 1 & \text{if } n \text{ is not a factor of } \nu, \\ -(n-1) & \text{if } n \text{ is a factor of } \nu. \end{cases}$$

3. Sequences and Series of Functions

1. General Remarks.

The terms of the infinite series hitherto considered have been constants; hence these series (when convergent) always represented definite numbers. But both in theory and in applications the series of outstanding importance are those in which the terms are functions of a variable, so that the sum of the series is also a function of the variable, as in the case of Taylor series.

We shall therefore consider a series

$$g_1(x) + g_2(x) + g_3(x) + \ldots,$$

in which all the functions $g_n(x)$ are functions defined in an interval $a \leqq x \leqq b$. The n-th partial sum of this series,

$$g_1(x) + g_2(x) + \ldots + g_n(x),$$

we denote by $f_n(x)$. Then the sum $f(x)$ of our series, where it exists, is simply the limit $\lim\limits_{n \to \infty} f_n(x)$.

We may therefore regard the *sum of an infinite series of functions* as the *limit of a sequence of functions* $f_1(x)$, $f_2(x)$, ..., $f_n(x)$, ... Conversely, for any such sequence of functions $f_1(x)$, $f_2(x)$, ... we can form an equivalent series by putting $g_1(x) = f_1(x)$ and $g_n(x) = f_n(x) - f_{n-1}(x)$ for $n > 1$. When it is convenient, therefore, we can pass from the consideration of series to that of sequences and conversely.

2. Limiting Processes with Functions and Curves.

We shall now state exactly what we mean by saying that a function $f(x)$ is the limit of a sequence $f_1(x)$, $f_2(x)$, ..., $f_n(x)$, ... in an interval $a \leqq x \leqq b$. The definition is as follows: the sequence $f_1(x)$, $f_2(x)$, ... converges in that interval to the limit function $f(x)$, if at each point x of the interval the values $f_n(x)$ converge in the usual sense to the value $f(x)$. In this case we shall write $\lim_{n \to \infty} f_n(x) = f(x)$. According to Cauchy's test (cf. p. 40) we can express the convergence of the sequence without necessarily knowing or stating the limit function $f(x)$. For our sequence of functions will converge to a limit function if, and only if, at each point x in our interval and for every positive number ϵ the quantity $|f_n(x) - f_m(x)|$ is less than ϵ, provided that the numbers n and m are chosen large enough, that is, larger than a certain number $N = N(\epsilon)$. This number $N(\epsilon)$ usually depends on ϵ and x and increases beyond all bounds as ϵ tends to zero.

We have frequently met with cases of limits of sequences of functions. We mention only the definition of the power x^a for irrational values of a by the equation

$$x^a = \lim_{n \to \infty} x^{r_n},$$

where r_1, r_2, ..., r_n, ... is a sequence of rational numbers tending to a; or the equation

$$e^x = \lim_{n \to \infty} \left(1 + \frac{x}{n}\right)^n,$$

where the functions $f_n(x)$ on the right are polynomials of degree n.

The graphical representation of functions by means of curves suggests that we can also speak of limits of sequences of curves, saying, for example, that the graphs of the above limit functions

x^a and e^x are to be regarded as the limit curves of the graphs of the functions x^{r_n} and $\left(1 + \dfrac{x}{n}\right)^n$ respectively. There is, however, a fine distinction between passages to the limit with functions and with curves. Until the middle of the nineteenth century this distinction was not sufficiently observed; only by having a clear idea of it can we avoid apparent paradoxes. We shall illustrate this point by an example.

For this purpose we consider the functions

$$f_n(x) = x^n \quad (n = 1, 2, \ldots)$$

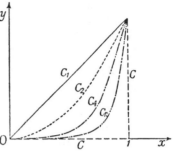

in the interval $0 \leq x \leq 1$. All these functions are continuous, and the limit function $\lim\limits_{n \to \infty} f_n(x) = f(x)$ exists. But this limit function is not continuous. On the contrary, since for all values of n the value of the function $f_n(1) = 1$, the limit

$$f(1) = 1;$$

Fig. 3.—Limit curve and limit function

while, on the other hand, for $0 \leq x < 1$, the limit $f(x) = \lim\limits_{n \to \infty} f_n(x) = 0$, as we saw in Chap. I, § 5 (p. 33). The function $f(x)$ is therefore a discontinuous function which at $x = 1$ has the value 1 and for all other values of x in the interval has the value 0.

This discontinuity becomes intelligible if we consider the graphs C_n of the functions $y = f_n(x)$. These (cf. fig. 16, p. 33) are continuous curves, all of which pass through the origin and the point $x = 1$, $y = 1$, and which draw in closer and closer to the x-axis as n increases. The *curves* possess a *limit curve* C which is not discontinuous at all, but consists (cf. fig. 3) of the portion of the x-axis between $x = 0$ and $x = 1$, and the portion of the line $x = 1$ between $y = 0$ and $y = 1$. The *curves* therefore converge to a *continuous* limit curve with a vertical portion, while the *functions* converge to a *discontinuous* limit function. We thus recognize that this discontinuity of the limit function expresses itself by the occurrence in the limit curve of a portion perpendicular to the x-axis. Such a portion *must* involve a discontinuity in the limit function, and, in fact, such a portion is always present when the limit function is discontinuous. This limit curve is *not* the graph of the limit function, nor can any curve with a vertical portion be the graph of any single-valued function $y = f(x)$; for corresponding to the value of x at which the vertical portion occurs the curve gives an infinite number of values of y and the function only one. Hence the limit of the graphs of the functions $f_n(x)$ is not the same as the graph of the limit $f(x)$ of these functions.

Corresponding statements, of course, hold for infinite series also.

4. Uniform and Non-uniform Convergence

1. General Remarks and Examples.

The distinction between the concept of the convergence of functions and that of the convergence of curves introduces a phenomenon which it is essential that the student should clearly recognize. This is the so-called *non-uniform convergence* of sequences or infinite series of functions. Since it is well known that beginners usually find difficulties here, we shall discuss the matter in some detail.

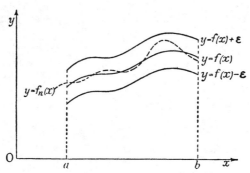

Fig. 4.—To illustrate uniform convergence

That a function $f(x)$ is the limit of a sequence $f_1(x), f_2(x), \ldots$ in an interval $a \leqq x \leqq b$ means only, by definition, that the usual limit relationship $f(x) = \lim_{n \to \infty} f(x)$ holds at each point x of the interval. From a naïve point of view one might expect that the following fact would automatically follow from this concept of convergence: if we assign an arbitrary degree of accuracy, say $\epsilon = \dfrac{1}{1000}$ or $\epsilon = \dfrac{1}{100}$, then from a certain index N onward all the functions $f_n(x)$ will lie between $f(x) + \epsilon$ and $f(x) - \epsilon$ for all values of x, so that their graphs $y = f_n(x)$ will lie entirely in the strip indicated in fig. 4. That is to say, for every positive ϵ there is a corresponding number $N = N(\epsilon)$, which, of course, will ordinarily increase beyond all bounds as $\epsilon \to 0$, such that for $n > N$ the difference $|f(x) - f_n(x)| < \epsilon$, no matter where x is chosen in the interval. (If this condition is fulfilled, then

$|f_n(x) - f_m(x)| < 2\epsilon$ for all values of x, provided that n and m are both greater than N.) If the accuracy of the approximation can be made at least equal to a pre-assigned number ϵ *everywhere* in the interval *at the same time*, that is, by everywhere choosing the same number $N(\epsilon)$ independent of x, we say that the approximation is *uniform*. One is at first astonished to find that the naïve assumption that convergence is necessarily uniform is entirely wrong; in other words, that convergence may very well be *non-uniform*.

Ex. 1. Non-uniform convergence occurs in the case of the sequence of functions just considered, $f_n(x) = x^n$; in the interval $0 \leqq x \leqq 1$ this sequence converges to the limit function $f(x) = 0$ for $0 \leqq x < 1$, $f(1) = 1$. Convergence occurs at every point in the interval; that is, if ε is any positive number, and if we select any definite fixed value $x = \xi$, the inequality $|\xi^n - f(\xi)| < \varepsilon$ certainly holds if n is sufficiently large. Yet this approximation is not uniform. For, if we choose $\varepsilon = \frac{1}{2}$, then no matter how large the number n is chosen, we can find a point $x = \eta \neq 1$ at which $|\eta^n - f(\eta)| = \eta^n > \frac{1}{2}$; this is, in fact, true for all points $x = \eta$ where $1 > \eta > \sqrt[n]{\frac{1}{2}}$. It is therefore impossible to choose the number n so large that the difference between $f(x)$ and $f_n(x)$ is less than $\frac{1}{2}$ *throughout the whole interval*.

This behaviour becomes intelligible if we refer to the graphs of these functions (fig. 3, p. 385). We see that no matter how large a value of n we choose, for values of ξ only a little less than 1 the value of the function $f_n(\xi)$ will be very near 1, and therefore cannot be a good approximation to $f(\xi)$, which is 0.

Similar behaviour is exhibited by the functions

$$f_n(x) = \frac{1}{1 + x^{2n}}$$

in the neighbourhood of the points $x = 1$ and $x = -1$; this can easily be established. (Compare also the discussion in Chap. I, § 8 (p. 52)).

Ex. 2. In the two examples above the non-uniformity of the convergence is connected with the fact that the limit function is discontinuous. Yet it is also easy to construct a sequence of continuous functions which do converge to a continuous limit function, but not uniformly. We restrict our attention to the interval $0 \leqq x \leqq 1$ and make the following definitions for $n \geqq 2$:

$$f_n(x) = x n^\alpha \text{ for } 0 \leqq x \leqq \frac{1}{n},$$

$$f_n(x) = \left(\frac{2}{n} - x\right) n^\alpha \text{ for } \frac{1}{n} \leqq x \leqq \frac{2}{n},$$

$$f_n(x) = 0 \text{ for } \frac{2}{n} \leqq x \leqq 1,$$

where to begin with we can choose any value for α, but must then keep this value of α fixed for all terms of the sequence. Graphically our functions are represented by a roof-shaped figure made of two line segments lying over the interval $0 \leqq x \leqq 2/n$ of the x-axis, while from $x = 2/n$ onwards the graph is the x-axis itself (cf. fig. 5).

If $\alpha < 1$, the altitude of the highest point of the graph, which has in general the value $n^{\alpha-1}$, will tend to 0 as n increases; the curves will then tend towards the x-axis, and the functions $f_n(x)$ will converge uniformly to the limit function $f(x) = 0$.

If $\alpha = 1$, the peak of the graph will have the height 1 for every value of n. If $\alpha > 1$, the height of the peak will increase beyond all bounds as n increases.

But no matter how α is chosen, the sequence $f_1(x)$, $f_2(x)$, ... always tends to the limit function $f(x) = 0$. For, if x is positive, for all sufficiently

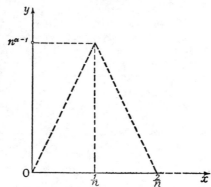

Fig. 5.—To illustrate non-uniform convergence

large values of n we have $2/n < x$, so that x is not under the roof-shaped part of the graph and $f_n(x) = 0$; for $x = 0$ all the functional values $f_n(x)$ are equal to 0, so that in either case $\lim_{n \to \infty} f_n(x) = 0$.

The convergence is certainly non-uniform, however, if $\alpha \geqq 1$; for it is plainly impossible to choose n so large that the expression $|f(x) - f_n(x)| = f_n(x)$ is less than $\frac{1}{2}$ *everywhere* in the interval.

Ex. 3. Exactly similar behaviour is exhibited by the sequence of functions (cf. fig. 6)

$$f_n(x) = x n^\alpha e^{-nx},$$

where, in contrast with the preceding case, each function of the sequence is represented by a single analytical expression. Here again the equation $\lim_{n \to \infty} f_n(x) = 0$ holds for every positive value of x, since as n increases the function e^{-nx} tends to 0 to a higher order than any power of $1/n$ (cf. Chap. III, § 9, p. 192). For $x = 0$, we have always $f_n(x) = 0$, and thus

$$f(x) = \lim_{n \to \infty} f_n(x) = 0$$

for every value of x in the interval $0 \leq x \leq a$, where a is an arbitrary positive number. But here again the convergence to the limit function is not uniform. For at the point $x = \dfrac{1}{n}$ (where $f_n(x)$ has its maximum) we have

$$f_n(x) = f_n\left(\frac{1}{n}\right) = \frac{n^{a-1}}{e},$$

and we thus recognize that if $\alpha \geq 1$, the convergence is non-uniform; for every curve $y = f_n(x)$, no matter how large n is chosen, will contain points (namely, the point $x = \dfrac{1}{n}$, which varies with n, and neighbouring points) at which $f_n(x) - f(x) = f_n(x) > \dfrac{1}{2e}$.

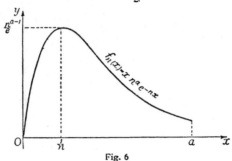

Fig. 6

Ex. 4. The concepts of uniform and non-uniform convergence may, of course, be extended to infinite series. We say that a series

$$g_1(x) + g_2(x) + \ldots$$

is uniformly convergent, or not, according to the behaviour of its partial sums $f_n(x)$. A very simple example of a non-uniformly convergent series is given by

$$f(x) = x^2 + \frac{x^2}{1 + x^2} + \frac{x^2}{(1 + x^2)^2} + \frac{x^2}{(1 + x^2)^3} + \ldots .$$

For $x = 0$ every partial sum $f_n(x) = x^2 + \ldots + \dfrac{x^2}{(1 + x^2)^{n-1}}$ has the value 0; therefore $f(0) = 0$. For $x \neq 0$ the series is simply a geometric series with the positive ratio $\dfrac{1}{1 + x^2} < 1$; we can therefore sum it by the elementary rules and thus obtain for every $x \neq 0$ the sum

$$\frac{x^2}{1 - 1/(1 + x^2)} = 1 + x^2.$$

The limit function $f(x)$ is thus given everywhere except at $x = 0$ by the expression $f(x) = 1 + x^2$, while $f(0) = 0$; it therefore has a somewhat artificial-looking discontinuity at the origin.

Here again we have non-uniform convergence in every interval containing the origin. For the difference $f(x) - f_n(x) = r_n(x)$ is always 0 for $x = 0$, while for all other values of x it is given by the expression $r_n(x) = \dfrac{1}{(1 + x^2)^{n-1}}$, as the reader may verify for himself. If we require this expression to be less than, say, $\frac{1}{2}$, then for each fixed value of x this can be attained by choosing n large enough. But we can find no value of n sufficiently large to ensure that $r_n(x)$ is everywhere less than $\frac{1}{2}$; for if we fix upon any value of n, no matter how large, we can make $r_n(x)$ greater than $\frac{1}{2}$ by taking x near enough to 0. A uniform approximation to within $\frac{1}{2}$ is therefore impossible. The matter becomes clear if we consider the approximating curves (cf. fig. 7). These curves, except near $x = 0$, lie

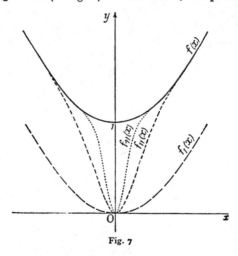

Fig. 7

nearer and nearer to the parabola $y = 1 + x^2$ as n increases; near $x = 0$, however, the curves send down a narrower and narrower extension to the origin, and as n increases this extension draws in closer and closer to a certain straight line, a portion of the y-axis, so that for limiting curve we have the parabola plus a linear extension reaching vertically down to the origin.

As a further example of non-uniform convergence we mention the series $\displaystyle\sum_{\nu=0}^{\infty} g_\nu(x)$, where $g_\nu(x) = x^\nu - x^{\nu-1}$ for $\nu \geq 1$, $g_0(x) = 1$, defined in the interval $0 \leq x \leq 1$. The partial sums of this series are the functions x^ν already considered in the first example (p. 387).

2. A Test of Uniform Convergence.

The preceding considerations show us that the uniform convergence of a sequence or series is a special property not possessed by all sequences and series. We shall now formulate the concept of uniform convergence again. The convergent series

$$g_1(x) + g_2(x) + \ldots$$

is said to be *uniformly convergent* in an interval if the sum $f(x)$ can be approximated to within ϵ (where ϵ is an arbitrarily small positive number) by taking a number of terms which is sufficiently large and which is the same throughout the interval.

We suppose first that the series $g_1(x) + g_2(x) + \ldots$ converges at every point of a certain interval to a limit function $f(x)$; by $f_n(x)$ we denote the n-th partial sum of the series, $f_n(x) = g_1(x) + \ldots + g_n(x)$, and by $R_n(x)$ the remainder of the series after n terms,

$$R_n(x) = f(x) - f_n(x).$$

The series $g_1(x) + g_2(x) + \ldots$ *is said to be uniformly convergent in the interval if to every positive number ϵ there corresponds a number N, dependent on ϵ alone and not on x, such that for $n > N$ the inequality $| R_n(x) | = | f(x) - f_n(x) | < \epsilon$ holds for all values of x in the interval.*

Expressed more pictorially, the partial sum $f_n(x)$ represents the sum $f(x)$ to within an error of less than ϵ everywhere in the interval at the same time, provided only that n is chosen large enough. By Cauchy's test we readily see that the series converges uniformly if, and only if, the difference $| f_n(x) - f_m(x) |$ can be made less than an arbitrary quantity ϵ everywhere in the interval by choosing n and m larger than a number N independent of x. For, firstly, if the convergence is uniform we can make $| f_n(x) - f(x) |$ and $| f_m(x) - f(x) |$ both less than $\epsilon/2$ by choosing n and m greater than a number N independent of x, from which it follows that $| f_n(x) - f_m(x) | < \epsilon$; and secondly, if $| f_n(x) - f_m(x) | < \epsilon$ for all values of x whenever n and m are greater than N, then on choosing any fixed value of $n > N$ and letting m increase beyond all bounds we have the relation

$$| f_n(x) - f(x) | = \lim_{m \to \infty} | f_n(x) - f_m(x) | \leqq \epsilon,$$

for every value of x, so that the convergence is uniform.

If we wish to speak of the uniform convergence of a sequence of functions we need make only trifling changes in the above definition; the sequence $f_1(x)$, $f_2(x)$, ... converges uniformly to $f(x)$ in an interval if the difference $|f(x) - f_n(x)|$ can be made less than ϵ everywhere in the interval by choosing n greater than a number N independent of x. As above, a necessary and sufficient condition for the uniform convergence of the sequence is that $|f_n(x) - f_m(x)| < \epsilon$ for all values of x when n and m are both greater than a certain number N dependent on ϵ but not on x.

We shall soon see that it is just this condition of uniform convergence that makes infinite series and other limiting processes with functions into convenient and useful tools of analysis. Fortunately, in the limiting processes usually encountered in the calculus and its applications, non-uniform convergence is a sort of exceptional phenomenon which will scarcely trouble us in our present applications of analysis.

In most cases the uniformity of convergence of a series is established by means of the following criterion:

If the terms of the series $\sum_{\nu=1}^{\infty} g_\nu(x)$ satisfy the condition $|g_\nu(x)| \leqq a_\nu$, where the numbers a_ν are constants which form a convergent series $\sum_{\nu=1}^{\infty} a_\nu$, then the series $\sum_{\nu=1}^{\infty} g_\nu(x)$ converges uniformly (and, we may incidentally remark, absolutely).

For we then have

$$\left| \sum_{\nu=n}^{m} g_\nu(x) \right| \leqq \sum_{\nu=n}^{m} |g_\nu(x)| \leqq \sum_{\nu=n}^{m} a_\nu,$$

and since by Cauchy's test the sum $\sum_{\nu=n}^{m} a_\nu$ can be made arbitrarily small by choosing n and $m > n$ large enough, this exactly expresses the necessary and sufficient condition for uniform convergence.

A first example is offered by the geometric series $1 + x + x^2 + \ldots$ where x is restricted to the interval $|x| \leq q$, q being any positive number less than 1. The terms of the series are then numerically less than or equal to the terms of the convergent geometric series $\sum q^\nu$.

A further example is given by the " trigonometric series "

$$\frac{c_1 \sin(x - \delta_1)}{1^2} + \frac{c_2 \sin(x - \delta_2)}{2^2} + \frac{c_3 \sin(x - \delta_3)}{3^2} + \cdots,$$

provided that $|c_n| < c$, where c is a positive constant independent of n. For then we have

$$g_n(x) = \frac{c_n \sin(x - \delta_n)}{n^2}, \text{ so that } |g_n(x)| < \frac{c}{n^2}.$$

Hence the uniform and absolute convergence of the trigonometric series follows from the convergence of the series $\sum\limits_{\nu=1}^{\infty} \dfrac{c}{n^2}$.

3. Continuity of the Sum of a Uniformly Convergent Series of Continuous Functions.

As we have already hinted, the significance of the uniform convergence of an infinite series lies in the fact that a uniformly convergent series in many respects behaves exactly like the sum of a finite number of functions. Thus, for example, the sum of a finite number of continuous functions is itself continuous, and correspondingly we have the following theorem:

If a series of continuous terms converges uniformly in an interval, its sum is also a continuous function.

The proof is quite simple. We subdivide the series

$$f(x) = g_1(x) + g_2(x) + \ldots$$

into the n-th partial sum $f_n(x)$ plus the remainder $R_n(x)$. As usual, $f_n(x) = g_1(x) + \ldots + g_n(x)$. If now any positive number ϵ is assigned, we can in virtue of the uniform convergence choose the number n so large that the remainder is less than $\epsilon/4$ throughout the whole interval, and hence

$$|R_n(x + h) - R_n(x)| < \frac{\epsilon}{2}$$

for every pair of numbers x and $x + h$ in the interval. The partial sum $f_n(x)$ consists of the sum of a finite number of continuous functions and is therefore continuous; for each point x in the interval, therefore, we can choose a positive δ so small that

$$|f_n(x + h) - f_n(x)| < \frac{\epsilon}{2}$$

provided $|h| < \delta$ and the points x and $x + h$ lie in the interval. It then follows that

$$|f(x + h) - f(x)| = |f_n(x + h) - f_n(x) + R_n(x + h) - R_n(x)|$$
$$\leq |f_n(x + h) - f_n(x)| + |R_n(x + h) - R_n(x)| < \epsilon,$$

which expresses the continuity of our function.

The significance of this theorem becomes clear when we recall that the sums of non-uniformly convergent series of continuous functions are not necessarily continuous, as our previous examples show. From the preceding theorem we may conclude that if the sum of a convergent series of continuous functions has a point of discontinuity, then in every neighbourhood of this point the convergence is non-uniform. Hence every representation of discontinuous functions by series of continuous functions is based on the use of non-uniformly convergent limiting processes.

4. Integration of Uniformly Convergent Series.

A sum of a finite number of continuous functions can be " integrated term by term "; that is, the integral of the sum can be found by integrating each term separately and adding the integrals. In the case of a convergent infinite series the same procedure is permissible, provided that the series converges uniformly in the interval of integration.

A series $\sum_{\nu=1}^{\infty} g_{\nu}(\mathrm{x}) = \mathrm{f}(\mathrm{x})$ which converges uniformly in an interval can be integrated term by term in that interval: or, more precisely, if a *and* x *are two numbers in the interval of uniform convergence, the series $\sum_{\nu=1}^{\infty} \int_{a}^{x} g_{\nu}(\mathrm{t})\,\mathrm{dt}$ converges, and, in fact, converges uniformly with respect to* x *for each fixed value of* a, *its sum being equal to* $\int_{a}^{x} \mathrm{f}(\mathrm{t})\,\mathrm{dt}.$

To prove this we write as before

$$f(x) = \sum_{\nu=1}^{\infty} g_{\nu}(x) = f_n(x) + R_n(x).$$

We have assumed that the separate terms of the series are continuous; hence by the previous sub-section the sum is also continuous and therefore integrable. Now if ϵ is any positive number, we can find a number N so large that for every $n > N$ the inequality $|R_n(x)| < \epsilon$ holds for every value of x in the interval. By the first mean value theorem of the integral calculus we have

$$\left| \int_{a}^{x} \left\{ f(t) - f_n(t) \right\} dt \right| \leq \epsilon l,$$

where l is the length of the interval of integration. Since the integration of the finite sum $f_n(x)$ can be performed term by term, this gives us

$$\left| \int_a^x f(t)\,dt - \sum_1^n \int_a^x g_\nu(t)\,dt \right| < \epsilon l.$$

But since ϵl can be made as small as we please, this states that

$$\sum_{\nu=1}^\infty \int_a^x g_\nu(t)\,dt = \lim_{n\to\infty} \sum_{\nu=1}^n \int_a^x g_\nu(t)\,dt = \int_a^x f(t)\,dt,$$

which was to be proved.

If, instead of infinite series, we wish to deal with sequences of functions, our result can be expressed in the following way:

If in an interval the sequence of functions $f_1(x)$, $f_2(x)$, . . . *tends uniformly to the limit function* $f(x)$, *then*

$$\int_a^b f(x)\,dx = \lim_{n\to\infty} \int_a^b f_n(x)\,dx$$

for every pair of numbers a *and* b *lying in the interval;* in other words, *we can then interchange the order of the operations of integration and passing to the limit.*

This fact is far from being a triviality. It is true that from a naïve point of view such as prevailed in the eighteenth century the interchangeability of the two processes is hardly to be doubted; but a glance at the examples in No. 1 of this section (p. 387) shows us that in the case of non-uniform convergence the above equation might not hold. We need only consider Ex. 2 (p. 387), in which the integral of the limit function is 0, while the integral of the function $f_n(x)$ over the interval $0 \le x \le 1$, that is to say, the area of the triangle in fig. 5, p. 388, has the value

$$\int_0^1 f_n(x)\,dx = n^{\alpha-2},$$

and when $\alpha \ge 2$ this does not tend to zero. Here we immediately see from the figure that the reason for the difference between $\int_0^1 f(x)\,dx$ and $\lim_{n\to\infty} \int_0^1 f_n(x)\,dx$ lies in the non-uniformity of the convergence.

On the other hand, by considering values of α such that $1 \le \alpha < 2$, we see that the equation $\lim_{n\to\infty} \int_0^1 f_n(x)\,dx = \int_0^1 f(x)\,dx$ can hold good although the convergence is non-uniform. As a further example, the series $\sum_0^\infty g_n(x)$, where $g_n(x) = x^n - x^{n-1}$ for $n \ge 1$ and $g_0(x) = 1$, can be integrated term by term between the limits 0 and 1, even though it does not converge uniformly. Thus while uniformity of convergence is a *sufficient* condition for term-by-term integrability, it is by no means a *necessary* condition. Neglect of this point may easily lead to misunderstanding.

5. Differentiation of Infinite Series.

The behaviour of uniformly convergent series or sequences with respect to differentiation is quite different from that with respect to integration. For example, the sequence of functions $f_n(x) = \dfrac{\sin n^2 x}{n}$ certainly converges uniformly to the limit function $f(x) = 0$, but the derivative $f_n'(x) = n \cos n^2 x$ certainly does not converge everywhere to the derivative of the limit function $f'(x) = 0$, as we see by considering $x = 0$. In spite of the uniformity of the convergence, therefore, we cannot change the order of the processes of differentiation and passing to the limit.

Corresponding statements of course hold for infinite series. For example, the series

$$\sin x + \frac{\sin 2^4 x}{2^2} + \frac{\sin 3^4 x}{3^2} + \dots$$

is absolutely and uniformly convergent, for its terms are numerically not greater than the terms of the convergent series $\dfrac{1}{1^2} + \dfrac{1}{2^2} + \dfrac{1}{3^2} + \dots$. If, however, we differentiate the series term by term, we obtain the series

$$\cos x + 2^2 \cos 2^4 x + 3^2 \cos 3^4 x + \dots ,$$

which plainly does not converge everywhere; for example, it diverges at $x = 0$.

The only useful criterion which assures us in special cases that term-by-term differentiation is permissible is given by the following theorem:

If, on differentiating a convergent infinite series $\sum\limits_{\nu=0}^{\infty} G_\nu(x) = F(x)$ term by term, we obtain a uniformly convergent series of continuous terms $\sum\limits_{\nu=0}^{\infty} g_\nu(x) = f(x)$, then the sum of this last series is equal to the *derivative of the sum of the first series.* This theorem, therefore, expressly requires that after differentiating the series term by term we must still investigate whether the result of the differentiation is a uniformly convergent series or not.

The proof of the theorem is almost trivial. For by the theorem in No. 4 (p. 394) we can integrate term by term the series obtained by differentiation. Recalling that $g_\nu(t) = G_\nu'(t)$, we obtain

$$\int_a^x f(t)\,dt = \int_a^x \left(\sum_{\nu=0}^\infty g_\nu(t)\right) dt = \sum_{\nu=0}^\infty \int_a^x g_\nu(t)\,dt = \sum_{\nu=0}^\infty (G_\nu(x) - G_\nu(a))$$

$$= F(x) - F(a).$$

This being true for every value of x in the interval of uniform convergence, it follows that

$$f(x) = F'(x),$$

which was to be proved.

EXAMPLES

1. Show by comparison with a series of constant terms that the following series converge uniformly in the intervals stated:

(a) $x - x^2 + x^3 - x^4 + \ldots$ $(-\tfrac{1}{2} \leq x \leq \tfrac{1}{2})$.

(b) $\tfrac{1}{2}\sqrt{1 - x^2} + \tfrac{1}{4}\sqrt{1 - x^4} + \tfrac{1}{8}\sqrt{1 - x^3} + \ldots + \dfrac{1}{2^n}\sqrt{1 - x^{2n}} + \ldots$ $(-1 \leq x \leq 1)$.

(c) $\dfrac{\sin x}{1^2} + \dfrac{\sin 2x}{2^2} + \ldots + \dfrac{\sin nx}{n^2} + \ldots$.

(d) $e^x + e^{2x} + \ldots + e^{nx} + \ldots$ $(-2 \leq x \leq -1)$.

2. Prove that $\lim f_n(x) = 0$, where $f_n(x) = \dfrac{nx}{1 + n^2 x^2}$, $-1 \leq x \leq 1$. Prove that the convergence is non-uniform.

3.* (a) Find $\lim\limits_{n \to \infty} f_n(x)$, where $f_n(x) = \dfrac{n^2 x^2}{1 + n^2 x^2}$, $-1 \leq x \leq 1$. Prove that the convergence is non-uniform. Prove that nevertheless

$$\lim_{n \to \infty} \int_{-1}^1 f_n(x)\,dx = \int_{-1}^1 \lim_{n \to \infty} f_n(x)\,dx.$$

(b) Discuss the behaviour of the sequence given by $f_n(x) = \dfrac{n^\alpha x^2}{1 + n^\alpha x^2}$ with regard to convergence, uniform convergence, and term-by-term integrability.

4.* Sketch the curves $y = f_n(x) = \dfrac{x^{2n}}{1 + x^{2n}}$, $-2 \leq x \leq 2$, for $n = 1$, 3, 10. Find $\lim\limits_{n \to \infty} f_n(x)$. Prove that the convergence is non-uniform.

5. Show that $\sum\limits_{\nu=-\infty}^{\infty} e^{-(x-\nu)^2}$ converges uniformly in any fixed interval $a \leqq x \leqq b$.

6. Show that in the interval $0 \leqq x \leqq \pi$ the following sequences converge, but not uniformly:

(a) $\sqrt[n]{\sin x}$.

(b) $(\sin x)^n$.

(c) $\sqrt[n]{x \sin x}$.

(d) $[f(x)]^n$, where $f(x) = \dfrac{\sin x}{x}$, $f(0) = 1$.

(e) $\sqrt[n]{f(x)}$, where $f(x) = \dfrac{\sin x}{x}$, $f(0) = 1$.

7. The sequence $f_n(x)$, $n = 1, 2, \ldots$, is defined in the interval $0 \leqq x \leqq 1$ by the equations

$$f_0(x) \equiv 1, \quad f_n(x) = \sqrt{x f_{n-1}(x)}.$$

(a) Prove that in the interval $0 \leqq x \leqq 1$ the sequence converges to a continuous limit.

(b)* Prove that the convergence is uniform.

8.* Let $f_0(x)$ be continuous in the interval $0 \leqq x \leqq a$. The sequence of functions $f_n(x)$ is defined by

$$f_n(x) = \int_0^x f_{n-1}(t)\, dt, \quad n = 1, 2, \ldots.$$

Prove that in any fixed interval $0 \leqq x \leqq a$ the sequence converges uniformly to 0.

9. Sketch the curves $x^{2n} + y^{2n} = 1$ for $n = 1, 2, 4$. To what limit do these curves tend as $n \to \infty$?

10.* Let $f_n(x)$, $n = 1, 2, \ldots$, be a sequence of functions with continuous derivatives in the interval $a \leqq x \leqq b$. Prove that if $f_n(x)$ converges at each point of the interval and the inequality $|f_n'(x)| < M$ (where M is a constant) is satisfied for all values of n and x, then the convergence is uniform.

5. Power Series

Among infinite series, *power series* occupy the chief place. By a power series we mean a series of the type

$$P(x) = c_0 + c_1 x + c_2 x^2 + \ldots = \sum_{\nu=0}^{\infty} c_\nu x^\nu$$

("power series in x"), or more generally

$$P(x) = c_0 + c_1(x - x_0) + c_2(x - x_0)^2 + \ldots = \sum_{\nu=0}^{\infty} c_\nu (x - x_0)^\nu$$

("power series in $(x - x_0)$"), where x_0 is a fixed number. If in the last series we introduce $\xi = x - x_0$ as a new variable, it becomes a power series $\overset{\infty}{\underset{\nu=0}{\Sigma}} c_\nu \xi^\nu$ in the new variable ξ, and we can therefore confine our attention to power series of the more special form $\overset{\infty}{\underset{\nu=0}{\Sigma}} c_\nu x^\nu$ without any loss of generality.

In Chap. VI (p. 320) we considered the approximate representation of functions by polynomials and were thus led to the expansion of functions in Taylor series, which are in fact power series. In this section we shall study power series in somewhat greater detail, and shall obtain the expansions of the most important functions in series in simpler and more convenient ways than before.

1. Convergence Properties of Power Series.

There are power series which converge for *no* value of x, except of course for $x = 0$, e.g. the series

$$x + 2^2 x^2 + 3^3 x^3 + \ldots + n^n x^n + \ldots .$$

For if $x \neq 0$, we can find an integer N such that $|x| > 1/N$. Then all the terms $n^n x^n$ for which $n > N$ will be greater than 1 in absolute value, and, in fact, as n increases $n^n x^n$ will increase beyond all bounds, so that the series fails to converge.

On the other hand, there are series which converge for *every* value of x; for example, the power series for the exponential function,

$$e^x = 1 + x + \frac{x^2}{2!} + \frac{x^3}{3!} + \ldots ,$$

whose convergence for every value of x follows at once from the ratio test (Criterion IIIa, p. 378). The $(n + 1)$-th term divided by the n-th term gives x/n, and, whatever number x is chosen, this ratio tends to zero as n increases.

The behaviour of power series with regard to convergence is expressed in the following fundamental theorem:

If a power series in x *converges for a value* x $= \xi$, *it converges absolutely for every value* x *such that* $|$x$| < |\xi|$, *and the convergence is uniform in every interval* $|$x$| \leq \eta$, *where* η *is any positive number less than* $|\xi|$. *Here* η *may lie as near* $|\xi|$ *as we please.*

The proof is simple. If the series $\overset{\infty}{\underset{\nu=0}{\Sigma}} c_\nu \xi^\nu$ converges, its terms tend to 0 as n increases. From this follows the weaker statement that the terms all lie below a bound M independent of ν, that is, $|\,c_\nu \xi^\nu\,| < M$. If now q is any number such that $0 < q < 1$, and if we restrict x to the interval $|\,x\,| \leqq q\,|\,\xi\,|$, then $|\,c_\nu x^\nu\,| \leqq |\,c_\nu \xi^\nu\,|\,q^\nu < Mq^\nu$. In this interval, therefore, the terms of our series $\overset{\infty}{\underset{0}{\Sigma}} c_\nu x^\nu$ are smaller in absolute value than the terms of the convergent geometric series ΣMq^ν. Hence from the theorem on p. 392 the absolute and uniform convergence of the series in the interval $-q\,|\,\xi\,| \leqq x \leqq q\,|\,\xi\,|$ follows.

If a power series does not converge everywhere, that is, if there is a value $x = \xi$ for which it diverges, it must diverge for every value of x such that $|\,x\,| > |\,\xi\,|$. For if it were convergent for such a value of x, by the theorem above it would have to converge for the numerically smaller value ξ.

From this we recognize that a power series which converges for at least one value of x other than 0 and which diverges for at least one value of x has an *interval of convergence*; that is, a definite positive number ρ exists such that for $|\,x\,| > \rho$ the series diverges and for $|\,x\,| < \rho$ the series converges. For $|\,x\,| = \rho$ no general statement can be made. The limiting cases, that in which the series converges only for $x = 0$ and that in which it converges everywhere, are expressed symbolically by writing $\rho = 0$ and $\rho = \infty$ respectively.*

For example, for the geometric series $1 + x + x^2 + \ldots$ we have $\rho = 1$; at the end-points of the interval of convergence the series diverges. Similarly, for the series for the inverse tangent (p. 319),

$$\text{arc tan}\, x = x - x^3/3 + x^5/5 - + \ldots ,$$

we have $\rho = 1$, and at both the end-points $x = \pm 1$ of the interval of convergence the series converges, as we recognize at once from Leibnitz's test (p. 370).

* It is possible to find this interval of convergence directly from the co-efficients c_ν of the series. If the limit $\lim\limits_{n \to \infty} \sqrt[n]{|c_n|}$ exists, then

$$\rho = \frac{1}{\lim\limits_{n \to \infty} \sqrt[n]{|c_n|}}.$$

In general ρ is given by the formula

$$\rho = \frac{1}{\overline{\lim}\limits_{n \to \infty} \sqrt[n]{|c_n|}}$$

where $\overline{\lim}$ is the symbol for the upper limit, as defined in the appendix to Chap. I (p. 62).

From the uniform convergence we derive the important fact that within its interval of convergence (if such an interval exists) the power series represents a continuous function.

2. Integration and Differentiation of Power Series.

On account of the uniformity of convergence *it is always permissible to integrate a power series*

$$f(x) = \sum_{\nu=0}^{\infty} c_\nu x^\nu$$

term by term over any closed interval lying entirely within the interval of convergence. We thus obtain the function

$$F(x) = c + \sum_{\nu=0}^{\infty} \frac{c_\nu}{\nu+1} x^{\nu+1},$$

for which $\qquad\qquad F'(x) = f(x).$

Further, since $\left| \dfrac{c_\nu}{\nu+1} \right| \leq |c_\nu|$ for all values of ν, the series obtained by integration converges more rapidly than the original series.

We can also differentiate a power series term by term within its interval of convergence, thus obtaining the equation

$$f'(x) = \sum_{\nu=1}^{\infty} \nu c_\nu x^{\nu-1}.$$

In order to prove this statement we need only show that the series on the right converges uniformly if x is restricted to an interval lying entirely within the interval of convergence. Suppose then that ξ is a number, lying as close to ρ as we please, for which $\sum_{\nu=1}^{\infty} c_\nu \xi^\nu$ converges; then, as we have seen before, the numbers $|c_\nu \xi^\nu|$ all lie below a bound M independent of ν, so that $|c_\nu \xi^{\nu-1}| < \dfrac{M}{|\xi|} = N$. Now let q be any number such that $0 < q < 1$; if we restrict x to the interval $|x| \leq q\,|\xi|$, the terms of the series under discussion are not greater than those of the series $\sum_{\nu=1}^{\infty} |\nu c_\nu q^{\nu-1} \xi^{\nu-1}|$, and therefore less than those of the

14

series $\sum\limits_{\nu=1} N\nu q^{\nu-1}$. But in this last series the ratio of the $(n+1)$-th term to the n-th term is $\dfrac{n+1}{n} q$, which tends to q as n increases. Since $0 < q < 1$, it follows (criterion IIIa, p. 378) that this series converges. Hence the series obtained by differentiation converges uniformly, and by the theorem at the end of last section (p. 396) represents the derivative $f'(x)$ of the function $f(x)$, which proves our statement.

If we apply this result again to the power series

$$f'(x) = \sum_{\nu=1}^{\infty} \nu c_\nu x^{\nu-1},$$

we find on differentiating term by term that

$$f''(x) = \sum_{\nu=2}^{\infty} \nu(\nu - 1) c_\nu x^{\nu-2},$$

and, continuing the process, we arrive at the theorem: *Every function represented by a power series can be differentiated as often as we please within the interval of convergence, and the differentiation can be performed term by term.**

3. Operations with Power Series.

The preceding theorems on the behaviour of power series are our justification for operating in the same way with power series as with polynomials. It is obvious that two power series can be added or subtracted by adding or subtracting the corresponding coefficients (see p. 376). It is also clear that a power series, like any other convergent series, can be multiplied by a constant factor by multiplying each term by that factor. On the other hand, the multiplication and division of two power series requires somewhat more detailed study, for

---

* As an explicit expression for the k-th derivative we obtain

$$f^{(k)}(x) = \sum_{\nu=k}^{\infty} \nu(\nu - 1) \ldots (\nu - k + 1) c_\nu x^{\nu-k},$$

or in a slightly different form,

$$\frac{f^{(k)}(x)}{k!} = \sum_{\nu=k}^{\infty} \binom{\nu}{k} c_\nu x^{\nu-k} = \sum_{\nu=0}^{\infty} \binom{k+\nu}{k} c_{k+\nu} x^{\nu}.$$

These two formulæ are frequently useful.

which we refer the reader to the appendix (p. 416). Here we merely mention without proof that two power series

$$f(x) = \sum_{\nu=0}^{\infty} a_{\nu} x^{\nu}$$

and
$$g(x) = \sum_{\nu=0}^{\infty} b_{\nu} x^{\nu}$$

can be multiplied together like polynomials. To be specific, we have the following theorem: throughout the common part of the intervals of convergence of these two series their product is given by the convergent power series $\sum_{\nu=0}^{\infty} c_{\nu} x^{\nu}$, where the coefficients c_{ν} are given by the formulæ

$$c_0 = a_0 b_0,$$
$$c_1 = a_0 b_1 + a_1 b_0,$$
$$c_2 = a_0 b_2 + a_1 b_1 + a_2 b_0,$$
$$\cdot \quad \cdot \quad \cdot \quad \cdot \quad \cdot \quad \cdot \quad \cdot \quad \cdot$$
$$c_n = a_0 b_n + a_1 b_{n-1} + \ldots + a_n b_0,$$
$$\cdot \quad \cdot \quad \cdot \quad \cdot \quad \cdot \quad \cdot \quad \cdot$$

(For the proof see the appendix, § 1, p. 416.)

4. Theorem of Uniqueness for Power Series.

In the theory of power series the following fact is of importance: if two power series $\sum_{\nu=0}^{\infty} a_{\nu} x^{\nu}$ and $\sum_{\nu=0}^{\infty} b_{\nu} x^{\nu}$ both converge in an interval which contains the point $x = 0$ in its interior, and if in that interval the two series represent the same function $f(x)$, then they are identical, that is, the equation $a_n = b_n$ is true for every value of n. In other words:

A function f(x) *can be represented by a power series in* x *in only one way, if at all.*

Briefly, the representation of a function by a power series is " unique ".

To prove this we need only notice that the difference of the two power series, that is, the power series $\phi(x) = \sum_{\nu=0}^{\infty} c_{\nu} x^{\nu}$ with coefficients $c_{\nu} = a_{\nu} - b_{\nu}$, represents the function

$$\phi(x) = f(x) - f(x) = 0$$

in the interval; that is, this last power series converges to the limit 0 everywhere in the interval. For $x = 0$, in particular, the sum of the series must be 0; that is, $c_0 = 0$, so that $a_0 = b_0$. We now differentiate the series in the interior of the interval, obtaining $\phi'(x) = \sum\limits_{\nu=1}^{\infty} \nu c_\nu x^{\nu-1}$. But $\phi'(x)$ is also 0 throughout the interval; hence for $x = 0$, in particular, we have $c_1 = 0$, or $a_1 = b_1$. Continuing this process of differentiating and then putting $x = 0$, we find successively that all the coefficients c_ν are equal to zero, which proves the theorem.

We see, in addition, that we can draw the following conclusion from the above discussion: if we take the ν-th derivative of a series $f(x) = \Sigma a_\nu x^\nu$ and then put $x = 0$, we at once obtain

$$a_\nu = \frac{1}{\nu!} f^{(\nu)}(0),$$

that is:

Every power series which converges for points other than $x = 0$ *is the Taylor series of the function which it represents.*

The uniqueness of the expansion is here expressed by the fact that the coefficients are uniquely determined by the function itself.

6. Expansion of Given Functions in Power Series. Method of Undetermined Coefficients. Examples

Within its interval of convergence every power series represents a continuous function with continuous derivatives of all orders. We shall now discuss the converse problem of the expansion of a given function in a power series. In theory we can always do this by means of Taylor's theorem; in practice we often meet with difficulties in the actual calculation of the n-th derivative and in the estimation of the remainder. But we can often reach our goal more simply by making use of the following device. We first write down the statement $f(x) = \sum\limits_{\nu=0}^{\infty} c_\nu x^\nu$, where the coefficients c_ν are unknown to begin with. Then by some known property of the function $f(x)$ we determine the coefficients, and then prove the convergence of the series. The series represents a function, and it only remains to prove that this function is identical with $f(x)$. Because of the uniqueness of the expansion in power series we know that no other series than the one just found can be the required expansion. We shall now consider some examples of this method. Actually, we have already obtained the series for $\arctan x$ and $\log(1 + x)$ by a method which forms part of the range of ideas of the present chapter. For we simply integrated term by term the series for the derivatives of these functions, which we knew to be geometric series.

1. The Exponential Function.

Our problem is to find a function $f(x)$ for which $f'(x) = f(x)$ and $f(0) = 1$. If we write down the series with undetermined coefficients

$$f(x) = c_0 + c_1 x + c_2 x^2 + \ldots \,,$$

and differentiate it, we obtain

$$f'(x) = c_1 + 2c_2 x + 3c_3 x^2 + \ldots \, .$$

Since by hypothesis these two power series must be identical, we have the equation

$$nc_n = c_{n-1},$$

true for all values of $n \geqq 1$. If we observe that because of the relation $f(0) = 1$ the coefficient c_0 must have the value 1, we can calculate all the coefficients successively, and obtain the power series

$$f(x) = 1 + \frac{x}{1!} + \frac{x^2}{2!} + \frac{x^3}{3!} + \ldots \, .$$

As we easily see by the ratio test, this series converges for all values of x and therefore represents a function for which the relations $f'(x) = f(x)$, $f(0) = 1$ are actually fulfilled. (Here we intentionally avoid making any use of what we have previously learned about the expansion of the exponential function).

Now the function e^x certainly possesses these properties; we readily deduce that the function $f(x)$ is identical with e^x. For if we form the quotient $\varphi(x) = f(x)/e^x$ and differentiate we have

$$\varphi'(x) = \frac{e^x f'(x) - e^x f(x)}{e^{2x}} = 0.$$

The function $\varphi(x)$ is therefore a constant, and since it has the value 1 for $x = 0$, it must be identically equal to 1, thus proving that our power series and the exponential function are identical (cf. the analogous discussion on p. 178).

2. The Binomial Series.

We can now return to the binomial series (Chap. VI, § 3, p. 329), this time making use of the method of undetermined coefficients. We wish to expand the function $f(x) = (1 + x)^\alpha$ in a power series, and therefore write

$$f(x) = (1 + x)^\alpha = c_0 + c_1 x + c_2 x^2 + \ldots,$$

the coefficients c_ν being undetermined. We now notice that our function obviously satisfies the relation

$$(1 + x)f'(x) = \alpha f(x) = \sum_{\nu=0}^{\infty} \alpha c_\nu x^\nu.$$

On the other hand, if we differentiate the series for $f(x)$ term by term and multiply by $(1 + x)$, we obtain

$$(1 + x)f'(x) = c_1 + (2c_2 + c_1)x + (3c_3 + 2c_2)x^2 + \ldots;$$

and since these two power series for $(1 + x)f'(x)$ must be identical,

$$\alpha c_0 = c_1, \quad \alpha c_1 = 2c_2 + c_1, \quad \alpha c_2 = 3c_3 + 2c_2, \ldots.$$

Now it is certain that $c_0 = 1$, since our series must have the value 1 for $x = 0$, and so we obtain in succession the expressions

$$c_1 = \alpha, \quad c_2 = \frac{(\alpha - 1)\alpha}{2}, \quad c_3 = \frac{(\alpha - 2)(\alpha - 1)\alpha}{3 \cdot 2}, \ldots,$$

for the coefficients, and in general, as is easily established, we have

$$c_\nu = \frac{(\alpha - \nu + 1)(\alpha - \nu + 2) \ldots (\alpha - 1)\alpha}{\nu(\nu - 1) \ldots 2 \cdot 1} = \binom{\alpha}{\nu}.$$

Substituting these values for the coefficients, we have the series $\sum_{\nu=0}^{\infty} \binom{\alpha}{\nu} x^\nu$; we have yet to investigate the convergence of this series and to show that it actually represents $(1 + x)^\alpha$.

By the ratio test we find that when α is not a positive integer, the series converges if $|x| < 1$ and diverges if $|x| > 1$; for then the ratio of the $(n + 1)$-th term to the n-th term is $\dfrac{\alpha - n + 1}{n} x$, and the absolute value of this expression tends to $|x|$ as n increases beyond all bounds.* Hence, if $|x| < 1$ our series represents a function $f(x)$ which satisfies the condition $(1 + x)f'(x) = \alpha f(x)$, as follows from the method of forming the

---

* Here we state, without proof, the exact conditions under which this series converges. If the index α is an integer ≥ 0, the series terminates and is therefore valid for all values of x (becoming the ordinary binomial theorem). For all other values of α the series is absolutely convergent for $|x| < 1$ and divergent for $|x| > 1$. For $x = +1$ the series converges absolutely if $\alpha > 0$, converges conditionally if $-1 < \alpha < 0$, and diverges if $\alpha \leq -1$. Finally, at $x = -1$ the series is absolutely convergent if $\alpha > 0$, divergent if $\alpha < 0$.

coefficients. Moreover, $f(0) = 1$. But these two conditions ensure that the function $f(x)$ is identical with $(1 + x)^a$. For on putting

$$\varphi(x) = f(x)/(1 + x)^a$$

we find that

$$\varphi'(x) = \frac{(1 + x)^a f'(x) - \alpha(1 + x)^{a-1} f(x)}{(1 + x)^{2a}} = 0;$$

$\varphi(x)$ is therefore a constant, and, in fact, is always equal to 1, since $\varphi(0) = 1$. We have therefore proved that when $|x| < 1$

$$(1 + x)^a = \sum_{\nu=0}^{\infty} \binom{\alpha}{\nu} x^\nu,$$

which is the binomial series.

Here we quote the following special cases of the binomial series: the geometric series

$$\frac{1}{1 + x} = (1 + x)^{-1} = 1 - x + x^2 - x^3 + x^4 - + \ldots$$

$$= \sum_{\nu=0}^{\infty} (-1)^\nu x^\nu;$$

the series

$$\frac{1}{(1 + x)^2} = (1 + x)^{-2} = 1 - 2x + 3x^2 - 4x^3 + - \ldots$$

$$= \sum_{\nu=0}^{\infty} (-1)^\nu (\nu + 1) x^\nu,$$

which may also be obtained from the geometric series by differentiation; and the series

$$\sqrt{(1 + x)} = (1 + x)^{\frac{1}{2}} = 1 + \frac{1}{2} x - \frac{1}{2 \cdot 4} x^2 + \frac{1 \cdot 3}{2 \cdot 4 \cdot 6} x^3$$

$$- \frac{1 \cdot 3 \cdot 5}{2 \cdot 4 \cdot 6 \cdot 8} x^4 + - \ldots,$$

$$\frac{1}{\sqrt{(1 + x)}} = (1 + x)^{-\frac{1}{2}} = 1 - \frac{1}{2} x + \frac{1 \cdot 3}{2 \cdot 4} x^2 - \frac{1 \cdot 3 \cdot 5}{2 \cdot 4 \cdot 6} x^3$$

$$+ \frac{1 \cdot 3 \cdot 5 \cdot 7}{2 \cdot 4 \cdot 6 \cdot 8} x^4 - + \ldots,$$

the first two or three terms of which form useful approximations.

3. The Series for arc sin x.

This series can be obtained very easily by expanding the expression $1/\sqrt{(1 - t^2)}$ according to the binomial series,

$$(1 - t^2)^{-\frac{1}{2}} = 1 + \frac{1}{2} t^2 + \frac{1 \cdot 3}{2 \cdot 4} t^4 + \ldots .$$

This series converges if $|t| \leqq 1$, and so converges uniformly if $|t| \leqq q < 1$. On integrating term by term between 0 and x, we obtain

$$\text{arc } \sin x = x + \frac{1}{2}\frac{x^3}{3} + \frac{1.3}{2.4}\frac{x^5}{5} + \dots ;$$

by the ratio test we find that this converges if $|x| < 1$, and diverges if $|x| > 1$.

The deduction of this series from Taylor's theorem would be decidedly less convenient, owing to the difficulty of estimating the remainder.

4. The Series for ar $\sinh x = \log\left\{x + \sqrt{(1 + x^2)}\right\}$.

We obtain this expansion by a similar method. Using the binomial theorem we write down the series for the derivative of ar $\sinh x$,

$$\frac{1}{\sqrt{(1 + x^2)}} = 1 - \frac{1}{2}x^2 + \frac{1.3}{2.4}x^4 - \frac{1.3.5}{2.4.6}x^6 + - \dots ,$$

and then integrate term by term. We thus obtain the expansion

$$\text{ar } \sinh x = x - \frac{1}{2}\frac{x^3}{3} + \frac{1.3}{2.4}\frac{x^5}{5} - + \dots ,$$

whose interval of convergence is $-1 \leqq x \leqq 1$.

5. Example of Multiplication of Series.

The expansion of the function

$$\frac{\log(1 + x)}{1 + x}$$

is a simple example of the application of the rule for the multiplication of power series. We have only to multiply the logarithmic series

$$\log(1 + x) = x - \frac{x^2}{2} + \frac{x^3}{3} - \frac{x^4}{4} + - \dots$$

by the geometric series

$$\frac{1}{1 + x} = 1 - x + x^2 - x^3 + x^4 - + \dots ;$$

as the reader may verify for himself, we obtain the remarkable expansion

$$\frac{\log(1 + x)}{1 + x} = x - \left(1 + \frac{1}{2}\right)x^2 + \left(1 + \frac{1}{2} + \frac{1}{3}\right)x^3$$
$$- \left(1 + \frac{1}{2} + \frac{1}{3} + \frac{1}{4}\right)x^4 + - \dots$$

for $|x| < 1$.

6. Example of Term-by-Term Integration (Elliptic Integral).

In previous applications we have met with the elliptic integral

$$K = \int_0^{\pi/2} \frac{d\varphi}{\sqrt{(1 - k^2 \sin^2 \varphi)}} \quad (k^2 < 1)$$

(the period of oscillation of a pendulum (p. 302)). In order to evaluate the integral we can first expand the integrand by the binomial theorem, thus obtaining

$$\frac{1}{\sqrt{(1 - k^2 \sin^2 \varphi)}} = 1 + \frac{1}{2} k^2 \sin^2 \varphi + \frac{1 \cdot 3}{2 \cdot 4} k^4 \sin^4 \varphi$$
$$+ \frac{1 \cdot 3 \cdot 5}{2 \cdot 4 \cdot 6} k^6 \sin^6 \varphi + \dots$$

Since $k^2 \sin^2 \varphi$ is never greater than k^2 this series converges uniformly for all values of φ, and we may integrate term by term:

$$K = \int_0^{\pi/2} \frac{d\varphi}{\sqrt{(1 - k^2 \sin^2 \varphi)}} = \int_0^{\pi/2} d\varphi + \frac{1}{2} k^2 \int_0^{\pi/2} \sin^2 \varphi \, d\varphi$$
$$+ \frac{1 \cdot 3}{2 \cdot 4} k^4 \int_0^{\pi/2} \sin^4 \varphi \, d\varphi + \dots$$

The integrals occurring here have already been calculated (cf. Chap. IV, § 4, p. 223). If we substitute their values we have

$$K = \int_0^{\pi/2} \frac{d\varphi}{\sqrt{(1 - k^2 \sin^2 \varphi)}} = \frac{\pi}{2} \left\{ 1 + \left(\frac{1}{2}\right)^2 k^2 + \left(\frac{1 \cdot 3}{2 \cdot 4}\right)^2 k^4 \right.$$
$$\left. + \left(\frac{1 \cdot 3 \cdot 5}{2 \cdot 4 \cdot 6}\right)^2 k^6 + \dots \right\}.$$

For further examples on the theory of series we refer the reader to the appendix (p. 415).

Examples

Determine the intervals of convergence of the series $\sum\limits_{n=1}^{\infty} a_n x^n$, where a_n is given by the formulæ in Ex. 1–20:

1. $\dfrac{1}{n}$.

2. n.

3. $\dfrac{1}{\sqrt{n}}$.

4. $\sqrt{n}$.

5. $\dfrac{1}{n^2}$.

6. $\dfrac{n}{n!}$.

7. $\dfrac{1}{a + n}$.

8. $\dfrac{1}{an + b}$.

9. $\dfrac{1}{\log(n + 1)}$.

10. $\dfrac{1}{\log \log 10n}$.

11. $\dfrac{1}{\sqrt[n]{n}}$.

12. a^n.

13. $a^{\sqrt{n}}$.

14. $a^{\log n}$.

15. $(\sqrt[n]{n} - 1)^n$.

16. $\dfrac{(n!)^2}{(2n)!}$.

17. $\dfrac{n + \sqrt{n}}{n^2 - n}$.

18. $\dfrac{1}{1 + a^n}$.

19. $\dfrac{1}{\sqrt{n}} + \dfrac{(-1)^n}{n}$.

20. $\dfrac{1}{n^{1+1/n}}$.

14*

Expand the functions in Ex. 21-26 in power series:

21. a^x.

22. $\dfrac{x + \log(1 - x)}{x^2}$.

23. $\sin^2 x$.

24. $\cos^2 x$.

25. $\sin^6 x$.

26. arc $\sin x^3$.

27. Using the binomial series, calculate $\sqrt{2}$ to four decimal places.

28. Obtain approximations in series for the following integrals by expanding the integrand in a power series and integrating:

(a) $\displaystyle\int_0^1 \frac{\sin x}{x}\, dx$.

(b) $\displaystyle\int_0^{\frac{1}{2}} \frac{dx}{\sqrt{(1 - x^4)}}$.

(c) $\displaystyle\int_0^1 \frac{\log(1 + x)}{x}\, dx$.

(d) $\displaystyle\int_5^{10} \frac{dx}{\sqrt{(1 + x^4)}}$.

29. By multiplication of power series obtain the expansions of the following up to the terms in x^4:

(a) $e^x \sin x$.

(b) $[\log(1 + x)]^2$.

(c) $\dfrac{\text{arc } \sin x}{\sqrt{1 - x}}$.

(d) $\sin^2 x$.

30.* By multiplication of power series prove that

(a) $e^x e^y = e^{x+y}$.

(b) $\sin 2x = 2 \sin x \cos x$.

31. If the interval of convergence of the power series $\Sigma a_n x^n$ is $|x| < \rho$, and that of $\Sigma b_n x^n$ is $|x| < \rho'$, where $\rho < \rho'$, what is the interval of convergence of $\Sigma(a_n + b_n) x^n$?

32. Using the method of undetermined coefficients, find the function $f(x)$ which satisfies the following conditions:

$$(a)\ f(0) = 3; \quad (b)\ f'(x) = f(x) + x.$$

7. Power Series with Complex Terms

1. Introduction of Complex Terms into Power Series.

The similarity between certain power series representing functions which are apparently unrelated led Euler to set up a purely formal connexion between them, found by giving complex values, in particular, pure imaginary values, to the variable x. We shall first do this formally, unhindered by questions of rigour, and shall investigate the results of the process.

The first striking relation of this sort is obtained if we replace the quantity x in the series for e^x by a pure imaginary $i\phi$, where ϕ is a real number. If we recall the fundamental equation for the imaginary unit i, that is, $i^2 = -1$, from which it follows that $i^3 = -i$, $i^4 = 1$, $i^5 = i, \ldots$, then on separating the real and the imaginary terms of the series, we obtain

$$e^{i\phi} = (1 - \frac{\phi^2}{2!} + \frac{\phi^4}{4!} - \frac{\phi^6}{6!} + - \ldots)$$

$$+ i(\phi - \frac{\phi^3}{3!} + \frac{\phi^5}{5!} - \frac{\phi^7}{7!} + - \ldots),$$

or, in another form,

$$e^{i\phi} = \cos\phi + i\sin\phi.$$

This is the well-known and important "Euler's formula"; as yet it is purely formal. It is consistent with De Moivre's theorem (p. 74), which is expressed by the equation

$$(\cos\phi + i\sin\phi)(\cos\psi + i\sin\psi) = \cos(\phi + \psi) + i\sin(\phi + \psi).$$

In virtue of Euler's formula this equation merely states that the relation

$$e^x \cdot e^y = e^{x+y}$$

continues to hold for pure imaginary values $x = i\phi$, $y = i\psi$.

If we replace the variable x in the power series for $\cos x$ by the pure imaginary ix we at once obtain the series for $\cosh x$; this relation can be expressed by the equation

$$\cosh x = \cos ix.$$

In the same way we obtain

$$\sinh x = \frac{1}{i}\sin ix.$$

Since Euler's formula also gives $e^{-i\phi} = \cos\phi - i\sin\phi$, we arrive at the exponential expressions for the trigonometric functions,

$$\sin x = \frac{e^{ix} - e^{-ix}}{2i}, \quad \cos x = \frac{e^{ix} + e^{-ix}}{2}.$$

These are exactly analogous to the exponential expressions for the hyperbolic functions and are, in fact, transformed into them by the relations $\cosh x = \cos ix$, $\sinh x = \frac{1}{i}\sin ix$.

Corresponding formal relations can, of course, be obtained for the functions $\tan x$, $\tanh x$, $\cot x$, $\coth x$, which are connected by the equations $\tanh x = \dfrac{1}{i} \tan ix$, $\coth x = i \cot ix$.

Finally, similar relations can also be found for the inverse trigonometric and hyperbolic functions. For example, from

$$y = \tan x = \frac{e^{ix} - e^{-ix}}{i(e^{ix} + e^{-ix})} = \frac{e^{2ix} - 1}{i(e^{2ix} + 1)}$$

we immediately find that

$$e^{2ix} = \frac{1 + iy}{1 - iy}.$$

If we take the logarithms of both sides of this equation and then write x instead of y and arc $\tan x$ instead of x, we obtain the equation

$$\mathrm{arc}\,\tan x = \frac{1}{2i} \log \frac{1 + ix}{1 - ix},$$

which expresses a remarkable connexion between the inverse tangent and the logarithm. If in the known power series for $\dfrac{1}{2} \log \dfrac{1 + x}{1 - x}$ (p. 318) we replace x by ix, we actually obtain the power series for arc $\tan x$,

$$\mathrm{arc}\,\tan x = \frac{1}{i}\left(ix + \frac{(ix)^3}{3} + \frac{(ix)^5}{5} + \dots\right)$$

$$= x - \frac{x^3}{3} + \frac{x^5}{5} - + \dots.$$

The above relations are as yet of a purely formal character, and naturally call for a more exact statement as to the meaning they are intended to convey. In the next sub-section we shall indicate how this can be given with the help of function theory.

For later use, however, we shall only need Euler's formula $e^{i\phi} = \cos\phi + i \sin\phi$ and, this being so, we can avoid a thorough analysis. We need only regard the symbol $e^{i\phi}$ as a *formal abbreviation* for the right-hand side $\cos\phi + i \sin\phi$, in which case De Moivre's formula $e^{i\phi} \cdot e^{i\psi} = e^{i(\phi+\psi)}$ appears merely as a *consequence of the elementary addition theorems of trigonometry*. From this formal point of view, in order to make the relation $e^x . e^y = e^{x+y}$ remain valid for any complex arguments, we set up the further definition

$$e^x = e^\xi (\cos\eta + i\sin\eta),$$

where $x = \xi + i\eta (\xi, \eta$ being real).

2. A Glance at the General Theory of Functions of a Complex Variable.

Although the purely formal point of view indicated above is in itself free from objection, it is still desirable to recognize in the above formulæ something more than a mere formal connexion. To follow out this aim leads us into the general theory of functions, as (for the sake of brevity) we call the general theory of the so-called analytic functions of a complex variable. In this we may use as our starting-point a general discussion of the theory of power series with complex variables and complex coefficients. The construction of such a theory of power series offers no difficulty once we define the concept of limit in the domain of complex numbers; in fact, it follows the theory of real power series almost exactly. But as we shall not make any use of these matters in what follows we shall content ourselves here by stating certain facts, omitting the proofs. It is found that the following generalization of the theorem of § 5, No. 1 (p. 400), holds for complex power series:

If a power series converges for any complex value $x = \xi$ *whatever, then it converges absolutely for every value* x *for which* $|x| < |\xi|$; *if it diverges for a value* $x = \xi$, *then it diverges for every value* x *for which* $|x| > |\xi|$. *A power series which does not converge everywhere, but does converge for some other point in addition to* $x = 0$, *possesses a circle of convergence, that is, there exists a number* $\rho > 0$ *such that the series converges absolutely for* $|x| < \rho$ *and diverges for* $|x| > \rho$.

Having once established the concept of functions of a complex variable represented by power series, and having developed the rules for operating with such functions, we can think of the functions e^x, $\sin x$, $\cos x$, arc $\tan x$, &c., of the *complex* variable x as simply *defined* by the power series which represent them for real values of x. Then all the above formal relationships reduce to trivialities.

We shall merely indicate by two examples how this introduction of complex variables helps us to understand the elementary functions.   The geometric series for $1/(1 + x^2)$ ceases to converge when x leaves the interval $-1 \leqq x \leqq 1$, and so does the series for arc $\tan x$, although there are no peculiarities in the behaviour of these functions at the ends of the interval of convergence;   in fact, they and all their· derivatives are continuous for all real values of x.   On the other hand, we can readily understand that the series for $1/(1 - x^2)$ and $\log(1 - x)$ cease to converge as x passes through the value 1, since they become infinite there.   But this divergence of the series for the inverse tangent   and   the   series $\sum\limits_{\nu=0}^{\infty} (-1)^\nu x^{2\nu}$ for $|x| > 1$ immediately becomes clear if we consider complex values of x also.   For we find that when $x = i$ the sum-functions become infinite and so cannot be represented by a convergent series.   Hence by our theorem about the circle of convergence the series must diverge for all values of x such that $|x| > |i|$;   in particular, for real values of x the series diverge outside the interval $-1 \leqq x \leqq 1$.

Another example is given us by the function $f(x) = e^{-1/x^2}$ for $x \neq 0, f(0) = 0$ (see pp. 196, 336), which, in spite of its apparently regular behaviour, cannot be expanded in a Taylor series.   As a matter of fact, this function ceases to be continuous if we take pure imaginary values of $x = i\xi$ into account.   The function then takes the form e^{1/ξ^2} and increases beyond all bounds as $\xi \to 0$. It is therefore clear that no power series in x can represent this function for all complex values of x in a neighbourhood of the origin, no matter how small a neighbourhood we choose.

These remarks on the theory of functions and power series of a complex variable must suffice for us here.

Appendix to Chapter VIII

1. MULTIPLICATION AND DIVISION OF SERIES

1. Multiplication of Absolutely Convergent Series.

Let
$$A = \sum_{\nu=0}^{\infty} a_\nu, \quad B = \sum_{\nu=0}^{\infty} b_\nu$$

be two absolutely convergent series. Together with these we consider the corresponding series of absolute values

$$\bar{A} = \sum_{\nu=0}^{\infty} |a_\nu| \quad \text{and} \quad \bar{B} = \sum_{\nu=0}^{\infty} |b_\nu|.$$

We further put

$$A_n = \sum_{\nu=0}^{n} a_\nu, \quad B_n = \sum_{\nu=0}^{n} b_\nu, \quad \bar{A}_n = \sum_{\nu=0}^{n} |a_\nu|, \quad \bar{B}_n = \sum_{\nu=0}^{n} |b_\nu|$$

and
$$c_n = a_0 b_n + a_1 b_{n-1} + \ldots + a_n b_0.$$

We assert that the series $\sum_{\nu=0}^{\infty} c_n$ is absolutely convergent, and that its sum is equal to AB.

To prove this, we write down the series

$$a_0 b_0 + a_1 b_0 + a_1 b_1 + a_0 b_1 + a_2 b_0 + a_2 b_1$$
$$+ a_2 b_2 + a_1 b_2 + a_0 b_2 + \ldots + a_n b_0 + a_n b_1$$
$$+ \ldots + a_n b_n + \ldots + a_1 b_n + a_0 b_n + \ldots,$$

the n^2-th partial sum of which is $A_n B_n$, and we assert that it converges absolutely. For the partial sums of the corresponding series with absolute values increase monotonically; the n^2-th partial sum is equal to $\bar{A}_n \bar{B}_n$, which is less than $\bar{A}\bar{B}$ (and which tends to $\bar{A}\bar{B}$). The series with absolute values therefore converges, and the series written down above converges absolutely. The sum of the series is obviously AB, since its n^2-th partial sum is $A_n B_n$, which tends to AB as $n \to \infty$. We now interchange the order of the terms, which is permissible for absolutely convergent series, and bracket successive terms together. In a convergent series we may bracket successive terms together in as many places as we desire without disturbing the convergence or altering the sum of the series, for if we bracket together, say, all the terms $(a_{n+1} + a_{n+2} + \ldots + a_m)$, then when we form the partial sums we shall omit those partial sums that originally fell between s_n and s_m, which does not affect the convergence or change the value of the limit. Also, if the series was absolutely convergent before the brackets were inserted, it remains absolutely convergent. Since the series

$$\sum_{\nu=0}^{\infty} c_\nu = (a_0 b_0) + (a_0 b_1 + a_1 b_0) + (a_0 b_2 + a_1 b_1 + a_2 b_0) + \ldots$$

is formed in this way from the series written down above, the required proof is complete.

2. Multiplication and Division of Power Series.

The principal use of our theorem is found in the theory of power series. The following assertion is an immediate consequence of it: the product of the two power series

$$\sum_{\nu=0}^{\infty} a_\nu x^\nu \quad \text{and} \quad \sum_{\nu=0}^{\infty} b_\nu x^\nu$$

is represented in the interval of convergence common to the two power series by a third power series $\sum_{\nu=0}^{\infty} c_\nu x^\nu$, whose coefficients are given by

$$c_\nu = a_0 b_\nu + a_1 b_{\nu-1} + \ldots + a_\nu b_0.$$

As for the division of power series, we can likewise represent the quotient of the two power series above by a power series $\sum_{\nu=0}^{\infty} q_\nu x^\nu$, provided b_0, the constant term in the denominator, does not vanish. (In the latter case such a representation is in general impossible; for it could not converge at $x = 0$ on account of the vanishing of the denominator, while on the other hand every power series must converge at $x = 0$.) The coefficients of the power series

$$\sum_{\nu=0}^{\infty} q_\nu x^\nu$$

can be calculated by remembering that $\overset{\infty}{\underset{\nu=0}{\Sigma}} q_\nu x^\nu \cdot \overset{\infty}{\underset{\nu=0}{\Sigma}} b_\nu x^\nu = \overset{\infty}{\underset{\nu=0}{\Sigma}} a_\nu x^\nu$, so that the following equations must be true:

$$a_0 = q_0 b_0,$$
$$a_1 = q_0 b_1 + q_1 b_0,$$
$$a_2 = q_0 b_2 + q_1 b_1 + q_2 b_0,$$
$$\cdots \cdots \cdots \cdots \cdots$$
$$a_\nu = q_0 b_\nu + q_1 b_{\nu-1} + \ldots + q_\nu b_0.$$

From the first of these equations q_0 is readily found, from the second we find the value q_1, from the third (by using the values of q_0 and q_1) we find the value q_2, &c. In order to give strict justification for the expression of the quotient of two power series by the third power series we have still to investigate the convergence of the formally-calculated power series $\overset{\infty}{\underset{\nu=0}{\Sigma}} q_\nu x^\nu$. We shall pass over this general investigation, of whose result we shall make no further use, and shall content ourselves with the statement that the series for the quotient does actually converge, provided x remains within a sufficiently small interval, in which the denominator does not vanish and both numerator and denominator are convergent series.

2. INFINITE SERIES AND IMPROPER INTEGRALS

The infinite series and the concepts developed in connexion with them have simple applications and analogies in the theory of improper integrals (cf. Chap. IV, § 8, p. 249). Here we confine ourselves to the case of a convergent integral with an infinite interval of integration, say an integral of the form $\int_0^\infty f(x)\,dx$. If we divide up the interval of integration by a sequence of numbers $x_0 = 0, x_1, \ldots$ tending monotonically to $+\infty$, we can write the improper integral in the form

$$\int_0^\infty f(x)\,dx = a_1 + a_2 + \ldots,$$

where each term of our infinite series is an integral;

$$a_1 = \int_0^{x_1} f(x)\,dx, \quad a_2 = \int_{x_1}^{x_2} f(x)\,dx, \ldots,$$

and so on. This is true no matter how we choose the points x_ν. We can therefore reduce the idea of a convergent improper integral to that of an infinite series in many ways.

It is especially convenient to choose the points x_ν in such a way that the integrand does not change sign within any individual sub-interval. The series $\sum\limits_{\nu=1}^{\infty} |a_\nu|$ will then correspond to the integral of the absolute value of our function,

$$\int_0^\infty |f(x)|\, dx.$$

We are thus naturally led to the following concept: *an improper integral $\int_0^\infty f(x)\,dx$ is said to be absolutely convergent if the integral $\int_0^\infty |f(x)|\,dx$ exists. Otherwise, if our integral exists at all, we say that it is conditionally convergent.*

Some of the integrals considered earlier (pp. 250–251), such as

$$\int_0^\infty \frac{1}{1+x^2}\,dx, \quad \int_0^\infty e^{-x^2}\,dx, \quad \Gamma(x) = \int_0^\infty e^{-t} t^{x-1}\,dt,$$

are absolutely convergent. On the other hand, the integral

$$\int_0^\infty \frac{\sin x}{x}\,dx = \lim_{A \to \infty} \int_0^A \frac{\sin x}{x}\,dx,$$

studied on p. 251, is a simple example of a conditionally convergent integral. In order to give a proof of the convergence of this integral which is independent of the former proof, we subdivide the interval from 0 to A at the points $x_\nu = \nu\pi (\nu = 0, 1, 2, \ldots, \mu_A)$ where μ_A is the largest possible integer for which $\mu_A \pi \leq A$. We therefore divide the integral into terms of the form $a_\nu = \int_{(\nu-1)\pi}^{\nu\pi} \frac{\sin x}{x}\,dx\,(\nu = 1, 2, \ldots)$, and a remainder R_A of the form

$$\int_{\mu_A \pi}^A \frac{\sin x}{x}\,dx \quad (0 \leq A - \mu_A \pi < \pi).$$

It is clear that the quantities a_ν have alternating signs, since $\sin x$ is alternately positive and negative in consecutive intervals. Moreover, $|a_{\nu+1}| < |a_\nu|$; for on applying the transformation $x = \xi - \pi$, we have

$$|a_\nu| = \int_{(\nu-1)\pi}^{\nu\pi} \frac{|\sin x|}{x}\,dx = \int_{\nu\pi}^{(\nu+1)\pi} \frac{|\sin(\xi - \pi)|}{\xi - \pi}\,d\xi = \int_{\nu\pi}^{(\nu+1)\pi} \frac{|\sin \xi|}{\xi - \pi}\,d\xi,$$

$$> \int_{\nu\pi}^{(\nu+1)\pi} \frac{|\sin \xi|}{\xi}\,d\xi = |a_{\nu+1}|.$$

Hence by Leibnitz's test we see that Σa_ν converges. Moreover, the remainder R_A has the absolute value

$$|R_A| = \left| \int_{\mu_A\pi}^{A} \frac{\sin x}{x} \, dx \right| \leq \int_{\mu_A\pi}^{(\mu_A+1)\pi} \frac{|\sin x|}{x} \, dx$$

$$\leq \frac{1}{\mu_A\pi} \int_{\mu_A\pi}^{(\mu_A+1)\pi} |\sin x| \, dx = \frac{2}{\mu_A\pi},$$

and this tends to 0 as A increases. Thus if we let A tend to ∞ in the equation

$$\int_0^A \frac{\sin x}{x} \, dx = a_1 + a_2 + a_3 + \ldots + a_{\mu_A} + R_A,$$

the right-hand side tends to Σa_ν as a limit, and our integral is convergent. But the convergence is not absolute; for

$$|a_\nu| > \int_{(\nu-1)\pi}^{\nu\pi} \frac{|\sin x|}{\nu\pi} \, dx = \frac{2}{\nu\pi}, \text{ so that } \Sigma |a_\nu| \text{ diverges.}$$

3. Infinite Products

In the introduction to this chapter (p. 366), we called attention to the fact that infinite series are only *one* way, although a particularly important way, of representing numbers or functions by infinite processes. As an example of another such process, we shall introduce the infinite product. No proofs will be given.

On p. 223 we met with Wallis's product,

$$\frac{\pi}{2} = \frac{2}{1} \cdot \frac{2}{3} \cdot \frac{4}{3} \cdot \frac{4}{5} \cdot \frac{6}{5} \cdot \frac{6}{7} \cdots$$

in which the number $\pi/2$ is expressed as an " infinite product ". By the value of the infinite product

$$\prod_{\nu=1}^{\infty} a_\nu = a_1 \cdot a_2 \cdot a_3 \cdot a_4 \ldots$$

we mean the limit of the sequence of partial products

$$a_1, \quad a_1 \cdot a_2, \quad a_1 \cdot a_2 \cdot a_3, \quad a_1 \cdot a_2 \cdot a_3 \cdot a_4, \quad \ldots,$$

provided it exists.

The factors a_1, a_2, a_3, ..., of course, may also be functions of a variable x. An especially interesting example is the "infinite product" for the function $\sin x$,

$$\sin \pi x = \pi x \left(1 - \frac{x^2}{1^2}\right)\left(1 - \frac{x^2}{2^2}\right)\left(1 - \frac{x^2}{3^2}\right) \cdots,$$

which we shall obtain in § 4 of the next chapter (p. 445).

The infinite product for the "zeta function" plays a very important part in the theory of numbers. In order to retain the notation usual in the theory of numbers we here denote the independent variable by s, and we define the zeta function for $s > 1$ by the expression

$$\zeta(s) = \sum_{n=1}^{\infty} \frac{1}{n^s}.$$

We know (cf. § 2, p. 380 *et seq.*) that the series on the right converges if $s > 1$. If p is any number greater than 1, we obtain the equation

$$\frac{1}{1 - \dfrac{1}{p^s}} = 1 + \frac{1}{p^s} + \frac{1}{p^{2s}} + \frac{1}{p^{3s}} + \cdots$$

by expanding the geometric series. If we imagine this series written down for all the prime numbers p_1, p_2, p_3, ... in increasing order of magnitude, and all the equations thus formed multiplied together, we obtain on the left a product of the form

$$\frac{1}{1 - p_1^{-s}} \cdot \frac{1}{1 - p_2^{-s}} \cdots$$

If, without stopping to justify the process in any way, we multiply together the series on the right-hand sides of our equations, and in addition remember that by an elementary theorem each integer $n > 1$ can be expressed in one and only one way as a product of powers of different prime numbers, we find that the product on the right is again the function $\zeta(s)$, and so we obtain the remarkable "product form"

$$\zeta(s) = \frac{1}{1 - p_1^{-s}} \cdot \frac{1}{1 - p_2^{-s}} \cdot \frac{1}{1 - p_3^{-s}} \cdots$$

This "product form", the derivation of which we have only briefly sketched here, is actually an expression of the zeta function as an infinite product, since the number of prime numbers is infinite.

In the general theory of infinite products we usually exclude the case where the product $a_1 a_2 \ldots a_n$ has the limit zero. Hence it is specially important that none of the factors a_n should vanish. In order that the product may converge, the factors a_n must accordingly tend to 1 as n increases. Since we can if necessary omit a finite number of factors (this has no bearing on the question of convergence), we may take it that $a_n > 0$. The following theorem applies to this case: a necessary and sufficient condition for the convergence of the product $\prod\limits_{\nu=1}^{\infty} a_\nu$, where $a_\nu > 0$, is that the series $\sum\limits_{\nu=1}^{\infty} \log a_\nu$ should converge. For it is clear that the partial sums $\sum\limits_{\nu=1}^{n} \log a_\nu = \log(a_1 a_2 \ldots a_n)$ of this series will tend to a definite limit if, and only if, the partial products $a_1 a_2 \ldots a_n$ possess a positive limit.

In studying convergence we usually apply the following criterion (a *sufficient* condition), where we put $a_\nu = 1 + \alpha_\nu$. The

product
$$\prod_{\nu=1}^{\infty} (1 + \alpha_\nu)$$

converges, if the series
$$\sum_{\nu=1}^{\infty} |\alpha_\nu|$$

converges and no factor $(1 + \alpha_\nu)$ is zero. In the proof we may assume, after omission of a finite number of factors if necessary, that each $|\alpha_\nu| < \dfrac{1}{2}$. Then we have $1 - |\alpha_\nu| > \dfrac{1}{2}$. By the mean value theorem $\log(1+h) = \log(1+h) - \log 1 = h\,\dfrac{1}{1+\theta h}$ for $0 < \theta < 1$. Therefore

$$|\log(1 + \alpha_\nu)| = \left| \frac{\alpha_\nu}{1 + \theta\alpha_\nu} \right| \leqq \frac{|\alpha_\nu|}{1 - |\alpha_\nu|} \leqq 2\,|\alpha_\nu|,$$

and so the convergence of the series $\sum\limits_{\nu=1}^{\infty} \log(1 + \alpha_\nu)$ follows from the convergence of $\sum\limits_{\nu=1}^{\infty} |\alpha_\nu|$.

From our criterion it follows that the infinite product given above for $\sin \pi x$ converges for all values of x except for $x = 0, \pm1, \pm2, \ldots$, where factors of the product are zero. Moreover, for $p \geqq 2$ and $s > 1$ we readily find that

$$\frac{1}{1 - p^{-s}} = 1 + \frac{1}{p^s - 1}, \quad 0 < \frac{1}{p^s - 1} < \frac{2}{p^s}.$$

Now if we let p assume all prime values, the series $\Sigma \dfrac{1}{p^s}$ must converge, since its terms form only a part of the convergent series $\overset{\infty}{\underset{\nu=1}{\Sigma}} \dfrac{1}{\nu^s}$. The convergence of the product $\Pi \dfrac{1}{1-p^{-s}}$ for $s > 1$ is thus proved.

4. Series involving Bernoulli's Numbers

So far we have given no expansions in power series for certain elementary functions, e.g. $\tan x$. The reason is that the numerical coefficients which occur are not of any very simple form. We can express these coefficients, and those in the series for a number of other functions, in terms of the so-called *Bernoulli's numbers*. These numbers are certain rational numbers, with a not very simple law of formation, which occur in many parts of analysis. We arrive at them most simply by expanding the function

$$\frac{x}{e^x - 1} = \frac{1}{1 + \dfrac{x}{2!} + \dfrac{x^2}{3!} + \ldots}$$

in a power series of the form

$$\frac{x}{e^x - 1} = \overset{\infty}{\underset{\nu=0}{\Sigma}} \frac{B_\nu}{\nu!} x^\nu.$$

If we write this equation in the form

$$x = (e^x - 1) \overset{\infty}{\underset{\nu=0}{\Sigma}} \frac{B_\nu}{\nu!} x^\nu$$

and substitute on the right the power series for $e^x - 1$, we obtain, as on p. 417, a recurrence relation which enables all the numbers B_ν to be calculated. These numbers are called Bernoulli's numbers.* They are rational, since in their formation only rational operations are concerned; as we easily recognize, they vanish for all odd indices other than $\nu = 1$. The first few are

$$B_0 = 1, \quad B_1 = -\frac{1}{2}, \quad B_2 = \frac{1}{6}, \quad B_4 = -\frac{1}{30}, \quad B_6 = \frac{1}{42},$$

$$B_8 = -\frac{1}{30}, \quad B_{10} = \frac{5}{66}, \ldots$$

---

* In some works a slightly different notation is used, the basic formula being written

$$\frac{x}{e^x - 1} = 1 - \frac{1}{2}x + \overset{\infty}{\underset{\nu=1}{\Sigma}} (-1)^{\nu+1} \frac{B_\nu}{(2\nu)!} x^{2\nu}.$$

We must content ourselves with a brief hint as to how these numbers are involved in the power series in question. First, by making use of the transformation

$$1 + \frac{B_2}{2!} x^2 + \ldots = \frac{x}{e^x - 1} + \frac{x}{2} = \frac{x}{2} \cdot \frac{e^x + 1}{e^x - 1} = \frac{x}{2} \cdot \frac{e^{\frac{1}{2}x} + e^{-\frac{1}{2}x}}{e^{\frac{1}{2}x} - e^{-\frac{1}{2}x}}$$

we obtain

$$\frac{x}{2} \coth \frac{x}{2} = \sum_{\nu=0}^{\infty} \frac{B_{2\nu}}{(2\nu)!} x^{2\nu}.$$

If we replace x by $2x$, we have the series

$$x \coth x = \sum_{\nu=0}^{\infty} \frac{2^{2\nu} B_{2\nu}}{(2\nu)!} x^{2\nu},$$

for $|x| < \pi$, from which, by replacing x by $-ix$, we obtain

$$x \cot x = \sum_{\nu=0}^{\infty} (-1)^\nu \frac{2^{2\nu} B_{2\nu}}{(2\nu)!} x^{2\nu}, \qquad |x| < \pi.$$

By means of the equation $2 \cot 2x = \cot x - \tan x$ we now obtain the series

$$\tan x = \sum_{\nu=1}^{\infty} (-1)^{\nu-1} \frac{2^{2\nu}(2^{2\nu} - 1)}{(2\nu)!} B_{2\nu} x^{2\nu-1},$$

which holds for $|x| < \frac{\pi}{2}$.

For further information we must refer the reader to more detailed treatises,* owing to the lengths of the proofs involved.

EXAMPLES

1. Prove that the power series for $\sqrt{(1 - x)}$ still converges when $x = 1$.

2. Prove that for every positive ε there is a polynomial in x which represents $\sqrt{(1 - x)}$ in the interval $0 \leqq x \leqq 1$ with an error less than ε.

3. Prove that for every positive ε there is a polynomial in t which represents $|t|$ in the interval $-1 \leqq t \leqq 1$ with an error less than ε.

4. *Weierstrass' Approximation Theorem.* Prove that if $f(x)$ is continuous in $a \leqq x \leqq b$, then for every positive ε there exists a polynomial $P(x)$ such that $|f(x) - P(x)| < \varepsilon$ for all values of x in the interval $a \leqq x \leqq b$.

5. Prove that the following infinite products converge:

$$\prod_{n=1}^{\infty} (1 + (\tfrac{1}{2})^{2n}); \qquad \prod_{n=2}^{\infty} \frac{n^3 - 1}{n^3 + 1}; \qquad \prod_{n=1}^{\infty} \left(1 - \frac{z^n}{n}\right), \text{ if } |z| < 1.$$

* See e.g. K. Knopp, *Theory and Application of Infinite Series*, p. 183 (Blackie & Son, Ltd., 1928).

6. Prove by the methods of the text that $\prod\limits_{n=1}^{\infty} \left(1 + \dfrac{1}{n}\right)$ diverges.

7. Using the identity

$$\sum_{n=1}^{\infty} \frac{1}{n^s} = \prod_{i=1}^{\infty} \left(\frac{1}{1 - p_i^{-s}}\right) \quad \text{(where } p_i \text{ is the } i\text{-th prime)}$$

prove that the number of primes is infinite.

8. Prove the identity

$$\prod_{\nu=1}^{\infty} (1 + x^{2^\nu}) = \frac{1}{1 - x}$$

for $|x| < 1$.

CHAPTER IX

Fourier Series

In addition to the power series there is another class of infinite series which plays a particularly important part both in pure mathematics and in applications. These are the Fourier series, in which the individual terms are trigonometric functions and the sum is a periodic function.

1. PERIODIC FUNCTIONS

1. General Remarks.

Periodic functions of the time, that is, functions which repeat their course after a definite interval of time, are met with in many applications. In most machines a periodic process takes place in rhythm with the rotation of a flywheel, e.g. the alternating current developed by a dynamo. Periodic functions are also associated with all vibration phenomena.

A periodic function with period 2l is represented by the equation

$$f(x + 2l) = f(x),$$

true for all values of x. We specially call attention to the fact that $2l$ is called the *period*.* It is worth notice that in addition to the period $2l$, the function $f(x)$ necessarily has the period $4l$

---

* In representing periodic functions it is often convenient to denote the independent variable x by a point on the circumference of a circle instead of the usual point on a straight line. If a function $f(x)$ has the period 2π, say, that is, if the equation

$$f(x + 2\pi) = f(x)$$

is true for all values of x, and if we denote by x the angle at the centre of a circle of unit radius which is included between an arbitrary initial radius and the radius to a variable point on the circumference, then the periodicity of the function $f(x)$ is expressed simply by the fact that to each point on the circumference there corresponds just one value of the function. In the case of a machine, for example, the periodicity may be expressed in terms of the position of a point on the flywheel.

also, since $f(x+4l) = f(x+2l) = f(x)$; $f(x)$ likewise has periods $6l$, $8l$, ...; and it is also possible (though not necessarily true) that $f(x)$ may have shorter periods, such as l or $l/5$. Graphically, in any two consecutive intervals of length $2l$ the graph of the function has exactly the same form. In order to have available a second interpretation which some readers may prefer, we may think of the variable x as the time (and accordingly we sometimes write t instead of x), the function $f(x)$ then representing a periodic process or, as we shall also say, a *vibration* (or *oscillation*). The period $2l = T$ is then called the *period of vibration* (or *oscillation*).

If any arbitrary function f(x) *is given in a definite interval, say,* $-l \leqq x \leqq l$, *it can always be extended as a periodic function*; we have only to define $f(x)$ outside the interval by the equation $f(x + 2nl) = f(x)$, where n is an arbitrary positive or negative integer. Here we must point out that if $f(x)$ is continuous in the interval $-l \leqq x \leqq l$, but $f(-l) \neq f(+l)$, our extended periodic function will be discontinuous at the points $\pm l$, $\pm 3l$, ... , (cf. figs. 7 and 8 (pp. 441, 442), in which $l = \pi$). Further, in this case the extension fails to give us a single-valued function $f(x)$ at the points $x = \pm l$, $\pm 3l$, ... since e.g. we have defined $f(3l)$ as $f(l + 2l)$, which gives $f(3l) = f(l)$, and we have also defined it as $f(-l + 4l)$, which gives $f(3l) = f(-l)$. We avoid this difficulty by extending, not the function as defined for $-l \leqq x \leqq l$, but the function as defined either for $-l < x \leqq l$ or $-l \leqq x < l$; that is, we discard either the original value $f(-l)$ or the original value $f(+l)$.

Here we would point out a general fact relating to periodic functions, which is expressed by the equation

$$\int_{-l-a}^{l-a} f(x)\, dx = \int_{-l}^{l} f(x)\, dx,$$

or, in words: the integral of a periodic function over an interval whose length is one period $T = 2l$ always has the same value, no matter where the interval lies. To prove this we need only notice that in virtue of the equation $f(\xi - 2l) = f(\xi)$ the substitution $x = \xi - 2l$ gives us

$$\int_{a}^{\beta} f(x)\, dx = \int_{a+2l}^{\beta+2l} f(\xi)\, d\xi = \int_{a+2l}^{\beta+2l} f(x)\, dx.$$

In particular, for $\alpha = -l - a$ and $\beta = -l$ it follows that

$$\int_{-l-a}^{-l} f(x)\, dx = \int_{l-a}^{l} f(x)\, dx,$$

and hence

$$\int_{-l-a}^{l-a} f(x)\, dx = \int_{-l-a}^{-l} f(x)\, dx + \int_{-l}^{l-a} f(x)\, dx$$

$$= \int_{l-a}^{l} f(x)\, dx + \int_{-l}^{l-a} f(x)\, dx = \int_{-l}^{l} f(x)\, dx,$$

which proves our statement. If we recall the geometrical meaning of the integral, the statement is made obvious by fig. 1.

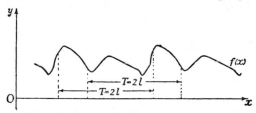

Fig. 1.—To illustrate the integral over a whole period

The simplest periodic functions, from which we shall later build up the most general periodic functions, are the functions $a \sin \omega x$ and $a \cos \omega x$, or more generally $a \sin \omega(x - \xi)$ and $a \cos \omega(x - \xi)$, where $a(\geqq 0)$, $\omega(>0)$, and ξ are constants. The processes represented by these functions * we call *sinusoidal vibrations* or *simple harmonic vibrations* (or *oscillations*). The period of vibration is $T = 2\pi/\omega$. The number ω is called the *circular frequency of the vibration*;† since $1/T$ is the number of vibrations in unit time, or *frequency*, ω is the *number of vibrations in time* 2π. The number a is called the *amplitude* of the vibration; it represents the maximum value of the function $a \sin \omega(x - \xi)$ or $a \cos \omega(x - \xi)$, since both sine and cosine have the maximum value 1. The number $\omega(x - \xi)$ is called the *phase* and the number $\omega \xi$ the *epoch* or *phase displacement*.

---

* Either of these formulæ taken by itself (for all values of a and ξ) represents the class of all sinusoidal vibrations; and the two formulæ are equivalent to one another, since $a \sin \omega(x - \xi) = a \cos \omega\{x - (\xi + \pi/2\omega)\}$.

† The reader should take care to distinguish between the *frequency* and the *circular frequency* (Ger., *Kreisfrequenz*).

We obtain these functions graphically by stretching the sine curve in the ratios $1 : \omega$ along the x-axis and $a : 1$ along the y-axis, and then translating the curve a distance ξ in the positive direction along the x-axis (cf. fig. 2).

By the addition formulæ for the trigonometric functions we can also express sinusoidal vibrations in the form

$$a \cos\omega x + \beta \sin\omega x \quad \text{and} \quad \beta \cos\omega x - a \sin\omega x$$

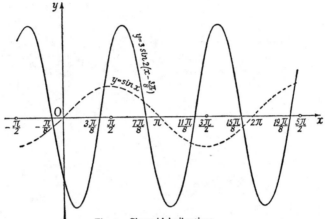

Fig. 2.—Sinusoidal vibrations

respectively, where $a = -a \sin\omega\xi$ and $\beta = a \cos\omega\xi$. Conversely, every function of the form

$$a \cos\omega x + \beta \sin\omega x$$

represents a sinusoidal vibration $a \sin\omega(x - \xi)$ with the amplitude $a = \sqrt{(a^2 + \beta^2)}$ and the phase displacement $\omega\xi$ given by the equations $a = -a \sin\omega\xi$, $\beta = a \cos\omega\xi$. By using the expression $a \cos\omega x + \beta \sin\omega x$ we see that the sum of two or more such functions with the same circular frequency ω always represents another sinusoidal vibration with the circular frequency ω.

2. Superposition of Sinusoidal Vibrations. Harmonics. Beats.

Although many vibrations are found to be sinusoidal (cf. Chap. V, § 4, p. 296), it is nevertheless true that most periodic motions have a more complicated character, being obtained by the superposition of sinusoidal vibrations. Mathematically this

simply means that the motion, e.g. the distance of a point from
its initial position as a function of the time, is given by a function
which is the sum of a number of pure periodic functions of the
above type. The sine waves of the function are then piled
up on top of one another (that is, their ordinates are added),
or, as we say, they are *superposed*. In this superposition
we assume that the circular frequencies (and, of course, the
periods) of the superposed vibrations are all different; for the
superposition of two sinusoidal vibrations with the same circular
frequency gives us another sinusoidal vibration with the same
circular frequency (but with a different amplitude and phase
displacement), as shown above.

If we consider the simplest instance, the superposition of
two sinusoidal vibrations with the circular frequencies ω_1 and ω_2,
we find that there are two fundamentally different cases, depending on whether the two circular frequencies have a rational ratio
or not, or, as we say, whether they are commensurable or incommensurable. We begin with the first case, and by way of an
example take the second circular frequency to be twice the first;
$\omega_2 = 2\omega_1$. The period of the second vibration will then be half
the period of the first, $2\pi/2\omega_1 = T_2 = T_1/2$, and so it will necessarily have not only the period T_2 but also the doubled period T_1,
since the function repeats itself after this double period; and the
function formed by superposing them will also have the period
T_1. The second vibration, with twice the circular frequency and
half the period of the first, is called a *first harmonic* of the first
vibration (the *fundamental*).

Corresponding statements hold if we introduce a further
vibration with the circular frequency $\omega_3 = 3\omega_1$. Here again the
vibration function $\sin 3\omega_1 x$ will necessarily repeat itself with the
period $2\pi/\omega_1 = T_1$. Such a vibration is called a *second harmonic*
of the given vibration. Likewise we can consider third, fourth, ... ,
$(n - 1)$-th harmonics with the circular frequencies $\omega_4 = 4\omega_1$,
$\omega_5 = 5\omega_1, \ldots, \omega_n = n\omega_1$, and, moreover, with any phase dis-
placements we please. Every such harmonic will necessarily
repeat itself after the period $T_1 = 2\pi/\omega_1$, and consequently every
function obtained by superposing a number of vibrations, each
of which is a harmonic of a given fundamental circular frequency
ω_1, will itself be a periodic function with the period $2\pi/\omega_1 = T_1$.
By superposing vibrations with circular frequencies ranging from

that of the fundamental to that of the $(n-1)$-th harmonic we obtain a periodic function of the form

$$S(x) = a + \sum_{\nu=1}^{n} (a_\nu \cos \nu\omega x + b_\nu \sin \nu\omega x).$$

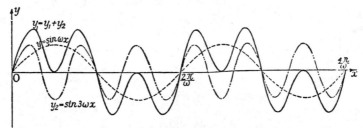

Fig. 3*.—Combination of vibrations

(The constant a, which we have introduced here in order to make the formula slightly more general, does not affect the periodicity, since it is periodic for any period.) Since this function contains

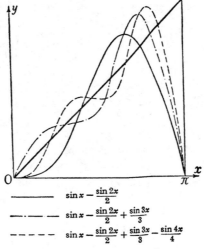

Fig. 4.—Combination of vibrations

$2n+1$ constants which we can choose arbitrarily, we are thus able to generate very complicated curves which are not at all like the original sine curves. Figs. 3–5 illustrate this graphically.

* The proportions of the figure correspond to the assumption $\omega = 1$.

The term "harmonic" originates in acoustics,* where we find that if a fundamental vibration with circular frequency ω corresponds to a note of a certain pitch, then the first, second, third, &c., harmonics correspond to the sequence of harmonics of the fundamental, that is, to the octave, octave + fifth, double octave, &c.

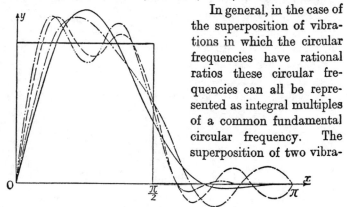

In general, in the case of the superposition of vibrations in which the circular frequencies have rational ratios these circular frequencies can all be represented as integral multiples of a common fundamental circular frequency. The superposition of two vibra-

Fig. 5†.—Combination of vibrations

tions with incommensurable circular frequencies ω_1 and ω_2, however, represents an intrinsically different type of phenomenon. Here the process resulting from the superposition of sinusoidal vibrations will no longer be periodic. We cannot go into the mathematical discussions that arise from this, but merely remark that such functions always have an approximately periodic character, or, as we say, are *almost periodic*. Such functions have just recently been studied in great detail.

A final remark on the superposition of sinusoidal vibrations is concerned with the phenomenon of so-called *beats*. If we superpose two vibrations both with unit amplitude but with different circular frequencies ω_1 and ω_2, and if for the sake of simplicity we take the same value of ξ (see p. 427) for both (the generalization to arbitrary phase can be left to the reader), then we are merely concerned with the behaviour of the function

$$y = \sin\omega_1 x + \sin\omega_2 x \qquad (\omega_1 > \omega_2 > 0).$$

* In acoustics the terms *overtone*, *(upper) partial* are also used.

† The curves drawn in the figure correspond to the trigonometrical polynomials obtained by taking 3, 5, 6 and 7 terms respectively of the series

$$\frac{\sin x}{1} + 2\frac{\sin 2x}{2} + \frac{\sin 3x}{3} + \frac{\sin 5x}{5} + 2\frac{\sin 6x}{6} + \frac{\sin 7x}{7} + \frac{\sin 9x}{9} + \ldots.$$

By a well-known trigonometrical formula we have

$$y = 2 \cos\tfrac{1}{2}(\omega_1 - \omega_2)x \sin\tfrac{1}{2}(\omega_1 + \omega_2)x.$$

This equation represents a phenomenon which we may think of as follows: we have a vibration with the circular frequency $\tfrac{1}{2}(\omega_1 + \omega_2)$ and the period $4\pi/(\omega_1 + \omega_2)$. This vibration, however, has not a constant amplitude; on the contrary, the " amplitude " is given by the expression $2 \cos\tfrac{1}{2}(\omega_1 - \omega_2)x$, which varies with a longer period $4\pi/(\omega_1 - \omega_2)$. This point of view is particularly useful and easy to interpret when the two circular frequencies ω_1 and ω_2 are relatively large while their difference

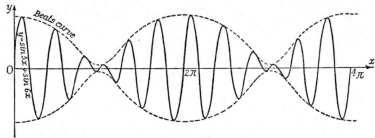

Fig. 6.—Beats

$(\omega_1 - \omega_2)$ is small compared to them. Then the amplitude $2 \cos\tfrac{1}{2}(\omega_1 - \omega_2)x$ of the vibration with period $4\pi/(\omega_1 + \omega_2)$ will vary only slowly compared with the period of vibration, and this change of amplitude will repeat itself periodically with the long period $4\pi/(\omega_1 - \omega_2)$. These rhythmic changes of amplitude are called *beats*. Everyone is acquainted with this phenomenon in acoustics and perhaps also in wireless telegraphy. In wireless telegraphy the circular frequencies ω_1 and ω_2 are as a rule far above those which the ear can detect, while the difference $\omega_1 - \omega_2$ falls in the range of audible notes. The beats then cause an audible note, while the original vibrations remain imperceptible to the ear.

An example of beats is illustrated graphically in fig. 6.

2. Use of Complex Notation

1. General Remarks.

The investigation of vibration phenomena and periodic functions gains in formal simplicity if we make use of complex numbers, combining each pair of trigonometric functions $\cos \omega x$ and $\sin \omega x$ to form an expression of the type $\cos \omega x + i \sin \omega x = e^{i\omega x}$ (cf. Chap. VIII, § 7, p. 411). Here we must bear in mind that one equation between complex quantities is equivalent to *two* equations between real quantities, and that our results must always be interpreted and made intelligible in the real domain.

If we everywhere replace the trigonometric functions by exponential functions in accordance with the formulæ

$$2 \cos \theta = e^{i\theta} + e^{-i\theta}, \quad 2i \sin \theta = e^{i\theta} - e^{-i\theta},$$

we express sinusoidal vibrations in terms of the complex quantities $e^{i\omega x}$, $e^{-i\omega x}$, or

$$ae^{i\omega(x-\xi)}, \quad ae^{-i\omega(x-\xi)}$$

respectively, where a, ω, and $\omega \xi$ are the real quantities amplitude, circular frequency, and phase displacement. The real vibrations are obtained from this complex expression simply by taking real and imaginary parts.

The convenience of this mode of representation for many purposes is due to the fact that the derivatives of the real vibrations with respect to the time x are obtained by differentiating the complex exponential function just as if i were a real constant, as is expressed by the formula

$$\frac{d}{dx} a\{ \cos\omega(x - \xi) + i \sin\omega(x - \xi) \}$$
$$= a\omega\{ -\sin\omega(x - \xi) + i \cos\omega(x - \xi) \}$$
$$= ia\omega\{ \cos\omega(x - \xi) + i \sin\omega(x - \xi) \},$$

or
$$\frac{d}{dx} ae^{i\omega(x-\xi)} = ia\omega e^{i\omega(x-\xi)}.$$

2. Application to the Study of Alternating Currents.

We shall now illustrate these matters by means of an important example. Here we shall denote the independent variable, the time, by t instead of x.

15

We consider an electric circuit with resistance R and inductance L, on which an external electromotive force (voltage) E is impressed. In the case of direct current E is constant, and the current I is given by Ohm's law,

$$E = RI.$$

If, however, we are dealing with alternating current, E is a function of the time t, and consequently so is I, and Ohm's law takes the form (cf. p. 182)

$$E - L\frac{dI}{dt} = RI.$$

In the simplest case, to which we restrict ourselves here, the external electromotive force E is sinusoidal with circular frequency ω. Now instead of taking this oscillation in the form $a\cos\omega t$ or $a\sin\omega t$, we combine both possibilities formally in the complex form

$$E = \varepsilon e^{i\omega t} = \varepsilon\cos\omega t + i\varepsilon\sin\omega t,$$

where $\varepsilon(>0)$ represents the amplitude. We shall operate with this " complex voltage " as if i were a real parameter, and we thus obtain a complex current I. Then the significance of the relation thus found between the complex quantities E and I is that the current corresponding to an electromotive force $\varepsilon\cos\omega t$ is the real part of I, while the current corresponding to an electromotive force $\varepsilon\sin\omega t$ is the imaginary part of I. The complex current can be calculated immediately if for I we write down an expression of the form

$$I = \alpha e^{i\omega t} = \alpha(\cos\omega t + i\sin\omega t);$$

that is, if we make the assumption that I is also sinusoidal with circular frequency ω. The derivative of I is then given formally by the expression

$$\frac{dI}{dt} = i\alpha\omega e^{i\omega t}$$

$$= \alpha\omega(-\sin\omega t + i\cos\omega t).$$

By substituting these quantities in the generalized form of Ohm's law and dividing out the factor $e^{i\omega t}$ we obtain the equation $\varepsilon - \alpha Li\omega = R\alpha$, or

$$\alpha = \frac{\varepsilon}{R + i\omega L},$$

so that $\qquad\qquad E = (R + i\omega L)I = WI.$

We may regard this last equation as Ohm's law for alternating currents in complex form, if we call the quantity

$$W = R + i\omega L$$

the *complex resistance* of the circuit. Ohm's law is then the same as for direct current: the current is equal to the voltage divided by the resistance.

If we write the complex resistance in the form

$$W = we^{i\delta} = w \cos \delta + iw \sin \delta,$$

where

$$w = \sqrt{(R^2 + L^2\omega^2)}, \quad \tan \delta = \frac{\omega L}{R},$$

we obtain

$$I = \frac{\varepsilon}{w} e^{i(\omega t - \delta)}.$$

According to this formula the current has the same period (and circular frequency) as the voltage; the amplitude a of the current is connected with the amplitude ε of the electromotive force by the equation

$$a = \frac{\varepsilon}{w},$$

and, in addition, there is a difference of phase between the current and the voltage. The current reaches its maximum, not at the same time as the voltage, but at a time δ/ω later, and the same is of course true for the minimum. In electrical engineering the quantity $w = \sqrt{(R^2 + L^2\omega^2)}$ is frequently called the *impedance* or *alternating current resistance* of the circuit for the circular frequency ω; the phase displacement, usually stated in degrees, is called the *lag*.

3. Complex Representation of the Superposition of Sinusoidal Vibrations.

So far the complex notation has been used to denote the combination of *two* sinusoidal vibrations. But a *single* vibration or a compound vibration of the type

$$S(x) = a + \sum_{\nu=1}^{n} (a_\nu \cos \nu x + b_\nu \sin \nu x)$$

(for simplicity we have taken $\omega = 1$) can also be reduced to complex form by substituting

$$\cos \nu x = \frac{1}{2} (e^{i\nu x} + e^{-i\nu x})$$

and

$$\sin \nu x = \frac{1}{2i} (e^{i\nu x} - e^{-i\nu x}).$$

The above expression then becomes an expression of the form

$$S(x) = \sum_{\nu=-n}^{n} a_\nu e^{i\nu x},$$

where the complex numbers a_ν are connected with the real numbers a, a_ν, and b_ν by the equations

$$a_\nu = a_\nu + a_{-\nu}, \quad a = a_0, \quad b_\nu = i(a_\nu - a_{-\nu}).$$

In order that the equation $a_\nu = a_\nu + a_{-\nu}$ shall formally include the case $\nu = 0$, we often put $a = a_0 = a_0/2$.

Conversely, we may regard any arbitrary expression of the form

$$\sum_{\nu=-n}^{n} a_\nu e^{i\nu x}$$

as a function representing the superposition of vibrations, written in complex form. In order that the result of this superposition may be real it is only necessary that $a_\nu + a_{-\nu}$ should be real and $a_\nu - a_{-\nu}$ a pure imaginary; that is, a_ν and $a_{-\nu}$ must be conjugate complex numbers.

4. Deduction of a Trigonometric Formula.

By using complex notation we obtain a very simple proof of a formula which we shall need later. This is the *trigonometric summation formula*

$$\sigma_n(\alpha) = \tfrac{1}{2} + \cos\alpha + \cos 2\alpha + \ldots + \cos n\alpha = \frac{\sin(n + \tfrac{1}{2})\alpha}{2\sin\tfrac{1}{2}\alpha},$$

which is true for all values of α except the values 0, $\pm 2\pi$, $\pm 4\pi$, ...

To prove this we replace the cosine function by its exponential expression and thus bring the sum $\sigma_n(\alpha)$ into the form

$$\sigma_n(\alpha) = \tfrac{1}{2} \sum_{\nu=-n}^{n} e^{i\nu\alpha}.$$

On the right we have a geometric progression with the common ratio $q = e^{i\alpha} \neq 1$. Using the ordinary formula for the sum, we have

$$\sigma_n(\alpha) = \frac{1}{2} e^{-in\alpha} \cdot \frac{1 - q^{2n+1}}{1 - q} = \frac{1}{2} \frac{e^{-in\alpha} - e^{(n+1)i\alpha}}{1 - e^{i\alpha}}.$$

On multiplying numerator and denominator by $e^{-i\alpha/2}$ we obtain

$$\sigma_n(\alpha) = \frac{\sin(n + \tfrac{1}{2})\alpha}{2\sin\tfrac{1}{2}\alpha},$$

as was stated.

1. Sketch the curves $y = \sum\limits_{n=1}^{N} \dfrac{\sin nx}{n}$ for $N = 3, 5, 6$.

2. Sketch the curves $y = \sum\limits_{n=1}^{N} \dfrac{\cos t}{n^4}$ for $N = 3, 6, 8$.

3. Evaluate the sum $\sin\alpha + \sin 2\alpha + \ldots + \sin n\alpha$.

4. If $s_m(\alpha) = \dfrac{\sigma_0(\alpha) + \sigma_1(\alpha) + \ldots + \sigma_m(\alpha)}{m+1}$, where $\sigma_n(\alpha)$ has the value
$\sigma_n(\alpha) = \tfrac{1}{2} + \cos\alpha + \cos 2\alpha + \ldots + \cos n\alpha$, prove that

$$s_m(\alpha) = \frac{1}{m+1}\left[\frac{\sin\dfrac{(m+1)\alpha}{2}}{\sin\dfrac{\alpha}{2}} \right]^2.$$

(The expression s_m is called the " Fejér kernel ", and is of great importance in the more advanced study of Fourier series.)

5. Show that
$$\frac{1}{\pi}\int_{-\pi}^{\pi} s_m(\alpha)\,d\alpha = 1,$$

where $s_m(\alpha)$ is the Fejér kernel of Ex. 4.

3. FOURIER SERIES

The function
$$S(x) = a + \sum_{\nu=1}^{n}(a_\nu \cos\nu x + b_\nu \sin\nu x)$$

resulting from the superposition of sinusoidal vibrations contains $2n + 1$ arbitrary constants a, a_ν, b_ν. The question now arises whether these constants can be so chosen that in the interval $-\pi \leqq x \leqq \pi$ the sum $S(x)$ shall approximate to a given function $f(x)$, and if so, how they are to be found. More precisely, we inquire whether the given function $f(x)$ can be expanded in an infinite series

$$f(x) = a + \sum_{\nu=1}^{\infty}(a_\nu \cos\nu x + b_\nu \sin\nu x).$$

If we assume for the moment that this expansion of the function $f(x)$ is actually possible and that the series converges uniformly in the interval $-\pi \leqq x \leqq \pi$, we readily obtain a simple relation between the function $f(x)$ and the coefficients $a = \tfrac{1}{2}a_0$, a_ν, and b_ν. (We shall soon see that the notation $a = \tfrac{1}{2}a_0$

is justified by its convenience.) We multiply the hypothetical expansion above by $\cos \nu x$ and integrate term by term, as is permissible on account of the uniform convergence. In virtue of the orthogonality relations

$$\int_{-\pi}^{+\pi} \sin mx \, \sin nx \, dx = \begin{cases} 0, & \text{if } m \neq n, \\ \pi, & \text{if } m = n \neq 0. \end{cases}$$

$$\int_{-\pi}^{+\pi} \sin mx \, \cos nx \, dx = 0,$$

$$\int_{-\pi}^{+\pi} \cos mx \, \cos nx \, dx = \begin{cases} 0, & \text{if } m \neq n, \\ \pi, & \text{if } m = n, \end{cases}$$

proved in Chap. IV, § 3 (p. 217), we at once obtain the formula

$$a_\nu = \frac{1}{\pi} \int_{-\pi}^{+\pi} f(x) \cos \nu x \, dx$$

for the coefficients. Similarly, by multiplying the series by $\sin \nu x$ and integrating, we have

$$b_\nu = \frac{1}{\pi} \int_{-\pi}^{+\pi} f(x) \sin \nu x \, dx.$$

These formulæ assign a definite sequence of coefficients a_ν and b_ν, usually called the Fourier coefficients, to every function $f(x)$ which is defined and continuous in the interval $-\pi \leq x \leq \pi$, or has only a finite number of jump discontinuities there. If the function $f(x)$ is given, we can use these quantities a_ν, b_ν to form the Fourier partial sums

$$S_n(x) = \tfrac{1}{2}a_0 + \sum_{\nu=1}^{n} (a_\nu \cos \nu x + b_\nu \sin \nu x),$$

and we may also formally write down the corresponding infinite " Fourier series ". Our problem is to distinguish simple classes of functions $f(x)$ for which this Fourier series does actually converge and does represent the function.

In order to formulate the result which we wish to prove, we introduce the following definition. A function $f(x)$ is said to be *sectionally smooth* * in an interval if it is itself sectionally continuous † (that is, continuous in the interval except for a

---

* Ger. *stückweise glatt*.　　† Ger. *stückweise stetig*.

finite number of jump discontinuities) and if in addition its first derivative $f'(x)$ is sectionally continuous.

We shall imagine the function $f(x)$ originally defined in the interval $-\pi \leqq x \leqq \pi$ to be periodically extended.

At each point at which the function $f(x)$ has a jump discontinuity we shall alter the function, if necessary, and shall assign to it the value which is the arithmetic mean of the left-hand limit and the right-hand limit of $f(x)$; that is, we write

$$f(x) = \tfrac{1}{2}(f(x - 0) + f(x + 0)),$$

where $f(x - 0)$ and $f(x + 0)$ are simply the limits of $f(\bar{x})$ as $\bar{x}$ approaches x from the left and from the right respectively. This equation is obviously true for every point x at which $f(x)$ is continuous.

Our goal is the following theorem:

If the function f(x) *is sectionally smooth and satisfies the above equation, then its Fourier series converges at every point* x *and represents the function.**

Further, we shall prove the following theorem:

In every closed interval in which the function f(x) *(imagined periodically extended) is continuous as well as sectionally smooth, the Fourier series converges uniformly.*

Finally:

If the function f(x) *is sectionally smooth and has* **no** *discontinuities, the Fourier series converges absolutely.*

The proofs of these theorems will be postponed until § 5 (p. 447). Here we merely wish to emphasize that the functions which can be expanded according to these theorems have a very high degree of arbitrariness; it is by no means necessary that the functions should be given by a single analytical expression.

In the next section we shall display the extraordinary fertility of the Fourier expansion by discussing a number of examples.

---

* It may be remarked incidentally that this theorem can be proved for more general classes of functions. The result formulated here, however, suffices for all applications.

4. Examples of Fourier Series

1. Preliminary Remarks.

We shall assume that our functions $f(x)$ have the period 2π and are defined in the interval $-\pi < x < \pi$. Beyond this interval to the left and the right, they are to be extended periodically, as on p. 426.

Before going into details we remark that if $f(x)$ is an *even* function (cf. p. 20), then clearly $f(x) \sin \nu x$ is odd and $f(x) \cos \nu x$ is even, so that

$$b_\nu = \frac{1}{\pi} \int_{-\pi}^{+\pi} f(x) \sin \nu x \, dx = 0; \quad a_\nu = \frac{2}{\pi} \int_0^\pi f(x) \cos \nu x \, dx.$$

We thus obtain a " cosine series ". If, on the other hand, the function $f(x)$ is an odd function, then

$$a_\nu = \frac{1}{\pi} \int_{-\pi}^{+\pi} f(x) \cos \nu x \, dx = 0; \quad b_\nu = \frac{2}{\pi} \int_0^\pi f(x) \sin \nu x \, dx.$$

We therefore obtain a " sine series ".*

2. Expansion of the Functions $\psi(x) = x$ and $\varphi(x) = x^2$.

For the odd function x we have $b_\nu = \frac{2}{\pi} \int_0^\pi x \sin \nu x \, dx$, and on integration by parts

$$\frac{\pi}{2} b_\nu = \frac{-x \cos \nu x}{\nu} \Big|_0^\pi + \frac{1}{\nu} \int_0^\pi \cos \nu x \, dx = (-1)^{\nu+1} \frac{\pi}{\nu}.$$

Hence for the periodic function $\psi(x)$ which in the interval $-\pi < x < \pi$ is equal to x (cf. fig. 7) we obtain the expansion

$$\psi(x) = 2 \left(\frac{\sin x}{1} - \frac{\sin 2x}{2} + \frac{\sin 3x}{3} - + \dots \right).$$

If we put $x = \pi/2$ we obtain Gregory's series,

$$\frac{\pi}{4} = 1 - \frac{1}{3} + \frac{1}{5} - + \dots,$$

with which we are already familiar (p. 319). The function $\psi(x)$ represented by this series is not a continuous function; on the contrary, it jumps by

---

* Consequently, if the function $f(x)$ is initially given only in the interval $0 < x < \pi$, then we can extend it in the interval $-\pi < x < 0$ either as an odd function or as an even function, and correspondingly expand the function in the interval $0 < x < \pi$ either in a sine series or in a cosine series.

an amount 2π at the points $x = k\pi$, $k = \pm 1$, ± 3. $\pm 5, \ldots$. At these points of discontinuity, that is, at the points $x = k\pi$, $k = \pm 1$, ± 3, $\pm 5, \ldots$, each term of the series is zero, and hence the function itself is zero. Hence at the points of discontinuity the series represents the arithmetic mean of the left-hand and right-hand limits.

If ξ is any fixed number between $-\pi$ and π, and if we replace x in the above series by $(x - \xi)$, we obtain the series

$$\psi(x - \xi) = 2 \left(\frac{\sin(x - \xi)}{1} - \frac{\sin 2(x - \xi)}{2} + \frac{\sin 3(x - \xi)}{3} - + \ldots \right)$$

$$= -\frac{2}{1} \sin\xi \cos x + \frac{2}{1} \cos\xi \sin x + \frac{2}{2} \sin 2\xi \cos 2x$$

$$-\frac{2}{2} \cos 2\xi \sin 2x - \frac{2}{3} \sin 3\xi \cos 3x + \frac{2}{3} \cos 3\xi \sin 3x + \ldots .$$

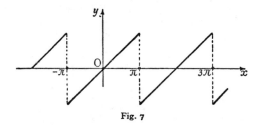

Fig. 7

This may also be written in the form of a Fourier series with coefficients

$$a_0 = 0, \quad a_n = 2\frac{(-1)^n}{n} \sin n\xi, \quad b_n = 2\frac{(-1)^{n-1}}{n} \cos n\xi,$$

which tend to zero as n increases; this series represents a function having the discontinuities described above at the points $x = \xi \pm \pi$, $x = \xi \pm 3\pi, \ldots$.

For the even function $\varphi(x) = x^2$ we find, on integrating by parts twice, that

$$a_\nu = \frac{2}{\pi} \int_0^\pi x^2 \cos \nu x \, dx = (-1)^\nu \frac{4}{\nu^2} \quad (\nu > 0),$$

$$a_0 = \frac{2\pi^2}{3},$$

so that we obtain the expansion

$$\varphi(x) = \frac{\pi^2}{3} - 4 \left(\frac{\cos x}{1^2} - \frac{\cos 2x}{2^2} + \frac{\cos 3x}{3^2} - + \ldots \right).$$

By differentiating this series term by term and dividing by 2 we *formally* recover the series for $\psi(x) = x$.

15$\bullet$ (E 798)

3. Expansion of the Function $x \cos x$.

For this odd function we have

$$a_\nu = 0, \quad b_\nu = \frac{2}{\pi} \int_0^\pi x \cos x \sin \nu x \, dx.$$

By using the formula

$$\int_0^\pi x \sin \mu x \, dx = (-1)^{\mu+1} \frac{\pi}{\mu} \quad (\mu = 1, 2, \ldots)$$

found in the previous sub-section, we evaluate b_ν:

$$b_\nu = \frac{2}{\pi} \int_0^\pi x \cos x \sin \nu x \, dx = \frac{1}{\pi} \int_0^\pi x(\sin(\nu+1)x + \sin(\nu-1)x) \, dx$$

$$= \frac{(-1)^{\nu+2}}{\nu+1} + \frac{(-1)^\nu}{\nu-1} = (-1)^\nu \frac{2\nu}{\nu^2-1} \quad (\nu = 2, 3, \ldots),$$

$$b_1 = -\frac{1}{2}.$$

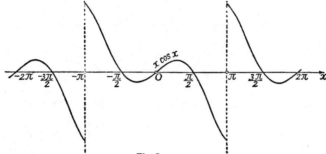

Fig. 8

We therefore obtain the series

$$x \cos x = -\frac{1}{2} \sin x + 2 \sum_{\nu=2}^\infty \frac{(-1)^\nu \nu}{\nu^2-1} \sin \nu x,$$

and if we add the series found on p. 440 this becomes

$$x(1 + \cos x) = \frac{3}{2} \sin x + 2 \left(\frac{\sin 2x}{1 \cdot 2 \cdot 3} - \frac{\sin 3x}{2 \cdot 3 \cdot 4} + \frac{\sin 4x}{3 \cdot 4 \cdot 5} - + \ldots \right).$$

When the function which is equal to $x \cos x$ in the interval $-\pi < x < \pi$ is extended periodically beyond this interval, the same discontinuities (cf. fig. 8) occur as are exhibited by the function $\psi(x)$ considered in No. 2. On the other hand, if the function $x(1 + \cos x)$ is periodically extended it remains continuous at the end-points of the intervals, and in fact its derivative also remains continuous, since the discontinuities are eliminated by the factor $1 + \cos x$, which together with its derivative vanishes at the end-points.

4. The Function $f(x) = |x|$.

This function is even; consequently $b_\nu = 0$ and

$$a_\nu = \frac{2}{\pi} \int_0^\pi x \cos \nu x \, dx,$$

and on integration by parts we readily obtain

$$\int_0^\pi x \cos \nu x \, dx = \frac{1}{\nu} x \sin \nu x \Big|_0^\pi - \frac{1}{\nu} \int_0^\pi \sin \nu x \, dx$$

$$= \begin{cases} 0, \text{ if } \nu \text{ is even and } \neq 0, \\ -\dfrac{2}{\nu^2}, \text{ if } \nu \text{ is odd.} \end{cases}$$

Consequently

$$f(x) = \frac{\pi}{2} - \frac{4}{\pi}\left(\cos x + \frac{\cos 3x}{3^2} + \frac{\cos 5x}{5^2} + \dots\right).$$

If we put $x = 0$, we obtain the remarkable formula

$$\frac{\pi^2}{8} = 1 + \frac{1}{3^2} + \frac{1}{5^2} + \dots.$$

5. Example.

The function defined by the equations

$$f(x) = \begin{cases} -1, \text{ for } -\pi < x < 0, \\ 0, \text{ for } x = 0, \\ +1, \text{ for } 0 < x < \pi, \end{cases}$$

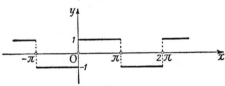

Fig. 9

as indicated in fig. 9, is an odd function. Hence $a_\nu = 0$ and

$$b_\nu = \frac{2}{\pi} \int_0^\pi \sin \nu x \, dx = \begin{cases} 0 \text{ if } \nu \text{ is even,} \\ \dfrac{4}{\pi\nu} \text{ if } \nu \text{ is odd,} \end{cases}$$

so that the Fourier series for the function is

$$f(x) = \frac{4}{\pi}\left(\frac{\sin x}{1} + \frac{\sin 3x}{3} + \dots\right).$$

For $x = \frac{\pi}{2}$, in particular, this again yields Gregory's series.

This series can be *formally* derived from that for $|x|$ by term-by-term differentiation.

6. The Function $f(x) = |\sin x|$.

The even function $f(x) = |\sin x|$ can be expanded in a cosine series, the coefficients a_ν being given by the following calculation:

$$\frac{\pi}{2}\, a_\nu = \int_0^\pi \sin x \cos \nu x \, dx = \frac{1}{2} \int_0^\pi \{\sin(\nu+1)x - \sin(\nu-1)x\} \, dx$$

$$= \begin{cases} 0 \text{ if } \nu \text{ is odd,} \\ \dfrac{-2}{\nu^2 - 1} \text{ if } \nu \text{ is even.} \end{cases}$$

We thus obtain

$$f(x) = |\sin x| = \frac{2}{\pi} - \frac{4}{\pi} \sum_{\mu=1}^{\infty} \frac{\cos 2\mu x}{4\mu^2 - 1}.$$

7. Expansion of the Function $\cos \mu x$. Resolution of the Cotangent into Partial Fractions. The Infinite Product for the Sine.

Let $f(x) = \cos \mu x$ for $-\pi < x < \pi$, where μ is *not* an integer. Since $f(x)$ is even we again obtain $b_\nu = 0$, while

$$\frac{\pi}{2}\, a_\nu = \int_0^\pi \cos \mu x \cos \nu x \, dx = \frac{1}{2} \int_0^\pi \{\cos(\mu+\nu)x + \cos(\mu-\nu)x\} \, dx$$

$$= \frac{1}{2} \left\{ \frac{\sin(\mu+\nu)\pi}{\mu+\nu} + \frac{\sin(\mu-\nu)\pi}{\mu-\nu} \right\}$$

$$= \frac{\mu(-1)^\nu}{\mu^2 - \nu^2} \sin \mu \pi.$$

We thus have

$$\cos \mu x = \frac{2\mu \sin \mu \pi}{\pi} \left(\frac{1}{2\mu^2} - \frac{\cos x}{\mu^2 - 1^2} + \frac{\cos 2x}{\mu^2 - 2^2} - + \cdots \right).$$

This function remains continuous at the points $x = \pm\pi$. If we put $x = \pi$, divide both sides of the equation by $\sin \mu \pi$, and then write x instead of μ, we obtain the equation

$$\cot \pi x = \frac{2x}{\pi} \left(\frac{1}{2x^2} + \frac{1}{x^2 - 1^2} + \frac{1}{x^2 - 2^2} + \cdots \right).$$

This is the so-called *resolution of the cotangent into partial fractions*, a very important formula frequently discussed in analysis. We now write this series in the form

$$\cot \pi x - \frac{1}{\pi x} = -\frac{2x}{\pi} \left\{ \frac{1}{1^2 - x^2} + \frac{1}{2^2 - x^2} + \cdots \right\}.$$

If x lies in an interval $0 \le x \le q < 1$, the n-th term on the right is less in absolute value than $\dfrac{2}{\pi}\, \dfrac{1}{n^2 - q^2}$. Hence the series converges uniformly in this interval, and can be integrated term by term. We thus obtain

$$\pi \int_0^x \left(\cot \pi t - \frac{1}{\pi t} \right) dt = \log \frac{\sin \pi x}{\pi x} - \lim_{a \to 0} \log \frac{\sin \pi a}{\pi a} = \log \frac{\sin \pi x}{\pi x}$$

on the left, and

$$\log \left(1 - \frac{x^2}{1^2} \right) + \log \left(1 - \frac{x^2}{2^2} \right) + \ldots = \lim_{n \to \infty} \sum_{\nu=1}^{n} \log \left(1 - \frac{x^2}{\nu^2} \right)$$

on the right, multiplying both sides by π. If we pass from the logarithm to the exponential function we have

$$\frac{\sin \pi x}{\pi x} = e^{\lim\limits_{n \to \infty} \sum\limits_{\nu=1}^{n} \log(1 - x^2/\nu^2)} = \lim_{n \to \infty} e^{\sum\limits_{\nu=1}^{n} \log(1 - x^2/\nu^2)}$$

$$= \lim_{n \to \infty} \prod_{\nu=1}^{n} \left(1 - \frac{x^2}{\nu^2} \right).$$

Hence

$$\sin \pi x = \pi x \left(1 - \frac{x^2}{1^2} \right) \left(1 - \frac{x^2}{2^2} \right) \left(1 - \frac{x^2}{3^2} \right) \ldots .$$

We have thus obtained the famous expression for the sine as an infinite product.* From this, by putting $x = \frac{1}{2}$, we obtain Wallis's product

$$\frac{\pi}{2} = \prod_{\nu=1}^{\infty} \frac{2\nu}{2\nu - 1} \cdot \frac{2\nu}{2\nu + 1} = \frac{2}{1} \cdot \frac{2}{3} \cdot \frac{4}{3} \cdot \frac{4}{5} \ldots ,$$

as on p. 224.

8. Further Examples.

By brief calculations similar to the preceding we obtain the following further examples of expansions in series.

The function $f(x)$ which is defined by the equation $f(x) = \sin \mu x$ for $-\pi < x < \pi$ can be expanded in the series

$$f(x) = \sin \mu x = -\frac{2 \sin \mu \pi}{\pi} \left(\frac{\sin x}{\mu^2 - 1^2} - \frac{2 \sin 2x}{\mu^2 - 2^2} + \frac{3 \sin 3x}{\mu^2 - 3^2} - + \ldots \right).$$

If we put $x = \frac{\pi}{2}$ and use the relation $\sin \mu \pi = 2 \sin \frac{\mu \pi}{2} \cos \frac{\mu \pi}{2}$, this gives us the resolution of the secant into partial fractions, that is, of the function $\dfrac{1}{\cos \mu \frac{\pi}{2}}$; this expansion is

$$\pi \sec \pi x = \frac{\pi}{\cos \pi x} = 4 \sum_{\nu=1}^{\infty} \frac{(-1)^\nu (2\nu - 1)}{4x^2 - (2\nu - 1)^2},$$

where for $\mu/2$ we have written x.

* This formula is particularly interesting because it shows directly that the function $\sin \pi x$ vanishes at the points $x = 0, \pm 1, \pm 2, \ldots$. In this respect it corresponds to the factorization of a polynomial when its zeros are known.

The series for the hyperbolic functions $\cosh \mu x$ and $\sinh \mu x$ $(-\pi < x < \pi)$ are

$$\cosh \mu x = \frac{2\mu}{\pi} \sinh \mu\pi \left(\frac{1}{2\mu^2} - \frac{\cos x}{\mu^2 + 1^2} + \frac{\cos 2x}{\mu^2 + 2^2} - \frac{\cos 3x}{\mu^2 + 3^2} + - \ldots \right),$$

$$\sinh \mu x = \frac{2}{\pi} \sinh \mu\pi \left(\frac{\sin x}{\mu^2 + 1^2} - \frac{2 \sin 2x}{\mu^2 + 2^2} + \frac{3 \sin 3x}{\mu^2 + 3^2} - + \ldots \right).$$

EXAMPLES

1. Find the Fourier expansions for the functions which are periodic with period 2π and which in $-\pi < x \leqq \pi$ are defined by the formulæ

(a) e^{ax}.   (b) $(x^2 - \pi^2)^2$.   (c) $\sin ax(1 + \cos x)$.

(d) $f(x) = 1 \, (a \leqq x \leqq b)$, $f(x) = 0 \, (-\pi < x < a)$, $f(x) = 0 \, (b < x \leqq \pi)$.

2. The function $f(t)$ is periodic with period 1, and in $0 \leqq x < 1$ it is given by $f(t) = t$. Prove that $f(t) = \dfrac{1}{2} - \dfrac{1}{\pi} \sum\limits_{n=1}^{\infty} \dfrac{\sin 2n\pi t}{n}$.

3. The polynomials $B_n(t)$ (Bernoulli polynomials) are defined by the relations: (a) $B_1(t) = t - \frac{1}{2}$; (b) $B_n'(t) = nB_{n-1}(t)$; (c) $\displaystyle\int_0^1 B_n(t) \, dt = 0$. Find $B_2(t)$, $B_3(t)$, $B_4(t)$.

(*Note*.—The numbers $B_n(0)$ are rational, and are, in fact, the same as Bernoulli's numbers B_n; cf. pp. 422, 423.)

4. Verify the Fourier expansions for the Bernoulli polynomials:

$$B_1(t) = -\frac{1}{\pi} \left\{ \sum_{n=1}^{\infty} \frac{\sin 2n\pi t}{n} \right\}, \qquad B_3(t) = \frac{3}{2\pi^3} \left\{ \sum_{n=1}^{\infty} \frac{\sin 2n\pi t}{n^3} \right\},$$

$$B_2(t) = \frac{1}{\pi^2} \left\{ \sum_{n=1}^{\infty} \frac{\cos 2n\pi t}{n^2} \right\}, \qquad B_4(t) = -\frac{3}{\pi^4} \left\{ \sum_{n=1}^{\infty} \frac{\cos 2n\pi t}{n^4} \right\}.$$

5. Prove that $\displaystyle\sum_{n=1}^{\infty} \frac{1}{n^2} = \frac{\pi^2}{6}$, $\displaystyle\sum_{n=1}^{\infty} \frac{1}{n^4} = \frac{\pi^4}{90}$.

6. Prove that $\dfrac{1}{1^3} - \dfrac{1}{3^3} + \dfrac{1}{5^3} - \dfrac{1}{7^3} + - \ldots = \dfrac{\pi^3}{32}$.

7. Prove that   (a) $1 + \dfrac{1}{3^2} + \dfrac{1}{5^2} + \dfrac{1}{7^2} + \ldots = \dfrac{\pi^2}{8}$.

(b) $1 - \dfrac{1}{2^2} + \dfrac{1}{3^2} - \dfrac{1}{4^2} + \ldots = \dfrac{\pi^2}{12}$.

(c) $1 - \dfrac{1}{2^4} + \dfrac{1}{3^4} - \dfrac{1}{4^4} + \ldots = \dfrac{7\pi^2}{720}$.

8. Obtain the infinite product for the cosine from the relation

$$\cos \pi x = \frac{\sin 2\pi x}{2 \sin \pi x}.$$

5. The Convergence of Fourier Series

We now proceed to establish rigorously the theorems which were stated in § 3 (p. 439) and illustrated in § 4 (p. 440).

1. The Convergence of the Fourier Series of a Sectionally Smooth Function.

We first recall that if $f(x)$ is any function which is defined and sectionally continuous (that is, continuous except for a finite number of jump discontinuities at most) in the interval $-\pi \leqq x \leqq \pi$, we can form the Fourier coefficients of $f(x)$ according to the formulæ

$$a_\nu = \frac{1}{\pi} \int_{-\pi}^{+\pi} f(t) \cos \nu t \, dt, \quad b_\nu = \frac{1}{\pi} \int_{-\pi}^{+\pi} f(t) \sin \nu t \, dt,$$

and we can formally write down the series

$$\tfrac{1}{2} a_0 + \sum_{\nu=1}^{\infty} (a_\nu \cos \nu x + b_\nu \sin \nu x).$$

This series is called the Fourier series corresponding to $f(x)$, irrespective of whether it converges or not. We are now about to find the conditions which must be imposed on $f(x)$ in order to ensure that the Fourier series corresponding to $f(x)$ does converge and does represent $f(x)$. We assume that $f(x)$ is extended periodically beyond the interval $-\pi < x \leqq \pi$.

The theorem which we shall now prove is as follows:

If the function f(x) *is sectionally smooth** * and at each point of discontinuity* (x) *satisfies the equation* f(x) $= \tfrac{1}{2}\{$f(x$-$0)$+$f(x$+$0)$\}$, *then the Fourier series corresponding to* f(x) *converges at every point and represents the function.*

To prove this theorem we consider the partial sums

$$S_n(x) = \tfrac{1}{2} a_0 + \sum_{\nu=1}^{n} (a_\nu \cos \nu x + b_\nu \sin \nu x).$$

If for the coefficients we substitute the integral expressions found above and then interchange the order of integration and summation we obtain

$$S_n(x) = \frac{1}{\pi} \int_{-\pi}^{+\pi} f(t) \left\{ \frac{1}{2} + \sum_{\nu=1}^{n} (\cos \nu t \cos \nu x + \sin \nu t \sin \nu x) \right\} dt,$$

---

* That is, $f(x)$ is sectionally continuous and its derivative $f'(x)$ is sectionally continuous also.

or, using the addition theorem for the cosine,

$$S_n(x) = \frac{1}{\pi} \int_{-\pi}^{+\pi} f(t) \left\{ \frac{1}{2} + \sum_{\nu=1}^{n} \cos \nu (t - x) \right\} dt.$$

If we now apply the summation formula obtained on p. 436, this becomes

$$S_n(x) = \frac{1}{2\pi} \int_{-\pi}^{+\pi} f(t) \frac{\sin (n + \frac{1}{2})(t - x)}{\sin \frac{1}{2}(t-x)} dt.$$

Finally, making the transformation $\tau = (t - x)$ and noting the periodicity of the integrand, we obtain

$$S_n(x) = \frac{1}{2\pi} \int_{-\pi}^{+\pi} f(x + \tau) \frac{\sin (n + \frac{1}{2})\tau}{\sin \frac{1}{2}\tau} d\tau.$$

Starting with this form of the partial sum $S_n(x)$ we can prove, by means of the following lemma, that $S_n(x)$ tends to $f(x)$.

Lemma. *If the function* s(x) *is sectionally continuous in the interval* a $\leq$ x $\leq$ b, *then the integral*

$$I = \int_a^b s(t) \sin \lambda t \, dt$$

tends to 0 *as* λ *increases.*

In the proof we may take $s(x)$ to be continuous in the whole interval, since otherwise we need only carry out the argument for each sub-interval in which $s(x)$ is continuous.

As in the similar argument on p. 418 *et seq.*, we notice that if λ is positive the function $\sin \lambda t$ is alternately positive and negative in successive intervals of length π/λ. For large values of λ the contributions to the integral from adjacent intervals almost cancel one another, since on account of continuity the values of $s(x)$ in two such adjacent intervals differ from one another only slightly. We make use of this circumstance by transforming the integral I by the substitution $t = \tau + h$, where $h = \pi/\lambda$; then $\sin \lambda t = -\sin \lambda \tau$, and we obtain

$$I = -\int_{a-h}^{b-h} s(\tau + h) \sin \lambda \tau \, d\tau.$$

If we again write the letter t instead of τ and then add the two expressions for I, we have

$$2I = -\int_{a-h}^{a} s(t+h)\sin\lambda t\, dt + \int_{a}^{b-h}\{s(t)-s(t+h)\}\sin\lambda t\, dt$$
$$+\int_{b-h}^{b} s(t)\sin\lambda t\, dt.$$

If M is an upper bound for the absolute value of $s(x)$, that is, if for all values of x in the interval under consideration $|s(x)| \leqq M$, then from this expression for I the inequality

$$2|I| \leqq 2Mh + \int_{a}^{b-h} |s(t)-s(t+h)|\, dt$$

follows at once. Now let ϵ be any positive number; if we choose λ so large that in the whole interval $a \leqq t \leqq b-h$ the expression $|s(t)-s(t+h)|$ remains less than $\epsilon/(b-a)$ and also $Mh = \dfrac{M\pi}{\lambda} < \dfrac{\epsilon}{2}$, then $|I| < \epsilon$; and consequently,* since ϵ can be chosen as small as we please, $\lim_{\lambda\to\infty} I = 0$.

Besides this lemma we need the integration formula

$$\int_{0}^{\pi} \frac{\sin(n+\tfrac{1}{2})t}{2\sin\tfrac{1}{2}t}\, dt = \frac{\pi}{2},$$

which is true for every positive integer n. This we readily establish by using our summation formula for the cosine, since

$$\int_{0}^{\pi} \frac{\sin(n+\tfrac{1}{2})t}{2\sin\tfrac{1}{2}t}\, dt = \int_{0}^{\pi} (\tfrac{1}{2} + \sum_{1}^{n}\cos\nu t)\, dt = \frac{\pi}{2}.$$

Proof of the Main Theorem.—By means of the lemma it is easy to prove our main theorem, i.e. to prove the formula

$$\lim_{n\to\infty} S_n(x) = \lim_{n\to\infty} \frac{1}{2\pi} \int_{-\pi}^{+\pi} f(x+t) \frac{\sin(n+\tfrac{1}{2})t}{2\sin\tfrac{1}{2}t}\, dt = f(x).$$

---

* If we assume that $s(x)$, besides being continuous, has a sectionally continuous derivative $s'(x)$, the proof of this lemma follows simply on integration by parts. For

$$\int_{a}^{b} s(t)\sin\lambda t\, dt = \frac{1}{\lambda}\left\{s(a)\cos\lambda a - s(b)\cos\lambda b + \int_{a}^{b} s'(t)\cos\lambda t\, dt\right\}.$$

Here we see at once that as λ increases the right-hand side tends to zero.

We begin by subdividing the interval of integration at the origin. For fixed values of x the function *

$$s(t) = \frac{f(x+t) - f(x+C)}{2 \sin \frac{1}{2} t}$$

is sectionally continuous in the interval $0 \leqq t \leqq \pi$. For this is obvious when $0 < t \leqq \pi$, while the continuity when $t = 0$ follows from the assumed existence of the right-hand derivative

$$\lim_{t \to 0,\, t > 0} \frac{f(x+t) - f(x+0)}{t} = \lim_{t \to +0} \frac{f(x+t) - f(x+0)}{2 \sin \frac{1}{2} t} \cdot \frac{2 \sin \frac{1}{2} t}{t}$$

$$= \lim_{t \to +0} \frac{f(x+t) - f(x+0)}{2 \sin \frac{1}{2} t}.$$

Hence, as $\lambda = n + \frac{1}{2}$ increases, the integral

$$\frac{1}{\pi} \int_0^\pi s(t) \sin \lambda t \, dt$$

$$= \frac{1}{2\pi} \int_0^\pi f(x+t) \frac{\sin \lambda t}{\sin \frac{1}{2} t} \, dt - \frac{1}{2\pi} \int_0^\pi f(x+0) \frac{\sin \lambda t}{\sin \frac{1}{2} t} \, dt$$

tends to zero.

Since, however, the factor $f(x+0)$ can be taken out of the second integral on the right, and since for $\lambda = n + \frac{1}{2}$ the integral $\int_0^\pi \frac{\sin \lambda t}{2 \sin \frac{1}{2} t} \, dt$ is equal to $\frac{\pi}{2}$, we immediately obtain the equation †

$$\lim_{\lambda \to \infty} \frac{1}{2\pi} \int_0^\pi f(x+t) \frac{\sin \lambda t}{\sin \frac{1}{2} t} \, dt = \frac{1}{2} f(x+0).$$

In the same way we obtain

$$\lim_{\lambda \to \infty} \frac{1}{2\pi} \int_{-\pi}^0 f(x+t) \frac{\sin \lambda t}{\sin \frac{1}{2} t} \, dt = \frac{1}{2} f(x-0),$$

for the interval $-\pi \leqq t \leqq 0$, and by addition

$$\lim_{\lambda \to \infty} \frac{1}{2\pi} \int_{-\pi}^{+\pi} f(x+t) \frac{\sin \lambda t}{\sin \frac{1}{2} t} \, dt = f(x).$$

---

* For this notation see p. 439.

† By putting $x = 0$, $f(t) = (\sin \frac{1}{2} t)/t$ in this equation, and then replacing t by u/λ, we obtain the important relation (cf. pp. 251–253)

$$\lim_{\lambda \to \infty} \int_0^{\lambda \pi} \frac{\sin u}{u} \, du = \frac{\pi}{2}.$$

2. Further Investigation of the Convergence.

In the neighbourhood of those points where the function $f(x)$ is discontinuous the Fourier series does not converge uniformly; for according to Chap. VIII, § 4 (p. 393), a uniformly convergent series of continuous functions possesses a continuous sum. Nevertheless, we have the following important theorem:

If a sectionally smooth periodic function has no discontinuities, its Fourier series converges absolutely and uniformly. The convergence of the Fourier series for any sectionally smooth function whatever is uniform in every closed interval which contains no point of discontinuity of the function.

In order to prove this theorem we start from a fundamental inequality satisfied by the Fourier coefficients of any function $f(x)$ which is sectionally continuous (note that $f(x)$ is not assumed to be sectionally *smooth*). This so-called *Bessel's inequality* states that for all values of n

$$\frac{1}{2} a_0^2 + \sum_{\nu=1}^{n} (a_\nu^2 + b_\nu^2) \leqq \frac{1}{\pi} \int_{-\pi}^{+\pi} \{f(x)\}^2 \, dx.$$

The proof follows from the fact that the expression

$$\int_{-\pi}^{+\pi} \left\{ f(x) - \tfrac{1}{2} a_0 - \sum_{\nu=1}^{n} (a_\nu \cos \nu x + b_\nu \sin \nu x) \right\}^2 dx$$

is always positive or zero. If we evaluate the integral by expanding the bracket under the integral sign and recalling the orthogonality relations and the definitions of the Fourier coefficients, we at once obtain Bessel's inequality in the form

$$\int_{-\pi}^{+\pi} \{f(x)\}^2 \, dx - \pi \left\{ \frac{1}{2} a_0^2 + \sum_{\nu=1}^{n} (a_\nu^2 + b_\nu^2) \right\} \geqq 0.$$

In addition to Bessel's inequality we make use of Schwarz's inequality (p. 13): if $u_1, u_2, \ldots, u_n$ and $v_1, v_2, \ldots, v_n$ are arbitrary real numbers, it is always true that

$$\left(\sum_{\nu=1}^{n} u_\nu v_\nu \right)^2 \leqq \sum_{\nu=1}^{n} u_\nu^2 \cdot \sum_{\nu=1}^{n} v_\nu^2,$$

the sign of equality occurring only when the sequence u is proportional to the sequence v.

We now assume that the periodic function $f(x)$ is sectionally smooth, and also continuous. The derivative $g(x) = f'(x)$ is sectionally continuous, and we easily show that c_ν and d_ν, the Fourier coefficients of $g(x)$, satisfy the relations

$$c_0 = 0,$$
$$\left.\begin{array}{l} c_\nu = \nu b_\nu \\ d_\nu = -\nu a_\nu \end{array}\right\} (\nu \geqq 1);$$

for on integration by parts we have

$$c_\nu = \frac{1}{\pi} \int_{-\pi}^{+\pi} g(x) \cos \nu x \, dx$$

$$= \frac{1}{\pi} f(x) \cos \nu x \Big|_{-\pi}^{+\pi} + \frac{\nu}{\pi} \int_{-\pi}^{+\pi} f(x) \sin \nu x \, dx = \nu b_\nu,$$

a similar proof holding for the other statements.

Bessel's inequality applied to the function $g(x)$ therefore gives us

$$\sum_{\nu=1}^{n} \nu^2 (a_\nu^2 + b_\nu^2) = \sum_{\nu=1}^{n} (c_\nu^2 + d_\nu^2) \leqq \frac{1}{\pi} \int_{-\pi}^{+\pi} \{g(x)\}^2 \, dx.$$

If for brevity we denote the right-hand side of this inequality by M^2, and apply Schwarz's inequality, we find that when $m > n$

$$\sum_{\nu=n+1}^{m} | a_\nu \cos \nu x + b_\nu \sin \nu x | \leqq \sum_{\nu=n+1}^{m} \sqrt{(a_\nu^2 + b_\nu^2)}$$

$$= \sum_{\nu=n+1}^{m} \left\{ \frac{1}{\nu} (\nu \sqrt{(a_\nu^2 + b_\nu^2)}) \right\} \leqq M \sqrt{\left(\sum_{\nu=n+1}^{m} \frac{1}{\nu^2} \right)},$$

since $\sqrt{(a_\nu^2 + b_\nu^2)}$ is the amplitude of the periodic function $a_\nu \cos \nu x + b_\nu \sin \nu x$.

But owing to the convergence of $\sum_{\nu=1}^{\infty} \frac{1}{\nu^2}$ the right-hand side, which is independent of x, can be made as small as we please by choosing n and m large enough, which proves the absolute and uniform convergence of the series.*

In order to prove the above theorem for sectionally smooth functions which are discontinuous, we first consider a *special* function $\psi(x)$ of this type.

---

* The same considerations show incidentally that for periodic functions with continuous derivatives of the $(h-1)$-th order and derivatives of the h-th order which are at least sectionally continuous, the sum $\sum_{}^{n} \nu^{2h}(a_\nu^2 + b_\nu^2)$ remains below a fixed bound. This gives us a definite statement about the order to which the Fourier coefficients vanish. For such a function the Fourier series of the derivatives up to the order $(h-1)$ converge absolutely and uniformly.

In the interval $-\pi < x < \pi$ we define $\psi(x)$ as equal to x; outside this interval $\psi(x)$ is extended periodically. According to p. 440 its Fourier series is

$$2\left(\frac{\sin x}{1} - \frac{\sin 2x}{2} + \frac{\sin 3x}{3} - + \cdots\right).$$

This series cannot be uniformly convergent, for its sum is the discontinuous function $\psi(x)$. We shall show, however, that the convergence is uniform in every interval $-l \leqq x \leqq l$ for which $0 < l < \pi$.

The proof is based on a special artifice.* We observe that in the interval $-l \leqq x \leqq l$ the function $\cos\dfrac{x}{2}$ is never less than the positive quantity $\cos\dfrac{l}{2} = \kappa$. If we multiply the absolute value of the difference between the m-th and n-th partial sums of the above series $(m > n)$, that is, the expression

$$|S_m(x) - S_n(x)|$$
$$= 2\left|\frac{\sin(n+1)x}{n+1} - \frac{\sin(n+2)x}{n+2} + - \cdots \pm \frac{\sin mx}{m}\right|$$

by the function $\cos\dfrac{x}{2}$, then in accordance with the well-known trigonometric formula $2\sin u \cos v = \sin(u+v) + \sin(u-v)$, we obtain the absolute value of the expression

$$2\cos\frac{x}{2}\left\{\frac{\sin(n+1)x}{n+1} - \frac{\sin(n+2)x}{n+2} + - \cdots \pm \frac{\sin mx}{mx}\right\}$$
$$= \frac{\sin(n+\frac{3}{2})x}{n+1} - \frac{\sin(n+\frac{5}{2})x}{n+2} + - \cdots \pm \frac{\sin(m+\frac{1}{2})x}{m}$$
$$+ \frac{\sin(n+\frac{1}{2})x}{n+1} - \frac{\sin(n+\frac{3}{2})x}{n+2} + \frac{\sin(n+\frac{5}{2})x}{n+3} - + \cdots.$$

* We are led to this artifice naturally by observing that the function $2y\cos y$, when extended periodically beyond the interval $-\dfrac{\pi}{2} \leqq y \leqq \dfrac{\pi}{2}$, remains continuous, so that according to the first part of the theorem its Fourier series must converge uniformly and must represent the function. This series, however, is obtained if we multiply the Fourier series for $2y$ by $\cos y$. If we now put $y = x/2$, this multiplication leads to the steps in the text.

If we combine the terms on the right which have the same numerators, we obtain the expression

$$\frac{\sin (n + \frac{1}{2})x}{n + 1} \pm \frac{\sin (m + \frac{1}{2})x}{m}$$

$$+ \frac{\sin (n + \frac{3}{2})x}{(n + 1)(n + 2)} - \frac{\sin (n + \frac{5}{2})x}{(n + 2)(n + 3)} + - \ldots \mp \frac{\sin (m - \frac{1}{2})x}{(m - 1)m},$$

and since $\cos \frac{x}{2} \geqq \kappa$ and $|\sin u| \leqq 1$, we obtain the estimate

$$| S_m (x) - S_n (x) |$$

$$\leqq \frac{1}{\kappa} \left[\frac{1}{n + 1} + \frac{1}{m} + \frac{1}{(n + 1)(n + 2)} + \ldots + \frac{1}{(m - 1)m} \right].$$

But the expression on the right does not depend on x, and in virtue of the convergence of the series $\sum\limits_{\nu = 1}^{\infty} \dfrac{1}{\nu (\nu + 1)}$ it can be made as small as we please by choosing n and m large enough. This implies the uniform convergence of the Fourier series, which we asserted.

Now that we have obtained the expansion for a particular discontinuous function, we can (cf. p. 441) transfer the discontinuity to any arbitrary point in the interval by translation of the curve or of the co-ordinate system. In fact, the function

$$\psi (x - \xi) = 2 \left(\frac{\sin (x - \xi)}{1} - \frac{\sin 2 (x - \xi)}{2} + \frac{\sin 3 (x - \xi)}{3} - + \ldots \right)$$

is continuous except at the points $(2k + 1)\pi + \xi$, where k is an integer. On passing these points, however, the function jumps by an amount -2π, from the value π to the value $-\pi$, while at these points themselves the value of the function is zero.

If now $f(x)$ is any sectionally smooth function which in the interval $-\pi \leqq x \leqq \pi$ is discontinuous only at the points $\xi_1, \xi_2, \ldots, \xi_m$, and if on passing these points from left to right the function jumps by the amounts $\delta_1, \delta_2, \ldots, \delta_m$ respectively, then the function

$$f(x) + \frac{\delta_1}{2\pi} \psi (x + \pi - \xi_1) + \frac{\delta_2}{2\pi} \psi (x + \pi - \xi_2) + \ldots$$

$$+ \frac{\delta_m}{2\pi} \psi (x + \pi - \xi_m)$$

will be continuous and sectionally smooth, and hence by the previous proof can be expanded in a uniformly convergent Fourier series. We now obtain the Fourier series of the function $f(x)$ by adding the finite number of Fourier series corresponding to the functions $-\dfrac{\delta_1}{2\pi}\,\psi(x+\pi-\xi_1),\ \ldots,\ -\dfrac{\delta_m}{2\pi}\,\psi(x+\pi-\xi_m)$ term by term. Our theorem is thus proved.

This result is quite adequate for most mathematical investigations and for applications. We would, however, point out that the investigation of Fourier series has been pushed much further. The conditions for expansion in Fourier series which we have found here are *sufficient*, but by no means *necessary*. Functions with far fewer continuity properties than those discussed here can be represented by Fourier series. There is an extensive literature devoted to these questions and to the general problem of the expansibility of a function in a Fourier series. As a remarkable result of such investigations we mention the fact that there are continuous functions whose Fourier series do not converge in any interval, no matter how small. Such a result does not in any way impugn the usefulness of Fourier series; on the contrary, it must be regarded as evidence that the concept of a continuous function involves fairly complicated possibilities, as has already been shown by the example of continuous functions which nowhere have a derivative.

Appendix to Chapter IX

Integration of Fourier Series

One of the remarkable properties of Fourier series is their term-by-term integrability. In general, a series can be integrated term by term if it is uniformly convergent; otherwise term-by-term integration may lead to false results. In contrast with this, for Fourier series, we have the theorem:

If $f(x)$ *is sectionally continuous in* $-\pi \leqq x \leqq \pi$, *and if the series* $\frac{1}{2}a_0+\Sigma(a_\nu\cos x + b_\nu\sin x)$ *is the Fourier series corresponding to* $f(x)$, *then this series can be integrated term by term between any* *two limits ξ and x lying in the interval $-\pi \leqq x \leqq \pi$; in symbols,*

$$\int_{\xi}^{x} f(x)\, dx = \int_{\xi}^{x} \tfrac{1}{2}a_0\, dx + \sum_{1}^{\infty} \left(\int_{\xi}^{x} a_\nu \cos \nu x\, dx + \int_{\xi}^{x} b_\nu \sin \nu x\, dx \right).$$

Moreover, for every fixed value of ξ the series on the right converges uniformly in x. The remarkable feature of this theorem is that not only do we not require that the Fourier series for $f(x)$ shall be uniformly convergent, but we need not even assume that it converges at all.

To prove this, let the function $F(x)$ be defined by the equation $F(x) = \int_{-\pi}^{x} \{f(x) - \tfrac{1}{2}a_0\}\, dx$. This function is sectionally smooth, and by the definition of a_0 we have $F(\pi) = F(-\pi) = 0$, so that $F(x)$ can be extended periodically and continuously. The Fourier series $\tfrac{1}{2}A_0 + \sum_{\nu=1}^{\infty} (A_\nu \cos \nu x + B_\nu \sin \nu x)$ of the function $F(x)$ therefore converges uniformly to $F(x)$. We now investigate the coefficients A_ν and B_ν. By integration by parts as on p. 449, we find that, for $\nu > 0$, $A_\nu = -b_\nu/\nu$ and $B_\nu = a_\nu/\nu$. Hence for any values ξ and x in the interval $-\pi \leqq x \leqq \pi$, we have

$$F(x) - F(\xi) = \sum_{\nu=1}^{\infty} \{A_\nu (\cos \nu x - \cos \nu \xi) + B_\nu (\sin \nu x - \sin \nu \xi)\}$$

$$= \sum_{\nu=1}^{\infty} \left\{ \frac{a_\nu}{\nu} (\sin \nu x - \sin \nu \xi) - \frac{b_\nu}{\nu} (\cos \nu x - \cos \nu \xi) \right\},$$

converging uniformly in x. If we replace $F(x)$ by its definition, this becomes

$$\int_{\xi}^{x} f(x)\, dx - \tfrac{1}{2}a_0 \int_{\xi}^{x} dx = \sum_{\nu=1}^{\infty} (a_\nu \int_{\xi}^{x} \cos \nu x\, dx + b_\nu \int_{\xi}^{x} \sin \nu x\, dx),$$

which was to be proved.

It is easy to see that if $f(x)$ is periodic and sectionally continuous the term-by-term integration can be performed over any interval whatever.

CHAPTER X

A Sketch of the Theory of Functions of Several Variables

Up to this point we have concerned ourselves exclusively with functions of a single independent variable. We must now go on to consider functions of several independent variables. Even the applications of the calculus force us to take this step. In almost all the relationships which occur in nature, in fact, the functions in question do not depend on a single independent variable; on the contrary, the dependent variable is usually determined by two, three, or more independent variables. Thus, for example, the volume of an ideal gas is a function of a single variable, the pressure, if we keep the temperature constant, but not otherwise. As a rule the temperature also varies, and the volume depends upon a pair of values, namely, the value of the pressure and that of the temperature; it is therefore a function of two independent variables.

From the point of view of pure mathematics also, the need for detailed study of functions of several independent variables is urgent. Here we shall be able to take advantage of what we have previously learned, so that in many cases we have only to make simple extensions of our arguments.

It is usually sufficient to consider the case of only two independent variables x and y, so long as no essentially new considerations are required for an extension to functions of three or more variables. In order to keep our statements and notation simple, therefore, we shall consider only two independent variables as a rule.

A systematic presentation of the differential and integral calculus for functions of several variables is impossible within the compass of this volume, but will be given in Vol. II

of this treatise. All that can be done here is to give the reader
a preliminary view of some of the most important new concepts
and operations. We shall frequently rely on intuitive plausi-
bility, the full proofs being developed subsequently in Vol. II.

1. The Concept of Function in the Case of Several Variables

1. Functions and their Ranges of Definition.

Equations of the form

$$u = x^2 + y^2, \quad u = x - y, \quad u = xy, \quad \text{or} \quad u = \sqrt{(1 - x^2 - y^2)}$$

assign a *functional value* u to each pair of values (x, y). In the
first three of our examples this correspondence holds for every
system of values (x, y), while in the last case the correspondence
has a meaning only for those pairs of values (x, y) for which the
inequality $x^2 + y^2 \leqq 1$ is true.

In these cases we say that u is a *function* of the *independent
variables x and y*. This expression we use in general whenever
some law assigns a value of u as *dependent variable*, corresponding to each pair of values (x, y) belonging to a certain
specified set. The relation between x, y and u may be stated in
terms of a "functional equation", as above, or by means of a verbal description such as "u is the area of the rectangle with sides
x and y", or it may follow from physical observations, as, for instance, in the case of the magnetic declination for different lati-
tudes and longitudes. The essential thing is that a *correspondence*
exists. Similarly, u is said to be a function of the three independent variables x, y, z if for each triad of values (x, y, z) of a
certain set there exists a corresponding value of u given by some
definite law; and similarly for the general case of functions of n
independent variables $x_1, x_2, \ldots, x_n$.

The set of values which the pair (x, y) can assume is called
the *range of definition* of the function $u = f(x, y)$. For the purposes of this chapter we shall restrict our attention to the simplest types of range of definition. We shall consider that (x, y)
is limited either to a so-called *rectangular region* (*domain*)

$$a \leqq x \leqq b, \quad c \leqq y \leqq d,$$

or else to a circle determined by an inequality of the form

$$(x - a)^2 + (y - b)^2 \leqq r^2.$$

In the case of functions of three variables x, y, z we shall again consider only rectangular regions

$$a \leqq x \leqq b, \quad c \leqq y \leqq d, \quad e \leqq z \leqq f$$

and spherical regions

$$(x - a)^2 + (y - b)^2 + (z - c)^2 \leqq r^2.$$

In dealing with more than three independent variables geometrical intuition fails us, but it is often convenient to extend geometrical phraseology to this case also. Thus for functions of n variables $x_1, \ldots, x_n$ we shall consider regions

$$a_1 \leqq x_1 \leqq b_1, \quad a_2 \leqq x_2 \leqq b_2, \ldots, \quad a_n \leqq x_n \leqq b_n$$

and also regions

$$(x_1 - a_1)^2 + (x_2 - a_2)^2 + \ldots + (x_n - a_n)^2 \leqq r^2,$$

which we call rectangular regions and spherical regions respectively.

2. The Simplest Types of Functions.

Just as in the case of functions of one variable, the simplest functions are the *rational integral functions* or *polynomials*. The most general polynomial of the first degree (linear function) is of the form

$$u = ax + by + c,$$

where a, b, and c are constants. The general polynomial of the second degree has the form

$$u = ax^2 + bxy + cy^2 + dx + ey + f.$$

The general polynomial is a sum of terms of the form $a_{mn} x^m y^n$, where the quantities a_{mn} are arbitrary constants.

Rational fractional functions are quotients of polynomials; to this class belongs e.g. the *linear fractional* function

$$u = \frac{ax + by + c}{a'x + b'y + c'}.$$

By extraction of roots we pass from the rational functions to certain *algebraic* functions,* e.g.

$$u = \sqrt{\left(\frac{x-y}{x+y}\right)} + \sqrt[3]{\left\{\frac{(x+y)^2}{x^3+xy}\right\}}.$$

In the construction of more complicated functions of several variables we almost always fall back on the well-known functions of one variable;† e.g.

$$u = \sin(xy) \quad \text{or} \quad u = \log(y^2 + \cos\tfrac{1}{2}x).$$

3. Geometrical Representation of Functions.

Just as we represent functions of one variable by means of curves, we seek to represent functions of two variables geometrically by means of *surfaces*; henceforward we shall consider only those functions which can actually be represented in this way. We achieve this representation very simply by considering a rectangular co-ordinate system in space, with co-ordinates x, y, and u, and marking off above each point (x, y) of the range (R) of definition of the function the point P with the third co-ordinate $u = f(x, y)$. As the point (x, y) ranges over the region R the point P describes a surface in space. This surface we take as the geometrical representation of the function.

Conversely, in analytical geometry surfaces in space are represented by functions of two variables, so that between such surfaces and functions of two variables there is a reciprocal relation.

For example, to the function

$$u = \sqrt{(1 - x^2 - y^2)}$$

there corresponds the *hemisphere* lying above the x, y plane, with unit radius and centre at the origin. To the function $u = x^2 + y^2$ there corresponds a so-called *paraboloid of revolution*, obtained by rotating the parabola $u = x^2$ about the u-axis (fig. 1). To the functions $u = x^2 - y^2$ and $u = xy$ there correspond *hyperbolic paraboloids* (fig. 2). The linear function $u = ax + by + c$ has for its graph a *plane* in space.‡

---

* For an accurate definition of the term " algebraic function " see p. 485.

† Cf. also the section on compound functions (p. 472).

‡ If in the function $u = f(x, y)$ one of the independent variables, say y, does not occur, so that u depends on x only, say $u = g(x)$, the function is represented in xyu-space by a cylindrical surface obtained by erecting the perpendiculars to the ux-plane at the points of the curve $u = g(x)$.

This representation by means of rectangular co-ordinates has, however, two disadvantages. Firstly, intuition fails us whenever we have to deal with three or more independent variables. Secondly, even in the case of two independent variables it is often more convenient to confine the discussion to the xy-plane alone, since in the plane we can sketch and make geometrical constructions without difficulty. From this point of view another geometrical representation of the function, by means of *contour lines*, is to be preferred. In the xy-plane we take all the points for which $u = f(x, y)$ has a constant value, say $u = k$. These

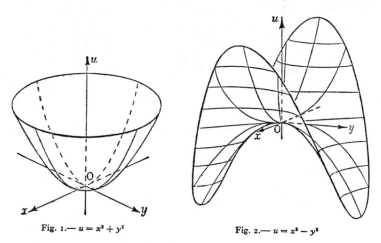

Fig. 1.— $u = x^2 + y^2$ Fig. 2.— $u = x^2 - y^2$

points will usually lie on a curve or curves, the so-called contour line for the given constant value of the function. We can also obtain these curves by cutting the surface $u = f(x, y)$ by the plane $u = k$ parallel to the xy-plane and projecting the curves of intersection perpendicularly on to the xy-plane. The system of these contour lines, marked with the corresponding values k_1, k_2, ... of the height k, gives us a representation of the function. As a rule k is assigned values in arithmetic progression, say $k = \nu h$, where $\nu = 1, 2, \ldots$. The distance between the contour lines then gives us a measure of the steepness of the surface $u = f(x, y)$; for between every two neighbouring lines the value of the function changes by the same amount. Where the contour lines are close together the function rises or falls steeply; where the lines are far apart the surface is flattish. This is the principle on which contour maps such as those of the Ordnance Survey and the U.S. Geological Survey are constructed.

In this method the linear function $u = ax + by + c$ is represented by a system of parallel straight lines $ax + by + c = k$. The function $u = x^2 + y^2$ is represented by a system of concentric circles (cf. fig. 3). The function $u = x^2 - y^2$, whose surface has a *saddle point* at the origin (fig. 2) is represented by the system of hyperbolas shown in fig. 4.

The method of representing the function $u = f(x, y)$ by contour lines has the advantage of being capable of extension to functions of three independent variables also. Instead of the

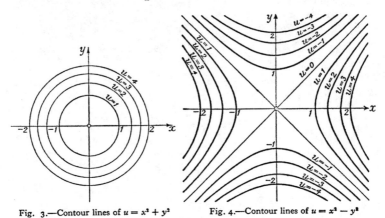

Fig. 3.—Contour lines of $u = x^2 + y^2$ Fig. 4.—Contour lines of $u = x^2 - y^2$

contour lines we then have the *level surfaces* $f(x, y, z) = k$, where k is a constant to which we can assign any suitable sequence of values. For example, the level surfaces for the function $u = x^2 + y^2 + z^2$ are concentric spheres about the origin of the co-ordinate system.

EXAMPLE

1. For each of the following functions sketch the contour lines corresponding to $z = -2, -1, 0, 1, 2, 3$:

(a) $z = x^2y$.

(b) $z = x^2 + y^2 - 1$.

(c) $z = x^2 - y^2$.

(d) $z = y^2$.

(e) $z = y\left(1 - \dfrac{1}{x^2 + y^2}\right)$.

2. Continuity

1. Definition.

As in the case of functions of one variable, the basic requirement that functions should be capable of being represented geometrically leads to the analytic condition of continuity. Here again the concept of continuity is given by the following definition: *a function* u = f(x, y), *defined in a region* R, *is said to be continuous at a point* (ξ, η) *of* R *if for all points* (x, y) *near* (ξ, η) *the value of the function* f(x, y) *differs but little from* $f(\xi, \eta)$, *the difference being arbitrarily small if only* (x, y) *is near enough to* (ξ, η).

More precisely: *the function* f(x, y), *defined in the region* R, *is continuous at the point* (ξ, η) *of* R, *provided that for every positive number* ϵ *it is possible to find a positive distance* $\delta = \delta(\epsilon)$ *(in general depending on* ϵ *and tending to 0 with* ϵ*) such that for all points of the region whose distance from* (ξ, η) *is less than* δ *(that is, for which the inequality*

$$(x - \xi)^2 + (y - \eta)^2 \leqq \delta^2$$

holds) the relation

$$\mid f(x, y) - f(\xi, \eta) \mid \leqq \epsilon$$

is satisfied. Or, in other words, the relation

$$\mid f(\xi + h, \eta + k) - f(\xi, \eta) \mid \leqq \epsilon$$

is to hold for all pairs of values (h, k) such that $h^2 + k^2 \leqq \delta^2$ and $(\xi + h, \eta + k)$ belongs to the region R.

If a function is continuous at every point of a region R, we say that it is *continuous in* R.

In the definition of continuity we can replace the distance condition $h^2 + k^2 \leqq \delta^2$ by the following equivalent condition:

To every $\epsilon > 0$ *there shall correspond two positive numbers* δ_1 *and* δ_2 *such that*

$$\mid f(\xi + h, \eta + k) - f(\xi, \eta) \mid \leqq \epsilon$$

whenever $\qquad \mid h \mid \leqq \delta_1 \quad and \quad \mid k \mid \leqq \delta_2.$

The two conditions are equivalent. For if the original condition is fulfilled, so is the second if we take $\delta_1 = \delta_2 = \delta/\sqrt{2}$;

and conversely, if the second condition is fulfilled so is the first, if for δ we take the smaller of the two numbers δ_1 and δ_2.

The following facts are almost obvious:

The sum, difference, and product of continuous functions are also continuous. The quotient of continuous functions is continuous except where the denominator vanishes. Continuous functions of continuous functions are themselves continuous (cf. the note on pp. 473–474). In particular, *all polynomials are continuous, and all rational fractional functions are also continuous except where the denominator vanishes.**

2. Examples of Discontinuities.

In the case of functions of one variable we met with three kinds of discontinuities: infinite discontinuities, jump discontinuities, and discontinuities at which no limit is approached from one side or both. With functions of two or more variables no such simple classification is possible. In particular, the situation is made more complicated by the fact that discontinuities may occur not merely at isolated points but also along whole curves.

Thus for the function $u = \dfrac{1}{x - y}$ the line $x = y$ is a line of infinite discontinuity. As we approach the line from one side or the other the values of u increase numerically beyond all bounds through positive or through negative values. The function $u = \dfrac{1}{(x - y)^2}$ has the same line of discontinuity but tends to $+\infty$ as we approach the line from either side. The function $u = \dfrac{1}{x^2 + y^2}$ has the single point of discontinuity $x = 0$, $y = 0$. The function $u = \sin \dfrac{1}{\sqrt{(x^2 + y^2)}}$ tends to no limit as we approach the origin; the surface which it represents is obtained by rotating the graph of the function $u = \sin \dfrac{1}{x}$ about the u-axis.

Another instructive example of a discontinuous function is given by the rational function $u = \dfrac{2xy}{x^2 + y^2}$. In the first instance the function is

---

* Another obvious fact, which, however, is worth stating, is as follows: *if a function* f(x, y) *is continuous in a region* R *and is different from zero at an interior point* P *of the region, it is possible to mark off about* P *a neighbourhood, say a circle, belonging entirely to* R, *in which* f(x, y) *is nowhere equal to zero.* For if the value of the function at P is a, we can mark off about P a circle so small that the value of the function within the circle differs from a by less than $a/2$, and therefore is certainly not zero.

undefined at $x = 0$, $y = 0$, and we supplement the definition by assuming that $u(0, 0) = 0$. This function has a peculiar type of discontinuity at the origin. If we put $x = 0$, that is, if we move along the y-axis, the function becomes $u(0, y) = 0$, which has the constant value 0 for all values of y. Along the x-axis we likewise have $u(x, 0) = 0$. Thus at the origin the function $u(x, y)$ is continuous in x if we keep y at the constant value 0 and is continuous in y if we keep x at the constant value 0. Nevertheless, the function is discontinuous when considered as a function of the two variables x and y. For at every point of the line $y = x$ we find that $u = 1$, so that arbitrarily near the origin we can find points at which u assumes the value 1. The function is therefore discontinuous at the origin,* and cannot be defined at the origin in such a way as to make it continuous.

The above example shows that a function can be continuous in x for every fixed value of y and continuous in y for every fixed value of x and yet discontinuous when considered as a function of the two variables. The essential point in the definition of continuity is that the value of the function at a point P must be arbitrarily close to the value of the function at a point Q, provided only that Q is near enough to P; it is not permissible to restrict the position of Q relative to P in any other way.

EXAMPLES

1. Examine the continuity of the function $z = \dfrac{x^2 + y}{\sqrt{x^2 + y^2}}$. Sketch the level lines $z = k$ ($k = -4, -2, 0, 2, 4$). Exhibit (on one graph) the behaviour of z as a function of x alone for $y = -2, -1, 0, 1, 2$. Similarly, exhibit the behaviour of z as a function of y alone for $x = 0, \pm 1, \pm 2$. Finally, exhibit the behaviour of z as a function of r alone when θ is constant (r, θ being polar co-ordinates).

2. Show that the following functions are continuous:

(a) $\sin(x^2 + y)$.

(b) $\dfrac{\sin xy}{\sqrt{x^2 + y^2}}$.

(c) $\dfrac{x^3 + y^3}{x^2 + y^2}$.

(d) $x^2 \log(x^2 + y^2)$.

---

* More generally, on the straight line $y = x \tan a$ inclined at the angle a to the x-axis we have $u = 2 \tan a/(1 + \tan^2 a) = 2 \sin a \cos a = \sin 2a$. The surface corresponding to the function $u = 2xy/(x^2 + y^2)$ is therefore formed by rotating a straight line at right angles to the u-axis about that axis until it coincides with the x-axis, and simultaneously raising or lowering it so that the height $\sin 2a$ is associated with the angle a. As a increases up to 45° the straight line rises to the height 1, and subsequently falls to the level of the y-axis and below it to the depth 1, thereafter rising again to the level of the x-axis. The surface enveloped by the moving straight line is known as the *cylindroid*; it is of importance in mechanics.

3. Find whether or not the following functions are continuous, and if not, where they are discontinuous:

(a) $\sin \dfrac{y}{x}$.　　(b) $\dfrac{x^3 + y^2}{x^2 + y^2}$.　　(c) $\dfrac{x^3 + y^2}{x^3 + y^3}$.　　(d) $\dfrac{x^3 + y^2}{x^2 + y}$.

3. THE DERIVATIVES OF A FUNCTION OF SEVERAL VARIABLES

1. Definition.　Geometrical Representation.

If in a function of several variables we assign definite numerical values to all but one of the variables and allow only that one variable, say x, to vary, the function becomes a function of one

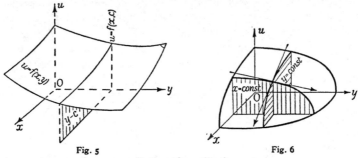

Fig. 5　　　　　　　　　　　　Fig. 6

Sections of $u = f(x, y)$

variable. We consider e.g. a function $u = f(x, y)$ of the two variables x and y and give y a definite fixed value $y = y_0 = c$. The function $u = f(x, y_0)$ of the single variable x which is thus formed may be simply represented geometrically by letting the surface $u = f(x, y)$ be cut by the plane $y = y_0$ (cf. figs. 5 and 6). The curve of intersection thus formed in the plane is represented by the equation $u = f(x, y_0)$. If we differentiate this function in the usual way at the point $x = x_0$ (we assume that the derivative exists), we obtain the *partial derivative of* f(x, y) *with respect to* x at the point (x_0, y_0). According to the usual definition of the derivative this is the limit *

$$\lim_{h \to 0} \frac{f(x_0 + h, \, y_0) - f(x_0, \, y_0)}{h}.$$

* If (x_0, y_0) is a point on the boundary of the region of definition we make the restriction that in the passage to the limit the point $(x + h, y_0)$ must always remain in the region.

Geometrically this partial derivative denotes the tangent of the angle between a parallel to the x-axis and the tangent line to the curve $u = f(x, y_0)$. It is therefore the *slope of the surface* u = f(x, y) *in the direction of the x-axis.*

To represent these partial derivatives several different notations are used, of which we mention the following:

$$\lim_{h \to 0} \frac{f(x_0 + h, y_0) - f(x_0, y_0)}{h} = f_x(x_0, y_0) = u_x(x_0, y_0).$$

If we wish to emphasize that the partial derivative is the limit of a difference quotient we denote it by

$$\frac{\partial f}{\partial x} \quad \text{or} \quad \frac{\partial}{\partial x} f.$$

Here we use a special round letter ∂, instead of the ordinary d used in the differentiation of functions of one variable, in order to show that we are dealing with a function of several variables and differentiating with respect to one of them.

It is sometimes convenient to use Cauchy's symbol D, mentioned on p. 90, and write

$$\frac{\partial f}{\partial x} = D_x f;$$

but we shall seldom use this symbol.

In exactly the same way we define the partial derivative of $f(x, y)$ with respect to y at the point (x_0, y_0) by the relation

$$\lim_{k \to 0} \frac{f(x_0, y_0 + k) - f(x_0, y_0)}{k} = f_y(x_0, y_0) = \frac{\partial f}{\partial y} = D_y f(x_0, y_0).$$

This represents the slope of the curve of intersection of the surface $u = f(x, y)$ with the plane $x = x_0$ perpendicular to the x-axis.

Let us now think of the point (x_0, y_0), hitherto considered fixed, as variable, and accordingly omit the suffixes 0. In other words, we think of the differentiation as carried out at any point (x, y) of the region of definition of $f(x, y)$. Then the two derivatives are themselves functions of x and y:

$$u_x(x, y) = f_x(x, y) = \frac{\partial f(x, y)}{\partial x} \quad \text{and} \quad u_y(x, y) = f_y(x, y) = \frac{\partial f(x, y)}{\partial y}.$$

For example, the function $u = x^2 + y^2$ has the partial derivatives $u_x = 2x$ (in differentiating with respect to x the term y^2 is regarded as a constant and so has the derivative 0) and $u_y = 2y$. The partial derivatives of $u = x^3 y$ are $u_x = 3x^2 y$ and $u_y = x^3$.

We similarly make the following definition for any number (n) of independent variables

$$\frac{\partial f(x_1, x_2, \ldots, x_n)}{\partial x_1}$$

$$= \lim_{h \to 0} \frac{f(x_1 + h, x_2, \ldots, x_n) - f(x_1, x_2, \ldots, x_n)}{h}$$

$$= f_{x_1}(x_1, x_2, \ldots, x_n) = D_{x_1} f(x_1, x_2, \ldots, x_n),$$

it being assumed that the limit exists.

Of course we can also form *higher partial derivatives* of $f(x, y)$ by again differentiating the partial derivatives of the "first order", $f_x(x, y)$ and $f_y(x, y)$, with respect to one of the variables, and repeating this process. We indicate the order of differentiation by the order of the suffixes or by the order of the symbols ∂x and ∂y in the "denominator", from right to left,* and use the following symbols for the second partial derivatives:

$$\frac{\partial}{\partial x}\left(\frac{\partial f}{\partial x}\right) = \frac{\partial^2 f}{\partial x^2} = f_{xx} = D^2_{xx} f,$$

$$\frac{\partial}{\partial x}\left(\frac{\partial f}{\partial y}\right) = \frac{\partial^2 f}{\partial x \partial y} = f_{xy} = D^2_{xy} f,$$

$$\frac{\partial}{\partial y}\left(\frac{\partial f}{\partial x}\right) = \frac{\partial^2 f}{\partial y \partial x} = f_{yx} = D^2_{yx} f,$$

$$\frac{\partial}{\partial y}\left(\frac{\partial f}{\partial y}\right) = \frac{\partial^2 f}{\partial y^2} = f_{yy} = D^2_{yy} f.$$

We likewise denote the third partial derivatives by

$$\frac{\partial}{\partial x}\left(\frac{\partial^2 f}{\partial x^2}\right) = \frac{\partial^3 f}{\partial x^3} = f_{xxx},$$

$$\frac{\partial}{\partial y}\left(\frac{\partial^2 f}{\partial x^2}\right) = \frac{\partial^3 f}{\partial y \partial x^2} = f_{yxx},$$

$$\frac{\partial}{\partial x}\left(\frac{\partial^2 f}{\partial x \partial y}\right) = \frac{\partial^3 f}{\partial x^2 \partial y} = f_{xxy}, \&c.;$$

* In Continental usage, on the other hand, $\frac{\partial}{\partial x}\left(\frac{\partial f}{\partial y}\right)$ is written $\frac{\partial^2 f}{\partial y \partial x}$.

and in general the n-th derivatives by

$$\frac{\partial}{\partial x}\left(\frac{\partial^{n-1}f}{\partial x^{n-1}}\right) = \frac{\partial^n f}{\partial x^n} = f_{x^n},$$

$$\frac{\partial}{\partial y}\left(\frac{\partial^{n-1}f}{\partial x^{n-1}}\right) = \frac{\partial^n f}{\partial y\,\partial x^{n-1}} = f_{yx^{n-1}},\ \&c.$$

Finally, we shall study a few examples of the actual calculation of partial derivatives. According to the definition all the independent variables are to be kept constant except the one with respect to which we are differentiating. We therefore have merely to regard the other variables as constants and carry out the differentiation according to the rules by which we differentiate functions of a single independent variable.

Thus for example we have:

1. Function $\qquad f(x, y) = xy;$

first derivatives, $\qquad f_x = y,\ f_y = x;$

second derivatives, $\quad f_{xx} = 0,\ f_{xy} = f_{yx} = 1,\ f_{yy} = 0.$

2. Function $\qquad f(x, y) = \sqrt{(x^2 + y^2)};$

first derivatives, $\quad f_x = \dfrac{x}{\sqrt{(x^2 + y^2)}},\ f_y = \dfrac{y}{\sqrt{(x^2 + y^2)}}.$

(Thus for the radius vector $r = \sqrt{(x^2 + y^2)}$ from the origin to the point (x, y) the partial derivatives with respect to x and to y are given by the cosine, $\cos\varphi = x/r$, and the sine, $\sin\varphi = y/r$, of the angle φ, which the radius vector makes with the positive direction of the x-axis.)

Second derivatives,

$$f_{xx} = \frac{\sqrt{(x^2 + y^2)} - \dfrac{x^2}{\sqrt{(x^2 + y^2)}}}{x^2 + y^2} = \frac{y^2}{\sqrt{(x^2 + y^2)^3}} = \frac{\sin^2\varphi}{r},$$

$$f_{xy} = f_{yx} = -\frac{xy}{\sqrt{(x^2 + y^2)^3}} = -\frac{\sin\varphi\cos\varphi}{r},$$

$$f_{yy} = \frac{\sqrt{(x^2 + y^2)} - \dfrac{y^2}{\sqrt{(x^2 + y^2)}}}{x^2 + y^2} = \frac{x^2}{\sqrt{(x^2 + y^2)^3}} = \frac{\cos^2\varphi}{r}.$$

3. Reciprocal of the radius vector in three dimensions:

$$f(x, y, z) = \frac{1}{\sqrt{(x^2 + y^2 + z^2)}} = \frac{1}{r};$$

first derivatives,

$$f_x = -\frac{x}{\sqrt{(x^2 + y^2 + z^2)^3}} = -\frac{x}{r^3},$$

$$f_y = -\frac{y}{\sqrt{(x^2 + y^2 + z^2)^3}} = -\frac{y}{r^3},$$

$$f_z = -\frac{z}{\sqrt{(x^2 + y^2 + z^2)^3}} = -\frac{z}{r^3};$$

second derivatives,

$$f_{xx} = -\frac{1}{r^3} + \frac{3x^2}{r^5}, \quad f_{yy} = -\frac{1}{r^3} + \frac{3y^2}{r^5}, \quad f_{zz} = -\frac{1}{r^3} + \frac{3z^2}{r^5},$$

$$f_{xy} = f_{yx} = \frac{3xy}{r^5}, \quad f_{yz} = f_{zy} = \frac{3yz}{r^5}, \quad f_{zx} = f_{xz} = \frac{3zx}{r^5}.$$

From this we see that for the function $f = \dfrac{1}{\sqrt{(x^2 + y^2 + z^2)}}$ the equation

$$f_{xx} + f_{yy} + f_{zz} = -\frac{3}{r^3} + \frac{3(x^2 + y^2 + z^2)}{r^5} = 0$$

holds for all values of x, y, z except 0, 0, 0; as we say, the equation

$$f_{xx} + f_{yy} + f_{zz} = 0$$

is satisfied identically in x, y, z by the function $f(x, y, z) = 1/r$.

4. Function $\qquad f(x, y) = \dfrac{1}{\sqrt{y}} e^{-(x-a)^2/4y};$

first derivatives,

$$f_x = \frac{1}{\sqrt{y}} \frac{-(x - a)}{2y} e^{-(x-a)^2/4y},$$

$$f_y = \left(\frac{-1}{2y^{3/2}} + \frac{(x - a)^2}{4y^{5/2}} \right) e^{-(x-a)^2/4y};$$

second derivatives,

$$f_{xx} = \left(\frac{-1}{2y^{3/2}} + \frac{(x - a)^2}{4y^{5/2}} \right) e^{-(x-a)^2/4y},$$

$$f_{xy} = f_{yx} = \left(\frac{3}{4} \frac{x - a}{y^{5/2}} - \frac{(x - a)^3}{8y^{7/2}} \right) e^{-(x-a)^2/4y},$$

$$f_{yy} = \left(\frac{3}{4} \frac{1}{y^{5/2}} - \frac{3}{4} \frac{(x - a)^2}{y^{7/2}} + \frac{(x - a)^4}{16y^{9/2}} \right) e^{-(x-a)^2/4y}.$$

The equation

$$f_{xx} - f_y = 0$$

is therefore satisfied identically in x and y.

Just as in the case of one independent variable, the possession of derivatives is a special property * of a function. All the same, this property is possessed by all functions of practical importance, except perhaps at isolated exceptional points.

In contrast with functions of one variable, the possession of derivatives does *not* imply the continuity of the function. This is clearly shown by the example $u = \dfrac{2xy}{x^2 + y^2}$ already considered on pp. 464–465; for the partial derivatives exist everywhere, and yet the function is discontinuous at the origin. But, as is stated by the following theorem, the possession of *bounded* derivatives does imply continuity:

If a function f(x, y) *has partial derivatives* f_x *and* f_y *everywhere in a region* R, *and these derivatives everywhere satisfy the inequalities*

$$\left| f_x(x, y) \right| < M, \quad \left| f_y(x, y) \right| < M,$$

where M *is independent of* x *and* y, *then* f(x, y) *is continuous everywhere in* R.

In particular, if f_x and f_y are continuous they are necessarily bounded, so that $f(x, y)$ is also continuous.

The proof of this theorem we shall leave for Vol. II.

The reader will have noticed that in all our examples the equation $f_{xy} = f_{yx}$ is satisfied. In other words, it made no difference whether we differentiated first with respect to x and then with respect to y or vice versa. This is no accidental occurrence. In fact, we have the following theorem:

If the " mixed " partial derivatives f_{xy} *and* f_{yx} *of a function* f(x, y) *are continuous in a region* R, *then the equation*

$$f_{yx} = f_{xy}$$

holds everywhere in the interior of this region; that is, the order of differentiation with respect to x *and* y *is immaterial.*

---

* The expression " differentiable " implies *more* than that the partial derivatives with respect to x and to y exist. Cf. Vol. II.

By applying this theorem to f_x and f_y, then to f_{xx}, f_{xy}, f_{yy}, and so on, we find that

$$f_{xxy} = f_{xyx} = f_{yxx},$$
$$f_{xyy} = f_{yxy} = f_{yyx},$$
$$f_{xxyy} = f_{xyxy} = f_{xyyx} = f_{yxxy} = f_{yxyx} = f_{yyxx}, \text{&c.},$$

and in general we have the following result:

In repeated differentiation of a function of two variables the order of differentiation can be changed arbitrarily, provided only that the derivatives in question are continuous functions.

For the proof of this theorem we refer the reader to Vol. II.

EXAMPLES

1. Find the first partial derivatives of the following:

    (a) $\sqrt[3]{x^2 + y^2}$.   (d) $\dfrac{1}{\sqrt{(1 + x + y^2 + z^2)}}$.

    (b) $\sin(x^2 - y)$.   (e) $y \sin(xz)$.

    (c) e^{x-y}.   (f) $\log \sqrt{(1 + x^2 + y^2)}$.

2. Find all the first and second partial derivatives of the following:

    (a) xy.   (d) x^y.

    (b) $\log xy$.   (e) $e^{(x^y)}$.

    (c) $\tan(\text{arc} \tan x + \text{arc} \tan y)$.

3.* Find a function $f(x, y)$ which is a function of $(x^2 + y^2)$ and is also a product of the form $\psi(x)\psi(y)$; that is, solve the equations

$$f(x, y) = \varphi(x^2 + y^2) = \psi(x)\psi(y)$$

for the unknown functions.

4. The Chain Rule and the Differentiation of Inverse Functions

1. Functions of Functions (Compound Functions).

It often happens that a function u of the independent variables x, y is stated in the form

$$u = f(\xi, \eta, \dots),$$

where the arguments ξ, η, ... of the function f are themselves functions of x and y:

$$\xi = \phi(x, y), \quad \eta = \psi(x, y), \dots.$$

We then say that

$$u = f(\xi, \eta, \ldots) = f(\phi(x, y), \psi(x, y), \ldots) = F(x, y)$$

is given as a compound function of x and y.

For example, the function

$$u = e^{x^2 y}(x + y)^3$$

may be written as a compound function by means of the relations

$$u = e^\xi \eta^3 = f(\xi, \eta); \quad \xi = x^2 y, \eta = x + y.$$

Similarly, the function

$$u = \log(x + 1) \cdot \text{arc cos } \sqrt{(4 - x^2 - y^2)}$$

can be expressed in the form

$$u = \eta \text{ arc cos } \xi = f(\xi, \eta); \quad \xi = \sqrt{(4 - x^2 - y^2)}, \quad \eta = \log(x + 1).$$

In order to make this concept more precise, we assume to begin with that the functions $\xi = \phi(x, y)$, $\eta = \psi(x, y)$, . . . are defined in a certain region R of the independent variables x, y. Then to every point (x, y) of R there corresponds a point $(\xi, \eta, \ldots)$ in the space with co-ordinates $\xi, \eta, \ldots$. As the point (x, y) ranges over R, the point $(\xi, \eta, \ldots)$ will range over a certain set of values. We assume that the point $(\xi, \eta, \ldots)$ always lies within a region S in which $f(\xi, \eta, \ldots)$ is defined. The function $u = f(\phi(x, y), \psi(x, y), \ldots) = F(x, y)$ is then defined in the region R.

Referring to our examples, in the first we find that ξ and η are defined for every x, y and $f(\xi, \eta)$ is defined for every ξ, η, so that our region R can be taken to be the whole xy-plane. In the second example, however, the region S is restricted by the inequality $|\xi| \leq 1$, since for $|\xi| > 1$ the function arc cos ξ is undefined. Secondly, the region R is restricted by the inequalities $x + 1 > 0$ and $x^2 + y^2 \leq 4$, since for other values ξ and η are not both defined. Thirdly, the region R must be further limited by the inequality $3 \leq x^2 + y^2$ in order that the point with co-ordinates ξ, η shall fall in S; that is, the restriction $|\xi| \leq 1$ implies that $x^2 + y^2 \geq 3$. Hence R consists of the part of the ring $3 \leq x^2 + y^2 \leq 4$ lying to the right of the line $x = -1$.

The following theorem on compound functions is an immediate consequence of the definitions:

If the function $u = f(\xi, \eta, \ldots)$ *is continuous in* S *and the functions* $\xi = \phi(x, y)$, $\eta = \psi(x, y)$, . . . *are continuous in* R, *then*

16 •   (E 793)

the compound function $u = F(x, y)$ *is continuous in* R. The reader will be able to prove this for himself.

2. The Chain Rule.

We now turn our attention to compound functions of the type $u = f(\xi, \eta, \ldots)$, where $\xi, \eta, \ldots$ depend on the single variable x:

$$\xi = \phi(x), \quad \eta = \psi(x), \ldots.$$

For such functions we have the important theorem known as the *chain rule*:

If the function $u = f(\xi, \eta, \ldots)$ *has continuous partial derivatives of the first order in* S, *and the functions* $\xi = \phi(x)$, $\eta = \psi(x)$, *... have continuous first derivatives in the interval* R, $a \leq x \leq b$, *then* $u = f\{\phi(x), \psi(x), \ldots\} = F(x)$ *has a continuous derivative in* R, *and*

$$F'(x) = f_\xi \phi'(x) + f_\eta \eta'(x) + \ldots.$$

The right-hand side of this equation is an abbreviation for

$$f_\xi\{\phi(x), \psi(x), \ldots\} \phi'(x) + \ldots.$$

To simplify the notation we shall assume that f is a function of the three arguments ξ, η, ζ. We shall denote by x_0 an arbitrary fixed point of the interval $a \leq x \leq b$, by ξ_0, η_0, ζ_0 the corresponding values $\xi_0 = \phi(x_0)$, $\eta_0 = \psi(x_0)$, $\zeta_0 = \chi(x_0)$, and by ξ, η, ζ the values $\phi(x), \psi(x), \chi(x)$ corresponding to a variable point $x = x_0 + h$. We first write down the identity

$$
\begin{aligned}
F(x) &- F(x_0) \\
&= f(\xi, \eta, \zeta) - f(\xi_0, \eta_0, \zeta_0) \\
&= \{f(\xi, \eta, \zeta) - f(\xi_0, \eta, \zeta)\} + \{f(\xi_0, \eta, \zeta) - f(\xi_0, \eta_0, \zeta)\} \\
&\qquad + \{f(\xi_0, \eta_0, \zeta) - f(\xi_0, \eta_0, \zeta_0)\}.
\end{aligned}
$$

In each bracket on the right we observe that only one of the independent variables changes its value. Hence to each bracket we can apply the mean value theorem for functions of one variable, and obtain

$$
\begin{aligned}
F(x) &- F(x_0) \\
&= (\xi - \xi_0) f_\xi(\bar{\xi}, \eta, \zeta) + (\eta - \eta_0) f_r(\xi_0, \bar{\eta}, \zeta) + (\zeta - \zeta_0) f_\zeta(\xi_0, \eta_0, \zeta),
\end{aligned}
$$

where $\bar{\xi}$ lies between ξ_0 and ξ, $\bar{\eta}$ between η_0 and η, and $\bar{\zeta}$ between ζ_0 and ζ. Further, by the mean value theorem, we have

$$\xi - \xi_0 = \phi(x) - \phi(x_0) = (x - x_0)\,\phi'(x_1),$$
$$\eta - \eta_0 = \psi(x) - \psi(x_0) = (x - x_0)\,\psi'(x_2),$$
$$\zeta - \zeta_0 = \chi(x) - \chi(x_0) = (x - x_0)\,\chi'(x_3),$$

where x_1, x_2, and x_3 all lie between x_0 and x. Substituting these values in the last equation and dividing by $x - x_0$, we have

$$\frac{F(x) - F(x_0)}{x - x_0}$$
$$= f_\xi(\bar{\xi}, \eta, \zeta)\phi'(x_1) + f_\eta(\xi_0, \bar{\eta}, \zeta)\psi'(x_2) + f_\zeta(\xi_0, \eta_0, \bar{\zeta})\chi'(x_3).$$

We now let x tend to x_0. Owing to the continuity of $\phi(x)$, $\psi(x)$, $\chi(x)$ the quantities ξ, η, ζ tend to ξ_0, η_0, ζ_0 respectively, and a fortiori $\bar{\xi}$, $\bar{\eta}$, $\bar{\zeta}$ do likewise. Also x_1, x_2, and x_3 tend to x_0. Since all the functions on the right are continuous, we have

$$\lim_{x \to x_0} \frac{F(x) - F(x_0)}{x - x_0} = F'(x_0)$$
$$= f_\xi(\xi_0, \eta_0, \zeta_0)\phi'(x_0) + f_\eta(\xi_0, \eta_0, \zeta_0)\psi'(x_0) + f_\zeta(\xi_0, \eta_0, \zeta_0)\chi'(x_0),$$

thus establishing the formula for $F'(x)$.

The continuity of $F'(x)$ follows immediately from the formula, since ϕ', ψ', and χ' are continuous by hypothesis and f_ξ, f_η, and f_ζ are continuous functions of continuous functions.

This theorem may be extended to compound functions of two or more variables, as follows:

If the function $u = f(\xi, \eta, \ldots)$ *has continuous partial derivatives of the first order in the region* S, *and the functions* $\xi = \phi(x, y)$, $\eta = \psi(x, y), \ldots$ *have continuous partial derivatives of the first order in* R, *then* $u = F(x, y) = f\{\phi(x, y), \psi(x, y), \ldots\}$ *has continuous partial derivatives of the first order in* R, *and these derivatives are given by the formulæ*

$$F_x = f_\xi \phi_x + f_\eta \psi_x + \ldots,$$
$$F_y = f_\xi \phi_y + f_\eta \psi_y + \ldots.$$

These formulæ are often written in the abbreviated form

$$u_x = u_\xi \xi_x + u_\eta \eta_x + \ldots,$$
$$u_y = u_\xi \xi_y + u_\eta \eta_y + \ldots.$$

To establish them we temporarily introduce the notation $g(x) = \phi(x, y_0)$, $h(x) = \psi(x, y_0)$, ..., where y_0 is a fixed value of y. By the definition of the partial derivatives it follows that $g'(x) = \phi_x(x, y_0)$, $h'(x) = \psi_x(x, y_0)$, ... Similarly, if we write $H(x) = F(x, y_0)$, we have $H'(x) = F_x(x, y_0)$. We now apply the theorem just proved to the function $u = H(x) = f(\xi, \eta, ...)$ $= f\{g(x), h(x), ...\}$, and obtain

$$H'(x_0) = f_\xi\, g'(x_0) + f_\eta h'(x_0) + \ldots.$$

Returning to the original symbols, we have

$$F_x(x_0, y_0) = f_\xi \phi_x(x_0, y_0) + f_\eta \psi_x(x_0, y_0) + \ldots.$$

The other formula is proved in a similar way.

If we wish to calculate the derivatives of higher order we need only differentiate the right-hand side of our formulæ again with respect to x and y, regarding $f_\xi, f_\eta, \ldots$ as compound functions. Thus for $u = f(\xi, \eta) = f\{\phi(x, y), \psi(x, y)\}$, we have

$$u_{xx} = f_{\xi\xi}\phi_x{}^2 + 2f_{\xi\eta}\phi_x\psi_x + f_{\eta\eta}\psi_x{}^2 + f_\xi\phi_{xx} + f_\eta\psi_{xx},$$
$$u_{xy} = f_{\xi\xi}\phi_x\phi_y + f_{\xi\eta}(\phi_x\psi_y + \phi_y\psi_x) + f_{\eta\eta}\psi_x\psi_y + f_\xi\phi_{xy} + f_\eta\psi_{xy},$$
$$u_{yy} = f_{\xi\xi}\phi_y{}^2 + 2f_{\xi\eta}\phi_y\psi_y + f_{\eta\eta}\psi_y{}^2 + f_\xi\phi_{yy} + f_\eta\psi_{yy}.$$

3. Examples.*

1. $u = e^{x \tan y + y \cos x}$.

Here we put $\xi = x \tan y$, $\eta = y \cos x$, so that $\xi_x = \tan y$, $\xi_y = \dfrac{x}{\cos^2 y}$, $\eta_x = -y \sin x$, $\eta_y = \cos x$. Since $u = e^{\xi+\eta}$, $u_\xi = u_\eta = e^{\xi+\eta}$, and

$$u_x = e^{x \tan y + y \cos x}(\tan y - y \sin x),$$
$$u_y = e^{x \tan y + y \cos x}\left(\frac{x}{\cos^2 y} + \cos x\right).$$

2. An example of a compound function of a single variable is

$$u = \{g(x)\}^{h(x)} = \xi^\eta = f(\xi, \eta),$$

where we put $\xi = g(x)$, $\eta = h(x)$. We immediately obtain

$$\frac{du}{dx} = f_\xi \xi' + f_\eta \eta' = \eta\xi^{\eta-1}\xi' + \xi^\eta \log \xi \cdot \eta'$$
$$= \{g(x)\}^{h(x)}\left\{h(x)\frac{g'(x)}{g(x)} + h'(x) \log g(x)\right\}.$$

We have already dealt with a special case of this by rather artificial methods (p. 203).

* We would emphasize that the following differentiations can also be carried out directly, without using the chain rule.

4. Change of the Independent Variables.

A particularly important type of compound function occurs in the process of changing the independent variables. For example, let $u = f(\xi, \eta)$ be a function of ξ and η, which we interpret as rectangular co-ordinates in the $\xi\eta$-plane. If we rotate the axes in the $\xi\eta$-plane through an angle θ we obtain a new system of co-ordinates x, y, related to the co-ordinates ξ, η by the equations:

$$\xi = x \cos\theta - y \sin\theta, \qquad \eta = x \sin\theta + y \cos\theta,$$

or $\qquad x = \xi \cos\theta + \eta \sin\theta, \qquad y = -\xi \sin\theta + \eta \cos\theta.$

The function $u = f(\xi, \eta)$ can then be expressed as a function of the new variables x, y:

$$u = f(\xi, \eta) = F(x, y).$$

Then the chain rule immediately gives

$$u_x = u_\xi \cos\theta + u_\eta \sin\theta, \quad u_y = -u_\xi \sin\theta + u_\eta \cos\theta.$$

Thus the partial derivatives are transformed by the same formulæ as the independent variables. This is true for rotation of the axes in space also.

Another important type of change of co-ordinates is the change from rectangular co-ordinates x, y to polar co-ordinates r, θ. This is done by means of the equations

$$x = r \cos\theta, \qquad y = r \sin\theta,$$

$$r = \sqrt{(x^2 + y^2)}, \qquad \theta = \arctan\frac{y}{x}.$$

We then find that for an arbitrary function $u = f(x, y)$ with continuous partial derivatives of the first order we have

$$u = f(x, y) = f(r \cos\theta, r \sin\theta) = F(r, \theta),$$

$$u_x = u_r r_x + u_\theta \theta_x = u_r \frac{x}{r} - u_\theta \frac{y}{r^2} = u_r \cos\theta - u_\theta \frac{\sin\theta}{r},$$

$$u_y = u_r r_y + u_\theta \theta_y = u_r \frac{y}{r} + u_\theta \frac{x}{r^2} = u_r \sin\theta + u_\theta \frac{\cos\theta}{r}.$$

From this we obtain the equation

$$u_x{}^2 + u_y{}^2 = u_r{}^2 + \frac{1}{r^2}u_\theta{}^2,$$

which is often useful.

In general, let us consider a pair of functions $\xi = \phi(x, y)$, $\eta = \psi(x, y)$ which are continuous and have continuous derivatives in a region R of the xy-plane. To each point (x, y) in R these equations assign a point $\xi = \phi(x, y)$, $\eta = \psi(x, y)$ in the $\xi\eta$-plane. As (x, y) ranges over R, the corresponding point (ξ, η) will range over some set of values S in the $\xi\eta$-plane. It is of course possible that several distinct points (x, y) will give the same values for ξ, η, so that to several points (x, y) there corresponds only one point (ξ, η). We shall assume that this is *not* the case, but instead that to one point $Q(\xi, \eta)$ in S there corresponds *exactly* one point $P(x, y)$ in R. We may therefore look at the correspondence from either point of view—saying that Q corresponds to P or that P corresponds to Q. The latter point of view can be expressed thus: to each point (ξ, η) in S there corresponds one x and one y, namely, the co-ordinates of P, or, in equations, there are two functions $x = g(\xi, \eta)$, $y = h(\xi, \eta)$, defined in S, which represent the correspondence inverse to $\xi = \phi(x, y)$, $\eta = \psi(x, y)$.

It often happens that the functions $g(\xi, \eta)$, $h(\xi, \eta)$ are by no means easy to calculate, even when they do exist. Hence we shall now find how to obtain the partial derivatives g_ξ, g_η, h_ξ, h_η directly from the partial derivatives ϕ_x, ϕ_y, ψ_x, ψ_y, without calculating g and h themselves at all. For this purpose we observe that if we choose any point $Q(\xi, \eta)$, find the corresponding point $P\{g(\xi, \eta), h(\xi, \eta)\}$ in R, and then find the point in S corresponding to P, which is $\phi\{g(\xi, \eta), h(\xi, \eta)\}$, $\psi\{g(\xi, \eta), h(\xi, \eta)\}$, we have simply returned to the point Q. That is, the equations $\xi = \phi\{g(\xi, \eta), h(\xi, \eta)\}$, $\eta = \psi\{g(\xi, \eta), h(\xi, \eta)\}$ are identities in ξ and η. We now differentiate * both sides of both equations with respect to ξ and to η. We have

$$1 = \phi_x g_\xi + \phi_y h_\xi, \quad 0 = \phi_x g_\eta + \phi_y h_\eta,$$
$$0 = \psi_x g_\xi + \psi_y h_\xi, \quad 1 = \psi_x g_\eta + \psi_y h_\eta.$$

---

* If an equation expresses an *identical* relationship, differentiation with respect to any independent variable in it yields an identity, as follows immediately from the definition.

Solving these equations, we find that

$$g_\xi = \frac{\psi_v}{D}, \quad g_\eta = -\frac{\phi_v}{D}, \quad h_\xi = -\frac{\psi_x}{D}, \quad h_\eta = \frac{\phi_x}{D},$$

or

$$x_\xi = \frac{\eta_v}{D}, \quad x_\eta = -\frac{\xi_v}{D}, \quad y_\xi = -\frac{\eta_x}{D}, \quad y_\eta = \frac{\xi_x}{D},$$

where by D we mean the determinant

$$D = \xi_x \eta_v - \xi_v \eta_x = \begin{vmatrix} \dfrac{\partial \xi}{\partial x} & \dfrac{\partial \xi}{\partial y} \\ \dfrac{\partial \eta}{\partial x} & \dfrac{\partial \eta}{\partial y} \end{vmatrix},$$

which we assume is not zero.

This determinant D, called the *functional determinant* or *Jacobian* of (ξ, η) with respect to (x, y), occurs so frequently that a special symbol is often used for it:

$$D = \frac{\partial(\xi, \eta)}{\partial(x, y)}.$$

EXAMPLES

1. Calculate the partial derivatives of the first order for:

(a) $f = \dfrac{1}{\sqrt{(x^2 + y^2 + 2xy \cos z)}}.$ (c) $f = x^2 + y \log(1 + x^2 + y^2 + z^2).$

(b) $f = \arcsin \dfrac{x}{z + y^2}.$ (d) $f = \arctan \sqrt{(x + yz)}.$

2. Calculate the derivatives of (a) $f = x^{(x^z)}$, (b) $f = \left(\left(\dfrac{1}{x} \right)^{1/x} \right)^{1/x}.$

3. Prove that if $f(x, y)$ satisfies " Laplace's equation "

$$\frac{\partial^2 f}{\partial x^2} + \frac{\partial^2 f}{\partial y^2} = 0,$$

so does $\varphi(x, y) = f\left(\dfrac{x}{x^2 + y^2}, \dfrac{y}{x^2 + y^2} \right).$

4. Prove that the functions

(a) $f(x, y) = \log \sqrt{(x^2 + y^2)}.$ (b) $g(x, y, z) = \dfrac{1}{\sqrt{(x^2 + y^2 + z^2)}}.$

(c) $h(x, y, z, w) = \dfrac{1}{x^2 + y^2 + z^2 + w^2},$

satisfy the respective Laplace's equations:

(a) $f_{xx} + f_{yy} = 0.$ (b) $g_{xx} + g_{yy} + g_{zz} = 0.$

(c) $h_{xx} + h_{yy} + h_{zz} + h_{ww} = 0.$

5. Given $z = r^2 \cos \theta$, where r and θ are polar co-ordinates, find z_x and z_y at the point $\theta = \dfrac{\pi}{4}, r = 2.$

Express z_r and z_θ in terms of z_x and z_y.

6. By the transformation $\xi = a + \alpha x + \beta y, \; \eta = b - \beta x + \alpha y,$ in which a, b, α, β are constants and $\alpha^2 + \beta^2 = 1$, the function $u(x, y)$ is transformed into a function $U(\xi, \eta)$ of ξ and η. Prove that

$$U_{\xi\xi} U_{\eta\eta} - U_{\xi\eta}{}^2 = u_{xx} u_{yy} - u_{xy}{}^2.$$

7. Find the Jacobians of the following transformations:

(a) $\xi = ax + by, \; \eta = cx + dy;$ (b) $r = \sqrt{(x^2 + y^2)}, \; \theta = \arctan\dfrac{y}{x};$

(c) $\xi = x^2, \; \eta = y^2.$

8. If $x = x(u, v), y = y(u, v)$ and $u = u(\xi, \eta), v = v(\xi, \eta)$, prove that

$$\frac{\partial(x, y)}{\partial(\xi, \eta)} = \frac{\partial(x, y)}{\partial(u, v)} \cdot \frac{\partial(u, v)}{\partial(\xi, \eta)}.$$

9. As a corollary to Ex. 8, prove that

$$\frac{\partial(x, y)}{\partial(u, v)} = \frac{1}{\dfrac{\partial(u, v)}{\partial(x, y)}}.$$

10. Using Ex. 9, find the Jacobians of the transformations which are the inverses of those in Ex. 7.

5. IMPLICIT FUNCTIONS

In the study of functions of several variables we have as yet had no analogue to the inverse function. We can regard the inverse function of $y = f(x)$ as the function obtained if we solve the equation $y - f(x) = 0$ for x. In this section we shall seek more generally to solve equations $F(x, y) = 0$ for x or for y, and to discuss functions of several variables in a corresponding way.

Even in elementary analytical geometry curves are frequently represented, not by equations $y = f(x)$ or $x = \phi(y)$, but by an equation involving x and y in the form $F(x, y) = 0$. For example, we have the circle $x^2 + y^2 - 1 = 0$, the ellipse $\dfrac{x^2}{a^2} + \dfrac{y^2}{b^2} - 1 = 0$, and the lemniscate $(x^2 + y^2)^2 - 2a^2(x^2 - y^2) = 0$. In order to obtain y as a function of x, or x as a function of y, we must solve the equation for y or for x. We then say that the function $y = f(x)$ or $x = \phi(y)$ found in this way is defined *implicitly* by the equation $F(x, y) = 0$, and that the solution of this equation gives us the function *explicitly*. In the examples cited and in many others the solution can be carried out and the solutions stated explicitly in terms of the elementary functions. In other cases the solution can be obtained in terms of an infinite series or other limiting process; that is, we can approximate to the solution $y = f(x)$ or $x = \phi(y)$ as closely as we desire.

For many purposes, however, it is more convenient to base our discussion on the implicit definition $F(x, y) = 0$, instead of resorting to an exact or approximate solution of the equation.

The idea that every function $F(x, y)$ yields a function $y = f(x)$ or $x = \phi(y)$ given implicitly by means of the equation $F(x, y) = 0$ is erroneous. On the contrary, it is easy to give examples of functions $F(x, y)$ which, when equated to zero, permit of no solution in terms of functions of one variable. Thus for example the equation $x^2 + y^2 = 0$ is satisfied by the single pair of values $x = 0$, $y = 0$ only, while the equation $x^2 + y^2 + 1 = 0$ is satisfied by no (real) values at all. It is therefore necessary to investigate the matter more closely in order to find out whether an equation $F(x, y) = 0$ can actually be solved, and what properties the solution has. Such an investigation we cannot undertake in detail here, but content ourselves with a geometrical interpretation which suggests the required results, the rigorous proofs being left for Volume II.

1. Geometrical Interpretation of Implicit Functions.

To discuss this problem geometrically we represent the function $u = F(x, y)$ by a surface in three-dimensional space. Finding values (x, y) which satisfy the equation $F(x, y) = 0$ is the same thing as finding values (x, y) which satisfy two equations $F(x, y) = u$, $u = 0$; in other words, we wish to find the intersection of the surface $u = F(x, y)$ and the plane $u = 0$, which is the xy-plane. We then suppose that we have a definite point (x_0, y_0) which satisfies the equation $F(x_0, y_0) = 0$; that is, at

(x_0, y_0) the surface $u = F(x, y)$ has a point in common with the plane $u = 0$. (If no such point exists, there is no intersection, and the equation $F(x, y) = 0$ cannot be solved.) If the tangent plane to the surface $u = F(x, y)$ at the point (x_0, y_0) is *not horizontal*, it cuts the plane $u = 0$ in a single straight line. Intuition then tells us that the surface $u = F(x, y)$, lying near the tangent plane, likewise cuts the plane $u = 0$ in a single well-defined curve. How *far* this curve extends does not at present concern us. The tangent plane will be horizontal if the two curves $u = F(x_0, y)$ and $u = F(x, y_0)$ both have horizontal tangent lines at (x_0, y_0); that is, if $F_x(x_0, y_0) = 0$ and $F_y(x_0, y_0) = 0$. Thus if either $F_x(x_0, y_0) \neq 0$ or $F_y(x_0, y_0) \neq 0$ the tangent plane is not horizontal, and, as we have just seen, we may expect that a solution in the form $y = f(x)$ or $x = \phi(y)$ will exist.

If, on the other hand, both $F_x(x_0, y_0)$ and $F_y(x_0, y_0)$ have the value 0, we readily see that there is no guarantee that the solution is possible.

For example, for $F = 1 - \sqrt{(1 - x^2 - y^2)}$ the corresponding spherical surface $u = 1 - \sqrt{(1 - x^2 - y^2)}$ has the point $(0, 0)$ in common with the xy-plane. The partial derivatives $F_x(0, 0)$ and $F_y(0, 0)$ are both zero; and we find that no point other than $(0, 0)$ satisfies the equation $F = 0$. For the function $F(x, y) = xy$ we find that $F(0, 0) = 0$, while $F_x(0, 0) = F_y(0, 0) = 0$. Here all the points on the x-axis and all the points on the y-axis satisfy the equation $F(x, y) = 0$; and in the neighbourhood of the origin we have no unique solution $x = \varphi(y)$ or $y = f(x)$. Thus we see that when $F_x(x_0, y_0) = F_y(x_0, y_0) = 0$ we cannot be sure that a solution exists.

If we accordingly return to the case in which one of the partial derivatives—say $F_y(x_0, y_0)$, to be specific—is not zero, the graphical suggestion that a smooth surface should be cut by a non-tangent plane in a smooth curve leads us to suspect that the following theorem is true:

If the function $F(x, y)$ *has continuous derivatives* F_x *and* F_y *and if at the point* (x_0, y_0) *the equation* $F(x_0, y_0) = 0$ *is satisfied, while* $F_y(x_0, y_0)$ *is not zero, then we can mark off about the point* (x_0, y_0) *a rectangle* $x_1 \leq x \leq x_2$, $y_1 \leq y \leq y_2$ *such that for every* x *in the interval* $x_1 \leq x \leq x_2$ *the equation* $F(x, y) = 0$ *determines just one value* $y = f(x)$ *lying in the interval* $y_1 \leq y \leq y_2$. *This function* $y = f(x)$ *satisfies the equation* $y_0 = f(x_0)$, *and the equation*

$$F\{x, f(x)\} = 0$$

is satisfied for every x *in the interval. Moreover, the function* y = f(x) *is continuous and has a continuous derivative.*

This can actually be rigorously proved, and will be proved in Vol. II. Assuming it to be true, we can add the following:

The derivative of the function y = f(x) *is given by the equation*

$$y' = f'(x) = -\frac{F_x}{F_y}.$$

This follows immediately by using the chain rule. For $\frac{d}{dx} F\{x, f(x)\} = F_x \frac{dx}{dx} + F_y \frac{df}{dx} = F_x + F_y f'$. But since $F\{x, f(x)\}$ is identically zero, its derivative is also zero; hence $F_x + F_y f' = 0$, and the formula is established.

If we regard the right-hand side of the formula as a compound function of x and differentiate according to the chain rule, replacing y' by $-F_x/F_y$, we have

$$y'' = -\frac{F_y(F_{xx} + F_{yx}y') - F_x(F_{xy} + F_{yy}y')}{F_y^2}$$

$$= -\frac{F_{xx}F_y^2 - 2F_{xy}F_xF_y + F_{yy}F_x^2}{F_y^3}.$$

Continuing the process, we may calculate y''', y^{iv}, &c.

By using this formula we can usually find the derivative of a function given in implicit form much more easily than by solving first and then differentiating.

For example, for the circle

$$F(x, y) = x^2 + y^2 - 1 = 0$$

we have

$$y' = -\frac{F_x}{F_y} = -\frac{x}{y}.$$

This is easily verified. For on solving the equation of the circle for y we obtain two solutions, namely, $y = \sqrt{(1 - x^2)}$ and $y = -\sqrt{(1 - x^2)}$, giving the upper and lower semicircles respectively. For the upper we have

$$y' = \frac{-x}{\sqrt{(1 - x^2)}},$$

for the lower

$$y' = \frac{+x}{\sqrt{(1 - x^2)}},$$

so that in either case $y' = -\frac{x}{y}$.

As another example, we have $F(x, y) = e^{x+y} + y - x = 0$. We find that $F_x\left(\dfrac{1}{2}, -\dfrac{1}{2}\right) = 0$, while $F_y\left(\dfrac{1}{2}, -\dfrac{1}{2}\right) = 2$. Thus the equation has a solution $y = f(x)$; but the actual explicit calculation of the function $f(x)$ would not be simple. Nevertheless, we have

$$y' = -\frac{F_x}{F_y} = -\frac{e^{x+y} - 1}{e^{x+y} + 1}.$$

In order that the function $f(x)$ may have a maximum or a minimum we must have $y' = 0$, that is, $e^{x+y} - 1 = 0$, whence $y = -x$. Substitution of $y = -x$ in the equation $F(x, y) = 0$ gives $1 - 2x = 0$, whence $x = \dfrac{1}{2}$, $y = -\dfrac{1}{2}$. If we calculate $f''(x)$ for $x = \dfrac{1}{2}$, we find it to be negative, so that $-\dfrac{1}{2}$ is the maximum value of y.

An extension of this theorem for implicit functions to functions of a greater number of independent variables readily suggests itself. The extension is as follows:

Let $F(x, y, \ldots, z, u)$ *be a continuous function of the independent variables* $x, y, \ldots, z, u$ *with continuous partial derivatives* $F_x, F_y, \ldots, F_z, F_u$. *For the system of values* $(x_0, y_0, \ldots, z_0, u_0)$ *let* $F(x_0, y_0, \ldots, z_0, u_0) = 0$ *and* $F_u(x_0, y_0, \ldots, z_0, u_0) \neq 0$. *Then we can mark off an interval* $u_1 \leqq u \leqq u_2$ *about* u_0 *and a region* R *containing* $(x_0, y_0, \ldots, z_0)$ *such that for every* $(x, y, \ldots, z)$ *in* R *the equation* $F(x, y, \ldots, z, u) = 0$ *is satisfied by just one value of* u *in the interval* $u_1 \leqq u \leqq u_2$. *This value of* u, *which we denote by* $u = f(x, y, \ldots, z)$, *is a continuous function of* $x, y, \ldots, z$ *and possesses continuous partial derivatives* $f_x, f_y, \ldots, f_z$, *and*

$$u_0 = f(x_0, y_0, \ldots, z_0).$$

The derivatives of f *are given by the equations*

$$F_x + F_u f_x = 0,$$
$$F_y + F_u f_y = 0,$$
$$\cdots \cdots \cdots$$
$$F_z + F_u f_z = 0.$$

For the proof of the existence and continuity of u we again refer the reader to Vol. II. The formulæ for f_{xy}, &c., follow immediately from the chain rule.

Incidentally, the concept of an implicit function enables us to give a general definition of the term " algebraic function ".

We say that $u = f(x, y, \ldots, z)$ is an *algebraic function* of the independent variables $x, y, \ldots, z$ if u can be defined implicitly by an equation $F(x, y, \ldots, z, u) = 0$, where F is a polynomial in $x, y, \ldots, z, u$; that is, if u satisfies an " algebraic equation". Functions which do not satisfy any algebraic equation are called *transcendental* (p. 24).

As an example of our differentiation formula we consider the ellipsoid

$$\frac{x^2}{a^2} + \frac{y^2}{b^2} + \frac{u^2}{c^2} - 1 = 0.$$

For the partial derivatives we have

$$u_x = -\frac{2x}{a^2} \cdot \frac{c^2}{2u} = -\frac{c^2}{a^2} \cdot \frac{x}{u},$$

$$u_y = -\frac{2y}{b^2} \cdot \frac{c^2}{2u} = -\frac{c^2}{b^2} \cdot \frac{y}{u};$$

and by differentiating again

$$u_{xx} = -\frac{c^2}{a^2} \cdot \frac{1}{u} + \frac{c^2}{a^2} \cdot \frac{x}{u^2} u_x = -\frac{c^2 a^2 u^2 + c^4 x^2}{a^4 u^3},$$

$$u_{xy} = +\frac{c^2}{a^2} \cdot \frac{x}{u^2} u_y = -\frac{c^4}{a^2 b^2} \frac{xy}{u^3},$$

$$u_{yy} = -\frac{c^2}{b^2} \cdot \frac{1}{u} + \frac{c^2}{b^2} \cdot \frac{y}{u^2} u_y = -\frac{c^2 b^2 u^2 + c^4 y^2}{b^4 u^3}.$$

EXAMPLES

1. Prove that the following equations have unique solutions for y near the points indicated:

(a) $x^2 + xy + y^2 = 7$ $(2, 1)$.

(b) $x \cos xy = 0$ $\left(1, \dfrac{\pi}{2}\right)$.

(c) $xy + \log xy = 1$ $(1, 1)$.

(d) $x^5 + y^5 + xy = 3$ $(1, 1)$.

2. Find the first derivatives of the solutions in Ex. 1.

3. Find the second derivatives of the solutions in Ex. 1.

4. Find the maximum and minimum values of the function $y = f(x)$ defined by the equation $x^2 + xy + y^2 = 27$.

5. Show that the equation $x + y + z = \sin xyz$ can be solved for z near $(0, 0, 0)$. Find the partial derivatives of the solution.

6. Multiple and Repeated Integrals

1. Multiple Integrals.

We consider a function $u = f(x, y)$ which is defined and continuous in a rectangle $R(a \leq x \leq b,\ c \leq y \leq d)$, and which takes positive values only. We wish to assign a volume to the portion of three-dimensional space bounded by the rectangle R, the surface $u = f(x, y)$, and the four planes $x = a$, $x = b$, $y = c$, $y = d$ perpendicular to the xy-plane. Moreover, the volume should be defined so as to satisfy certain elementary conditions: (1) if the three-dimensional region is a prism—i.e. if the function u is a constant k—the volume should be the product of the base by the altitude, $V = (b - a)(d - c)k$;

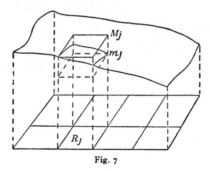

Fig. 7

(2) if we divide the rectangle R into smaller rectangles R_1 and R_2 by drawing straight lines, then the volume over R should be equal to the volume over R_1 plus the volume over R_2; (3) if the three-dimensional region R_1 completely includes R_2, the volume of R_1 should be at least as great as that of R_2.

These considerations lead us to a method of defining V which is an immediate extension of the method of defining area in Chap. II (p. 77 *et seq.*). By constructing lines parallel to the sides we subdivide the rectangle R into smaller rectangles $R_1, R_2, \ldots,$ R_n, whose areas we denote by $\Delta R_1, \Delta R_2, \ldots, \Delta R_n$. In each rectangle R_j the function has a least value m_j and a greatest value M_j. Therefore a prism whose base is R_j and whose height is M_j completely includes the portion of our region over R_j, while this portion of the region contains the prism with base R_j and height m_j (cf. fig. 7). We see, therefore, that the volume of the portion in question lies between m_jR, and M_jR_j. Thus the total volume V should be such that

$$\sum_{=1}^{n} m_j \Delta R_j \leqq V \leqq \sum_{j=1}^{n} M_j \Delta R_j.$$

Suppose now that the number n of rectangles increases beyond all bounds in such a way that the length of the longest diagonal tends to zero. Intuition leads us to expect that the two sums $\Sigma m_j \Delta R_j$ and $\Sigma M_j \Delta R_j$ will both converge and will tend to the same limit. This limit we therefore call the *volume V*.

The reader will have observed that we have carried out an immediate generalization of the discussion in Chap. II (p. 78). As in Chap. II, we call the common limit of the sums $\Sigma m_j \Delta R_j$ and $\Sigma M_j \Delta R_j$ *the integral of the function* u = f(x, y) *over the rectangle* R, and we denote it by the symbol

$$\iint_R f(x, y)\, dr.$$

It is at once clear that if in each rectangle R_j we choose a point (ξ_j, η_j) and find the corresponding value of the function $f(\xi_j, \eta_j)$ then the limiting relation

$$\lim_{n \to \infty} \Sigma f(\xi_j, \eta_j) \Delta R_j = \iint_R f(x, y)\, dr$$

must hold; for the sum $\Sigma f(\xi_j, \eta_j) \Delta R_j$ lies between $\Sigma m_j \Delta R_j$ and $\Sigma M_j \Delta R_j$, both of which approach the integral as a limit.

As a particular method of subdividing R into smaller rectangles, we may divide the side $a \leqq x \leqq b$ into n intervals of length $\Delta x = (b - a)/n$ and the side $c \leqq y \leqq d$ into m intervals of length $\Delta y = (d - c)/m$, and then draw parallels to the axes through the points of division thus marked. The area of each rectangle R_j is then $\Delta R_j = \Delta x \Delta y$. Choosing a point (ξ_j, η_j) arbitrarily in each rectangle R_j, we form the sum

$$\Sigma_j f(\xi_j, \eta_j) \Delta R_j = \Sigma_j f(\xi_j, \eta_j) \Delta x \Delta y.$$

As n and m both increase without limit, this sum approaches the integral as a limit. This type of subdivision suggests a second notation for the integral, which has been in common use since the time of Leibnitz, namely,

$$\iint_R f(x, y)\, dx\, dy.$$

The proof that such a limit exists if $u = f(x, y)$ is continuous can be carried out as in the appendix to Chap. II (p. 131 *et seq.*). We shall, however, assume without proof an even stronger statement, namely, the following:

If the function f(x, y) *is continuous except along a finite number of smooth* * *curves* y = f(x) *or* x = ϕ(y) *along which* f(x, y) *has jump discontinuities, then the double integral*

$$\iint_R f(x, y) \, dr$$

exists.

The proof of this we leave for Vol. II. It depends essentially on the fact that as the number of rectangles increases the total area of the rectangles having points in common with the curves of discontinuity tends to zero. Thus even though M_j and m_j may differ considerably for such rectangles, they give rise to little difference between the sums $\Sigma M_j \Delta R_j$ and $\Sigma m_j \Delta R_j$.

With this assumption we can find the area under surfaces $u = f(x, y)$ for which (x, y) ranges over quite complicated regions R. For suppose that the region R is bounded by a finite number of curves $x = \phi(y)$ or $y = \psi(x)$ with continuous derivatives, and that $f(x, y)$ is continuous in R. We enclose R in a rectangle R', and at the points of R' which do not belong to R we assign to $f(x, y)$ the value 0. Then we take the integral $\iint_{R'} f(x, y) \, dr$, taken over the region R', as the volume under the surface $u = f(x, y)$, where (x, y) is in R. This integral is usually denoted by $\iint_R f(x, y) \, dr$.

Certain simple but important theorems relating to these double integrals follow directly from the definition. Here we simply state the theorems; the reader will be able to prove them without any trouble.

If f(x, y) *and* g(x, y) *are integrable over a rectangle, then so are* f $\pm$ g, *and* cf, *where* c *is a constant:*

$$\iint_R \{f(x, y) \pm g(x, y)\} \, dr = \iint_R f(x, y) \, dr \pm \iint_R g(x, y) \, dr,$$

$$\iint_R c f(x, y) \, dr = c \iint_R f(x, y) \, dr.$$

---

* By smooth curves we mean, as before, curves with continuous derivatives.

If $f(x, y) \geqq g(x, y)$ *in R, then*

$$\iint_R f(x, y)\, dr \geqq \iint_R g(x, y)\, dr.$$

I, R is the sum of two regions R_1 *and* R_2, *then*

$$\iint_R f(x, y)\, dr = \iint_{R_1} f(x, y)\, dr + \iint_{R_2} f(x, y)\, dr.$$

2. Reduction of Double Integrals to Repeated Simple Integrals.

We now have a definition of the double integral, with its interpretation as a volume and with the many possibilities of usefulness which our experience with the single integral suggests; but as yet we do not possess a method for evaluating such integrals. In this section we shall see how the calculation of a double integral can be reduced to that of two single integrals.

We suppose that $u = f(x, y)$ is a function which is defined and continuous in a rectangle R, $a \leqq x \leqq b$, $c \leqq y \leqq d$. If we fix upon any value x_0 in the interval $a \leqq x \leqq b$, the function $f(x_0, y)$ is a continuous function of the remaining variable y. Hence the integral

$$\int_c^d f(x_0, y)\, dy$$

exists, and can be evaluated by the methods of earlier chapters. This integral has a definite value for each value of x_0 that we may choose; in other words, the integral is a function $\phi(x_0)$ of the quantity x_0;

$$\int_c^d f(x, y)\, dy = \phi(x).$$

For example, suppose that $u = f(x, y) = x^2y^3$, $0 \leqq x \leqq 1$, $0 \leqq y \leqq 3$. For each fixed x in the interval $0 \leqq x \leqq 1$ the integral $\int_0^3 x^2y^3\, dy$ can be evaluated, and is, in fact, $\dfrac{81}{4}x^2$; that is, it is a function of x. Or if $f(x, y) = e^{xy}$, $1 \leqq x \leqq 2$, $1 \leqq y \leqq 4$, we have $\int_1^4 e^{xy}\, dy = \dfrac{1}{x}(e^{4x} - e^x)$.

Having thus found the function $\phi(x)$, we can prove that it is continuous; this is a simple consequence of the uniform continuity of $f(x, y)$. It is therefore possible to integrate $\phi(x)$ between the limits a and b, thus obtaining the " repeated integral "

$$\int_a^b \phi(x) \, dx = \int_a^b \left(\int_c^d f(x, y) \, dy \right) dx.$$

By reversing the order of the process, first calculating the function of y defined by $\int_a^b f(x, y) \, dx$ and then integrating from c to d, we obtain the other repeated integral

$$\int_c^d \left(\int_a^b f(x, y) \, dx \right) dy.$$

These integrals, as we have seen, are obtained by a double application of the ordinary simple integration which we have studied in previous chapters. Their importance lies in the following fact:

For continuous functions f(x, y), *and for functions* f(x, y) *having at most jump discontinuities on a finite number of smooth curves, the repeated integrals are equal to the double integral:*

$$\iint_R f(x, y) \, dr = \int_a^b \left(\int_c^d f(x, y) \, dy \right) dx$$

$$= \int_c^d \left(\int_a^b f(x, y) \, dx \right) dy.$$

We shall content ourselves with an intuitive discussion of the case where $f(x, y)$ is continuous. In our original discussion of the double integral regarded as the volume lying above the rectangle $a \leq x \leq b$, $c \leq y \leq d$ and below the surface $u = f(x, y)$, we obtained this volume by subdividing the solid into vertical columns and then letting the diagonals of the bases of these columns approach zero. Instead of this we can divide the solid into slices of breadth $k = (d - c)/n$ by drawing the lines $y = c + vk$ ($v = 0, 1, \ldots, n$) parallel to the x-axis and then constructing a plane perpendicular to the xy-plane through each iine (cf. fig. 8). These planes cut the solid into n slices which grow thinner as n increases, and whose total volume is equal to the double integral. We now see that the volume of each slice is approximately (but of course not as a rule exactly) equal to the product of the thickness k by the area of the left-hand face, that is, equal to

$$k \int_a^b f(x, c + vk)\, dx.$$

Therefore, if we write

$$\phi(y) = \int_a^b f(x, y)\, dx$$

the desired volume is represented approximately by

$$\sum_{v=0}^{n-1} k\phi(c + vk).$$

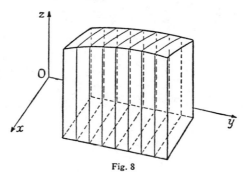

Fig. 8

As $n \to \infty$ these sums tend to

$$\int_c^d \phi(y)\, dy.$$

It is therefore reasonable to expect that the volume or double integral is exactly equal to

$$\int_c^d \phi(y)\, dy = \int_c^d \left(\int_a^b f(x, y)\, dx \right) dy,$$

which is the statement made above. A similar discussion makes it equally plausible that the statement

$$\int_a^b \left(\int_c^d f(x, y)\, dy \right) dx = \int\!\!\int_R f(x, y)\, dr$$

is also true.

3. Examples and Remarks.

A few examples will serve to illustrate how this theorem may be used to evaluate double integrals. For the function $u = f(x, y) = x^3y$, $0 \leqq x \leqq 1$, $0 \leqq y \leqq 2$, we have

$$\int\int_R x^3 y \, dr = \int_0^1 \left(\int_0^2 x^3 y \, dy \right) dx = \int_0^1 \left(\frac{1}{2} x^3 y^2 \Big|_0^2 \right) dx$$

$$= \int_0^1 2x^3 \, dx = \frac{1}{2} x^4 \Big|_0^1 = \frac{1}{2}.$$

The above example belongs to a general class of functions whose integration is often simplified by the following theorem:

If the function u = f(x, y), a $\leqq$ x $\leqq$ b, c $\leqq$ y $\leqq$ d, *can be represented as the product of a function of* x *alone, and a function of* y *alone,*

$$f(x, y) = \phi(x) \, \psi(y),$$

then the double integral of f *is the product of two simple integrals:*

$$\int\int_R f(x, y) \, dr = \left(\int_a^b \phi(x) \, dx \right) \left(\int_c^d \psi(y) \, dy \right).$$

For on integration with respect to y the function $\phi(x)$ can be treated as a constant and placed in front of the integral sign, while, on integration with respect to x, $\int_c^d \psi(y) \, dy$ is a constant; hence

$$\int_a^b \left(\int_c^d \phi(x) \, \psi(y) \, dy \right) dx = \int_a^b \left(\phi(x) \int_c^d \psi(y) \, dy \right) dx$$

$$= \left(\int_c^d \psi(y) \, dy \right) \left(\int_a^b \phi(x) \, dx \right).$$

For the function $u = \sin(x + y)$, $0 \leqq x \leqq \pi/2$, $0 \leqq y \leqq \pi/2$, we have

$$\int\int_R \sin(x + y) \, dr = \int_0^{\pi/2} \left(\int_0^{\pi/2} \sin(x + y) \, dy \right) dx$$

$$= \int_0^{\pi/2} \left(-\cos\left(x + \frac{\pi}{2} \right) + \cos x \right) dx = \int_0^{\pi/2} (\sin x + \cos x) \, dx$$

$$= (-\cos x + \sin x) \Big|_0^{\pi/2} = 1 + 1 = 2.$$

Again, let us calculate the volume V of the vertical prism whose base in the xy-plane is bounded by the co-ordinate axes and the line $x + y = 1$, and which lies below the plane $u = 2x + 3y$. We first extend the function $u = f(x, y)$ to the square $0 \leq x \leq 1$, $0 \leq y \leq 1$ by equating it to 0 outside the triangle (the base of the prism). Then for each x in the interval the function $f(x, y)$ is different from 0 for $0 \leq y \leq 1 - x$ only; hence

$$\int_0^1 f(x, y)\, dy = \int_0^{1-x} f(x, y)\, dy = \int_0^{1-x} (2x + 3y)\, dy$$

$$= 2x(1 - x) + \frac{3}{2}(1 - x)^2 = -\frac{1}{2}x^2 - x + \frac{3}{2},$$

and $\qquad V = \int \int_R f(x, y)\, dr = \int_0^1 \left(-\frac{1}{2}x^2 - x + \frac{3}{2}\right) dx = \frac{5}{6}.$

The device just used is capable of extension to any function $u = f(x, y)$ which is defined in a region R bounded above and below by curves $y = \psi(x)$ and $y = \phi(x)$. For suppose R is defined by the inequalities $a \leq x \leq b$, $\phi(x) \leq y \leq \psi(x)$. We mark off a rectangle R', $a \leq x \leq b$, $c \leq y \leq d$, completely containing R, and outside R we put $f = 0$. Then

$$\int_c^d f(x, y)\, dy = \int_{\phi(x)}^{\psi(x)} f(x, y)\, dy$$

for every x in the interval $a \leq x \leq b$, so that

$$\int \int_R f(x, y)\, dr = \int \int_{R'} f(x, y)\, dr = \int_a^b \left(\int_c^d f(x, y)\, dy\right) dx$$

$$= \int_a^b \left(\int_{\phi(x)}^{\psi(x)} f(x, y)\, dy\right) dx.$$

Thus to find the volume V of the ellipsoid $\frac{x^2}{a^2} + \frac{y^2}{b^2} + \frac{u^2}{c^2} - 1 = 0$, we notice that $\frac{1}{2}V$ is the volume under $u = f(x, y) = c\sqrt{\left(1 - \frac{x^2}{a^2} - \frac{y^2}{b^2}\right)}$, this function $f(x, y)$ being defined only inside an ellipse

$$\frac{x^2}{a^2} + \frac{y^2}{b^2} \leq 1, \quad \text{or} \quad -b\sqrt{\left(1 - \frac{x^2}{a^2}\right)} \leq y \leq b\sqrt{\left(1 - \frac{x^2}{a^2}\right)}, \quad -a \leq x \leq a.$$

Calculating the repeated integral we first have

$$\int_{-b}^b f(x, y)\, dy = \int_{-b\sqrt{(1-x^2/a^2)}}^{b\sqrt{(1-x^2/a^2)}} c\sqrt{\left(1 - \frac{x^2}{a^2} - \frac{y^2}{b^2}\right)} dy$$

$$= -\frac{1}{2}c\left(b - \frac{bx^2}{a^2}\right) \text{arc cos} \frac{y}{\sqrt{(b^2 - b^2 x^2/a^2)}} + \frac{cy}{2}\sqrt{\left(1 - \frac{x^2}{a^2} - \frac{y^2}{b^2}\right)}\Big|_{-b\sqrt{(1-x^2/a^2)}}^{+b\sqrt{(1-x^2/a^2)}}$$

$$= -\frac{c}{2}\left(b - \frac{bx^2}{a^2}\right)(0 - \pi) + 0 = \frac{c\pi}{2}b\left(1 - \frac{x^2}{a^2}\right).$$

Proceeding with the integration, we have

$$\frac{1}{2} V = \int_{-a}^{a} \left(\int_{-b}^{b} f(x, y)\, dy \right) dx = \int_{-a}^{a} \frac{\pi}{2} cb \left(1 - \frac{x^2}{a^2} \right) dx = \frac{\pi c}{2} b \left(x - \frac{x^3}{3a^2} \right) \Big|_{-a}^{a}$$

$$= \frac{\pi}{2} cb \left(\frac{4}{3} a \right) = \frac{2\pi}{3} abc,$$

so that
$$V = \frac{4}{3} \pi abc.$$

4. Polar Co-ordinates.

In our definition of the double integral, the subdivision into rectangles was of course chosen simply because such a subdivision is most convenient in connexion with rectangular coordinates. As we already know, however, there are many applications in which polar co-ordinates are much more suitable than rectangular. If we are considering a function $f(\rho, \phi)$ where ρ and ϕ are polar co-ordinates, the most convenient subdivision is not into rectangles, but into regions bounded by arcs of circles $\rho = $ constant and radii $\phi = $ constant. Suppose, then, that our function $f(\rho, \phi)$ is defined in such a region R, specified by the inequalities $a \leqq \rho \leqq b$, $\alpha \leqq \phi \leqq \beta$. (If $f(\rho, \phi)$ is originally defined in a region R' not of this type, we enclose R' in a larger region R of the desired form and put $f(\rho, \phi) = 0$ outside R'.) Then, just as on p. 486, we can insert points of subdivision $\rho_0 = a$, $\rho_1, \rho_2, \ldots, \rho_n = b$, $\phi_0 = \alpha$, $\phi_1, \phi_2, \ldots, \phi_m = \beta$ and construct the corresponding radii and arcs of circles, thus dividing R into regions R_{ij}, of area ΔR_{ij}. In each R_{ij} we choose a point (ρ_{ij}, ϕ_{ij}) and form the sum $\Sigma f(\rho_{ij}, \phi_{ij}) \Delta R_{ij}$, and then let m and n increase without limit. Then the sum will again tend to the volume under the surface $u = f(\rho, \phi)$, and we may denote this by the integral

$$\iint_{R} f(\rho, \phi)\, dr.$$

So far we have encountered nothing essentially new. The point of importance is to learn how to evaluate these integrals by reducing them either to repeated integrals or to integrals in terms of rectangular co-ordinates. For this purpose we mark off a pair of rectangular axes in a new plane, the $\rho\phi$-plane, and call them the ρ-axis and the ϕ-axis respectively. Corresponding to the point in R with *polar* co-ordinates ρ, ϕ we plot the point in the $\rho\phi$-plane with rectangular co-ordinates ρ, ϕ. Thus the region R, $a \leqq \rho \leqq b$, $a \leqq \phi \leqq \beta$ is represented in the $\rho\phi$-plane by a rectangle R', $a \leqq \rho \leqq b$, $a \leqq \phi \leqq \beta$, and each small region R_{ij}, $\rho_{i-1} \leqq \rho \leqq \rho_i$, $\phi_{j-1} \leqq \phi \leqq \phi_j$ is represented by a small rectangle R_{ij}'. But the area $\Delta R_{ij}'$ of the rectangle R_{ij}' is *not* the same as the area ΔR_{ij} of R_{ij}. The relation between them is easily found. The area $\Delta R_{ij}'$ is simply $(\phi_j - \phi_{j-1})(\rho_i - \rho_{i-1})$, while the area ΔR_{ij} is given by the formula

$$\Delta R_{ij} = \tfrac{1}{2}(\phi_j - \phi_{j-1})(\rho_i{}^2 - \rho_{i-1}{}^2)$$
$$= \tfrac{1}{2}(\rho_i + \rho_{i-1})(\phi_j - \phi_{j-1})(\rho_i - \rho_{i-1}) = \tfrac{1}{2}(\rho_i + \rho_{i-1})\Delta R_{ij}'.$$

In each region R_{ij} let us now choose the point $\bar{\rho}_i = \tfrac{1}{2}(\rho_i + \rho_{i-1})$, $\bar{\phi}_j = \tfrac{1}{2}(\phi_j + \phi_{j-1})$. Then by definition

$$\iint_R f(\rho,\ \phi)\,dr = \lim \Sigma f(\bar{\rho}_i,\ \bar{\phi}_j)\,\Delta R_{ij}.$$

But
$$\Sigma f(\bar{\rho}_i,\ \bar{\phi}_j)\Delta R_j = \Sigma f(\bar{\rho}_i,\ \bar{\phi}_j)\bar{\rho}_i\Delta R_{ij}',$$

and the latter expression is just the sum which we form in defining the double integral of the function $f(\rho,\ \phi)\rho$ over the rectangle R' in the $\rho\phi$-plane. Hence as the fineness of the subdivision increases the sum approaches this integral and

$$\iint_R f(\rho,\ \phi)\,dr = \iint_{R'} f(\rho,\ \phi)\rho\,dr' = \iint_{R'} f(\rho,\ \phi)\rho\,d\rho\,d\phi$$
$$= \int_a^b \left(\int_a^\beta f(\rho,\ \phi)\rho\,d\phi \right) d\rho = \int_a^\beta \left(\int_a^b f(\rho,\ \phi)\rho\,d\rho \right) d\phi.$$

As an example, let us calculate the volume V of the sphere of radius a. The upper hemisphere is given by the equation $u = \sqrt{(a^2 - \rho^2)}$, $0 \leqq \rho \leqq a$, $0 \leqq \varphi \leqq 2\pi$. Thus

$$\tfrac{1}{2} V = \int_0^{2\pi} \left(\int_0^a \sqrt{(a^2 - \rho^2)}\ \rho\,d\rho \right) d\varphi = \int_0^{2\pi} \left(-\tfrac{1}{3}(a^2 - \rho^2)^{3/2} \Big|_0^a \right) d\varphi$$
$$= \frac{1}{3} \int_0^{2\pi} a^3\,d\varphi = \frac{2\pi a^3}{3},$$

so that $V = \dfrac{4}{3}\pi a^3$.

5. Evaluation of $\int_{-\infty}^{\infty} e^{-x^2} dx$.

The formulæ of the preceding sub-section enable us to calculate the area under the curve $y = e^{-x^2}$, $-\infty < x < \infty$, which frequently occurs in the theory of probability. This integration is especially interesting, in that we can evaluate the definite integral from $-\infty$ to ∞ of a function for which we cannot find a primitive function or an indefinite integral at all. Let us first consider the integral I_a of the function $e^{-(x^2+y^2)} = e^{-\rho^2}$ over the circle $0 \leqq \rho \leqq a$. This is given by

$$I_a = \int_0^{2\pi} \left(\int_0^a e^{-\rho^2} \rho \, d\rho \right) d\phi = \int_0^{2\pi} \left(\frac{1}{2} - \frac{1}{2} e^{-a^2} \right) d\phi = \pi(1 - e^{-a^2}).$$

The square $-a \leqq x \leqq a$, $-a \leqq y \leqq a$ contains the circle $0 \leqq \rho \leqq a$ and is contained in the circle $0 \leqq \rho \leqq 2a$, and the integrand $e^{-x^2-y^2}$ is everywhere positive; hence

$$\pi(1 - e^{-a^2}) = I_a \leqq \int_{-a}^a \left(\int_{-a}^a e^{-x^2-y^2} dy \right) dx \leqq I_{2a} = \pi(1 - e^{-4a^2}).$$

The integral can be written in the form

$$\int_{-a}^a e^{-x^2} \left(\int_{-a}^a e^{-y^2} dy \right) dx = \left(\int_{-a}^a e^{-x^2} dx \right)^2,$$

hence

$$\pi(1 - e^{-a^2}) \leqq \left(\int_{-a}^a e^{-x^2} dx \right)^2 \leqq \pi(1 - e^{-4a^2}).$$

If we now let a increase without limit, this gives the equation

$$\int_{-\infty}^{\infty} e^{-x^2} dx = \sqrt{\pi},$$

and our integral is evaluated.

6. Moments and Centre of Mass; Moments of Inertia.

In Chap. V, § 2 (p. 283) we saw that the moment of a system of points P_1, P_2, $\ldots$, P_n with co-ordinates (x_1, y_1), (x_2, y_2), $\ldots$, (x_n, y_n) and masses m_1, m_2, $\ldots$, m_n about the x-axis is given by $\sum_{\nu=1}^n m_i y_i$, and that the ordinate of its centre of mass is given by the equation

$$\eta = \frac{1}{M} \sum_{\nu=1}^n m_\nu y_\nu, \quad \text{where} \quad M = \sum_{\nu=1}^n m_i,$$

with analogous expressions for the moment about the y-axis and the abscissa of the centre of mass. We now extend these ideas to masses distributed uniformly over a region R. We suppose that a mass is distributed with density 1 over the region R; that is, that each portion of R with area ΔR has also mass ΔR. Then the total mass M of R is the same as the area of R,

$$M = \int\int_R dr.$$

Let us now divide R into portions $R_1, \ldots, R_n$ with areas $\Delta R_1, \ldots, \Delta R_n$, and in each portion R_ν choose a point (ξ_ν, η_ν). If we imagine that the total mass ΔR_ν of the portion R_ν is concentrated at the point (ξ_ν, η_ν), the moment of the resulting system of points with respect to the x-axis will be $\Sigma \eta_\nu \Delta R_\nu$, and the ordinate of the centre of mass will be

$$\frac{\Sigma \eta_\nu \Delta R_\nu}{\Sigma \Delta R_\nu} = \frac{\Sigma \eta_\nu \Delta R_\nu}{M}.$$

If we now let $n \to \infty$ and let the diameter of the greatest R_ν tend to 0, these sums tend to the integrals

$$T_x = \int\int_R y\,dr, \qquad \eta = \frac{\int\int_R y\,dr}{M}$$

respectively. These expressions we take as the *definitions* of the moment T_x of R about the x-axis and of the ordinate η of its centre of mass. Similarly, the moment about the y-axis and the abscissa ξ of the centre of mass are respectively given by

$$T_y = \int\int_R x\,dr, \quad \xi = \frac{\int\int_R x\,dr}{M}, \quad \text{where} \quad M = \int\int_R dr.$$

For example, the moment of the semicircle R, $-\rho \leqq x \leqq \rho$, $0 \leqq y \leqq \sqrt{(\rho^2 - x^2)}$, about the x-axis is

$$T_x = \int\int_R y\,dr = \int_{-\rho}^{\rho} \left(\int_0^{\sqrt{(\rho^2 - x^2)}} y\,dy \right) dx$$

$$= \int_{-\rho}^{\rho} \frac{1}{2}(\rho^2 - x^2)\,dx = \frac{1}{2}\left(\rho^2 x - \frac{x^3}{3}\right)\Big|_{-\rho}^{\rho} = \frac{2}{3}\rho^3,$$

and since $M = \int\int_R dr = \text{area } R = \frac{1}{2}\pi\rho^2$,

$$\eta = \frac{2}{3}\rho^3 \div \left(\frac{1}{2}\pi\rho^2\right) = \frac{4}{3}\frac{\rho}{\pi}.$$

(E 798)

By a similar argument, starting from the definition of the moment of inertia I_x of a system of particles,

$$I_x = \Sigma m_\nu y_\nu^2,$$

we arrive at the expression for the moment of inertia of the region R about the x-axis,

$$I_x = \int\int_R y^2 \, dr,$$

and similarly we obtain the moment of inertia with respect to the y-axis,

$$I_y = \int\int_R x^2 \, dr.$$

Analogous formulæ hold for three-dimensional regions R; the co-ordinates ξ, η, ζ of the centre of mass are given by

$$\xi = \frac{\int\int\int_R x \, dr}{M}, \quad \eta = \frac{\int\int\int_R y \, dr}{M}, \quad \zeta = \frac{\int\int\int_R z \, dr}{M},$$

where $M = \int\int\int_R 1 \, dr = $ volume of R. To find the moments of inertia I_x, I_y, I_z of R about the x, y, z-axes respectively, we must remember that the distance of the point (x, y, z) from the x-axis is $\sqrt{(y^2 + z^2)}$; hence for a system of particles the moment of inertia about the x-axis is $\Sigma m_\nu \sqrt{(y_\nu^2 + z_\nu^2)^2} = \Sigma m_\nu (y_\nu^2 + z_\nu^2)$, and on dividing R into sub-regions and passing to the limit as before we obtain the formula

$$I_x = \int\int\int_R (y^2 + z^2) \, dr.$$

Similarly
$$I_y = \int\int\int_R (x^2 + z^2) \, dr,$$

$$I_z = \int\int\int_R (x^2 + y^2) \, dr.$$

Thus the moment of inertia of the cube $-h \leqq x \leqq h$, $-h \leqq y \leqq h$, $-h \leqq z \leqq h$ about the z-axis is

$$I_z = \int_{-h}^{h} \left\{ \int_{-h}^{h} \left(\int_{-h}^{h} (x^2 + y^2) \, dz \right) dy \right\} dx$$

$$= \int_{-h}^{h} \left\{ \int_{-h}^{h} 2h(x^2 + y^2) \, dy \right\} dx = \int_{-h}^{h} 2h \left(2x^2 h + \frac{2}{3} h^3 \right) dx$$

$$= \frac{4h}{3} (x^3 h + xh^3) \Big|_{-h}^{h} = \frac{4h}{3} (4h^4) = \frac{16}{3} h^5.$$

The significance of the moment of inertia, as we have already remarked in Chap. V (p. 286), lies in the fact that in rotatory motion it plays the part taken by the mass in translatory motion. For example, if the region R rotates about the x-axis with angular velocity ω, its kinetic energy is $\frac{1}{2}I_x\omega^2$. This, however, is not the only application of the concept of moment of inertia; for example, it is also important in structural engineering, where it is found that the stiffness of a beam of a given material is proportional to the moment of inertia of the cross-section taken about a line through its centre of mass. The reader will find further information about this in any textbook on strength of materials.

7. Further Applications.

The student should not assume that the applications already discussed exhaust the possibilities of the double integral. For instance, we have not proved the important theorem that the area A of the surface $z = f(x, y)$, where (x, y) is in R, is given by the integral

$$A = \int\int_R \sqrt{\left(1 + \left(\frac{\partial f}{\partial x}\right)^2 + \left(\frac{\partial f}{\partial y}\right)^2\right)}\, dr$$

provided $\dfrac{\partial f}{\partial x}$ and $\dfrac{\partial f}{\partial y}$ are continuous; and we have left many other interesting fields untouched. These further developments, however, do not come within the scope of the present book and must be left for Vol. II.

EXAMPLES

1. Perform the following integrations:

(a) $\displaystyle\int_0^a \int_0^b xy\,(x^2 - y^2)\,dy\,dx.$

(b) $\displaystyle\int_0^\pi \int_0^\pi \cos(x + y)\,dy\,dx.$

(c) $\displaystyle\int_1^e \int_1^2 \frac{1}{xy}\,dy\,dx.$

(d) $\displaystyle\int_0^a \int_0^b xe^{xy}\,dy\,dx.$

(e) $\displaystyle\int_0^1 \int_0^{\sqrt{1-x^2}} y^2 \, dy \, dx.$

(f) $\displaystyle\int_0^2 \int_0^{2-x} y \, dy \, dx.$

2. Find the volume between the xy-plane and the paraboloid $z = 2 - x^2 - y^2$.

3. Find the volume common to the two cylinders $x^2 + z^2 = 1$ and $y^2 + z^2 = 1$.

4. By integration, find the volume of the smaller of the two portions into which a sphere of radius r is cut by a plane whose perpendicular distance from the centre is h ($< r$).

5. For the following figures find the area, the centre of gravity, the moments about the x- and y-axes, and the moments of inertia about the x- and y-axes:

(a) the semicircle $0 \leq y \leq \sqrt{(r^2 - x^2)}$;

(b) the rectangle $0 \leq x \leq a$, $0 \leq y \leq b$;

(c) the rectangle $-a \leq x \leq a$, $-b \leq y \leq b$;

(d) the ellipse $|y| \leq b\sqrt{\left(1 - \dfrac{x^2}{a^2}\right)}$;

(e) the triangle with vertices $(0, 0)$, $(a, 0)$, $(0, b)$:

6. For the following figures find the volume, the centre of gravity, and the moments of inertia about the x-, y-, and z-axes:

(a) the parallelepiped $0 \leq x \leq a$, $0 \leq y \leq b$, $0 \leq z \leq c$;

(b) the hemisphere $0 \leq z \leq \sqrt{(a^2 - x^2 - y^2)}$;

(c) the triangular prism with vertices $(0, 0, 0)$, $(a, 0, 0)$, $(0, b, 0)$, $(0, 0, c)$.

CHAPTER XI

The Differential Equations
for the Simplest Types of Vibration

On several occasions we have already met with differentiaɩ equations, that is, equations from which an unknown function is to be determined and which involve not only this function itself but also its derivatives.

The simplest problem of this type is that of finding the indefinite integral of a given function $f(x)$. This problem requires us to find a function $y = F(x)$ which satisfies the differential equation $y' - f(x) = 0$. Further, we solved a problem of the same type in Chap. III, § 7 (p. 178), where we showed that an equation of the form $y' = ay$ is satisfied by an exponential function $y = ce^{ax}$. As we saw in Chap. V (p. 294), differential equations arise in connexion with the problems of mechanics, and indeed many branches of pure mathematics and most of applied mathematics depend on differential equations. In this chapter, without going into the general theory, we shall consider the differential equations of the simplest types of vibration. These are not only of theoretical value, but are also extremely important in applied mathematics.

It will be convenient to bear the following general ideas and definitions in mind. By a *solution* of a differential equation we mean a function which, when substituted in the differential equation, satisfies the equation for all values of the independent variable that are being considered. Instead of *solution* the term *integral* is often used: in the first place because the problem is more or less a generalization of the ordinary problem of integration; and in the second place because it frequently happens that the solution is actually found by integration.

1. Vibration Problems of Mechanics and Physics

1. The Simplest Mechanical Vibrations.

The simplest type of mechanical vibration has already been considered in Chap. V, § 4 (p. 295). We there considered a particle of mass m which is free to move on the x-axis and which is brought back to its initial position $x = 0$ by a restoring force. The magnitude of this restoring force we took to be proportional to the displacement x; in fact, we equated it to $-kx$, where k is a positive constant and the negative sign expresses the fact that the force is always directed towards the origin. We shall now assume that there is a frictional force present also and that this frictional force is proportional to the velocity $dx/dt = \dot{x}$ of the particle and opposed to it. This force is then given by an expression of the form $-r\dot{x}$, with a positive frictional constant r. Finally, we shall assume that the particle is also acted on by an external force which is a function $f(t)$ of the time t. Then by Newton's fundamental law the product of the mass m and the acceleration $\ddot{x}$ must be equal to the total force, that is, the elastic force plus the frictional force plus the external force. This is expressed by the equation

$$m\ddot{x} + r\dot{x} + kx = f(t).$$

This equation determines the motion of the particle. If we recall the previous examples of differential equations, such as the integration problem $\dot{x} = \dfrac{dx}{dt} = f(t)$ with its solution $x = \int f(t)\, dt + c$, or the solution of the particular differential equation $m\ddot{x} + kx = 0$ on p. 296, we observe that these problems have an infinite number of different solutions. Here too we shall find that there are an infinite number of solutions, which are expressed in the following way. It is possible to find a *general solution* or *complete integral* $x(t)$ of the differential equation, depending not only on the independent variable t, but also on two parameters c_1 and c_2, called the *constants of integration*. If we assign special values to these constants, we obtain a particular solution, and *every* solution can be found by assigning special values to these constants. The complete integral is then the totality of all particular solutions.

This fact is quite understandable (cf. also Chap. V, § 4, p. 298). We cannot expect that the differential equation alone will determine the motion completely. On the contrary, it is plausible that at a given instant, say at the time $t = 0$, we should be able to choose the initial position $x(0) = x_0$ and the initial velocity $\dot{x}(0) = \dot{x}_0$ (in short, the *initial state*) arbitrarily; in other words, at time $t = 0$ we should be able to start the particle from any initial position with any velocity. This being done, we may expect the rest of the motion to be definitely determined. The two arbitrary constants c_1 and c_2 in the general solution are just enough to enable us to select the particular solution which fits these initial conditions. In the next section (p. 508) we shall see that this can be done in one way only.

If no external force is present, that is, if $f(t) = 0$, the motion is called a *free motion*. The differential equation is then said to be *homogeneous*. If $f(t)$ is not equal to zero for all values of t, we say that the motion is *forced* and that the differential equation is *non-homogeneous*. The term $f(t)$ is also occasionally referred to as the *perturbation term*.

2. Electrical Oscillations.

A mechanical system of the simple type described can actually be realized only approximately. An approximation is offered by the pendulum, provided its oscillations are small. The oscillations of a magnetic needle, the oscillations of the centre of a telephone or microphone diaphragm, and other mechanical vibrations can be represented to within a certain degree of accuracy by systems such as we have described. But there is another type of phenomenon which corresponds far more exactly to our differential equation. This is the oscillatory electrical circuit.

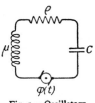

Fig. 1.—Oscillatory electrical circuit

We consider the circuit sketched in fig. 1, having inductance μ, resistance ρ and capacity $C = 1/\kappa$. We also suppose that the circuit is acted upon by an external electromotive force $\phi(t)$ which is known as a function of the time t, such as the voltage supplied by a dynamo or the voltage due to electric waves. In order to describe the process taking place in the circuit we denote the voltage across the condenser by E and the charge in the condenser by Q. These quantities are then connected by the equation $CE = E/\kappa = Q$. The current I, which like the voltage

E is a function of the time, is defined as the rate of change of the charge per unit time, that is, as the rate at which the charge on the condenser diminishes: $I = -\dot{Q} = -dQ/dt = -\dot{E}/\kappa$. Ohm's law states that the product of the current and the resistance is equal to the electromotive force (voltage); that is, it is equal to the condenser voltage E minus the counter electromotive force due to self-induction plus the external electromotive force $\phi(t)$. We thus arrive at the equation $I\rho = E - \mu\dot{I} + \phi(t)$

or $-\dfrac{\rho}{\kappa}\dot{E} = E + \dfrac{\mu}{\kappa}\ddot{E} + \phi(t)$, that is, $\mu\ddot{E} + \rho\dot{E} + \kappa E = -\kappa\phi(t)$,

which is satisfied by the voltage in the circuit. We see, therefore, that we have obtained a differential equation of exactly the type considered in No. 1 (p. 502). Instead of the mass we have the inductance, instead of the frictional force the resistance, and instead of the elastic constant the reciprocal of the capacity, while the external electromotive force (apart from a constant factor) corresponds to the external force. If the electromotive force is zero, the differential equation is homogeneous.

If we multiply both sides of the differential equation by $-1/\kappa$ and differentiate with respect to the time, we obtain for the current I the corresponding equation

$$\mu\ddot{I} + \rho\dot{I} + \kappa I = \dot{\phi}(t),$$

which differs from the equation for the voltage on the right-hand side only, and for free oscillations ($\phi = 0$) has identically the same form.

2. Solution of the Homogeneous Equation. Free Oscillations

1. The Formal Solution.

We can easily obtain a solution of the homogeneous equation $m\ddot{x} + r\dot{x} + kx = 0$ on p. 502 in the form of an exponential expression, by seeking to determine a constant λ in such a way that the expression $e^{\lambda t} = x$ is a solution. If we substitute this and its derivatives $\dot{x} = \lambda e^{\lambda t}$, $\ddot{x} = \lambda^2 e^{\lambda t}$ in the differential equation and remove the common factor $e^{\lambda t}$, we obtain the quadratic equation

$$m\lambda^2 + r\lambda + k = 0$$

for λ. The roots of this equation are

$$\lambda_1 = -\frac{r}{2m} + \frac{1}{2m}\sqrt{(r^2 - 4mk)}, \quad \lambda_2 = -\frac{r}{2m} - \frac{1}{2m}\sqrt{(r^2 - 4mk)}.$$

Each of the two expressions $x = e^{\lambda_1 t}$ and $x = e^{\lambda_2 t}$ is, at least formally, a particular solution of the differential equation, as we see by carrying out the calculations in the reverse direction. Three different cases can now occur.

1. $r^2 - 4mk > 0$. The two roots λ_1 and λ_2 are then real, negative, and unequal, and we have two solutions of the differential equation, $x = u_1 = e^{\lambda_1 t}$ and $x = u_2 = e^{\lambda_2 t}$. With the help of these two solutions we can at once construct a solution in which two arbitrary constants are present. For on differentiation we see that

$$x = c_1 u_1 + c_2 u_2$$

is also a solution of the differential equation. On p. 508 we shall show that this expression is in fact the most general solution of the equation; that is, that we can obtain *every* solution of the equation by substituting suitable numerical values for c_1 and c_2.

2. $r^2 - 4mk = 0$. The quadratic equation has a double root. Thus to begin with we have, apart from a constant factor, only the one solution $x = w_1 = e^{-rt/2m}$. But we easily verify that in this case the function

$$x = w_2 = te^{-rt/2m}$$

is also a solution of the differential equation.* For we find that

$$\dot{x} = \left(1 - \frac{r}{2m}t\right)e^{-rt/2m}, \quad \ddot{x} = \left(\frac{r^2}{4m^2}t - \frac{r}{m}\right)e^{-rt/2m},$$

and by substitution we see that the differential equation

$$m\ddot{x} + r\dot{x} + \frac{r^2}{4m}x = m\ddot{x} + r\dot{x} + kx = 0$$

---

* We are led to this solution naturally by the following limiting process: if $\lambda_1 \neq \lambda_2$, then the expression $(e^{\lambda_1 t} - e^{\lambda_2 t})/(\lambda_1 - \lambda_2)$ also represents a solution. If we now let λ_1 tend to λ_2 and write λ instead of λ_1, λ_2, our expression becomes $\frac{d}{d\lambda}e^{\lambda t} = te^{\lambda t}$.

is satisfied. Then the expression

$$x = c_1 e^{-rt/2m} + c_2 t e^{-rt/2m}$$

again gives us a solution of the differential equation with two arbitrary constants of integration c_1 and c_2.

3. $r^2 - 4mk < 0$. We put $r^2 - 4mk = -4m^2\nu^2$ and obtain two solutions of the differential equation in complex form, given by the expressions $x = u_1 = e^{-rt/2m + i\nu t}$ and $x = u_2 = e^{-rt/2m - i\nu t}$. Euler's formula

$$e^{\pm i\nu t} = \cos \nu t \pm i \sin \nu t$$

gives us for the real and imaginary parts of the complex solution u_1, on the one hand the expressions

$$v_1 = e^{-rt/2m} \cos \nu t, \quad v_2 = e^{-rt/2m} \sin \nu t,$$

and on the other hand the representation

$$v_1 = \frac{u_1 + u_2}{2}, \quad v_2 = \frac{u_1 - u_2}{2i}.$$

From the second form of representation we see that v_1 and v_2 are (real) solutions of the differential equation. To verify this directly by differentiation and substitution forms a simple but valuable exercise.

From our two particular solutions we can again form a general solution

$$x = c_1 v_1 + c_2 v_2 = (c_1 \cos \nu t + c_2 \sin \nu t) e^{-rt/2m}$$

with two arbitrary constants c_1 and c_2. This may also be written in the form

$$x = a e^{-rt/2m} \cos \nu (t - \delta),$$

where we have put $c_1 = a \cos \nu\delta$, $c_2 = a \sin \nu\delta$, and a, δ are two new constants.

We recall that we have already come across this solution for the special case $r = 0$ (Chap. V, § 4, p. 296).

2. Physical Interpretation of the Solution.

In the two cases $r > 2\sqrt{mk}$ and $r = 2\sqrt{mk}$ the solution is given by the exponential curve, or by the graph of the function $te^{-rt/2m}$, which for large values of t resembles the exponential curve, or by the superposition of such curves. In these cases the process is aperiodic; that is, as the time increases the "distance" x approaches the value 0 asymptotically, without oscillating about the value $x = 0$. The motion, therefore, is not oscillatory. The effect of friction or *damping* is so great that it prevents the elastic force from setting up oscillatory motions.

It is quite different in the case $r < 2\sqrt{mk}$, where the damping is so small that complex roots λ_1, λ_2 occur. The expression $x = a \cos v(t - \delta)e^{-rt/2m}$ here gives us *damped harmonic oscillations*. These are oscillations which follow the sine law and have the circular frequency $v = \sqrt{\left(\dfrac{k}{m} - \dfrac{r^2}{4m^2}\right)}$, but whose amplitude, instead of being constant, is given by the expression $ae^{-rt/2m}$. That is, the amplitude diminishes exponentially; the greater the expression $r/2m$ is, the faster is the rate of decrease. In physical literature this damping factor is frequently called the *logarithmic decrement* of the damped oscillation, the term indicating that the logarithm of the amplitude decreases at the rate $r/2m$. A damped oscillation of this kind is

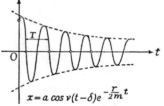

$x = a \cos v(t - \delta)e^{-\frac{r}{2m}t}$

Fig. 2.—Damped harmonic oscillations

illustrated in fig. 2. As before, we call the quantity $T = 2\pi/v$ the period of the oscillation and the quantity $v\delta$ the phase displacement. For the special case $r = 0$ we again obtain simple harmonic oscillations with the frequency $v_0 = \sqrt{k/m}$, the *natural frequency* of the undamped oscillatory system.

3. Fulfilment of Given Initial Conditions.   Uniqueness of the Solution.

We have still to show that the solution with the two constants c_1 and c_2 can be made to fit any pre-assigned initial state, and also that it represents all the possible solutions of the equation. Suppose that we have to find a solution which at time $t = 0$ satisfies the initial conditions $x(0) = x_0$, $\dot{x}(0) = \dot{x}_0$, where the numbers x_0 and $\dot{x}_0$ can have any values. Then in case 1 on p. 505 we must put

$$c_1 + c_2 = x_0,$$
$$c_1\lambda_1 + c_2\lambda_2 = \dot{x}_0.$$

For the constants c_1 and c_2 we accordingly have two linear equations, and these have the unique solutions

$$c_1 = \frac{\dot{x}_0 - \lambda_2 x_0}{\lambda_1 - \lambda_2}, \quad c_2 = \frac{\dot{x}_0 - \lambda_1 x_0}{\lambda_2 - \lambda_1}.$$

In case 2 (p. 505) the same process gives the two linear equations

$$c_1 = x_0,$$

$$\lambda c_1 + c_2 = \dot{x}_0 \quad \left(\lambda = -\frac{r}{2m}\right),$$

from which c_1 and c_2 can again be uniquely determined. Finally, in case 3 (p. 506) the equations determining the constants take the form

$$a \cos \nu \delta = x_0,$$

$$a\left(\nu \sin \nu \delta - \frac{r}{2m} \cos \nu \delta\right) = \dot{x}_0,$$

with the solutions

$$\delta = \frac{1}{\nu} \arccos \frac{x_0}{a}, \quad a = \frac{1}{\nu} \sqrt{\left\{\nu^2 x_0{}^2 + \left(\dot{x}_0 + \frac{r}{2m} x_0\right)^2\right\}}.$$

Thus we have shown that the general solution can be made to fit any arbitrary initial conditions. We have still to show that there is no other solution. For this we need only show that for a given initial state there can never be two different solutions.

If two such solutions $u(t)$ and $v(t)$ existed, for which $u(0) = x_0$, $\dot{u}(0) = \dot{x}_0$ and $v(0) = x_0$, $\dot{v}(0) = \dot{x}_0$, then their difference $w = u - v$ would also be a solution of the differential equation, and we should have $w(0) = 0$, $\dot{w}(0) = 0$. This solution would therefore correspond to an initial state of rest, that is, to a state in which at time $t = 0$ the particle is in its position of rest and has zero velocity. We must show that it can never set itself in motion. To do this we multiply both sides of the differential equation $m\ddot{w} + r\dot{w} + kw = 0$ by $2\dot{w}$ and recall that $2\dot{w}\ddot{w} = \frac{d}{dt}\dot{w}^2$ and $2w\dot{w} = \frac{d}{dt}w^2$. We thus obtain

$$\frac{d}{dt}(m\dot{w}^2) + \frac{d}{dt}(kw^2) + 2r\dot{w}^2 = 0.$$

If we integrate between the instants $t = 0$ and $t = \tau$ and use the initial conditions $w(0) = 0$, $\dot{w}(0) = 0$, we have

$$m\dot{w}^2(\tau) + kw^2(\tau) + 2r \int_0^\tau \left(\frac{dw}{dt}\right)^2 dt = 0.$$

This equation, however, would yield a contradiction if at any time $\tau > 0$ the function w were different from 0. For then the left-hand side of the equation would be positive, since we have taken m, k and r to be positive, while the right-hand side is zero. Hence $w = u - v$ is always equal to 0, which proves that the solution is unique.

For the equations in Ex. 1–5 find the general solution, and also the solution for which $x(0) = 0$, $\dot{x}(0) = 1$:

1. $\ddot{x} - 3\dot{x} + 2x = 0$.
2. $\ddot{x} + 3\dot{x} + 2x = 0$.
3. $2\ddot{x} + \dot{x} - x = 0$.
4. $\ddot{x} + 4\dot{x} + 4x = 0$.
5. $4\ddot{x} + 4\dot{x} + x = 0$.

6. Find the general solution, and also the solution for which $x(0) = 0$, $\dot{x}(0) = 1$, of the equation

$$\ddot{x} + \dot{x} + x = 0.$$

Determine the frequency (ν), the period (T), the amplitude (a), and the phase (δ) of the solution.

7. Find the solution of

$$2\ddot{x} + 2\dot{x} + x = 0$$

for which $x(0) = 1$, $\dot{x}(0) = -1$. Calculate the amplitude (a), the phase (δ), and the frequency (ν) of the solution.

3. The Non-homogeneous Equation.　Forced Oscillations

1. General Remarks.

Before proceeding to the solution of the problem when an external force $f(t)$ is present, that is, to the solution of the non-homogeneous equation, we make the following remark.

If w and v are two solutions of the non-homogeneous equation, the difference $u = w - v$ satisfies the homogeneous equation; this we see at once by substitution. Conversely, if u is a solution of the homogeneous equation and v a solution of the non-homogeneous equation, then $w = u + v$ is also a solution of the non-homogeneous equation. Therefore from one solution * of the non-homogeneous equation we obtain *all* its solutions by adding the complete integral of the homogeneous equation.† We therefore need only find a *single* solution of the non-homogeneous equation. Physically this means that if we have a forced oscillation due to an external force, and on it superpose an arbitrary free oscillation, represented by a solution of the homo-

---

* Often called the *particular integral.*
† Often called the *complementary function.*

geneous equation, we obtain a phenomenon which satisfies the same non-homogeneous equation as the original forced oscillation. If a frictional force is present, the free motion in the case of oscillatory motion will fade out as time goes on, because of the damping factor $e^{-rt/2m}$. Hence for a given forced vibration with friction it is immaterial what free vibration we superpose; the motion will always tend to the same final state as time goes on.

Secondly, we notice that the effect of a force $f(t)$ can be split up in the same way as the force itself. By this we mean the following: if $f_1(t), f_2(t)$, and $f(t)$ are three functions such that

$$f_1(t) + f_2(t) = f(t),$$

and if $x_1 = x_1(t)$ is a solution of the differential equation $m\ddot{x} + r\dot{x} + kx = f_1(t)$ and $x_2 = x_2(t)$ is a solution of the equation $m\ddot{x} + r\dot{x} + kx = f_2(t)$ then $x(t) = x_1(t) + x_2(t)$ is a solution of the differential equation $m\ddot{x} + r\dot{x} + kx = f(t)$. A corresponding statement of course holds if $f(t)$ consists of any number of terms. This simple but important fact is called the "principle of superposition". The proof follows from a glance at the equation itself. By subdividing the function $f(t)$ into two or more terms we can thus split the differential equation into several equations, which in certain circumstances may be easier to manipulate.

The most important case is that of a periodic external force $f(t)$. Such a periodic external force can be resolved into purely periodic components by expansion in a Fourier series, and can therefore * be approximated to as closely as we please by a sum of a finite number of purely periodic functions. It is therefore sufficient to find the solution of the differential equation subject to the assumption that the right-hand side has the form

$$a \cos \omega t \quad \text{or} \quad b \sin \omega t,$$

where a, b, and ω are arbitrary constants.

Instead of working with these trigonometric functions, we can obtain the solution more simply and neatly if we use complex notation. We put $f(t) = ce^{i\omega t}$, and the principle of superposition shows that we need only consider the differential equation

$$m\ddot{x} + r\dot{x} + kx = ce^{i\omega t},$$

* Provided that it is continuous and sectionally smooth (p. 439), which is the only case of importance in physics.

where by c we mean an arbitrary real or complex constant. Such a differential equation actually represents two real differential equations. For if we split the right-hand side into two terms, if e.g. we take $c = 1$ and write $e^{i\omega t} = \cos \omega t + i \sin \omega t$, then x_1 and x_2, the solutions of the two *real* differential equations $m\ddot{x} + r\dot{x} + kx = \cos \omega t$ and $m\ddot{x} + r\dot{x} + kx = \sin \omega t$, combine to form the solution $x = x_1 + ix_2$ of the *complex* differential equation. Conversely, if we first solve the differential equation in complex form, the real part of the solution gives us the function x_1 and the imaginary part the function x_2.

2. Solution of the Non-homogeneous Equation.

We solve the equation $m\ddot{x} + r\dot{x} + kx = ce^{i\omega t}$ by a device naturally suggested by intuition. We assume that c is real and (for the time being) that $r \neq 0$. We now make the guess that a motion will exist which has the same rhythm as the periodic external force, and we accordingly attempt to find a solution of the differential equation in the form

$$x = \sigma e^{i\omega t},$$

where we have only to determine the factor σ, which is independent of the time. If we substitute this expression and its derivatives $\dot{x} = i\omega\sigma e^{i\omega t}$, $\ddot{x} = -\omega^2\sigma e^{i\omega t}$ in the differential equation and remove the common factor $e^{i\omega t}$ we obtain the equation

$$-m\omega^2\sigma + ir\omega\sigma + k\sigma = c$$

or
$$\sigma = \frac{c}{-m\omega^2 + ir\omega + k}.$$

Conversely, we see that for this value of σ the expression $\sigma e^{i\omega t}$ is actually a solution of the differential equation. To express the meaning of this result clearly, however, we must perform a few transformations.

We begin by writing the complex factor σ in the form

$$\sigma = c\frac{k - m\omega^2 - ir\omega}{(k - m\omega^2)^2 + r^2\omega^2} = cae^{-i\omega\delta},$$

where the positive " distortion factor " a and the " phase displacement" $\omega\delta$ are expressed in terms of the given quantities m, r, k, by the equations

$$a^2 = \frac{1}{(k - m\omega^2)^2 + r^2\omega^2}, \quad \sin\omega\delta = r\omega a, \quad \cos\omega\delta = (k - m\omega^2)a.$$

With this notation our solution takes the form

$$x = cae^{i\omega(t-\delta)},$$

and the meaning of the result is as follows: to the force $c\cos\omega t$ there corresponds the "effect" $ca\cos\omega(t-\delta)$, and to the force $c\sin\omega t$ corresponds the effect $ca\sin\omega(t-\delta)$.

Hence we see that the effect is a function of the same type as the force, that is, an undamped oscillation. This oscillation differs from the oscillation representing the force in that the amplitude is increased in the ratio $a : 1$ and the phase is altered by the angle $\omega\delta$. Of course it is easy to obtain the same result without using the complex notation, but at the cost of somewhat longer calculations.

According to the remark at the beginning of this section (p. 509), by finding this one solution we have completely solved the problem; for by superposing any free oscillation we can obtain the most general forced oscillation.

Collecting the results, we have the following:
The complete integral of the differential equation

$$m\ddot{x} + r\dot{x} + kx = ce^{i\omega t}$$

(*where* $x \neq 0$) *is* $x = cae^{i\omega(t-\delta)} + u$, *where* u *is the complete integral of the homogeneous equation* $m\ddot{x} + r\dot{x} + kx = 0$ *and the quantities* a *and* δ *are defined by the equations*

$$a^2 = \frac{1}{(k - m\omega^2)^2 + r^2\omega^2}, \quad \sin\omega\delta = r\omega a, \quad \cos\omega\delta = (k - m\omega^2)a.$$

The constants in this general solution leave us the possibility of making the solution suit an arbitrary initial state, that is, for arbitrarily assigned values of x_0 and $\dot{x}_0$ the constants can be chosen in such a way that $x(0) = x_0$ and $\dot{x}(0) = \dot{x}_0$.

3. The Resonance Curve.

In order to acquire a grasp of the solution which we have obtained and of its significance in applications, we shall study the distortion factor a as a function of the " exciting frequency " ω, that is, the function

$$\phi(\omega) = \frac{1}{\sqrt{(k - m\omega^2)^2 + r^2\omega^2}}.$$

The reason for this detailed investigation is that for given constants k, m, r, or as we say for a given " oscillatory system ", we can think of the system as being acted on by periodic exciting forces of very different circular frequencies, and it is important to consider the solution of the differential equation for these widely different exciting forces. In order to describe the function conveniently we introduce the quantity $\omega_0 = \sqrt{k/m}$. This number ω_0 is the circular frequency which the system would have for free oscillations if the friction r were zero; or, briefly, the *natural frequency of the undamped system* (cf. p. 507). The actual frequency of the free system, owing to the friction r, is not equal to ω_0, but is instead

$$\nu = \sqrt{\left(\frac{k}{m} - \frac{r^2}{4m^2}\right)},$$

where we assume that $4km - r^2 > 0$. (If this is not the case the free system has no frequency; it is aperiodic.)

The function $\phi(\omega)$ tends asymptotically to the value 0 as the exciting frequency tends to infinity, and, in fact, it vanishes to the order $1/\omega^2$. Further, $\phi(0) = 1/k$; in other words, an exciting force of frequency zero and magnitude 1, that is, a constant force of magnitude 1, gives rise to a displacement of the oscillatory system amounting to $1/k$. In the region of positive values of ω the derivative $\phi'(\omega)$ cannot vanish except where the derivative of the expression $(k - m\omega^2)^2 + r^2\omega^2$ vanishes, that is, for a value $\omega = \omega_1 > 0$ for which the equation

$$-4m\omega(k - m\omega^2) + 2r^2\omega = 0$$

holds. In order that such a value may exist we must obviously have $2km - r^2 > 0$; in this case

$$\omega_1 = \sqrt{\left(\frac{k}{m} - \frac{r^2}{2m^2}\right)} = \sqrt{\left(\omega_0^2 - \frac{r^2}{2m^2}\right)}.$$

Since the function $\phi(\omega)$ is positive everywhere, increases monotonically for small values of ω, and vanishes at infinity, this value must give a maximum. We call the circular frequency ω_1 the " resonance frequency " of the system.

By substituting this expression for ω_1 we find that the value of the maximum is

$$\phi(\omega_1) = \frac{1}{r \sqrt{\left(\dfrac{k}{m} - \dfrac{r^2}{4m^2} \right)}}.$$

As $r \to 0$, this value increases beyond all bounds. For $r = 0$, that is, for an undamped oscillatory system, the function $\phi(\omega)$ has an infinite discontinuity at the value $\omega = \omega_1$. This is a limiting case to which we shall give special consideration later.

The graph of the function $\phi(\omega)$ is called the *resonance curve* of the system. The fact that for $\omega = \omega_1$ (and consequently for small values of r in the neighbourhood of the natural frequency) the distortion of amplitude $a = \phi(\omega)$ is particularly large is the mathematical expression of the " phenomenon of resonance ", which for fixed values of m and k is more and more evident as r becomes smaller and smaller.

In fig. 3 we have sketched a family of resonance curves, all corresponding to the values $m = 1$ and $k = 1$, and consequently to $\omega_0 = 1$, but with different values of $D = \frac{1}{2}r$. We see that for small values of D well-marked resonance occurs near $\omega = 1$; in the limiting case $D = 0$ there would be an infinite discontinuity of $\varphi(\omega)$ at $\omega = 1$, instead of a maximum. As D increases the maxima move towards the left, and for the value $D = 1/\sqrt{2}$ we have $\omega_1 = 0$. In this last case the point where the tangent is horizontal has moved to the origin, and the maximum has disappeared. If $D > 1/\sqrt{2}$ there is no zero of $\varphi'(\omega)$; the resonance curve no longer has a maximum, and resonance no longer occurs.

In general, the resonance phenomenon ceases as soon as the condition

$$2km - r^2 \leqq 0$$

becomes true. In the case of the equality sign, the resonance curve reaches its greatest height $\phi(0) = 1/k$ at $\omega_1 = 0$; its tangent is horizontal there, and after an initial course which is almost horizontal it diminishes towards zero.

4. Further Discussion of the Oscillation.

We cannot, however, rest content with the above discussion. In order that we may really understand the phenomenon of forced motion an additional point requires to be emphasized. The particular integral $cae^{i\omega(t-\delta)}$ is to be regarded as a *limiting state* which the complete integral

$$x(t) = cae^{i\omega(t-\delta)} + c_1u_1 + c_2u_2$$

approaches more and more closely *as time goes on*, since the free oscillation $c_1u_1 + c_2u_2$ superposed on the particular integral fades away with the passage of time. This fading away will take place slowly if r is small, rapidly if r is large.

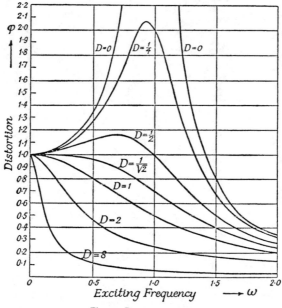

Fig. 3.—Resonance curves

Let us suppose, for example, that at the beginning of the motion, i.e. at time $t = 0$, the system is at rest, so that $x(0) = 0$ and $\dot{x}(0) = 0$. From this we can determine the constants c_1 and c_2, and we see at once that they are *not* both zero. Even when the exciting frequency is approximately or exactly equal to ω_1, so that resonance occurs, the relatively large amplitude $a = \phi(\omega_1)$ will not at first appear. On the contrary, it will be masked by the function $c_1u_1 + c_2u_2$, and will first make its appearance when this function fades away; that is, it will appear more slowly the smaller r is.

For the undamped system, that is, for $r = 0$, our solution fails when the exciting frequency is equal to the natural circular frequency $\omega_0 = \sqrt{k/m}$, for then $\phi(\omega_0)$ is infinite. We therefore cannot obtain a solution of the equation $m\ddot{x} + kx = e^{i\omega t}$ in the form $\sigma e^{i\omega t}$. We can, however, at once obtain a solution of the equation in the form $\sigma t e^{i\omega t}$. If we substitute this expression in the differential equation, remembering that

$$\dot{x} = \sigma e^{i\omega t}(1 + i\omega t), \quad \ddot{x} = \sigma e^{i\omega t}(2i\omega - t\omega^2),$$

we have

$$\sigma(2im\omega - m\omega^2 t + kt) = 1,$$

and, since $m\omega^2 = k$,

$$\sigma = \frac{1}{2im\omega}.$$

Thus *when resonance occurs in an undamped system we have the solution*

$$x = \frac{t}{2im\omega} e^{i\omega t} = \frac{t}{2i\sqrt{km}} e^{i\omega t}.$$

Using real notation, when $f(t) = \cos \omega t$ we have $x = \frac{1}{2} \frac{t}{\sqrt{km}} \sin \omega t$, and when $f(t) = \sin \omega t$ we have

$$x = -\frac{1}{2} \frac{t}{\sqrt{km}} \cos \omega t.$$

We thus see that we have found a function which may be referred to as an oscillation, but whose amplitude increases proportionally with the time. The superposed free oscillation does not fade away, since it is undamped; but it retains its original amplitude and becomes unimportant in comparison with the increasing amplitude of the special forced oscillation. The fact that in this case the solution oscillates backwards and forwards between positive and negative bounds which continually increase as time goes on represents the real meaning of the infinite discontinuity of the resonance function in the case of an undamped system.

5. Remarks on the Construction of Recording Instruments.

In a great variety of applications in physics and engineering the discussion in the previous sub-section is of the utmost importance. With many instruments, such as galvanometers, seismographs, oscillatory electrical circuits in radio receivers, and microphone diaphragms, the problem is to record an oscillatory displacement x due to an external periodic force. In such cases the quantity x satisfies our differential equation, at least to a first approximation.

If T is the period of oscillation of the external periodic force, we can expand the force in a Fourier series of the form

$$f(t) = \sum_{l=-\infty}^{\infty} \gamma_l e^{il(2\pi/T)t},$$

or, better still, we can think of it as represented with sufficient accuracy by a trigonometric sum $\sum_{l=-N}^{N} \gamma_l e^{il(2\pi/T)t}$ consisting of a finite number of terms only. By the principle of superposition (p. 510), the solution $x(t)$ of the differential equation, apart from the superposed free oscillation, will be represented by an infinite series * of the form

$$x(t) = \sum_{l=-\infty}^{\infty} \sigma_l e^{il(2\pi/T)t},$$

or approximately by a finite expression of the form

$$x(t) = \sum_{l=-N}^{N} \sigma_l e^{il(2\pi/T)t}.$$

In virtue of our previous results

$$\sigma_l = \gamma_l a_l e^{-i\delta_l(2\pi l/T)}$$

and

$$\alpha_l{}^2 = \frac{1}{\left(k - ml^2 \dfrac{4\pi^2}{T^2}\right)^2 + r^2 l^2 \dfrac{4\pi^2}{T^2}}, \quad \tan \frac{2\pi l}{T} \delta_l = \frac{2\pi l r}{T\left(k - m \dfrac{4\pi^2 l^2}{T^2}\right)}.$$

We can then describe the action of an arbitrary periodic external force in the following way: if we analyse the exciting force into purely periodic components, the individual terms of the Fourier series, then each component is subject to its own distortion of amplitude and phase displacement, and the separate effects are then superposed additively. If we are interested only in the distortion of amplitude (the phase displacement is only of secondary importance † in applications and, moreover, can be discussed in the same way as the distortion of amplitude), a study of the resonance curve gives us complete information about the way in which

---

* Questions of convergence will not be discussed here.
† Since e.g. it is imperceptible to the human ear.

the motions of the recording apparatus reproduce the external exciting force. For very large values of l or $\omega\left(=\dfrac{2\pi}{T}\,l\right)$ the effect of the exciting frequency on the displacement x will be hardly perceptible. On the other hand, all exciting frequencies in the neighbourhood of ω_1, the (circular) resonance frequency, will markedly affect the quantity x.

In the construction of physical measuring and recording apparatus the constants m, r, and k are at our disposal, at least within wide limits. These should be chosen so that the shape of the resonance curve is as well adapted as possible to the special requirements of the measurement in question. Here two considerations predominate. In the first place, it is desirable that the apparatus should be as sensitive as possible; that is, for all frequencies ω in question the value of α should be as large as possible. For small values of ω, as we have seen, α is approximately proportional to $1/k$, so that the number $1/k$ is a measure of the sensitiveness of the instrument for small exciting frequencies. The sensitiveness can therefore be increased by increasing $1/k$, that is, by weakening the restoring force.

The other important point is the necessity for *relative freedom from distortion*. Let us assume that the representation $f(t) = \sum\limits_{l=-N}^{N} \gamma_l e^{il(2\pi/T)t}$ is an adequate approximation to the exciting force. We then say that the apparatus records the exciting force $f(t)$ with relative freedom from distortion if for all circular frequencies $\omega \le N\,\dfrac{2\pi}{T}$ the distortion factor has approximately the same value. This condition is indispensable if we wish to derive conclusions about the exciting process directly from the behaviour of the apparatus; if, for example, a gramophone or wireless set is to reproduce both high and low musical notes with an approximately correct ratio of intensity. The requirement that the reproduction should be relatively " distortionless " can never be satisfied exactly, since no portion of the resonance curve is exactly horizontal. We can, however, attempt to choose the constants m, k, r, of the apparatus in such a way that no marked resonance occurs, and also in such a way that the curve has a horizontal tangent at the beginning, so that $\varphi(\omega) = \alpha$ remains approximately constant for small values of ω. As we have learned above, we can do this by putting

$$2km - r^2 = 0.$$

Given a constant m and a constant k, we can satisfy this requirement by adjusting the friction r properly, e.g. by inserting a properly chosen resistance in an electrical circuit. The resonance curve then shows us that from the frequency 0 to circular frequencies near the natural circular frequency ω_0 of the undamped system the instrument is nearly distortionless, and that above this frequency the damping is considerable. We therefore obtain relative freedom from distortion in a given interval of frequencies by first choosing m so small and k so large that the natural circular frequency ω_0 of the undamped system is greater than any of the exciting circular frequencies under consideration, and then choosing a damping factor r in accordance with the equation $2km - r^2 = 0$.

EXAMPLES

For the equations in Ex. 1–5 find the solution satisfying the initial conditions $x(0) = 0$, $\dot{x}(0) = 0$. For equations 1–4 state also the amplitude, the phase, and the value of ω for which the amplitude is a maximum:

1. $\ddot{x} + 3\dot{x} + 2x = \cos\omega t$.
2. $\ddot{x} + \dot{x} + x = \cos\omega t$.
3. $\ddot{x} + \dot{x} + x = \sin\omega t$.
4. $2\ddot{x} + 2\dot{x} + x = \cos\omega t$.
5. $\ddot{x} + 4\dot{x} + 4x = \cos\omega t$.

4. ADDITIONAL REMARKS ON DIFFERENTIAL EQUATIONS

A more systematic study of differential equations is made in Volume II, Chapter VI. Here only a few additions to the preceding special theory will be given.

1. Homogeneous Linear Differential Equations of Order n with Constant Coefficients.

More complicated vibration problems lead to a linear differential equation for the unknown function $x(t)$ of the independent variable, of the form

$$\frac{d^n x}{dt^n} + a_1 \frac{d^{n-1}x}{dt^{n-1}} + \ldots + a_n x = 0,$$

where $a_1, \ldots, a_n$ are constants and n is a positive integer. We can solve this by a method similar to that for the case $n = 2$ (p. 504).

Let $x = e^{\lambda t}$. If we substitute this function and its derivatives in the differential equation and remove the common factor $e^{\lambda t}$, we obtain the following equation of the n-th degree for λ:

$$f(\lambda) \equiv \lambda^n + a_1\lambda^{n-1} + \ldots + a_n = 0.$$

If λ is a root of this equation, $e^{\lambda t}$ satisfies the differential equation.

We shall now examine the various possibilities. Let $\lambda_1, \lambda_2, \ldots, \lambda_n$ be the roots of the equation $f(\lambda) = 0$, so that

$$f(\lambda) \equiv (\lambda - \lambda_1)(\lambda - \lambda_2)\ldots(\lambda - \lambda_n).$$

First assume that all the roots are different. If all the λ_n's are real, then we obtain n linearly independent solutions $e^{\lambda_n t}$, exactly as before. The *general solution* is any linear combination

$$c_1 e^{\lambda_1 t} + c_2 e^{\lambda_2 t} + \ldots + c_n e^{\lambda_n t}$$

of these. The constants c_n can be so determined that x and its first $n - 1$ derivatives take arbitrary pre-assigned values at time $t = 0$. To do this we must solve the following system of n linear equations *:

$$c_1 + c_2 + \ldots + c_n = x(0),$$
$$\lambda_1 c_1 + \lambda_2 c_2 + \ldots + \lambda_n c_n = x'(0),$$
$$\cdot \quad \cdot \quad \cdot \quad \cdot \quad \cdot \quad \cdot \quad \cdot \quad \cdot$$
$$\lambda_1{}^{n-1} c_1 + \lambda_2{}^{n-1} c_2 + \ldots + \lambda_n{}^{n-1} c_n = x^{(n-1)}(0).$$

If two of the roots are equal, say $\lambda_1 = \lambda_2$, then not only $e^{\lambda_1 t}$ but also $t e^{\lambda_1 t}$ is a solution. This can be verified as follows: since $f(\lambda) = 0$ has a double root $\lambda = \lambda_1 = \lambda_2$, by a well-known theorem in algebra it follows that

$$f'(\lambda) \equiv n\lambda^{n-1} + (n-1)a_1\lambda^{n-2} + \ldots + a_{n-1} = 0.$$

Now, by Leibnitz's rule for the derivative of a product (p. 202),

$$\frac{d^k}{dt^k}(te^{\lambda t}) = t\frac{d^k}{dt^k}e^{\lambda t} + k\frac{dt}{dt}\frac{d^{k-1}}{dt^{k-1}}e^{\lambda t} = t\lambda^k e^{\lambda t} + k\lambda^{k-1}e^{\lambda t}.$$

Substituting in the differential equation, we have

$$te^{\lambda t}(\lambda^n + a_1\lambda^{n-1} + \ldots + a_n) + e^{\lambda t}(n\lambda^{n-1} + (n-1)a_1\lambda^{n-2} + \ldots + a_{n-1})$$
$$= te^{\lambda t}f(\lambda) + e^{\lambda t}f'(\lambda) = 0,$$

since $f(\lambda) = 0$ and by the above remark on double roots $f'(\lambda) = 0$.

In the same way if $\lambda_1, \lambda_2, \ldots, \lambda_\mu$ are equal, we obtain the following linearly independent solutions:

$$e^{\lambda_1 t}, \; te^{\lambda_1 t}, \; \ldots t^{\mu-1}e^{\lambda_1 t},$$

which may be combined to give a general solution depending on $c_1, c_2, \ldots, c_n$. These parameters again enable us to adapt the solution to n pre-assigned conditions, so that for $t = 0$ we can fix the value of $x(0)$ and its first $n - 1$ derivatives.

---

* This set of equations always has a solution if the roots are unequal, for the determinant of the coefficients is not zero.

If the equation has complex roots, then by a theorem in algebra the roots occur in pairs, each one with its conjugate. Just as in the case $n = 2$, we obtain solutions of the form

$$\cos \beta t \cdot e^{at} \text{ and } \sin \beta t \cdot e^{at}, \text{ where } \lambda_1 = a + i\beta, \ \lambda_2 = a - i\beta.$$

A few examples will serve to illustrate the above.

Example 1.
$$\frac{d^3x}{dt^3} + 2\frac{d^2x}{dt^2} - \frac{dx}{dt} - 2x = 0,$$
$$f(\lambda) = \lambda^3 + 2\lambda^2 - \lambda - 2 = 0.$$

The general solution is $x = c_1 e^{-t} + c_2 e^{t} + c_3 e^{-2t}$.
A particular solution for which $x = 2$, $x' = 0$ at $t = 0$ is given by $x = e^{t} + e^{-t}$.

Example 2.
$$\frac{d^3x}{dt^3} - \frac{d^2x}{dt^2} - \frac{dx}{dt} + x = 0.$$

The general solution is $x = c_1 e^{t} + c_2 t e^{t} + c_3 e^{-t}$.

Example 3.
$$\frac{d^3x}{dt^3} - 2\frac{dx}{dt} + 4 = 0,$$
$$f(\lambda) = \lambda^3 - 2\lambda + 4 = (\lambda + 2)(\lambda - 1 + i)(\lambda - 1 - i).$$

The general solution is $x = c_1 e^{-2t} + c_2 e^{t} \cos t + c_3 e^{t} \sin t$.

2. Bernoulli's Equation.

An equation of the type

$$\frac{dx}{dt} + A(t)x = B(t),$$

where A and B are functions of t alone, is called a *linear equation*. In the case $B = 0$, if $x = \alpha(t)$, $x = \beta(t)$ are solutions, any linear combination of α and β is also a solution. We shall now consider the slightly more general type

$$\frac{dx}{dt} + A(t)x = B(t)x^n,$$

where n is a positive integer. This is known as Bernoulli's equation.

First consider the simpler case where B is zero, i.e. where

$$\frac{dx}{dt} + A(t)x = 0.$$

Rewriting the equation as $\dfrac{dx}{x} = -A(t)\,dt$, we see that it can be integrated immediately as follows:

$$\log x = -\int A(t)\,dt + c,$$

$$x = e^c e^{-\int A\,dt} = v e^{-\int A\,dt}$$

if we write $e^c = v$.

Let us now try to satisfy Bernoulli's equation by a function of the form $x = v e^{-\int A\,dt}$, where we assume that v is a variable, so that

$$\frac{dx}{dt} = \frac{dv}{dt} e^{-\int A\,dt} - vA e^{-\int A\,dt}.$$

Substituting, we have

$$\frac{dv}{dt} v^{-n} = B e^{-n\int A\,dt}\, e^{\int A\,dt},$$

which can be integrated at once, giving

$$x^{1-n} = (1-n)e^{(n-1)\int A\,dt}\left[\int B e^{(1-n)\int A\,dt}\,dt\right].$$

The above method is very important and may be applied in many cases. It is called the method of *variation of parameters*. (For further details, see Volume II, p. 445.) Note that our solution is expressed in terms of integrals which cannot in general be expressed in terms of the elementary functions.

Example.—Consider the equation

$$\frac{dx}{dt} - tx = t^3 x^2.$$

Let

$$x = v e^{\int t\,dt} = v e^{\frac{1}{2}t^2};$$

then

$$\frac{dx}{dt} - tx = \frac{dv}{dt} e^{\frac{1}{2}t^2} + vt e^{\frac{1}{2}t^2} - tv e^{\frac{1}{2}t^2} = \frac{dv}{dt} e^{\frac{1}{2}t^2},$$

and the equation becomes

$$\frac{dv}{dt} e^{\frac{1}{2}t^2} = t^3 v^2 e^{t^2}, \quad \text{or} \quad \frac{dv}{v^2} = t^3 e^{\frac{1}{2}t^2}\,dt.$$

By integration,

$$-\frac{1}{v} = (t^2 - 2)e^{\frac{1}{2}t^2} + c, \quad \text{or} \quad \frac{1}{x} = 2 - t^2 + c e^{-\frac{1}{2}t^2}.$$

This result could have been obtained by direct substitution in the formula given above, but actually to carry the method through is far more instructive.

3. Other Differential Equations of First Order Solvable by Simple Integration.

There are a few other types of differential equations of the first order which can be solved by integration (although in most cases the integration cannot be performed explicitly in terms of elementary functions).

The first method we shall consider is that of *separation of the variables*. If the differential equation can be brought into the form *

$$A(x)\,dx + B(y)\,dy = 0,$$

the variables are said to be *separable*. The solution obviously is

$$\int A(x)\,dx + \int B(y)\,dy + c = 0.$$

Example.—Consider the equation

$$yy' + xy^2 = x.$$

Here

$$y\,dy + x(y^2 - 1)\,dx = 0, \quad \text{or} \quad \frac{y\,dy}{y^2 - 1} + x\,dx = 0;$$

hence

$$\tfrac{1}{2}\log(y^2 - 1) + \tfrac{1}{2}x^2 = c, \quad \text{or} \quad (y^2 - 1)e^{x^2} = k.$$

Another type of equation which can be solved is of the form

$$M(x,\,y)\,dx + N(x,\,y)\,dy = 0,$$

where M and N are homogeneous functions of x and y of the same degree. In this case the fraction M/N is a function of y/x only, and we may write

$$\frac{dy}{dx} = f\left(\frac{y}{x}\right).$$

If we put $y = xv$, this becomes

$$x\frac{dv}{dx} + v = f(v).$$

The variables x, v are now separable as follows:

$$\frac{dx}{x} = \frac{dv}{f(v) - v}.$$

* That is, $y'B(y) + A(x) = 0$.

Integrating, we have

$$\log x = \int \frac{dv}{f(v) - v} + c.$$

Example.—Consider the equation

$$(2\sqrt{xy} - x)\,dy + y\,dx = 0.$$

Substituting $y = vx$, we have

$$(2v^{1/2} - 1)x\left(v + x\frac{dv}{dx}\right) + vx = 0,$$

$$v(2v^{1/2} - 1) + v + x(2v^{1/2} - 1)\frac{dv}{dx} = 0,$$

$$\frac{dx}{x} = -\frac{2v^{1/2} - 1}{2v^{3/2}}\,dv = -\frac{dv}{v} + \frac{dv}{2v^{3/2}}.$$

Integrating, we have

$$\log x = -\log v - v^{-\frac{1}{2}} + c$$

or

$$\log y + \sqrt{x/y} = c.$$

4. Differential Equations of the Second Order.

There are a few types of non-linear differential equations whose solutions can also be found by integration. One type has already been discussed implicitly in Chap. V (p. 297) when we studied the motion of a particle on a given curve. This type is as follows:

$$\frac{d^2x}{dt^2} = f(x).$$

Let $v = \dfrac{dx}{dt}$, so that

$$\frac{d^2x}{dt^2} = \frac{dv}{dt} = \frac{dv}{dx}\frac{dx}{dt} = v\frac{dv}{dx},$$

and our equation becomes

$$v\frac{dv}{dx} = f(x).$$

This may be regarded as an equation of the first order with v as dependent and x as independent variable. Separating the variables and integrating, we have

$$v\,dv = f(x)\,dx$$

$$v^2 = 2 \int f(x)\,dx + c \quad \text{or} \quad v = \sqrt{2 \int f(x)\,dx + c}.$$

Then

$$\frac{dx}{\sqrt{2 \int f(x)\,dx + c}} = dt,$$

which can be solved by integration (although in general it is impossible to carry out the integration explicitly).

This device aids us to solve equations of the following types:

$$\phi\left(\frac{d^2x}{dt^2}, \frac{dx}{dt}\right) = 0,$$

$$\psi\left(\frac{d^2x}{dt^2}, \frac{dx}{dt}, t\right) = 0,$$

$$\theta\left(\frac{d^2x}{dt^2}, \frac{dx}{dt}, x\right) = 0,$$

which reduce respectively, when we write $v = \dfrac{dx}{dt}$, to

$$\phi\left(\frac{dv}{dt}, v\right) = 0,$$

$$\psi\left(\frac{dv}{dt}, v, t\right) = 0,$$

$$\theta\left(v\frac{dv}{dx}, v, x\right) = 0.$$

These are equations of the first order which may be solvable by the preceding methods. This solution, after v has been replaced by $\dfrac{dx}{dt}$, will again be a differential equation of the first order, which must be solved for x. A few examples will make the process clear.

Example 1.
$$2a \frac{dy}{dx} \frac{d^2y}{dx^2} = 1.$$

Let $\frac{dy}{dx} = p$. The equation becomes

$$2ap \frac{dp}{dx} = 1.$$

Integrating by separating the variables, we have

$$ap^2 = x + c_1,$$

or

$$\sqrt{a} \frac{dy}{dx} = \sqrt{x + c_1}.$$

By integration,

$$\sqrt{a}(y + c_2) = \tfrac{2}{3}(x + c_1)^{3/2}.$$

Squaring, we have

$$a(y + c_2)^2 = \tfrac{4}{9}(x + c_1)^3.$$

Example 2.
$$(1 + x^2) \frac{d^2y}{dx^2} + x \frac{dy}{dx} = 0.$$

Let $\frac{dy}{dx} = p$.

$$(1 + x^2) \frac{dp}{dx} + xp = 0, \quad \text{or} \quad \frac{dp}{p} = - \frac{x\,dx}{1 + x^2}.$$

Integrating, we have

$$\log p = -\tfrac{1}{2} \log(1 + x^2) + c,$$

$$p = c_1(1 + x^2)^{-1/2},$$

or

$$\frac{dy}{dx} = \frac{c_1}{\sqrt{1 + x^2}},$$

whence

$$y = c_2 + c_1 \operatorname{ar\,sinh} x.$$

Example 3.
$$y \frac{d^2y}{dx^2} = 1 - \left(\frac{dy}{dx}\right)^2.$$

Let $\frac{dy}{dx} = p$, then $\frac{d^2y}{dx^2} = p \frac{dp}{dy}$. We have

$$py \frac{dp}{dy} = 1 - p^2, \quad \text{or} \quad \frac{p\,dp}{1 - p^2} = \frac{dy}{y}.$$

By integration,

$$-\tfrac{1}{2}\log(1 - p^2) = \log y + c,$$

that is,

$$y = c_1(1 - p^2)^{-1/2},$$

or

$$y^2(1 - p^2) = c_1^2,$$

and

$$\frac{dy}{dx} = p = \frac{\sqrt{y^2 - c_1^2}}{y}, \quad \text{or} \quad \frac{y\,dy}{\sqrt{y^2 - c_1^2}} = dx.$$

Integrating, we have

$$\sqrt{y^2 - c_1^2} = x + c_2,$$

that is,

$$y^2 = x^2 + c_3 x + c_4.$$

EXAMPLES

Solve the differential equations in Ex. 1–22.

1. $(1 + y^2)dx - (y - \sqrt{1 + y})(1 + x)^{3/2}\,dy = 0.$

2. $(x^3 + y^3)dy = 3x^2 y\,dx.$ 3. $y(\log x - \log y)\,dy - x\,dx = 0.$

4. $xy' + y = y^2 \log x.$ 5. $(1 + y^2)dx = (\arctan y - x)dy.$

6. $yy' + \tfrac{1}{2}y^2 = \sin x.$

7. $(x^3 y^3 + x^2 y^2 + xy + 1)y + (x^3 y^3 - x^2 y^2 - xy + 1)xy' = 0.$

8. $3y^2 y' + y^3 = x - 1.$ 9. $\sin x \cos y\,dx + \cos x \sin y\,dy = 0.$

10. $(1 + e^{x/y})dx + e^{x/y}\left(1 - \dfrac{x}{y}\right)dy = 0.$

11. $\dfrac{d^3x}{dt^3} - 3\dfrac{d^2x}{dt^2} + 3\dfrac{dx}{dt} - x = 0.$ 17. $\dfrac{d^2y}{dx^2} + \left(\dfrac{dy}{dx}\right)^2 + 1 = 0.$

12. $\dfrac{d^3x}{dt^3} - 6\dfrac{d^2x}{dt^2} + 9\dfrac{dx}{dt} = 0.$ 18. $\dfrac{d^4y}{dx^4} = \dfrac{d^2y}{dx^2}.$

13. $\dfrac{d^4y}{dx^4} + 2\dfrac{d^2y}{dx^2} + y = 0.$ 19. $(1 + x^2)\dfrac{d^2y}{dx^2} + 2x\dfrac{dy}{dx} = 0.$

14. $\dfrac{d^3y}{dx^3} - \dfrac{d^2y}{dx^2} + \dfrac{dy}{dx} = 0.$ 20. $(1 - y)\dfrac{d^2y}{dx^2} + 2\left(\dfrac{dy}{dx}\right) = 0.$

15. $\dfrac{d^8y}{dx^8} - 2\dfrac{d^4y}{dx^4} + y = 0.$ 21. $x\dfrac{d^2x}{dt^2} = 2\left(\dfrac{dx}{dt}\right)^2.$

16. $a\dfrac{d^2y}{dx^2} = \dfrac{dy}{dx}.$ 22. $(1 - t^2)\dfrac{d^2s}{dt^2} - t\dfrac{ds}{dt} = 2.$

23. Find the motion of a particle moving in a straight line under the attraction of a force varying as the inverse square of the distance from the origin.

SUMMARY OF IMPORTANT THEOREMS
AND FORMULÆ

1. HYPERBOLIC FUNCTIONS

(pp. 183–189)

$$\sinh x = \tfrac{1}{2}(e^x - e^{-x}).$$
$$\tanh x = \frac{\sinh x}{\cosh x} = \frac{e^x - e^{-x}}{e^x + e^{-x}}.$$

$$\cosh x = \tfrac{1}{2}(e^x + e^{-x}).$$
$$\coth x = \frac{1}{\tanh x} = \frac{e^x + e^{-x}}{e^x - e^{-x}}.$$

$$\cosh^2 x - \sinh^2 x = 1.$$
$$\cosh^2 x = \frac{1}{1 - \tanh^2 x}.$$

$$\cosh(x \pm y) = \cosh x \cosh y \pm \sinh x \sinh y.$$

$$\sinh(x \pm y) = \sinh x \cosh y \pm \cosh x \sinh y.$$

$$\cosh^2 x = \tfrac{1}{2}(\cosh 2x + 1). \qquad \sinh^2 x = \tfrac{1}{2}(\cosh 2x - 1).$$

$$\operatorname{ar\,sinh} x = \log\{x + \sqrt{(x^2 + 1)}\}.$$
$$\operatorname{ar\,cosh} x = \log\{x \pm \sqrt{(x^2 - 1)}\} \; (x \geqq 1).$$

$$\operatorname{ar\,tanh} x = \tfrac{1}{2} \log \frac{1+x}{1-x} \ (|\,x\,| < 1).$$

$$\operatorname{ar\,coth} x = \tfrac{1}{2} \log \frac{x+1}{x-1} \ (|\,x\,| > 1).$$

2. Convergence of Sequences and Series

1. Infinite Sequences (p. 38).

Cauchy's Convergence Test (p. 40).   A sequence of numbers a_n is convergent if, and only if, for every positive quantity ϵ there exists a number N such that

$$|\,a_n - a_m\,| < \epsilon$$

when $n > N,\ m > N$.

Operations with limits (pp. 41–42).   If $\lim\limits_{n\to\infty} a_n$ and $\lim\limits_{n\to\infty} b_n$ exist, then

$$\lim_{n\to\infty}(a_n \pm b_n) = \lim_{n\to\infty} a_n \pm \lim_{n\to\infty} b_n;$$

$$\lim_{n\to\infty}(a_n \cdot b_n) = \lim_{n\to\infty} a_n \cdot \lim_{n\to\infty} b_n;$$

$$\lim_{n\to\infty} \frac{a_n}{b_n} = \frac{\lim\limits_{n\to\infty} a_n}{\lim\limits_{n\to\infty} b_n}, \text{ provided } \lim_{n\to\infty} b_n \neq 0.$$

2. Infinite Series (p. 365 *et seq.*).

Cauchy's Convergence Test (p. 367).   The series Σa_n converges if, and only if, for every positive quantity ϵ there exists a number N such that

$$|\,a_n + a_{n+1} + \ldots + a_m\,| < \epsilon$$

when $m > n > N$.

Note.—All the following criteria are *sufficient* but not *necessary*.

Principle of comparison of series (p. 377). Σa_n converges if numbers b_n exist such that $b_n \geqq |\,a_n\,|$ for all values of n and Σb_n converges.

Ratio test and root test (p. 378). Σa_n converges if there is a number N, and also a number $q < 1$, such that

$$\left| \frac{a_{n+1}}{a_n} \right| < q \quad \text{or} \quad \sqrt[n]{|a_n|} < q$$

for all values of $n > N$; in particular, if there is a number $k < 1$ such that

$$\lim_{n \to \infty} \left| \frac{a_{n+1}}{a_n} \right| = k \quad \text{or} \quad \lim_{n \to \infty} \sqrt[n]{|a_n|} = k.$$

Σa_n diverges if there is a number $k > 1$ such that

$$\lim_{n \to \infty} \left| \frac{a_{n+1}}{a_n} \right| = k \quad \text{or} \quad \lim_{n \to \infty} \sqrt[n]{|a_n|} = k.$$

Leibnitz's Test (p. 370). Σa_n converges if the terms have alternating signs and $|a_n|$ tends monotonically to zero.

3. DIFFERENTIATION

1. **General Rules (Fundamental Ideas, p. 88 et seq.).**

$$\{f(x) \pm g(x)\}' = f'(x) \pm g'(x).$$
$$\{f(x)g(x)\}' = f'(x)g(x) + f(x)g'(x).$$

$$\left(\frac{f(x)}{g(x)} \right)' = \frac{f'(x)g(x) - f(x)g'(x)}{\{g(x)\}^2}, \quad g(x) \neq 0 \text{ (pp. 136–139).}$$

$$\{f(x)g(x)\}^{(n)} = f^{(n)}(x)g(x) + \binom{n}{1}f^{(n-1)}(x)g'(x)$$
$$+ \binom{n}{2}f^{(n-2)}(x)g''(x) + \ldots + f(x)g^{(n)}(x).$$

(Leibnitz's Rule, p. 202.)

Chain rule. If $f(x) = g\{\phi(x)\}$,

$$\frac{df}{dx} = \frac{dg}{d\phi} \frac{d\phi}{dx},$$

$$\frac{d^2f}{dx^2} = \frac{d^2g}{d\phi^2}\left(\frac{d\phi}{dx}\right)^2 + \frac{dg}{d\phi}\frac{d^2\phi}{dx^2}, \text{ and so on (pp. 153 et seq., 202).}$$

If $u = f(\xi, \eta, \zeta, \ldots)$, where $\xi = \xi(x, y)$, $\eta = \eta(x, y)$, $\ldots$,

$$u_x = f_\xi \xi_x + f_\eta \eta_x + f_\zeta \zeta_x + \ldots,$$
$$u_{xx} = f_{\xi\xi} \xi_x{}^2 + f_{\eta\eta} \eta_x{}^2 + f_{\zeta\zeta} \zeta_x{}^2 + \ldots$$
$$+ 2f_{\xi\eta} \xi_x \eta_x + 2f_{\xi\zeta} \xi_x \zeta_x + \ldots$$
$$+ \ldots\ldots\ldots\ldots\ldots\ldots$$
$$+ f_\xi \xi_{xx} + f_\eta \eta_{xx} + f_\zeta \zeta_{xx} + \ldots,$$

with corresponding formulæ for u_{xy} and u_{yy} (p. 476).

Implicit functions. If $F(x, y) = 0$,

$$\frac{dy}{dx} = -\frac{F_x}{F_y},$$

$$\frac{d^2y}{dx^2} = -\frac{F_{xx} F_y{}^2 - 2F_{xy} F_x F_y + F_{yy} F_x{}^2}{F_y{}^3} \qquad \text{(p. 483)}.$$

Functions expressed in terms of a parameter. If $x = x(t)$, $y = y(t)$,

$$\frac{dy}{dx} = \frac{dy}{dt} \Big/ \frac{dx}{dt} \qquad \text{(p. 262)}.$$

Inverse functions.

$$\frac{dy}{dx} = 1 \Big/ \frac{dx}{dy} \qquad \text{(p. 145)}.$$

If $\xi = \phi(x, y)$, $\eta = \psi(x, y)$,

$$\frac{\partial x}{\partial \xi} = \frac{\psi_y}{D}, \quad \frac{\partial x}{\partial \eta} = -\frac{\phi_y}{D}, \quad \frac{\partial y}{\partial \xi} = -\frac{\psi_x}{D}, \quad \frac{\partial y}{\partial \eta} = \frac{\phi_x}{D},$$

where

$$D = \frac{\partial(\xi, \eta)}{\partial(x, y)} = \begin{vmatrix} \phi_x & \phi_y \\ \psi_x & \psi_y \end{vmatrix} = \phi_x \psi_y - \phi_y \psi_x$$

(functional determinant or Jacobian) (p. 479).

2. **Special Formulæ** (pp. 94–96, 139–141, 149–150, 167 *et seq.*, 186–187).

$$(x^n)' = nx^{n-1}.$$

$$(\sin x)' = \cos x. \qquad\qquad (\text{arc } \sin x)' = \frac{1}{\sqrt{(1 - x^2)}}.$$

$$(\cos x)' = -\sin x. \qquad\qquad (\text{arc } \cos x)' = -\frac{1}{\sqrt{(1 - x^2)}}.$$

$$(\tan x)' = \frac{1}{\cos^2 x} = \sec^2 x. \qquad (\text{arc}\tan x)' = \frac{1}{1 + x^2}.$$

$$(\cot x)' = -\frac{1}{\sin^2 x} = -\text{cosec}^2 x. \quad (\text{arc}\cot x)' = -\frac{1}{1 + x^2}.$$

$$(\sinh x)' = \cosh x. \qquad (\text{ar}\sinh x)' = \frac{1}{\sqrt{(1 + x^2)}}.$$

$$(\cosh x)' = \sinh x. \qquad (\text{ar}\cosh x)' = \pm\frac{1}{\sqrt{(x^2 - 1)}}$$
$$(x > 1).$$

$$(\tanh x)' = \frac{1}{\cosh^2 x} = \text{sech}^2 x. \qquad (\text{ar}\tanh x)' = \frac{1}{1 - x^2} \quad (|x| < 1).$$

$$(\coth x)' = -\frac{1}{\sinh^2 x} = -\text{cosech}^2 x. \; (\text{ar}\coth x)' = \frac{1}{1 - x^2} \quad (|x| > 1).$$

$$(\log_a x)' = \frac{1}{x}\log_a e; \qquad\qquad (a^x)' = a^x \log_e a;$$

in particular,   in particular,

$$(\log x)' = \frac{1}{x}. \qquad\qquad (e^x)' = e^x.$$

$$(u^v)' = u^v(vu'/u + v'\log u).$$

4. INTEGRATION

1. **General Rules (Fundamental Ideas, p. 79 et seq.).**

$$\int_a^b f(x)\, dx + \int_b^c f(x)\, dx = \int_a^c f(x)\, dx.$$

$$\int_a^b f(x)\, dx = -\int_b^a f(x)\, dx.$$

$$\int_a^b \{f(x) + g(x)\}\, dx = \int_a^b f(x)\, dx + \int_a^b g(x)\, dx.$$

$$\int_a^b c f(x)\, dx = c\int_a^b f(x)\, dx \quad (\text{pp. 81 et seq., 141}).$$

Estimation of integrals. If $f(x) \geqq g(x)$, $b \geqq a$,

$$\int_a^b f(x)\, dx \geqq \int_a^b g(x)\, dx \qquad\qquad (\text{p. 126}).$$

Integration by Parts (pp. 218–219).

$$\int_a^b f(x)g'(x)\,dx = f(x)g(x)\,\Big|_a^b - \int_a^b f'(x)g(x)\,dx.$$

Method of substitution (pp. 207–212).

$$\int_a^b f(x)\,dx = \int_\alpha^\beta f\{\phi(u)\}\,\phi'(u)\,du,$$

where $a = \phi(\alpha)$, $b = \phi(\beta)$.

Connexion between differentiation and integration (p. 111 *et seq.*).

$$\frac{d}{dx}\int_a^x f(u)\,du = f(x).$$

Improper Integrals (pp. 197–254).

If $f(x)$ is continuous except at the point $x = b$ where it becomes infinite, $\int_a^b f(x)\,dx$ is (absolutely) convergent, if in the neighbourhood of $x = b$

$$|f(x)| \leqq \frac{M}{(b-x)^\nu},$$

where $\nu < 1$ (p. 248).

$\int_a^\infty f(x)\,dx$ converges (absolutely) if

$$|f(x)| \leqq \frac{M}{x^\nu},$$

where $\nu > 1$, for values of $x \geqq A$ (p. 250).

2. **Special Formulæ** (pp. 82–87, 128–130, 142 *et seq.*, 151, 168 *et seq.*, 206, 208–209, 210, 213–217, 220 *et seq.*).

$$\int x^n\,dx = \frac{x^{n+1}}{n+1}. \qquad\qquad \int \log x\,dx = x\log x - x.$$

$$\int \frac{dx}{x} = \log|x|. \qquad\qquad \int \frac{1}{x}\log x\,dx = \tfrac{1}{2}(\log x)^2.$$

$$\int a^x\,dx = \frac{a^x}{\log a}. \qquad\qquad \int \frac{1}{x\log x}\,dx = \log|\log x|.$$

$$\int x^a \log x\,dx = \frac{x^{a+1}}{a+1}\left(\log x - \frac{1}{a+1}\right); \quad a \neq -1.$$

$$\int \sin x \, dx = -\cos x. \qquad \int \sinh x \, dx = \cosh x.$$

$$\int \cos x \, dx = \sin x. \qquad \int \cosh x \, dx = \sinh x.$$

$$\int \tan x \, dx = -\log |\cos x|. \qquad \int \tanh x \, dx = \log \cosh x.$$

$$\int \cot x \, dx = \log |\sin x|. \qquad \int \coth x \, dx = \log |\sinh x|.$$

$$\int \text{arc} \sin x \, dx = x \, \text{arc} \sin x + \sqrt{(1 - x^2)}.$$

$$\int \text{arc} \cos x \, dx = x \, \text{arc} \cos x - \sqrt{(1 - x^2)}.$$

$$\int \text{arc} \tan x \, dx = x \, \text{arc} \tan x - \tfrac{1}{2} \log (1 + x^2).$$

$$\int \text{arc} \cot x \, dx = x \, \text{arc} \cot x + \tfrac{1}{2} \log (1 + x^2).$$

$$\int \text{ar} \sinh x \, dx = x \, \text{ar} \sinh x - \sqrt{(1 + x^2)}.$$

$$\int \text{ar} \cosh x \, dx = x \, \text{ar} \cosh x - \sqrt{(x^2 - 1)}.$$

$$\int \text{ar} \tanh x \, dx = x \, \text{ar} \tanh x + \tfrac{1}{2} \log (1 - x^2).$$

$$\int \text{ar} \coth x \, dx = x \, \text{ar} \coth x + \tfrac{1}{2} \log (x^2 - 1).$$

$$\int \frac{dx}{\sin x} = \log \left| \tan \frac{x}{2} \right|. \qquad \int \frac{dx}{\sinh x} = \log \left| \tanh \frac{x}{2} \right|.$$

$$\int \frac{dx}{\cos x} = \log \left| \tan \left(\frac{x}{2} + \frac{\pi}{4} \right) \right| \quad \int \frac{dx}{\cosh x} = 2 \, \text{arc} \tan \left(\tanh \frac{x}{2} \right).$$

$$= 2 \, \text{ar} \tanh \left(\tan \frac{x}{2} \right).$$

$$\int \frac{dx}{\sin x \cos x} = \log |\tan x|. \qquad \int \frac{dx}{\sinh x \cosh x} = \log |\tanh x|.$$

$$\int \frac{dx}{\sin^2 x} = -\cot x. \qquad \int \frac{dx}{\sinh^2 x} = -\coth x.$$

$$\int \frac{dx}{\cos^2 x} = \tan x. \qquad \int \frac{dx}{\cosh^2 x} = \tanh x.$$

$$\int \sin^2 x \, dx = \tfrac{1}{2}(x - \sin x \cos x).$$

$$\int \cos^2 x \, dx = \tfrac{1}{2}(x + \sin x \cos x).$$

$$\left.\begin{aligned}
\int \frac{dx}{a^2 \sin^2 x + b^2 \cos^2 x} &= \frac{1}{ab} \arctan\left(\frac{a}{b} \tan x\right) \\[2mm]
\int \frac{dx}{a^2 \sin^2 x - b^2 \cos^2 x} &= -\frac{1}{ab} \operatorname{ar\,tanh}\left(\frac{a}{b} \tan x\right)
\end{aligned}\right\} \; a,\, b \neq 0.$$

$$\int \frac{dx}{x^2 + a^2} = \frac{1}{a} \arctan \frac{x}{a}.$$

$$\int \frac{dx}{x^2 - a^2} = \begin{cases}
-\dfrac{1}{a} \operatorname{ar\,tanh} \dfrac{x}{a} = \dfrac{1}{2a} \log \dfrac{a-x}{a+x}, & \text{if } |x| < a. \\[4mm]
-\dfrac{1}{a} \operatorname{ar\,coth} \dfrac{x}{a} = \dfrac{1}{2a} \log \dfrac{x-a}{x+a}, & \text{if } |x| > a,\ a > 0.
\end{cases}$$

$$\int \frac{dx}{\sqrt{(a^2 - x^2)}} = \begin{cases} +\arcsin \dfrac{x}{a}. \\[2mm] -\arccos \dfrac{x}{a}. \end{cases} \qquad \int \frac{dx}{x\sqrt{(x^2 - a^2)}} = \begin{cases} -\dfrac{1}{a} \arcsin \dfrac{a}{x}. \\[2mm] +\dfrac{1}{a} \arccos \dfrac{a}{x}. \end{cases}$$

$$\int \frac{x \, dx}{\sqrt{(a^2 + x^2)}} = \sqrt{(a^2 + x^2)}.$$

$$\int \frac{x \, dx}{\sqrt{(a^2 - x^2)}} = -\sqrt{(a^2 - x^2)}.$$

$$\int \frac{dx}{\sqrt{(a^2 + x^2)}} = \operatorname{ar\,sinh} \frac{x}{a} = \log\{\pm x + \sqrt{(x^2 + a^2)}\}.$$

$$\int \frac{dx}{\sqrt{(x^2 - a^2)}} = \operatorname{ar\,cosh} \frac{x}{a} = \log\{x \pm \sqrt{(x^2 - a^2)}\}.$$

$$\int \frac{dx}{x\sqrt{(x^2 + a^2)}} = -\frac{1}{a} \operatorname{ar\,sinh} \frac{a}{x} = -\frac{1}{a} \log \frac{\pm a + \sqrt{(a^2 + x^2)}}{x}.$$

$$\int \frac{dx}{x\sqrt{(a^2 - x^2)}} = -\frac{1}{a} \operatorname{ar\,cosh} \frac{a}{x} = -\frac{1}{a} \log \frac{a \pm \sqrt{(a^2 - x^2)}}{x}.$$

$$\int \sqrt{(a^2 - x^2)}\, dx = -\frac{1}{2}\, a^2 \text{ arc cos}\, \frac{x}{a} + \frac{1}{2} x \sqrt{(a^2 - x^2)}.$$

$$\int \sqrt{(x^2 - a^2)}\, dx = -\frac{1}{2}\, a^2 \text{ ar cosh}\, \frac{x}{a} + \frac{1}{2} x \sqrt{(x^2 - a^2)}.$$

$$\int \sqrt{(x^2 + a^2)}\, dx = \frac{1}{2}\, a^2 \text{ ar sinh}\, \frac{x}{a} + \frac{1}{2} x \sqrt{(x^2 + a^2)}.$$

$$\int \frac{dx}{x^2 + 2bx + c} = -\frac{1}{\sqrt{(b^2 - c)}} \text{ ar tanh}\, \frac{x + b}{\sqrt{(b^2 - c)}}$$

$$= -\frac{1}{2\sqrt{(b^2 - c)}}\, \log\left|\frac{\sqrt{(b^2 - c)} - x - b}{\sqrt{(b^2 - c)} + x + b}\right|,$$

$$\text{if } c < b^2, \text{ i.e. } x^2 + 2bx + c = 0 \text{ has real roots.}$$

$$\int \frac{dx}{x^2 + 2bx + c} = \frac{1}{\sqrt{(c - b^2)}} \text{ arc tan}\, \frac{x + b}{\sqrt{(c - b^2)}},$$

$$\text{if } c > b^2, \text{ i.e. } x^2 + 2bx + c = 0 \text{ has imaginary roots.}$$

$$\int e^{ax} \sin bx\, dx = \frac{1}{a^2 + b^2}\, e^{ax}(a \sin bx - b \cos bx).$$

$$\int e^{ax} \cos bx\, dx = \frac{1}{a^2 + b^2}\, e^{ax}(a \cos bx + b \sin bx).$$

$$\int \sin^n x \cos x\, dx = \frac{\sin^{n+1} x}{n + 1}.$$

Recurrence Formulæ (p. 221 *et seq.*).

$$\int \cos^n x\, dx = \frac{1}{n} \cos^{n-1} x \sin x + \frac{n - 1}{n} \int \cos^{n-2} x\, dx.$$

$$\int \sin^n x\, dx = -\frac{1}{n} \sin^{n-1} x \cos x + \frac{n - 1}{n} \int \sin^{n-2} x\, dx.$$

$$\int x^n \cos x\, dx = x^n \sin x - n \int x^{n-1} \sin x\, dx.$$

$$\int x^n \sin x\, dx = -x^n \cos x + n \int x^{n-1} \cos x\, dx.$$

$$\int \sin^m x \cos^n x\, dx = \frac{\sin^{m+1} x \cos^{n-1} x}{m + n} + \frac{n - 1}{m + n} \int \sin^m x \cos^{n-2} x\, dx.$$

$$\int (\log x)^n \, dx = x (\log x)^n - n \int (\log x)^{n-1} dx.$$

$$\int x^n e^x \, dx = x^n e^x - n \int x^{n-1} e^x \, dx.$$

$$\int x^a (\log x)^n \, dx = \frac{x^{a+1}(\log x)^n}{a+1} - \frac{n}{a+1} \int x^a (\log x)^{n-1} dx \quad (a \neq -1).$$

$$\int \frac{dx}{(1+x^2)^n} = \frac{x}{2(n-1)(1+x^2)^{n-1}} + \frac{2n-3}{2(n-1)} \int \frac{dx}{(1+x^2)^{n-1}}.$$

3. Integration of Special Types of Functions.

(a) *Rational Functions.* These are reduced to the following three fundamental types by resolution into partial fractions (pp. 226–234):

$$\int \frac{dx}{(x-a)^n} = -\frac{1}{n-1} \frac{1}{(x-a)^{n-1}};$$

$$\int \frac{dx}{(x^2+2bx+c)^n} = \frac{1}{(c-b^2)^{n-\frac{1}{2}}} \int \frac{du}{(1+u^2)^n},$$

where $\quad c - b^2 > 0, \quad u = (x+b)/\sqrt{(c-b^2)},$

the integral on the right being evaluated by the last recurrence formula given above;

$$\int \frac{x \, dx}{(x^2+2bx+c)^n}$$
$$= -\frac{1}{2(n-1)} \frac{1}{(x^2+2bx+c)^{n-1}} - b \int \frac{dx}{(x^2+2bx+c)^n},$$

where the integral on the right is of the type immediately preceding.

In what follows R denotes a rational function.

(b) $\int R(\sin x, \cos x) \, dx$ (p. 237).

Substitution: $t = \tan \dfrac{x}{2}$, so that $\sin x = \dfrac{2t}{1+t^2}$, $\cos x = \dfrac{1-t^2}{1+t^2}$, $\dfrac{dx}{dt} = \dfrac{2}{1+t^2}.$

If, however, R is an even function or involves $\tan x$ alone, the following substitution is more convenient:

$$u = \tan x, \quad \sin^2 x = \frac{u^2}{1+u^2}, \quad \cos^2 x = \frac{1}{1+u^2}, \quad \frac{dx}{du} = \frac{1}{1+u^2}.$$

(c) $\int R(\cosh x, \sinh x)\, dx$ (p. 237).

Substitution: $t = \tanh \dfrac{x}{2}$, so that $\sinh x = \dfrac{2t}{1-t^2}$, $\cosh x = \dfrac{1+t^2}{1-t^2}$, $\dfrac{dx}{dt} = \dfrac{2}{1-t^2}$.

(d) $\int R(e^{mx})\, dx$.

Substitution: $t = e^{mx}$, $\dfrac{dx}{dt} = \dfrac{1}{mt}$.

(e) $\int R(x, \sqrt{(1-x^2)})\, dx$ (pp. 237–238).

Substitution:

$$t = \sqrt{\left(\frac{1-x}{1+x}\right)}, \quad x = \frac{1-t^2}{1+t^2}, \quad \sqrt{(1-x^2)} = \frac{2t}{1+t^2}, \quad \frac{dx}{dt} = -\frac{4t}{(1+t^2)^2}.$$

(f) $\int R(x, \sqrt{(x^2-1)})\, dx$ (p. 238).

Substitution:

$$t = \sqrt{\left(\frac{x-1}{x+1}\right)}, \quad x = \frac{1+t^2}{1-t^2}, \quad \sqrt{(x^2-1)} = \frac{2t}{1-t^2}, \quad \frac{dx}{dt} = \frac{4t}{(1-t^2)^2}.$$

(g) $\int R(x, \sqrt{(1+x^2)})\, dx$ (p. 238).

Substitution:

$$t = x + \sqrt{(x^2+1)}, \quad x = \frac{t^2-1}{2t}, \quad \sqrt{(x^2+1)} = \frac{1+t^2}{2t}, \quad \frac{dx}{dt} = \frac{t^2+1}{2t^2}.$$

(h) $\int R(x, \sqrt{(ax^2 + 2bx + c)})\, dx$ (p. 239).

The substitution $\xi = \dfrac{ax+b}{\sqrt{|\, ac - b^2\,|}}$ reduces this integral to one of the three preceding types.

(*i*) $\int R(x, \sqrt{(ax+b)}, \sqrt{(cx+d)})\, dx$ (p. 239).

Substitution: $\xi = \sqrt{(cx+d)}$ or $x = \dfrac{1}{c}(\xi^2 - d)$, $\dfrac{dx}{d\xi} = \dfrac{2\xi}{c}$.

(*k*) $\int R\left(x, \sqrt[n]{\left(\dfrac{ax+b}{cx+d}\right)}\right)\, dx$ (p. 240).

Substitution:

$$\xi = \sqrt[n]{\left(\frac{ax+b}{cx+d}\right)}, \quad x = -\frac{d\xi^n - b}{c\xi^n - a}, \quad \frac{dx}{d\xi} = \frac{ad - bc}{(c\xi^n - a)^2}\, n\xi^{n-1}.$$

5. Uniform Convergence and Interchange of Infinite Operations

For the definition of uniform convergence see p. 391.

A series which is uniformly convergent in a closed interval and whose terms are continuous functions represents a continuous function in that interval (p. 393).

If $|f_n(x)| \leqq a_n$ and Σa_n converges, $\Sigma f_n(x)$ converges uniformly (and absolutely) (p. 392).

Interchange of Summation and Differentiation pp. 396–397). Any convergent series of continuous functions may be differentiated term by term, provided the resulting series converges *uniformly*.

Interchange of Summation and Integration (p. 394). Any uniformly convergent series of continuous functions may be integrated term by term. The resulting series also converges uniformly.

6. Special Limits

Stirling's Formula (p. 361).

$$\lim_{n \to \infty} \frac{n!}{\sqrt{2\pi}\, n^{n+\frac{1}{2}} e^{-n}} = 1.$$

Wallis's Product (pp. 223–225, 363, 445).

$$\frac{\pi}{2} = \prod_{n=1}^{\infty} \left(\frac{2n}{2n-1}\, \frac{2n}{2n+1}\right).$$

$$\sqrt{\pi} = \lim_{n \to \infty} \frac{(n!)^2\, 2^{2n}}{(2n)!\, \sqrt{n}}.$$

(For infinite products, see pp. 419–422.)

$$e^x = \lim_{n \to \infty} \left(1 + \frac{x}{n}\right)^n \qquad \text{(p. 175)}.$$

$$\zeta(s) = \sum_{n=1}^{\infty} \frac{1}{n^s} = \prod_p \frac{1}{1 - p^{-s}}, \quad s > 1 \qquad \text{(p. 420)}.$$

$$\sin \pi x = \pi x \prod_{n=1}^{\infty} \left(1 - \frac{x^2}{n^2}\right) \qquad \text{(p. 445)}.$$

Definition of the Gamma Function (pp. 250–251).

$$\Gamma(x) = \int_0^{\infty} e^{-t} t^{x-1} dx \qquad (x \geqq 1):$$

$$\Gamma(x + 1) = x\Gamma(x);$$

if x is a positive integer n,

$$\Gamma(n) = (n - 1)!$$

Order of magnitude of functions (pp. 190–195).

$$\lim_{x \to \infty} \frac{e^{cx}}{x^a} = \infty, \text{ if } c > 0 \qquad \text{(p. 192)}.$$

$$\lim_{x \to \infty} \frac{\log x}{x^a} = 0, \text{ if } a > 0 \qquad \text{(p. 192)}.$$

$$\lim_{x \to 0} x^a \log x = 0, \text{ if } a > 0 \qquad \text{(p. 195)}.$$

7. Special Definite Integrals

Orthogonality relations of the trigonometric functions (p. 217).

$$\int_{-\pi}^{+\pi} \sin mx \, \sin nx \, dx = \begin{cases} 0, & \text{if } m \neq n. \\ \pi, & \text{if } m = n, \ n \neq 0. \end{cases}$$

$$\int_{-\pi}^{+\pi} \sin mx \, \cos nx \, dx = 0.$$

$$\int_{-\pi}^{+\pi} \cos mx \, \cos nx \, dx = \begin{cases} 0, & \text{if } m \neq n. \\ \pi, & \text{if } m = n, \ n \neq 0. \end{cases}$$

$$\int_0^{\infty} e^{-x^2} dx = \frac{1}{2}\sqrt{\pi} \qquad \text{(p. 496)}.$$

$$\int_0^{\infty} \frac{\sin x}{x} dx = \frac{1}{2}\pi \quad \text{(pp. 251–253, 418, 450)}.$$

8. Mean Value Theorems

Mean Value Theorem of the Differential Calculus (p. 103).

$$\frac{f(x+h)-f(x)}{h}=f'(x+\theta h),\ \ 0<\theta<1.$$

If $f(x)=f(x+h)=0$, this gives *Rolle's Theorem* (p. 105)· between two zeros of the function there is always a zero of the derivative.

Generalized Mean Value Theorem (pp. 135, 203).

$$\frac{f(b)-f(a)}{g(b)-g(a)}=\frac{f'(\xi)}{g'(\xi)},$$

where ξ is a value between a and b.

Taylor's Theorem (pp. 320–323).

$$f(x+h)=f(x)+\frac{h}{1!}f'(x)+\frac{h^2}{2!}f''(x)+\ .\ .\ .+\frac{h^n}{n!}f^{(n)}(x)+R_n,$$

with the remainder (pp. 323–324).

$$R_n=\frac{1}{n!}\int_0^h(h-\tau)^n f^{(n+1)}(x+\tau)\,d\tau$$

$$=\frac{h^{n+1}}{(n+1)!}f^{(n+1)}(x+\theta h)$$

$$=\frac{h^{n+1}}{n!}(1-\theta)^n f^{(n+1)}(x+\theta h)\qquad(0<\theta<1).$$

Mean Value Theorem of the Integral Calculus (p. 127).

$$\int_a^b f(x)\,dx=(b-a)\,f(\xi),\ \text{where}\ a\leqq\xi\leqq b.$$

$$\int_a^b f(x)\,p(x)\,dx=f(\xi)\int_a^b p(x)\,dx,\ \text{if}\ p(x)\geqq 0.$$

9. Expansions in Series: Taylor Series, Fourier Series

1. **Power Series** (for definition, see p. 398).

(a) *Power Series in General.*

Any power series $\sum\limits_{n=0}^{\infty} a_n x^n$

in one variable has a *radius of convergence* ρ (which may be zero or infinite); the series converges when $|x| < \rho$, and in fact it converges *uniformly* and absolutely in every interval $|x| \leq \eta$, where $\eta < \rho$; when $|x| > \rho$, the series diverges (p. 400).

If the remainder in Taylor's theorem tends to zero as n increases, we have the infinite power series (p. 325)

$$f(x + h) = f(x) + \frac{h}{1!} f'(x) + \frac{h^2}{2!} f''(x) + \ldots + \frac{h^n}{n!} f^{(n)}(x) + \ldots.$$

(b) *Special Taylor Series* (pp. 316–319, 326–330, 405–409, 422–423).

$$\log(1 + x) = x - \frac{x^2}{2} + \frac{x^3}{3} - \frac{x^4}{4} + - \ldots + (-1)^{n-1} \frac{x^n}{n} + \ldots$$
$$\text{for } -1 < x \leq 1.$$

$$e^x = 1 + \frac{x}{1!} + \frac{x^2}{2!} + \ldots + \frac{x^n}{n!} + \ldots$$

$$\sin x = x - \frac{x^3}{3!} + \frac{x^5}{5!} - + \ldots + (-1)^n \frac{x^{2n+1}}{(2n+1)!} + \ldots$$

$$\cos x = 1 - \frac{x^2}{2!} + \frac{x^4}{4!} - + \ldots + (-1)^n \frac{x^{2n}}{(2n)!} + \ldots \qquad \text{for all values of } x.$$

$$\sinh x = x + \frac{x^3}{3!} + \frac{x^5}{5!} + \ldots + \frac{x^{2n+1}}{(2n+1)!} + \ldots$$

$$\cosh x = 1 + \frac{x^2}{2!} + \frac{x^4}{4!} + \ldots + \frac{x^{2n}}{(2n)!} + \ldots$$

$$\tan x = \sum_{\nu=1}^{\infty} (-1)^{\nu-1} \frac{2^{2\nu}(2^{2\nu} - 1) B_{2\nu}}{(2\nu)!} x^{2\nu-1} \quad \text{for } -\frac{\pi}{2} < x < \frac{\pi}{2},$$

$$x \cot x = \sum_{\nu=0}^{\infty} (-1)^{\nu} \frac{2^{2\nu} B_{2\nu}}{(2\nu)!} x^{2\nu} \quad \text{for } -\pi < x < \pi,$$

where the quantities $B_{2\nu}$ are *Bernoulli's numbers* (p. 423).

$$\left.\begin{aligned}
\text{arc } \sin x &= x + \frac{1}{2}\frac{x^3}{3} + \frac{1.3}{2.4}\frac{x^5}{5} + \frac{1.3.5}{2.4.6}\frac{x^7}{7} + \cdots \\
\text{ar } \sinh x &= x - \frac{1}{2}\frac{x^3}{3} + \frac{1.3}{2.4}\frac{x^5}{5} - \frac{1.3.5}{2.4.6}\frac{x^7}{7} + - \cdots \\
\text{arc } \tan x &= x - \frac{x^3}{3} + \frac{x^5}{5} - + \cdots
\end{aligned}\right\} \begin{aligned}&\text{for}\\ &-1 \leqq x \leqq 1.\end{aligned}$$

$$\text{ar } \tanh x = x + \frac{x^3}{3} + \frac{x^5}{5} + \cdots \qquad\qquad \text{for } |x| < 1.$$

Binomial Series.

$$(1 + x)^a$$

$$= 1 + ax + \frac{a(a-1)}{2!}x^2 + \cdots + \frac{a(a-1)(a-2)\ldots(a-n+1)}{n!}x^n + \cdots$$

$$\begin{aligned}
&\text{for } -1 < x < 1,\\
&\text{if } a > -1 \text{ for } x = 1 \text{ also,}\\
&\text{if } a \geqq 0 \text{ for } x = -1 \text{ also;}
\end{aligned}$$

in particular,

$$\frac{1}{1 + x} = 1 - x + x^2 - x^3 + - \cdots,$$

$$\frac{1}{(1 + x)^2} = 1 - 2x + 3x^2 - 4x^3 + - \cdots,$$

$$\sqrt{(1 + x)} = 1 + \frac{1}{2}x - \frac{1}{2.4}x^2 + \frac{1.3}{2.4.6}x^3 - \frac{1.3.5}{2.4.6.8}x^4 + - \cdots,$$

$$\frac{1}{\sqrt{(1 + x)}} = 1 - \frac{1}{2}x + \frac{1.3}{2.4}x^2 - \frac{1.3.5}{2.4.6}x^3 + \frac{1.3.5.7}{2.4.6.8}x^4 - + \cdots.$$

Elliptic integral:

$$\int_0^{\pi/2} \frac{d\phi}{\sqrt{(1 - k^2 \sin^2 \phi)}}$$

$$= \frac{\pi}{2}\left\{1 + \left(\frac{1}{2}\right)^2 k^2 + \left(\frac{1.3}{2.4}\right)^2 k^4 + \left(\frac{1.3.5}{2.4.6}\right)^2 k^6 + \cdots\right\}.$$

2. Fourier Series.

If the function $f(x)$ is sectionally smooth in the interval $-\pi \leqq x \leqq \pi$, i.e. if its first derivative is sectionally continuous, the Fourier series

$$f(x) = \frac{1}{2} a_0 + \sum_{\nu=1}^{\infty} (a_\nu \cos \nu x + b_\nu \sin \nu x),$$

where $\quad a_\nu = \frac{1}{\pi} \int_{-\pi}^{+\pi} f(t) \cos \nu t \, dt, \; b_\nu = \frac{1}{\pi} \int_{-\pi}^{+\pi} f(t) \sin \nu t \, dt$

is absolutely convergent throughout the whole interval. If $f(x)$ has a finite number of jump discontinuities, while elsewhere $f'(x)$ is sectionally continuous, the series converges uniformly in every closed sub-interval which contains no discontinuities of $f(x)$. At every point at which $f(x)$ is continuous, the series represents the value of the function $f(x)$, while at every point of discontinuity of $f(x)$ it represents the arithmetic mean of the right-hand and left-hand limits of $f(x)$ (pp. 447–450).

10. Maxima and Minima

The following rule holds only for maxima and minima in the *interior* of the region under consideration.

In order that ξ may be an extreme value of the function $y = f(x)$, $f'(\xi)$ must vanish. When this condition is satisfied there is a maximum or minimum if the first non-vanishing derivative of $f(x)$ is of even order; if it is of odd order, there is neither a maximum nor a minimum. In the former case there is a maximum or a minimum according as the sign of the first non-zero derivative is negative or positive (p. 158 *et seq.*).

11. Curves

In what follows ξ, η are current co-ordinates.

Equation of the curve:

(a) $y = f(x)$, (b) $F(x, y) = 0$, (c) $x = \phi(t)$, $y = \psi(t)$.

Equation of the tangent at the point (x, y) (p. 263):

(a) $\eta - y = (\xi - x)f'(x)$,   (b) $(\xi - x)F_x + (\eta - y)F_y = 0$,

(c) $\{\xi - \phi(t)\}\psi'(t) - \{\eta - \psi(t)\}\phi'(t) = 0$.

Equation of the normal at the point (x, y) (p. 263):

(a) $\xi - x + (\eta - y)f'(x) = 0$,   (b) $(\xi - x)F_y - (\eta - y)F_x = 0$,

(c) $\{\xi - \phi(t)\}\phi'(t) + \{\eta - \psi(t)\}\psi'(t) = 0$.

Curvature (p. 281):

(a) $k = \dfrac{y''}{(1 + y'^2)^{\frac{3}{2}}}$,   (b) $k = -\dfrac{F_{xx}F_y^2 - 2F_{xy}F_xF_y + F_{yy}F_x^2}{(F_x^2 + F_y^2)^{\frac{3}{2}}}$,

(c) $k = \dfrac{\dot{\phi}\ddot{\psi} - \ddot{\phi}\dot{\psi}}{(\dot{\phi}^2 + \dot{\psi}^2)^{\frac{3}{2}}}$.

Radius of curvature (p. 282):

$$\rho = \frac{1}{|k|}.$$

Evolute (locus of centre of curvature) (pp. 283, 307–311):

(a) $\xi = x - y'\dfrac{1 + y'^2}{y''}$, $\eta = y + \dfrac{1 + y'^2}{y''}$;

(b) $\xi = x + F_x\dfrac{F_x^2 + F_y^2}{F_{xx}F_y^2 - 2F_{xy}F_xF_y + F_{yy}F_x^2}$,

$\eta = y + F_y\dfrac{F_x^2 + F_y^2}{F_{xx}F_y^2 - 2F_{xy}F_xF_y + F_{yy}F_x^2}$;

(c) $\xi = \phi - \dot{\psi}\dfrac{\dot{\phi}^2 + \dot{\psi}^2}{\dot{\phi}\ddot{\psi} - \ddot{\phi}\dot{\psi}}$, $\eta = \psi + \dot{\phi}\dfrac{\dot{\phi}^2 + \dot{\psi}^2}{\dot{\phi}\ddot{\psi} - \ddot{\phi}\dot{\psi}}$.

Involute (p. 309):

$$\xi = x + (a - s)\dot{x}, \quad \eta = y + (a - s)\dot{y},$$

where a is an arbitrary constant and s the length of arc measured from a given point.

Point of inflection (pp. 159, 266). Necessary condition for a point of inflection is

(a) $y'' = 0$,   (b) $F_{xx}F_y^2 - 2F_{xy}F_xF_y + F_{yy}F_x^2 = 0$,

(c) $\dot{x}\ddot{y} - \ddot{x}\dot{y} = 0$.

Angle between two curves (p. 264):

$$(b)\ \cos\omega = \frac{F_x G_x + F_y G_y}{\sqrt{(F_x^2 + F_y^2)}\sqrt{(G_x^2 + G_y^2)}},$$

$$(c)\ \cos\omega = \frac{\dot{x}\dot{x}_1 + \dot{y}\dot{y}_1}{\sqrt{(\dot{x}^2 + \dot{y}^2)}\sqrt{(\dot{x}_1^2 + \dot{y}_1^2)}}.$$

In particular, the curves are orthogonal if

$$(b)\ F_x G_x + F_y G_y = 0, \quad (c)\ \dot{x}\dot{x}_1 + \dot{y}\dot{y}_1 = 0;$$

the curves touch if

$$(b)\ F_x G_y - F_y G_x = 0, \quad (c)\ \dot{x}\dot{y}_1 - \dot{x}_1\dot{y} = 0.$$

Two curves $y = f(x)$, $y = g(x)$ have contact of order n at a point x, if

$$f(x) = g(x), \quad f'(x) = g'(x), \quad \ldots, \quad f^{(n)}(x) = g^{(n)}(x),$$
$$f^{(n+1)}(x) \neq g^{(n+1)}(x)$$

(pp. 331–333).

12. Length of Arc, Area, Volume

Length of Arc (pp. 276–280). Let a plane curve be given by the equations

$$(a)\ y = f(x), \quad (b)\ F(x, y) = 0, \quad (c)\ x = \phi(t),\ y = \psi(t),$$
$$(d)\ \text{(polar co-ordinates)}\ r = r(\theta).$$

The length of arc is

$$(a)\ s = \int_{x_0}^{x_1} \sqrt{(1 + y'^2)}\,dx, \qquad (c)\ s = \int_{t_0}^{t_1} \sqrt{(\dot{x}^2 + \dot{y}^2)}\,dt,$$

$$(b)\ s = \int_{x_0}^{x_1} \frac{1}{F_y}\sqrt{(F_x^2 + F_y^2)}\,dx, \qquad (d)\ s = \int_{\theta_0}^{\theta_1} \sqrt{(r^2 + r'^2)}\,d\theta.$$

Area of Plane Surface. The area bounded by the curve

$$r = r(\theta)$$

and two radii vectores θ_0, θ_1, where r, θ are polar co-ordinates, is given by

$$\frac{1}{2}\int_{\theta_0}^{\theta_1} r^2\,d\theta \qquad \text{(p. 275).}$$

The area enclosed by the curve

$$y = f(x),$$

the two ordinates $x = x_0$, $x = x_1$, and the x-axis, is

$$\int_{x_0}^{x_1} y \, dx \qquad \text{(p. 80)}.$$

Volume. The volume lying over the region R and bounded above by the surface with the equation

$$z = f(x, y)$$

is given by

$$V = \int \int_R f(x, y) \, dx \, dy \qquad \text{(p. 487)}.$$

MISCELLANEOUS EXAMPLES

CHAPTER I

1. Prove that if p and q are integers the expansion of p/q as a decimal either terminates or recurs from a certain point onward. Prove also that every terminating or recurring decimal represents a rational number.

2. Express 39 in the ternary scale (scale of 3).

3. How would the number one hundred and fifty-six be written if (a) the binary scale (scale of 2), (b) the scale of 4, were in common use?

4. Express the following numbers in the scale of 12: (a) 1076, (b) 10,000, (c) 20,736, (d) 1/6, (e) 1/64, (f) 1/5.

5. We can find $\sqrt{2}$ to one decimal place thus: $1^2 = 1 < 2, 2^2 = 4 > 2$, therefore $1 < \sqrt{2} < 2$. Next, $1 \cdot 3^2 = 1 \cdot 69 < 2$, $1 \cdot 4^2 = 1 \cdot 96 < 2$, $1 \cdot 5^2 = 2 \cdot 25 > 2$, therefore $1 \cdot 4 < \sqrt{2} < 1 \cdot 5$.
(a) Continue this process one step further.
(b) Calculate $\sqrt{7}$ to two decimal places by the same method.

6. For what values of x do the following inequalities hold?

(a) $x^2 + 3x + 1 \geqq 0$.   (c) $\left| x + \dfrac{1}{x} \right| \geqq 6$.

(b) $x^2 - x + 1 \geqq 0$.   (d) $3x - 2 \leqq x^3$.

7. Prove that the arithmetic mean $\dfrac{a + b}{2}$ of two positive quantities a, b is not less than the geometric mean $\sqrt{ab}$, i.e. that

$$\frac{a + b}{2} \geqq \sqrt{ab}.$$

State when the equality sign holds.

8. The quantity ξ defined by $\dfrac{1}{\xi} = \dfrac{1}{2}\left(\dfrac{1}{a} + \dfrac{1}{b} \right)$ is called the harmonic mean of the two positive quantities a, b. Prove that the geometric mean is not less than the harmonic mean, i.e. that $\sqrt{ab} \geqq \xi$.
When does the equality sign hold?

9.* Show that the following inequalities hold, if a, b, c are positive:
    (a) $a^2 + b^2 + c^2 \geqq ab + bc + ca$.
    (b) $(a + b)(b + c)(c + a) \geqq 8abc$.
    (c) $a^2b^2 + b^2c^2 + c^2a^2 \geqq abc\,(a + b + c)$.

10. The numbers x_1, x_2, x_3 and a_{ik} (i, $k = 1, 2, 3$) are all positive. In addition, $a_{ik} \leqq M$ and $x_1^2 + x_2^2 + x_3^2 \leqq 1$. Prove that

$$a_{11}x_1^2 + 2a_{12}x_1x_2 + \ldots + a_{33}x_3^2 \leqq 3M.$$

11.* Prove that if the numbers $a_1, a_2, \ldots, a_n$ and $b_1, b_2, \ldots, b_n$ satisfy the inequalities $a_1 \geqq a_2 \geqq \ldots \geqq a_n$, $b_1 \geqq b_2 \geqq \ldots \geqq b_n$, then

$$n \sum_1^n a_i b_i \geqq \left(\sum_{i=1}^n a_i \right) \left(\sum_{i=1}^n b_i \right).$$

12. Prove the following properties of the binomial coefficients:

(a) $1 - \binom{n}{1} + \binom{n}{2} - \binom{n}{3} + \ldots \pm \binom{n}{n} = 0.$

(b) $\binom{n}{1} + 2\binom{n}{2} + 3\binom{n}{3} + \ldots + n\binom{n}{n} = n \, 2^{n-1}.$

(c) $1 . 2\binom{n}{2} + 2 . 3\binom{n}{3} + \ldots + (n-1)n\binom{n}{n} = n(n-1)2^{n-2}.$

(d) $1 + \frac{1}{2}\binom{n}{1} + \frac{1}{3}\binom{n}{2} + \ldots + \frac{1}{n+1}\binom{n}{n} = \frac{2^{n+1} - 1}{n+1}.$

(e) $\binom{n}{0}^2 + \binom{n}{1}^2 + \ldots + \binom{n}{n}^2 = \binom{2n}{n}.$

13. By summing

$$\nu(\nu + 1)(\nu + 2) \ldots (\nu + k + 1) - (\nu - 1)\nu(\nu + 1) \ldots (\nu + k)$$

from $\nu = 1$ to $\nu = n$, show that

$$\sum_{\nu=1}^n \nu(\nu + 1)(\nu + 2) \ldots (\nu + k) = \frac{n(n+1) \ldots (n + k + 1)}{k + 2}.$$

14. Evaluate $1^3 + 2^3 + \ldots + n^3$ by using the relation

$$\nu^3 = \nu(\nu + 1)(\nu + 2) - 3\nu(\nu + 1) + \nu.$$

15. Evaluate

(a) $\dfrac{1}{1 . 2 . 3} + \dfrac{1}{2 . 3 . 4} + \ldots + \dfrac{1}{n(n+1)(n+2)};$

(b) $\dfrac{1}{1 . 3} + \dfrac{1}{2 . 4} + \dfrac{1}{3 . 5} + \ldots + \dfrac{1}{n(n+2)}.$

(c) $\dfrac{1}{1 . 2 . 4} + \dfrac{1}{2 . 3 . 5} + \ldots + \dfrac{1}{n(n+1)(n+3)}.$

16. Find a formula for the n-th term of the following arithmetic progressions:

(a) 1, 2, 4, 7, 11, 16, ...
(b) -7, -10, -9, 1, 25, 68, ...

17.* Show that the sum of the first n terms of an arithmetic progression of order k is

$$aS_k + bS_{k-1} + \ldots + pS_1 + qn,$$

where S_ν represents the sum of the first n ν-th powers, and $a, b, \ldots, p, q$ are independent of n. Evaluate the sums for the arithmetic progressions of Ex. 16.

18.* Prove the binomial theorem

$$(a + b)^n = a^n + \binom{n}{1}a^{n-1}b + \binom{n}{2}a^{n-2}b^2 + \ldots + b^n$$

by mathematical induction. (See also Chap. III, p. 201.)

19. Find

(a) $\displaystyle\lim_{n\to\infty}\left(\frac{1}{1 \cdot 2} + \frac{1}{2 \cdot 3} + \ldots + \frac{1}{n(n+1)}\right).$

(b) $\displaystyle\lim_{n\to\infty}\left(\frac{1}{1 \cdot 2 \cdot 3} + \frac{1}{2 \cdot 3 \cdot 4} + \ldots + \frac{1}{n(n+1)(n+2)}\right).$

(c) $\displaystyle\lim_{n\to\infty}\left(\frac{1}{\sqrt[4]{n}} + \frac{1}{\sqrt[4]{n+1}} + \ldots + \frac{1}{\sqrt[4]{2n}}\right).$

20. If $\displaystyle\sum_{i=0}^{k} a_i = 0$, prove that $\displaystyle\lim_{n\to\infty}\sum_{i=0}^{k} a_i\sqrt{n+i} = 0$.

21. Prove that $\displaystyle\lim_{n\to\infty}\frac{n^5}{2^n} = 0.$

22. Prove that $\displaystyle\lim_{n\to\infty}\frac{(n+1)^5}{2^n} = 0.$

23. Prove that $\displaystyle\lim_{n\to\infty}\sqrt[n]{n^2} = 0.$

24. Prove that $\displaystyle\lim_{n\to\infty}\sqrt[(2n+1)]{(n^2+n)} = 0.$

25. Use Cauchy's convergence test to show that the following sequences converge:

(a) $a_n = \dfrac{1}{n}.$

(b) $a_n = \dfrac{n+1}{n}.$

(c)* $a_n = 1 + \dfrac{1}{1!} + \dfrac{1}{2!} + \ldots + \dfrac{1}{n!}.$

(d)* $a_n = 1 - \dfrac{1}{1!} + \dfrac{1}{2!} - \dfrac{1}{3!} + \ldots \pm \dfrac{1}{n!}.$

26.* Show that the limits of the sequences (c), (d) of the previous example are reciprocals of one another (so that the limit of the sequence (d) is $1/e!$).

27.* Prove that the limit of the sequence

$$\sqrt{2}, \quad \sqrt{2 + \sqrt{2}}, \quad \sqrt{2 + \sqrt{2 + \sqrt{2}}}, \; \ldots$$

(a) exists, (b) is equal to 2.

28.* Prove that the limit of the sequence

$$a_n = \frac{1}{n} + \frac{1}{n + 1} + \ldots + \frac{1}{2n}$$

exists. Show that the limit is less than 1 but not less than $\frac{1}{2}$.

29. Prove that the limit of the sequence

$$a_n = \frac{1}{n + 1} + \ldots + \frac{1}{2n}$$

exists, is equal to the limit of the previous example, and is greater than $\frac{1}{2}$ but not greater than 1.

30. Obtain the following bounds for the limit L of the two previous examples: $\dfrac{37}{60} < L < \dfrac{57}{60}$.

31.* Let a_1, b_1 be any two positive numbers, and let $a_1 < b_1$. Let

$$a_2 = \frac{2a_1 b_1}{a_1 + b_1}, \quad b_2 = \sqrt{a_1 b_1},$$

and in general

$$a_n = \frac{2a_{n-1} b_{n-1}}{a_{n-1} + b_{n-1}}, \quad b_n = \sqrt{a_{n-1} b_{n-1}}.$$

Prove that the sequences $a_1, a_2, \ldots$ and $b_1, b_2, \ldots$ converge and have the same limit.

32.* If $a_n > 0$, and $\lim\limits_{n \to \infty} \dfrac{a_{n+1}}{a_n} = L$, then $\lim\limits_{n \to \infty} \sqrt[n]{a_n} = L$.

33. Use Ex. 32 to evaluate the limits of the following sequences:

$$(a) \; \sqrt[n]{n}, \quad (b) \; \sqrt[n]{(n^5 + n^4)}, \quad (c) \; \sqrt[n]{\left(\frac{n!}{n^n}\right)}.$$

34. Use Ex. 33(c) to show that

$$n! = n^n e^{-n} a_n,$$

where a_n is a number whose n-th root tends to 1. (See Chap. VII, Appendix, p. 363.)

35. Prove that $\lim\limits_{x \to 0} \dfrac{x + 2}{x + 1} = 2$. Find a δ such that for $|x| < \delta$ the difference between 2 and $\dfrac{x + 2}{x + 1}$ is, in absolute value, (a) less than $\frac{1}{10}$, (b) less than $\frac{1}{1000}$, (c) less than ε, $\varepsilon > 0$.

36. (a) Prove that $\lim\limits_{x\to 1}\dfrac{x+2}{x+1}=\dfrac{3}{2}$. Find a δ such that for $|1-x|<\delta$ the difference between $\dfrac{3}{2}$ and $\dfrac{x+2}{x+1}$ is, in absolute value, less than ε, $\varepsilon>0$.

Do the same for (b) $\lim\limits_{x\to 2}\sqrt{(1+x^3)}$; (c) $\lim\limits_{x\to 0}\dfrac{\sin x}{x}$.

37. Prove that (a) $\lim\limits_{x\to 0}\dfrac{\sqrt{(1+x)}-1}{x}=\dfrac{1}{2}$.

(b) $\lim\limits_{x\to\infty}\sqrt{x+\tfrac{1}{2}}(\sqrt{x+1}-\sqrt{x})=\tfrac{1}{2}$.

38. Prove that $\lim\limits_{m\to\infty}(\cos\pi x)^{2m}$ exists for each value of x and is equal to 1 or 0 according as x is an integer or not.

39.* Prove that $\lim\limits_{n\to\infty}[\lim\limits_{m\to\infty}(\cos\pi n!\,x)^{2m}]$ exists for each value of x and is equal to 1 or 0 according as x is rational or irrational.

40. Determine which of the following functions are continuous. For those which are discontinuous, find the points of discontinuity.

(a) $f(x)=\dfrac{x^5+5x^3+3x^2}{\sin x}$, $f(0)=0$.

(b) $f(x)=\dfrac{x^5+5x^3+3x}{\sin x}$, $f(0)=0$.

(c) $f(x)=\lim\limits_{m\to\infty}(\cos\pi x)^{2m}$.

(d) $f(x)=\lim\limits_{n\to\infty}[\lim\limits_{m\to\infty}(\cos\pi n!\,x)^{2m}]$.

41. Let $f(x)$ be continuous for $0\leq x\leq 1$. Suppose further that $f(x)$ assumes rational values only, and that $f(x)=\tfrac{1}{2}$ when $x=\tfrac{1}{2}$. Prove that $f(x)=\tfrac{1}{2}$ everywhere.

42. Has the function

$$f(x)=2\sin 3x+10\cos 5x$$

any real zeros?

43.* If $f(x)$ satisfies the functional equation

$$f(x+y)=f(x)+f(y)$$

for all values of x and y, find the values of $f(x)$ at the rational points and prove that, if $f(x)$ is continuous, $f(x)=cx$, where c is a constant.

44.* Prove a converse of the theorem of uniform continuity; namely, that if $f(x)$ is uniformly continuous in the half-open interval $a<x\leq b$, then $f(x)$ tends to a unique limit as $x\to a$ (which may be taken as the value of $f(a)$).

45. Plot the following graphs, and express the equations in Cartesian co-ordinates:

$$(a) \quad r = a + b \cos\theta \quad \text{(Limaçon)}.$$

$$(b) \quad r = \frac{2}{2 - \cos\theta} \quad \text{(Ellipse)}.$$

$$(c) \quad r = \frac{2a \sin^2\theta}{\cos\theta} \quad \text{(Cissoid)}.$$

$$(d) \quad r = \frac{3a \sin\theta \cos\theta}{\sin^3\theta + \cos^3\theta} \quad \text{(Folium of Descartes)}.$$

46.* Show that the equation of an ellipse with one focus at the origin is

$$r = \frac{k}{1 - e \cos(\theta - \theta_0)}.$$

47. Let c be the complex number $x + iy$ represented by a point in a Cartesian co-ordinate system. Plot the curves

$$(a) \quad \left| \frac{c - i}{c + i} \right| = 2.$$

$$(b)^* \quad \left| \frac{c - \alpha}{c - \beta} \right| = k, \quad \alpha, \beta \text{ complex constants}.$$

$$(c) \quad |c^2 - 1| = k.$$

48. Let c_1, c_2 be two complex numbers. Prove that

$$(a) \quad |c_1 \pm c_2| \leq |c_1| + |c_2|.$$

$$(b) \quad |c_1 \pm c_2| \geq |c_1| - |c_2|.$$

49. Prove the equality

$$|c_1 + c_2|^2 + |c_1 - c_2|^2 = 2|c_1|^2 + 2|c_2|^2$$

and state its geometrical interpretation.

50. Prove that $(\cos\theta + i \sin\theta)^n = \cos n\theta + i \sin n\theta$ by mathematical induction.

CHAPTER II

51.* Prove directly that the derivative of the function

$$f(x) = x^2 \sin\frac{1}{x}, \quad x \neq 0; \quad f(0) = 0$$

exists at every point and is equal to

$$-\cos\frac{1}{x} + 2x \sin\frac{1}{x}, \quad x \neq 0; \quad 0 \text{ at } x = 0.$$

Show that although $f'(x)$ is not continuous at $x = 0$, nevertheless the mean value theorem still applies and the property of Ex. 57 below holds good. (See pp. 199, 200 of the text.)

52. Draw the graph of the function

$$f(x) = x \sin \frac{1}{x}, \ x \neq 0; \ f(0) = 0$$

and find its derivative for $x \neq 0$. Show that its derivative does not exist at $x = 0$, but that the difference quotient $\dfrac{f(x) - f(0)}{x}$ as $x \to 0$ has the upper and lower limits 1 and -1 respectively. (See p. 199.)

53. Investigate the behaviour of the function

$$f(x) = x \sin \frac{1}{x} + x^2 \sin \frac{1}{x}, \ x \neq 0; \ f(0) = 0$$

with regard to differentiability.

54. Prove that the derivative of the function

$$f(x) = \frac{1}{x} \sin x, \ x \neq 0; \ f(0) = 1$$

exists at every point and is equal to

$$f'(x) = -\frac{1}{x^2} \sin x + \frac{1}{x} \cos x, \ x \neq 0; \ f'(0) = 0.$$

Show that $f'(x)$ is continuous, and find $f''(x)$.

55. If $f(x)$ is continuous and differentiable for $a \leq x \leq b$, show that if $f'(x) \leq 0$ for $a \leq x < \xi$ and $f'(x) \geq 0$ for $\xi < x \leq b$, the function is never less than $f(\xi)$.

56.* If the continuous function $f(x)$ has a derivative $f'(x)$ at each point x in the neighbourhood of $x = \xi$, and if $f'(x)$ approaches a limit L as $x \to \xi$, then $f'(\xi)$ exists and is equal to L.

57.* If $f(x)$ possesses a derivative $f'(x)$ (not necessarily continuous at each point x of $a \leq x \leq b$, and if $f'(x)$ assumes the values m and M) it also assumes every value μ between m and M.

58. If $f''(x) \geq 0$ for all values of x in $a \leq x \leq b$, the graph of $y = f(x)$ lies above the tangent line at any point $x = \xi$, $y = f(\xi)$ of the graph. (The curve is convex upwards.)

59. If $f''(x) \geq 0$ for all values of x in $a \leq x \leq b$, the graph of $y = y f(x)$ in the interval $x_1 \leq x \leq x_2$ lies below the line segment joining the two points of the graph for which $x = x_1$, $x = x_2$.

60. If $f''(x) \geq 0$, then $f\left(\dfrac{x_1 + x_2}{2}\right) \leq \dfrac{f(x_1) + f(x_2)}{2}$.

61. Given $f(x) = \frac{1}{3}x^3 - x^2 + 1$, find a number δ such that for every h less in absolute value than δ and every x in the interval $-\frac{1}{2} \leq x \leq \frac{1}{2}$ the following inequality holds:

$$\left| f'(x) - \frac{f(x + h) - f(x)}{h} \right| \leq \frac{1}{100}.$$

62. Differentiate directly and write down the corresponding integration formulæ: (a) $x^{1/2}$; (b) $\tan x$.

63. Evaluate

(a) $\lim\limits_{n \to \infty} \dfrac{1}{\sqrt{n}} \left(1 + \dfrac{1}{\sqrt{2}} + \ldots + \dfrac{1}{\sqrt{n}} \right)$.

(b) $\lim\limits_{n \to \infty} \dfrac{1}{n} \left(1 + \sec^2 \dfrac{\pi}{4n} + \sec^2 \dfrac{2\pi}{4n} + \ldots + \sec^2 \dfrac{n\pi}{4n} \right)$.

64. Prove that

(a) $\displaystyle\int_{-1}^{1} (x^3 - 1)^2 \, dx = \dfrac{16}{15}$; (b) $(-1)^n \displaystyle\int_{-1}^{1} (x^2 - 1)^n \, dx = \dfrac{2^{n+1}(n!)^2}{(2n+1)!}$.

65. Show that

$$\frac{1}{\nu + 1} < \int_{\nu}^{\nu+1} \frac{dx}{x} < \frac{1}{\nu}$$

and

$$\frac{1}{2} + \frac{1}{3} + \ldots + \frac{1}{n} < \int_{1}^{n} \frac{dx}{x} < 1 + \frac{1}{2} + \ldots + \frac{1}{n-1}.$$

Prove that the sequence $1 + \dfrac{1}{2} + \ldots + \dfrac{1}{\nu} - \displaystyle\int_{1}^{\nu} \dfrac{dx}{x}$, $\nu = 1, 2, \ldots$, is a decreasing sequence and is bounded below.

66.* Let $f(x)$ be a function such that $f''(x) \geqq 0$ for all values of x, and let $u = u(t)$ be an arbitrary continuous function. Then

$$\frac{1}{a} \int_{0}^{a} f(u(t)) \, dt \geqq f\left(\frac{1}{a} \int_{0}^{a} u(t) \, dt \right).$$

67.* If a particle traverses distance 1 in time 1, beginning and ending at rest, then at some point in the interval it must have been subjected to an acceleration $\geqq 4$.

CHAPTER III

68. Differentiate the following functions:

(a) $e^{\tan^2 x + \log \sin x}$.

(b) $(x + 2)^4 (1 - x^2)^{1/3} (x^2 + 1)^{5/7}$.

(c) $\dfrac{x^3 \sin x - x^5 \cos x}{x^2 \tan x}$.

69. What conditions must the coefficients α, β, a, b, c satisfy in order that

$$\frac{\alpha x + \beta}{\sqrt{(ax^2 + 2bx + c)}}$$

shall everywhere have a finite derivative which is never zero?

70. Sketch the graph of the function

$$y = (x^2)^x, \quad y(0) = 1.$$

Show that the function is continuous at $x = 0$. Has the function maxima, minima, or points of inflection?

71. Among all triangles with given base and given perimeter, the isosceles triangle has the maximum area.

72. Among all triangles with given base and given vertical angle, the isosceles triangle has the maximum area.

73. Among all triangles with given base and given area, the isosceles triangle has the maximum vertical angle.

74.* Among all triangles with given area, the equilateral triangle has the least perimeter.

75.* Among all triangles with given perimeter, the equilateral triangle has the maximum area.

76.* Among all triangles inscribed in a circle, the equilateral triangle has the maximum area.

77. Prove the following inequalities:

(a) $e^x > \dfrac{1}{1 + x}, \quad x > 0.$

(b) $e^x > 1 + \log(1 + x), \quad x > 0.$

(c) $e^x > 1 + (1 + x)\log(1 + x), \quad x > 0.$

78.* Let a, b be two positive numbers, p and q any non-zero numbers, $p < q$. Prove that

$$\frac{[\theta a^p + (1 - \theta)b^p]^{1/p}}{[\theta a^q + (1 - \theta)b^q]^{1/q}} \leq 1$$

for all values of θ in the interval $0 < \theta < 1$.

(This is Jensen's inequality, which states that the p-th power mean $[\theta a^p + (1 - \theta)b^p]^{1/p}$ of two positive quantities a, b is an increasing function of p.)

79. Show that the equality sign in the above inequality holds if, and only if, $a = b$.

80. Prove that $\lim\limits_{p \to 0} [\theta a^p + (1 - \theta)b^p]^{1/p} = a^\theta b^{1-\theta}.$

81. Defining the zero-th power mean of a, b as $a^\theta b^{1-\theta}$, show that Jensen's inequality applies to this case, and becomes $(a \neq b)$

$$a^\theta b^{1-\theta} \gtrless [\theta a^q + (1 - \theta)b^q]^{1/q} \text{ according as } q \lessgtr 0.$$

For $q = 1$, $\qquad a^\theta b^{1-\theta} \leq \theta a + (1 - \theta)b.$

82. Prove the inequality

$$a^\theta b^{1-\theta} \leq \theta a + (1 - \theta)b,$$

$a, b > 0, \ 0 < \theta < 1$, without reference to Jensen's inequality, and show that equality holds only if $a = b$. (This inequality states that the θ, $1 - \theta$ geometric mean is less than the corresponding arithmetic mean.)

83. If $\varphi(x) \to \infty$ as $x \to \infty$, show that $\log \varphi(x)$ is of a lower order and $e^{\phi(x)}$ of a higher order of magnitude than $\varphi(x)$.

84. If the order of magnitude of the positive function $f(x)$ as $x \to \infty$ is higher, the same, or lower than that of x^m, prove that $\int_a^x f(\xi)\,d\xi$ has the corresponding order of magnitude relative to x^{m+1}.

85. Compare the order of magnitude as $x \to \infty$ of $\int_a^x f(\xi)\,d\xi$ relative to $f(x)$ for the following functions $f(x)$:

(a) $\dfrac{e^{\sqrt{x}}}{\sqrt{x}}$.

(c) xe^{x^2}.

(b) e^x.

(d) $\log x$.

86. Prove that if $f(x)$ is continuous and

$$f(x) = \int_0^x f(t)\,dt,$$

then $f(x)$ is identically zero.

87. Prove that $\sum\limits_{i=1}^{n-1} ix^{i-1} = \dfrac{(n-1)x^n - nx^{n-1} + 1}{(x-1)^2}$.

88. Show that $\dfrac{d^n(e^{x^2/2})}{dx^n} = u_n(x)e^{x^2/2}$,

where $u_n(x)$ is a polynomial of degree n. Establish the recurrence relation

$$u_{n+1} = xu_n + u_n'.$$

89.* By applying Leibnitz's rule to

$$\frac{d}{dx}(e^{x^2/2}) = xe^{x^2/2},$$

obtain the recurrence relation

$$u_{n+1} = xu_n + nu_{n-1}.$$

90.* By combining the recurrence relations of Ex. 88, 89, obtain the differential equation

$$u_n'' + xu_n' - nu_n = 0$$

satisfied by $u_n(x)$.

91. Find the polynomial solution

$$u_n(x) = x^n + a_1 x^{n-1} + \ldots + a_n$$

of the differential equation $u_n'' + xu_n' - nu_n = 0$.

92.* If $P_n(x) = \dfrac{1}{2^n n!} \dfrac{d^n}{dx^n} (x^2 - 1)^n$, prove the relations

(a) $P_{n+1}' = \dfrac{x^2 - 1}{2(n+1)} P_n'' + \dfrac{(n+2)x}{x+1} P_n' + \dfrac{n+2}{2} P_n.$

(b) $P_{n+1}' = x P_n' + (n+1) P_n.$

(c) $\dfrac{d}{dx} ((x^2 - 1) P_n') - n(n+1) P_n = 0.$

93. Find the polynomial solution

$$P_n = \frac{(2n)!}{2^n (n!)^2} x^n + a_1 x^{n-1} + \ldots + a_n$$

of the differential equation

$$\frac{d}{dx} ((x^2 - 1) P_n') - n(n+1) P_n = 0.$$

94. Determine the polynomial $P_n(x) = \dfrac{1}{2^n n!} \dfrac{d^n}{dx^n} (x^2 - 1)^n$ by using the binomial theorem.

95.* Let $\lambda_{n,p}(x) = \dbinom{p}{n} x^n (1-x)^{p-n}$, $n = 0, 1, 2, \ldots, p$. Show that

$$1 = \sum_{n=0}^{p} \lambda_{n,p}(x).$$

$$x = \sum_{n=1}^{p} \frac{n}{p} \lambda_{n,p}(x).$$

$$x^k = \sum_{n=k}^{p} \frac{\dbinom{n}{k}}{\dbinom{p}{k}} \lambda_{n,p}(x).$$

$$\vdots \qquad \cdot \qquad \cdot \qquad \cdot \qquad \cdot \qquad \cdot$$

$$x^p = \lambda_{p,p}(x).$$

$$\cdot \qquad \cdot \qquad \cdot \qquad \cdot \qquad \cdot \qquad \cdot \qquad \cdot$$

CHAPTER IV

Perform the integrations in Ex. 96–101.

96. $\displaystyle\int \frac{1 + \sqrt[3]{x}}{1 + \sqrt{x}} \, dx.$

99. $\displaystyle\int \frac{x^2 - 1}{x^4 + x^2 + 1} \, dx.$

97. $\displaystyle\int \frac{e^{2x}}{\sqrt[4]{(e^x + 1)}} \, dx.$

100. $\displaystyle\int \frac{dx}{x \sqrt{(x^{2n} - 1)}}.$

98. $\displaystyle\int \frac{x \, dx}{\sqrt[3]{(1+x)} - \sqrt{(1+x)}}.$

101. $\displaystyle\int \frac{dx}{x(x+1) \ldots (x+n)}.$

Evaluate the integrals in Ex. 102–107.

102. $\displaystyle\int_0^{\pi/2} \cos^n x \, dx.$

103. $\displaystyle\int_0^{\pi/6} \cos^7 3\theta \sin^4 6\theta \, d\theta.$

104. $\displaystyle\int_0^1 \frac{x^{2n}\,dx}{\sqrt{(1-x^2)}}.$ **105.** $\displaystyle\int_0^1 \frac{x^{2n+1}\,dx}{\sqrt{(1-x^2)}}.$

106. $\displaystyle\int_0^1 x^3\sqrt{(1-x^2)}\,dx.$ **107.** $\displaystyle\int_0^1 x^2(1-x^2)^{3/2}\,dx.$

Obtain recurrence formulæ for the integrals of Ex. 108–112.

108. $\displaystyle\int x^a (\log x)^m\,dx.$ **111.** $\displaystyle\int e^{ax} \sinh bx\,dx.$

109. $\displaystyle\int x^n e^{ax} \sin bx\,dx.$ **112.** $\displaystyle\int e^{ax} \cosh bx\,dx.$

110. $\displaystyle\int x^n e^{ax} \cos bx\,dx.$

113. Integrate $\displaystyle\int \frac{dx}{\sqrt{(a^2-x^2)}}$ in three different ways and compare the results.

114.* Let $P_n(x) = \dfrac{1}{2^n n!}\,\dfrac{d^n}{dx^n}\,(x^2-1)^n.$ Show that

$$\int_{-1}^1 P_n(x)P_m(x)\,dx = 0, \text{ if } m \neq n.$$

115. Prove that $\displaystyle\int_{-1}^1 P_n{}^2(x)\,dx = \frac{2}{n+1}.$

116. Prove that $\displaystyle\int_1^{-1} x^m P_n(x)\,dx = 0,$ if $m < n.$

117. Evaluate $\displaystyle\int_{-1}^1 x^n P_n(x)\,dx.$

Test whether the improper integrals in Ex. 118–131 converge or diverge.

118. $\displaystyle\int_0^a \frac{dx}{\sqrt{(ax-x^2)}}.$ **125.** $\displaystyle\int_0^\pi x \log \sin x\,dx.$

119. $\displaystyle\int_1^\infty \frac{dx}{x\sqrt{(x^2-1)}}.$ **126.** $\displaystyle\int_{-\infty}^\infty e^{-x^2}\,dx.$

120. $\displaystyle\int_0^1 \left(\log\frac{1}{x}\right)^n dx.$ **127.** $\displaystyle\int_0^\infty x^{2n-1}e^{-x^2}\,dx.$

121. $\displaystyle\int_0^1 x^m\left(\log\frac{1}{x}\right)^n dx.$ **128.** $\displaystyle\int_0^{\pi/2} \frac{x^m\,dx}{(\sin x)^n}.$

122. $\displaystyle\int_0^\infty e^{-x}x^m(\log x)^n\,dx.$ **129.** $\displaystyle\int_0^\infty \frac{dx}{1+x^4\sin^2 x}.$

123. $\displaystyle\int_0^\pi \log \sin x\,dx.$ **130.** $\displaystyle\int_0^\infty \frac{x\,dx}{1+x^3\sin^2 x}.$

124. $\displaystyle\int_0^\pi \frac{1}{x}\log \sin x\,dx.$ **131.*** $\displaystyle\int_0^\infty \frac{x^a\,dx}{1+x^\beta\sin^2 x}.$

132.* If $\int_a^\infty \dfrac{f(x)}{x}\,dx$ converges for any positive value of a, and if $f(x)$ tends to a limit L as $x \to 0$, show that $\int_0^\infty \dfrac{f(\alpha x) - f(\beta x)}{x}\,dx$ converges and has the value $L \log \dfrac{\beta}{\alpha}$.

133. By reference to the previous example, show that

$$(a) \int_0^\infty \frac{e^{-\alpha x} - e^{-\beta x}}{x}\,dx = \log \frac{\beta}{\alpha}.$$

$$(b) \int_0^\infty \frac{\cos \alpha x - \cos \beta x}{x}\,dx = \log \frac{\beta}{\alpha}.$$

134.* If $\int_a^b \dfrac{f(x)}{x}\,dx$ converges for any positive values of a and b, and if $f(x)$ tends to a limit M as $x \to \infty$ and a limit L as $x \to 0$, show that

$$\int_0^\infty \frac{f(\alpha x) - f(\beta x)}{x}\,dx = (L - M) \log \frac{\beta}{\alpha}.$$

135. Obtain the following expressions for the gamma function:

$$\Gamma(n) = 2\int_0^\infty x^{2n-1}e^{-x^2}\,dx,$$

$$\Gamma(n) = \int_0^1 \left(\log \frac{1}{x}\right)^{n-1} dx.$$

CHAPTER V

136. Plot the following curves and find their equations in non-parametric form:

$$(a)\ x = \frac{5at^2}{1 + t^5}, \quad y = \frac{5at^3}{1 + t^5}.$$

(b) $x = at + b \sin t, \quad y = a - b \cos t.$

137.* Show that the two families of ellipses and hyperbolas,

$$\frac{x^2}{a^2 - \lambda} + \frac{y^2}{b^2 - \lambda} = 1, \quad \text{for } \lambda < b,$$

$$\frac{x^2}{a^2 - \tau} + \frac{y^2}{b^2 - \tau} = 1, \quad \text{for } a < \tau < b,$$

are confocal and intersect at right angles.

138. Find the pedal curves (see p. 267, Ex. 11) of the following:
(a) the ellipse $x = a \cos \theta,\ y = b \sin \theta$ with respect to the origin;
(b) the hyperbola $x = \cosh \theta,\ y = b \sinh \theta$ with respect to the origin;
(c) the parabola $y^2 = 4px$ with respect to the origin;
(d) the parabola $y^2 = 4px$ with respect to the focus.

139. Show that the tangent to an ellipse is equally inclined to the focal radii drawn to the point of contact.

140. Show that the tangent to a hyperbola is equally inclined to the focal radii drawn to the point of contact.

141. A constant length l is measured off along the normal to a parabola. Find the curve described by the extremity of this segment.

142. Find the area bounded by the loop of the curve

$$x^5 + y^5 - 5ax^2y^2 = 0.$$

143. Find the area enclosed by the curve

$$a^2(x^2 + y^2)^2(b^2x^2 + a^2y^2) = (a^2 - b^2)^2b^2x^4.$$

144. Find the length of arc of the epicycloid

$$x = (a + b)\cos t - b\cos\frac{a + b}{b}t$$

$$y = (a + b)\sin t - b\sin\frac{a + b}{b}t$$

reckoned from the initial point $t = 0$.

145. Prove that the radius of curvature at a point of the polar curve $r = f(\theta)$ is

$$\frac{\left[r^2 + \left(\frac{dr}{d\theta}\right)^2\right]^{3/2}}{r^2 - r\frac{d^2r}{d\theta^2} + 2\left(\frac{dr}{d\theta}\right)^2}.$$

146.* If the curvature of a curve in the xy-plane is a monotonic function of the length of arc, prove that the curve is not closed and that it has no double points.

147. Find the moment of inertia of a rod of length L

(*a*) with respect to its centre;
(*b*) with respect to one end;
(*c*) with respect to a point on the line of the rod at a distance d from the centre;
(*d*) with respect to any point at a distance d from the centre.

148. Find the equation of the curves which everywhere intersect the straight lines through the origin at the same angle α.

149. Find the equation of the curves whose normal is of constant length k. (The "length" of the normal is the length of the portion of the normal intercepted between the curve and the x-axis.)

150. Show that the only curves whose curvature is a fixed constant k are circles of radius $1/k$.

151. Find the equation of the curves whose centre of curvature lies on the x-axis and whose radius of curvature is therefore equal to the length of the normal.

152. Find the equation of the curves whose radius of curvature is equal to the length of the normal but whose centre of curvature does not lie on the x-axis.

153.* Obtain the formula for the length of a curve in polar co-ordinates.

CHAPTER VI

154. Deduce the integral formula for the remainder R_n by applying integration by parts to

$$f(x+h) - f(x) = \int_0^h f'(x+\tau)d\tau.$$

155. Integrate the formula

$$R_n = \frac{1}{n!}\int_0^h (h-\tau)^n f^{(n+1)}(x+\tau)d\tau,$$

and so obtain

$$R_n = f(x+h) - f(x) - hf'(x) - \ldots - \frac{h^n}{n!}f^{(n)}(x).$$

156.* Suppose that in some way a series for the function $f(x)$ has been obtained, namely

$$f(x) = a_0 + a_1 x + a_2 x^2 + \ldots + a_n x^n + R_n(x),$$

where $a_0, a_1, \ldots, a_n$ are constants, $R_n(x)$ is n times continuously differentiable, and $\dfrac{R_n(x)}{x^n} \to 0$ as $x \to 0$. Show that $a_k = \dfrac{f^k(0)}{k!}$ $(k = 0, \ldots, n)$, i.e. that the series is a Taylor series.

157.* Find the first three non-vanishing terms of the Taylor series for $\sin^2 x$ in the neighbourhood of $x = 0$ by multiplying the Taylor series for $\sin x$ by itself. Justify this procedure.

158.* Find the first three non-vanishing terms of the Taylor series for $\tan x$ in the neighbourhood of $x = 0$, by using the relation $\tan x = \dfrac{\sin x}{\cos x}$, and justify the procedure.

159.* Find the first three non-vanishing terms of the Taylor series for $\sqrt{\cos x}$ in the neighbourhood of $x = 0$, by applying the binomial theorem to the Taylor series for $\cos x$, and justify the procedure.

160. Find the first four non-vanishing terms of the Taylor series for the following functions in the neighbourhood of $x = 0$:

(a) $x \cot x$. (c) $\sec x$. (e) e^{e^x}.

(b) $\dfrac{\sqrt{\sin x}}{\sqrt{x}}$. (d) $e^{\sin x}$. (f) $\log \sin x - \log x$.

161. Find the Taylor series for arc $\sin x$ in the neighbourhood of $x = 0$ by using

$$\text{arc}\sin x = \int_0^x \frac{dt}{\sqrt{1 - t^2}}.$$

(Cf. p. 203, Ex. 5.)

162.* Find the Taylor series for $(\text{arc}\sin x)^2$. (Cf. p. 203, Ex. 5.)

163. Find the Taylor series for the following functions in the neighbourhood of $x = 0$:

$$(a)\ \sinh^{-1} x. \qquad (b)\ \int_0^x e^{-t^2}\, dt. \qquad (c)\ \int_0^x \frac{\sin t}{t}\, dt.$$

164.* Estimate the error involved in using the first n terms in the series in Ex. 163.

165.* Two oppositely charged particles $+e$, $-e$ situated at a small distance d apart form an electric dipole with moment $M = ed$. Show that the potential energy (a) at a point situated on the axis of the dipole at a distance r from the centre of the dipole is $\dfrac{M}{r^2}(1 + \varepsilon)$, where ε is approximately equal to $\dfrac{d^2}{4r^2}$;

(b) at a point situated on the perpendicular bisector of the dipole is 0;

(c) at a point with polar co-ordinates r, θ relative to the centre and axis of the dipole is $\dfrac{M \cos \theta}{r^3}(1 + \varepsilon)$, where ε is approximately equal to $\dfrac{d^2}{8r^2}(5\cos^2\theta - 3)$.

(The potential energy of a single charge q at a point at a distance r from the charge is q/r; the potential energy of several charges is the sum of the potential energies of the separate charges.)

166.* Find the first three terms of the Taylor series for $\left(1 + \dfrac{1}{x}\right)^x$ in powers of $\dfrac{1}{x}$.

167. Evaluate the following limits:

$$(a)\ \lim_{x \to \infty} x\left[\left(1 + \frac{1}{x}\right)^x - e\right].$$

$$(b)\ \lim_{x \to \infty} \frac{e}{2} x + x^2 \left[\left(1 + \frac{1}{x}\right)^x - e\right].$$

$$(c)^*\ \lim_{x \to \infty} x\left[\left(1 + \frac{1}{x}\right)^x - e \log\left(1 + \frac{1}{x}\right)^x\right].$$

$$(d)\ \lim_{x \to 0}\left(\frac{\sin x}{x}\right)^{1/x^2} \qquad (e)\ \lim_{x \to \infty}\left(\frac{\sin x}{x}\right)^{1/x^2}.$$

168.* Show that the osculating circle at a point where the radius of curvature is a maximum or minimum does not cross the curve.

169. Find the maxima and minima of the following functions: $(a)\ |x|$, $(b)\ x\sin(1/x)$.

CHAPTER VII

170. Show that the length of the ellipse $x = a\cos t$, $y = b\sin t$ is

$$s = 4a \int_0^{\pi/2} \sqrt{1 - e^2 \cos^2 t}\, dt, \quad \text{where } e^2 = \frac{a^2 - b^2}{a^2}.$$

Calculate the length of the ellipse for which $e = \frac{1}{2}$ to four significant figures, by using Simpson's rule with six divisions.

171. Expand the integral of Ex. 170 as a series, and estimate the number of terms necessary for accuracy to four significant figures.

172. Evaluate $\int_0^1 \frac{\log(1 + x)}{x}\, dx$, using Simpson's rule with $h = 0\cdot 1$.

173. The hypotenuse of a right-angled triangle is measured accurately as 40, and one angle is measured as 30° with a possible error of $\frac{1}{2}$°. Find the possible error in the lengths of each of the sides and in the area of the triangle.

174.* By considering $\int_{1/2}^{n+1/2} \log(\alpha + x)\, dx$, $\alpha > 0$, show that

$$\alpha(\alpha + 1) \ldots (\alpha + n) = a_n n!\, n^\alpha,$$

where a_n is bounded below by a positive number. Show that a_n is monotonically decreasing for sufficiently large values of n. (The limit of a_n as $n \to \infty$ is $1/\Gamma(\alpha)$.)

175. Find an approximate expression for $\log \dfrac{n_1!\, n_2! \ldots n_l!}{n!}$, where $n_1 + n_2 + \ldots + n_l = n$.

176. Show that the coefficient of x^n in the binomial expansion of $\dfrac{1}{\sqrt{1 - x}}$ is asymptotically given by $\dfrac{1}{\sqrt{\pi n}}$.

CHAPTER VIII

177. Prove that if $\sum_{\nu=1}^{\infty} a_\nu^2$ converges, so does $\sum_{\nu=1}^{\infty} \frac{a_\nu}{\nu}$.

178. If a_n is a monotonic increasing sequence with positive terms, when does the series $\dfrac{1}{a_1} + \dfrac{1}{a_1 a_2} + \dfrac{1}{a_1 a_2 a_3} + \ldots$ converge?

179.* If the series $\sum_{\nu=1}^{\infty} a_\nu$ with decreasing positive terms converges, then $\lim_{n \to \infty} na_n = 0$.

180. Show that the series $\sum_{\nu=1}^{\infty} \sin \frac{\pi}{\nu}$ diverges.

181.* Prove that if Σa_ν converges and if b_1, b_2, b_3, ... is a bounded monotonic sequence of numbers, then $\Sigma a_\nu b_\nu$ converges.

182.* Prove that if Σa_ν oscillates between finite bounds and if b_ν is a monotonic sequence tending towards zero, then $\Sigma a_\nu b_\nu$ converges.

183. Discuss the convergence or divergence of the following series:

(a) $\Sigma \dfrac{(-1)^\nu}{\nu}$. (b) $\Sigma \dfrac{(-1)^\nu \cos(\theta/\nu)}{\nu}$. (c) $\Sigma \dfrac{\cos \nu\theta}{\nu}$.

(d) $\Sigma \dfrac{\sin \nu\theta}{\nu}$. (e) $\Sigma \dfrac{(-1)^\nu \cos \nu\theta}{\nu}$. (f) $\Sigma \dfrac{(-1)^\nu \sin \nu\theta}{\nu}$.

184. Find the sums of the following derangements of the series $1 - \frac{1}{2} + \frac{1}{3} - \frac{1}{4} + \frac{1}{5} - \frac{1}{6} + \dots$ for $\log 2$:

(a) $1 - \frac{1}{2} - \frac{1}{4} + \frac{1}{3} - \frac{1}{6} - \frac{1}{8} + \frac{1}{5} - \frac{1}{10} - \frac{1}{12} + - - \dots$.

(b) $1 + \frac{1}{3} + \frac{1}{5} - \frac{1}{2} - \frac{1}{4} - \frac{1}{6} + + + \dots$.

185. For what values of α do the following series converge?

(a) $1 - \dfrac{1}{2^\alpha} + \dfrac{1}{3} - \dfrac{1}{4^\alpha} + \dfrac{1}{5} - \dfrac{1}{6^\alpha} + - \dots$.

(b) $1 + \dfrac{1}{3^\alpha} - \dfrac{1}{2^\alpha} + \dfrac{1}{5^\alpha} + \dfrac{1}{7^\alpha} - \dfrac{1}{4^\alpha} + + - \dots$.

186. Find whether the following series converge or diverge:

(a) $1 + \frac{1}{2} - \frac{1}{3} + \frac{1}{4} + \frac{1}{5} - \frac{1}{6} + \frac{1}{7} + \frac{1}{8} - \frac{1}{9} + + - \dots$.

(b) $1 + \frac{1}{2} - \frac{2}{3} + \frac{1}{4} + \frac{1}{5} - \frac{2}{6} + \frac{1}{7} + \frac{1}{8} - \frac{2}{9} + + - \dots$.

187. Show that

(a) $\displaystyle\sum_{\nu=1}^{\infty} \dfrac{\nu!}{(2\nu)!}$ converges.

(b) $\displaystyle\sum_{\nu=2}^{\infty} \dfrac{\log(\nu+1) - \log \nu}{(\log \nu)^2}$ converges.

(c) $\displaystyle\sum_{\nu=1}^{\infty} \dfrac{1 \cdot 2 \cdot 3 \dots \nu}{(\alpha+1)(\alpha+2)\dots(\alpha+\nu)}$ converges if $\alpha > 1$ and diverges if $\alpha \leqq 1$.

188.* By comparison with the series $\displaystyle\sum_{\nu=1}^{\infty} \dfrac{1}{\nu^\alpha}$, prove the following test:

If $\dfrac{\log(1/|a_n|)}{\log n} > 1 + \varepsilon$ for every sufficiently large n, the series Σa_ν converges absolutely; if $\dfrac{\log(1/|a_n|)}{\log n} < 1 - \varepsilon$ for every sufficiently large n, the series Σa_ν does not converge absolutely.

189. Show that the series $\displaystyle\sum_{\nu=1}^{\infty} \left(1 - \dfrac{1}{\sqrt{\nu}}\right)^\nu$ converges.

190. By comparison with the series $\Sigma \dfrac{1}{\nu (\log \nu)^a}$, prove the following test:

The series $\Sigma \,|\, a_\nu \,|$ converges or diverges according as

$$\frac{\log (1/n \,|\, a_n \,|)}{\log \log n}$$

is greater than $1 + \varepsilon$ or less than $1 - \varepsilon$ for every sufficiently large n.

191. Derive the n-th root test from the test of Ex. 188.

192.* Prove the following comparison test: if the series Σb_ν of positive terms converges, and

$$\left|\frac{a_{n+1}}{a_n}\right| < \frac{b_{n+1}}{b_n}$$

from a certain term onwards, the series Σa_ν is absolutely convergent; if Σb_ν diverges and

$$\left|\frac{a_{n+1}}{a_n}\right| > \frac{b_{n+1}}{b_n}$$

from a certain term onwards, the series Σa_ν is not absolutely convergent.

193. Obtain the ratio test by comparison with the geometric series.

194.* By comparison with $\displaystyle\sum_{\nu=1}^{\infty} \dfrac{1}{\nu^a}$, prove Raabe's test:

The series $\Sigma \,|\, a_\nu \,|$ converges or diverges according as

$$n\left(\frac{|\, a_n \,|}{|\, a_{n+1} \,|} - 1\right)$$

is greater than $1 + \varepsilon$ or less than $1 - \varepsilon$ for every sufficiently large n.

195. By comparison with $\Sigma \dfrac{1}{\nu (\log \nu)^a}$, prove the following test:

The series $\Sigma \,|\, a_\nu \,|$ converges or diverges according as

$$n \log n\left(\frac{|\, a_n \,|}{|\, a_{n+1} \,|} - 1 - \frac{1}{n}\right)$$

is greater than $1 + \varepsilon$ or less than $1 - \varepsilon$ for every sufficiently large n.

196. Prove Gauss's test:

If
$$\frac{|\, a_n \,|}{|\, a_{n+1} \,|} = 1 + \frac{\mu}{n} + \frac{R_n}{n^{1+\epsilon}},$$

where $|\, R_n \,|$ is bounded, then $\Sigma \,|\, a_\nu \,|$ converges if $\mu > 1$, diverges if $\mu \leqq 1$.

197. Test the following series for convergence or divergence:

(a) $\dfrac{\alpha}{\beta} + \dfrac{\alpha(\alpha + 1)}{\beta(\beta + 1)} + \dfrac{\alpha(\alpha + 1)(\alpha + 2)}{\beta(\beta + 1)(\beta + 2)} + \dots .$

(b) $1 + \dfrac{\alpha \cdot \beta}{1 \cdot \gamma} + \dfrac{\alpha(\alpha + 1) \cdot \beta(\beta + 1)}{1 \cdot 2 \cdot \gamma(\gamma + 1)} + \dfrac{\alpha(\alpha + 1)(\alpha + 2) \cdot \beta(\beta + 1)(\beta + 2)}{1 \cdot 2 \cdot 3 \cdot \gamma(\gamma + 1)(\gamma + 2)} + \dots .$

198. (a) Show that the series $\sum\limits_{\nu=1}^{\infty} \dfrac{1}{\nu^x}$ converges uniformly for $x \geqq 1 + \varepsilon$.

(b) Show that the derived series $-\sum \dfrac{\log \nu}{\nu^x}$ converges uniformly for $x \geqq 1 + \varepsilon$.

199.* Show that the series $\sum \dfrac{\cos \nu x}{\nu^a}$, $\alpha > 0$, converges uniformly for $\varepsilon \leqq x \leqq 2\pi - \varepsilon$.

200. The series

$$\frac{x-1}{x+1} + \frac{1}{3}\left(\frac{x-1}{x+1}\right)^3 + \frac{1}{5}\left(\frac{x-1}{x+1}\right)^5 + \cdots$$

converges uniformly for $\varepsilon \leqq x \leqq N$.

201. Find the regions in which the following series are convergent:

(a) $\sum x^{\nu!}$.

(e) $\sum \dfrac{a^\nu}{\nu^x}$, $a < 1$.

(b) $\sum \dfrac{(\nu!)^2 x^\nu}{(2\nu)!}$.

(f) $\sum \dfrac{a^\nu}{\nu^x}$, $a > 1$.

(c) $\sum \dfrac{1}{\nu^x}$.

(g) $\sum \dfrac{\log \nu}{\nu^x}$.

(d) $\sum \dfrac{(-1)^\nu}{\nu^x}$.

(h) $\sum \dfrac{x^\nu}{1 - x^\nu}$.

202.* Prove that if the series $\sum \dfrac{a^\nu}{\nu^x}$ converges for $x = x_0$, it converges for any $x > x_0$; if it diverges for $x = x_0$, it diverges for any $x < x_0$. Thus, there is an " abscissa of convergence " such that for any greater value of x the series converges, and for any smaller value of x the series diverges.

203. If $\sum \dfrac{a_\nu}{\nu^x}$ converges for $x = x_0$, the derived series $-\sum \dfrac{a_\nu \log \nu}{\nu^x}$ converges for any $x > x_0$.

204. If $a_\nu > 0$ and $\sum a_\nu$ converges, then

$$\lim_{x \to 1-0} \sum a_\nu x^\nu = \sum a_\nu.$$

205. If $a_\nu > 0$ and $\sum a_\nu$ diverges,

$$\lim_{x \to 1-0} \sum a_\nu x^\nu = \infty.$$

206.* Prove Abel's theorem:

If $\sum a_\nu X^\nu$ converges, then $\sum a_\nu x^\nu$ converges uniformly for $0 \leqq x \leqq X$.

207.* If $\sum a_\nu X^\nu$ converges, then $\lim\limits_{x \to X-0} \sum a_\nu x^\nu = \sum a_\nu X^\nu$.

208. Find the rational functions represented by the following Taylor series:

(a) $x + x^2 - x^3 - x^4 + x^5 + x^6 - - + + \cdots$.

(b) $1 + 2x - 4x^3 - 5x^4 + 7x^6 + 8x^7 - - + + \cdots$.

209. Show that

(a) $\dfrac{1}{2!} + \dfrac{2}{3!} + \dfrac{3}{4!} + \ldots = 1.$

(b) $\dfrac{1}{2} + \dfrac{1 \cdot 3}{2 \cdot 4 \cdot 6} + \dfrac{1 \cdot 3 \cdot 5 \cdot 7}{2 \cdot 4 \cdot 6 \cdot 8 \cdot 10} + \ldots = \dfrac{1}{2}\sqrt{2}.$

210. Let $z = re^{i\theta} = r(\cos\theta + i\sin\theta)$. From the expansion $\dfrac{1}{1-z} = \overset{\infty}{\underset{\nu=0}{\Sigma}} z^{\nu}$, show that

$$\frac{1 - r\cos\theta}{1 - 2r\cos\theta + r^2} = \overset{\infty}{\underset{\nu=0}{\Sigma}} r^{\nu}\cos\nu\theta$$

and

$$\frac{r\sin\theta}{1 - 2r\cos\theta + r^2} = \overset{\infty}{\underset{\nu=0}{\Sigma}} r^{\nu}\sin\nu\theta.$$

CHAPTER IX

211.* Using the expression for the cotangent in partial fractions, expand $\pi x \cot \pi x$ as a power series in x. By comparing this with the series given on p. 423, show that

$$\overset{\infty}{\underset{\nu=1}{\Sigma}} \frac{1}{\nu^{2m}} = (-1)^{m-1} \frac{(2\pi)^{2m}}{2 \cdot (2m)!} B_{2m}.$$

212. Show that

$$\overset{\infty}{\underset{\nu=1}{\Sigma}} \frac{1}{(2\nu-1)^{2m}} = \frac{(-1)^{m-1}(2^{2m}-1)\pi^{2m}}{2(2m)!} B_{2m}.$$

213. Show that

$$\overset{\infty}{\underset{\nu=1}{\Sigma}} \frac{(-1)^{\nu}}{\nu^{2m}} = \frac{(-1)^m (2^{2m}-2)\pi^{2m}}{2 \cdot (2m)!} B_{2m}.$$

214. Prove that

(a) $\displaystyle\int_0^1 \frac{\log x}{1-x}\, dx = -\frac{\pi^2}{6};$

(b) $\displaystyle\int_0^1 \frac{\log x}{1+x}\, dx = -\frac{\pi^2}{12}.$

215. Using the infinite products for the sine and cosine, show that

(a) $\log\left(\dfrac{\sin x}{x}\right) = -\overset{\infty}{\underset{\nu=1}{\Sigma}} \dfrac{(-1)^{\nu-1}2^{2\nu-1}B_{2\nu}}{(2\nu)!\,\nu} x^{2\nu};$

(b) $\log\cos x = -\overset{\infty}{\underset{\nu=1}{\Sigma}} \dfrac{(-1)^{\nu-1}2^{2\nu-1}(2^{2\nu}-1)B_{2\nu}}{(2\nu)!\,\nu} x^{2\nu}:$

19 *

(E 798)

216. Using the infinite products for the sine and cosine, evaluate

$$(a) \ \tfrac{2}{1} \cdot \tfrac{2}{3} \cdot \tfrac{6}{5} \cdot \tfrac{6}{7} \cdot \tfrac{10}{9} \cdot \tfrac{10}{11} \cdot \tfrac{14}{13} \cdots;$$

$$(b) \ 2 \cdot \tfrac{2}{3} \cdot \tfrac{4}{3} \cdot \tfrac{8}{9} \cdot \tfrac{10}{9} \cdot \tfrac{14}{15} \cdot \tfrac{16}{15}.$$

217. Express the hyperbolic cotangent in terms of partial fractions.

CHAPTER XI

218. Find the curves whose tangent is of constant length a. (The "length" of the tangent is the length of the portion of the tangent intercepted between the curve and the x-axis.)

219. Find the curves which are orthogonal to the family $y = ce^{kx}$.

220. If s denotes the length of arc of a chain measured from a point at which the tangent is horizontal, the form of the chain is determined by the differential equation

$$\frac{d}{dx}(\log s) = \frac{d}{dx}\left(\log \frac{dy}{dx}\right).$$

Show that the equation of the chain is $y = c \cosh \dfrac{x}{c} + a$.

221. Integrate the equation for the electric circuit

$$\mu \dot{I} + \rho I = E,$$

where $E = E_0 \sin \omega t$, and μ, ρ, E_0, ω are constants.

222. A particle falls towards a point which attracts inversely as the cube of the distance and directly as the mass. Find the motion and time of descent if $v = 0$ and $x = a$ at $t = 0$.

223.* Integrate $y = -xp + x^4 p^2$, where $p = \dfrac{dy}{dx}$.

224. Integrate $y = p + \log p$.

225.* Solve the difference equation

$$u_{n+2} + 2au_{n+1} + bu_n = 0,$$

where a, b are constants, by putting $u_n = \lambda^n$. Show that the solution can be expressed in the form $u_n = \alpha r_1{}^n + \beta r_2{}^n$, where r_1, r_2 are the roots (supposed distinct) of the equation $\lambda^2 + 2a\lambda + b = 0$. Show that the form of the solution when $b = a^2$ is $u_n = \alpha(-a)^n + \beta n(-a)^n$.

ANSWERS AND HINTS

CHAPTER I

§ 1, p. 13.

1. (d), (e). Show that x satisfies an equation of the type

$$x^6 + a_1 x^5 + \ldots + a_6 = 0,$$

where $a_1, \ldots, a_6$ are integers; prove that x is then either irrational or an integer.

2. Use the irrationality of $\sin 60° = \sqrt{3}/2$.

4. Write $ax^2 + 2bx + c$ as $a\left(x + \dfrac{b}{a}\right)^2 + \dfrac{ac - b^2}{a}$.

7. If $a > 0$ and $b^2 - ac \leq 0$, it is possible to make $ax^2 + 2bx + c = 0$ for some value of x if, and only if, $b^2 - ac = 0$; then use Example 6.

8. The cosine of the angle between two straight lines is ≤ 1 in absolute value.

9. Use Schwarz's inequality.

10. Square both sides and then use Schwarz's inequality. The sum of the lengths of two sides of a triangle is not less than the third side.

§§ 2, 3, p. 26.

2. (a), (d), (e), (g) odd; (b) even.

3. (b), (c), (h) monotonic; (a), (d), (e), (l), (m) even; (d) and (e) identical.

§ 4, p. 28.

2. $(n + 1)(2n + 1)(2n + 3)/3$.

3. (c) Expand $(1 + 1)^n$ by the binomial theorem.

4. (a) $n(n + 1)(n + 2)/3$.

(b) Sum $\dfrac{1}{\nu + 1} - \dfrac{1}{\nu}$ from $\nu = 1$ to $\nu = n$. $n/(n + 1)$.

(c) Sum $\dfrac{1}{(\nu + 1)^2} - \dfrac{1}{\nu^2}$ from $\nu = 1$ to $\nu = n$. $n(n + 2)/(n + 1)^2$.

5. 3; 193.

7. $\frac{1}{6}(2n^3 + 3n^2 - 11n + 30)$.

§ 5, p. 36.

   1. (a) 1; (b) 333; (c) 333,333.

   2. (a) 0; (b) ∞; (c) 6; (d) a_0/b_0; (e) 1/3.

   4. 19.

   5. (a) 6; (b) 10; (c) 14.

   6. (a) 25; (b) 2500; (6) 250,000.

   9. (a) 0; (b) no; (c) yes; (e) 30.

   15. The greatest of $a_1, \ldots, a_k$.

   16. 2.

   17. Use the fact that $n/2^n \to 0$.

§ 6, p. 45.

   1. (a) For every number M, no matter how large, there exists an n such that $|a_n| > M$.

   (b) There exists a positive number ε such that for every number M there exist numbers n, m greater than M for which $|a_n - a_m| \geqq \varepsilon$.

   5. Error is less than $\dfrac{1}{n(n!)}$; $e = 2 \cdot 71828 \ldots$.

§ 7, p. 49.

   1. (a) 6; (b) 15; (c) $\frac{1}{2}$; (d) 3.

   3. Limits (a) and (b) do not exist; limit (c) exists and is equal to 1.

§ 8, p. 55.

   3. (a) 1/60; 1/600; 1/6000.

     (b) $1/10(1 + 2|\xi - 1|)$, &c.

     (c) $1/120(1 + |\xi|)^3$, &c.

     (d) 1/100; 1/10000; 1/1000000.   (e) 1/10; 1/100; 1/1000.

   4. (a) 1/600; $\varepsilon/6$. (b) 1/400; $\varepsilon/4$. (c) 1/77600; $\varepsilon/776$. (d) 1/10000; ε^2. (e) 1/100; ε.

   5. (a), (b), (c), (d), (g) continuous;

     (e) discontinuous at $x = 2, 4$;

     (f)   ,,   ,, $x = 3$;

     (h), (k), (m) ,,   ,, $x = (n + \frac{1}{2})\pi$;

     (i), (j)   ,,   ,, $x = n\pi$;

     (l)   ,,   ,, $x = n\pi$, $n \neq 0$.

Appendix I, p. 70.

   1. (a) Upper bound $= \frac{324}{5}$,* lower $= 0$, upper limit $= 0$, lower $= 0$.

     (b)   ,, $= \frac{1}{2}$,*   ,, $= -1$,*   ,, $= 0$,*   ,, $= 0$.*

     (c)   ,, $= \frac{9}{10}$,   ,, $= -\frac{2}{3}$,*   ,, $= \frac{1}{2}$,   ,, $= \frac{1}{2}$.

     (d)   ,, $= \frac{19}{10}$,*   ,, $= -\frac{1}{3}$,*   ,, $= \frac{3}{2}$,   ,, $= \frac{1}{2}$.

     (e)   ,, $= 2$,*   ,, $= 0$,   ,, $= 1$,   ,, $= 0$.

The quantities marked * belong to the sequence

2. Divide the interval into a finite number of sub-intervals by points $a = x_0, x_1, x_2, \ldots, x_n = b$ so close that $|f(x) - f(\bar{x})| < \varepsilon$ if x and $\bar{x}$ lie in the same sub-interval. Join adjacent points $x = x_i$, $y = f(x_i)$ by straight lines.

3. The expression $-\dfrac{k}{2}|x - x_i| + \dfrac{k}{2}|x - x_{i-1}|$ has the slope zero outside the interval (x_{i-1}, x_i). Add suitable terms of this kind.

$$\tfrac{3}{2} + x - \tfrac{3}{2}|x - 2| + |x - 3| - \tfrac{1}{2}|x - 5|.$$

4. (a) $\varepsilon/6$; (b) ε/na^{n-1}, $n > 0$; (c) $\varepsilon^3/2$.

Appendix II, p. 75.

2. (a) $r = a$; (b) $r = 2a\cos(\varphi - \varphi_0)$; (c) $r = a/\cos(\varphi - \varphi_0)$.

3. $\cos 2\theta = \cos^2\theta - \sin^2\theta$, $\sin 2\theta = 2\sin\theta\cos\theta$;
$\cos 3\theta = 4\cos^3\theta - 3\cos\theta$, $\sin 3\theta = 3\sin\theta - 4\sin^3\theta$;
$\cos 5\theta = 16\cos^5\theta - 20\cos^3\theta + 5\cos\theta$, $\sin 5\theta = 16\sin^5\theta - 20\sin^3\theta + 5\sin\theta$.

4. (a) $-6i$; $\theta = \pi, r = 3$; $\theta = \pi/2, r = 2$; $\theta = 3\pi/2, r = 6$.

(b) $1 + \sqrt{3} + i(1 - \sqrt{3})$; $\theta = \pi/4, r = 4\sqrt{2}$; $\theta = \pi/3, r = \tfrac{1}{2}$;
$\theta = 7\pi/12, r = 2\sqrt{2}$.

(c) 2; $\theta = \pi/4, r = \sqrt{2}$; $\theta = 7\pi/4, r = \sqrt{2}$; $\theta = 2\pi, r = 2$.

(d) $2 - 2i\sqrt{3}$; $\theta = 5\pi/6, r = 2$; $\theta = 5\pi/3, r = 4$.

(e) ± 1; $\theta = 0, r = 1$; $\theta = 0, r = \pm 1$.

(f) $\pm\left(\dfrac{1}{\sqrt{2}} + \dfrac{i}{\sqrt{2}}\right)$; $\theta = \pi/2, r = 1$; $\theta = \pi/4, r = \pm 1$.

(g) $\{\sqrt{\sqrt{2}+1} + i\sqrt{\sqrt{2}-1}\}/\sqrt{2}$; $\theta = \pi/4, r = \sqrt{2}$;
$\theta = \pi/8, r = \sqrt[4]{2}$.

(h) $-\sqrt[3]{18}(\sqrt{3} + i)/2$; $\theta = 7\pi/4, r = 3\sqrt{2}$;
$\theta = 7\pi/6, r = (3\sqrt{2})^{2/3} = \sqrt[3]{18}$.

(k) $1, (-1 \pm i\sqrt{3})/2$; $\theta = 0, r = 1$; $\theta = 0, \dfrac{2\pi}{3}, \dfrac{4\pi}{3}, r = 1$.

(l) $\sqrt[4]{2}\{\sqrt{\sqrt{2}+1} + i\sqrt{\sqrt{2}-1}\}$; $\theta = \pi/2, r = 16$;
$\theta = \pi/8, r = \pm 2$.

5. Note that ε^ν satisfies the equation $x^n - 1 = 0$; then factorize $x^n - 1$.

CHAPTER II

§ 2, p. 87.

1. Use formulæ of § 2 and the basic rules: 70/3.

2. Required area may be regarded as the difference between the area under the line and the area under the parabola, taken between the points of intersection of curve and line: $10\sqrt{5}/3$.

3. $\sqrt{5}/6$.

4. $\frac{1}{6}(a^2 + 4b)^{3/2}$.

5. (a) $\{(1 + b)^{1+a} - (1 + a)^{1+a}\}/(1 + \alpha)$;   (b) $-(\cos \alpha b - \cos \alpha a)/\alpha$;
(c) $(\sin \alpha b - \sin \alpha a)/\alpha$.

7. $(b^4 - a^4)/4$.

8. $1/(n + 1)$.

§ 3, p. 109.

1. For every number a there exists an ε such that for every positive number δ there exists an x for which

$$|x - \xi| \leqq \delta \quad \text{and} \quad \left| a - \frac{f(x) - f(\xi)}{x - \xi} \right| \geqq \varepsilon.$$

2. (a) $-1/(x + 1)^2$;   (b) $-2x/(x^2 + 2)^2$;   (c) $-4x/(2x^2 + 1)^2$;
    (d) $-\cos x/\sin^2 x$;   (e) $3 \cos 3x$;   (f) $-a \sin ax$;   (g) $2 \sin x \cos x$;
    (h) $-2 \cos x \sin x$.

3. (a) ξ has any value;   (b) $\xi = (x_1 + x_2)/2$;   (c) $\xi = \sqrt{\left(\dfrac{x_2{}^2 + x_2 x_1 + x_1{}^2}{3} \right)}$.

$$\text{(e)} \quad \xi = \left(\frac{x_2{}^{2/3} + x_2{}^{1/3} x_1{}^{1/3} + x_1{}^{2/3}}{3} \right)^{3/2}.$$

§ 4, p. 119.

2. (a) $\frac{1}{2}$;   (b) $\frac{1}{2}$.

§ 5, p. 121.

1. $\dfrac{\pi}{4} = 0 \cdot 785$.

§§ 6, 7, p. 130.

1. (b) $\xi = \dfrac{a + b}{2}$;   (c) $\xi = \sqrt[n]{\left(\dfrac{a^n + a^{n-1}b + \ldots + b^n}{n + 1} \right)}$;   (d) $\xi = \sqrt{ab}$.

3. (a) $I_n = a^{1+1/n}/(1 + 1/n)$, $\lim\limits_{n \to \infty} I_n = a$;

    (b) $I_n = a^{n+1}/(n + 1)$, $\lim\limits_{n \to \infty} I_n = 0$ for $-1 \leqq a \leqq 1$, ∞ for $a > 1$.

4. $|F(x) - f(x)| \leqq \dfrac{1}{2\delta} \displaystyle\int_{-\delta}^{\delta} |f(x + t) - f(x)| \, dt$.   Use the uniform continuity of $f(x)$ in $a \leqq x \leqq b$. Also, we may write

$$F(x) = \frac{1}{2\delta} \left\{ \int_{x-\delta}^{c} f(t) \, dt + \int_{c}^{x+\delta} f(t) \, dt \right\},$$

where c is a fixed number.

5. Express the integrals as limits of sums, using equal subdivisions of $a \leqq x \leqq b$, and applying Schwarz's inequality (p. 12) to these sums. Another method is to integrate $\{f(x) + tg(x)\}^2 \geqq 0$ and use Ex. 6, p. 13.

Appendix, p. 135.

1. Let $\varphi(x) = f'(x)$ where $f'(x) \geqq 0$, $\varphi(x) = 0$ elsewhere. Let $\psi(x) = f'(x) - \varphi(x)$; then $\psi(x) \leqq 0$. Consider $\displaystyle\int_{a}^{x} \varphi(x) \, dx$, $\displaystyle\int_{a}^{x} \psi(x) \, dx$.

CHAPTER III

§§ 1, 2, p. 143.

1. $f'(1) = 1$, $f''(1) = 8$, $f'''(1) = 36$, $f^{iv}(1) = 96$, $f^{v}(1) = 120$, $f^{vi}(1) = 0$, $f^{vii}(1) = 0, \ldots$.

2. 0.

3. (a) a;   (b) $175\,cx^6$;   (c) $2(b + cx)$;   (d) $\dfrac{ad - bc}{(cx + d)^2}$;

(e) $\dfrac{2x^2(a\beta - \alpha b) + 2x(a\gamma - \alpha c) + 2(b\gamma - \beta c)}{(\alpha x^2 + 2\beta x + \gamma)^2}$;

(f) $\dfrac{4x(1 + x^4)}{(1 - x^2)^2(1 + x^2)^2}$;   (g) **0**.

4. (a) $F(x) = a_n x^n + (a_{n-1} + na_n)x^{n-1}$
$\qquad + \{a_{n-2} + (n-1)(a_{n-1} + na_n)\}x^{n-2} + \ldots$.

(b) $F(x) = \dfrac{a_n}{c_0} x^n + \left(a_{n-1} - na_n \dfrac{c_1}{c_0}\right)x^{n-1} +$

$\qquad \left\{\dfrac{a_{n-2}}{c_0} - (n-1)a_{n-1}\dfrac{c_1}{c_0^2} - n(n-1)a_n \dfrac{(c_0 c_2 - c_1^2)}{c_0^3}\right\} x^{n-2} + \ldots$.

5. (a) $2\cos 2x$;   (b) $-1/(1 + \sin 2x)$;   (c) $\tan x + x/\cos^2 x$;

(d) $-2/(1 - \sin 2x)$;   (e) $-\dfrac{\sin x}{x^2} + \dfrac{\cos x}{x}$.

6. $\sec^3 x + \sec x \tan^2 x$.   **7.** $24 \sec^5 x - 20 \sec^3 x + \sec x$.

8. $\cos x\,(\operatorname{cosec}^2 x - 6\operatorname{cosec}^4 x)$.

9. $24 \sec^5 x - 20 \sec^3 x + \sec x - \cos x$.   **10.** ∞.

11. $ax^2/2 + bx$.   **12.** $ax^3/3 + bx^2 + cx$.

13. $x^9 + x^7 + x^5 + x^3 + x$.   **14.** $-(1/x + 1/2x^2 + 1/3x^3)$.

15. $x^3/3 - 1/x$.   **16.** $a \sin x - b \cot x$.

17. $3x^2/2 - 7\cos x - 5/2x^2 - 9\tan x$.   **18.** $\sec x$.

§ 3, p. 152.

1. 4.   **3.** $\sqrt[n]{x}\{1 - (n-1)x\}/nx(1 + x)^2$.

4. $\cos^2 x/2\sqrt{x} - 2\sqrt{x}\sin x \cos x$.   **5.** $1/\sqrt{x}(1 - \sqrt{x})^2$.

6. $\dfrac{(1 - \tan x) + 3x(1 + \tan^2 x)}{3x^{2/3}(1 - \tan x)^2}$.

7. $(\operatorname{arc}\cos x - \operatorname{arc}\sin x)/\sqrt{(1 - x^2)}$.

8. $2/(1 + x^2)(1 - \operatorname{arc}\tan x)^2$.

9. $\dfrac{1}{\sqrt{1 - x^2}\,\operatorname{arc}\tan x} - \dfrac{\operatorname{arc}\sin x}{(1 + x^2)(\operatorname{arc}\tan x)^2}$.

10. $-\dfrac{5}{1 + x^2} + \dfrac{1}{\sqrt{1 - x^2}(\operatorname{arc}\cos x)^2}$.

11. 0·785.

§ 4, p. 157.

1. $3(x + 1)^2$. **2.** $6(3x + 5)$. **3.** $15x^{14}(3x^6 - 6x^3 - 1)(x^6 - 3x^3 - 1)^4$.

4. $-1/(1 + x)^2$. **5.** $2x/(1 - x^2)^2$. **6.** $an(ax + b)^{n-1}$.

7. $-\dfrac{1}{\sqrt{x^2 - 1}(x + \sqrt{x^2 - 1})}$.

8. $\dfrac{x^2(am - bl) + 2x(an - cl) + (bn - cm)}{2\sqrt{\{(ax^2 + bx + c)(lx^2 + mx + n)^3\}}}$.

9. $-\frac{5}{3}(1 - x)^{2/3}$. **10.** $\sin 2x$. **11.** $2x \cos(x^2)$.

12. $\sin x \cos x / \sqrt{(1 + \sin^2 x)}$. **13.** $2\left(x \sin\dfrac{1}{x^2} - \dfrac{1}{x}\cos\dfrac{1}{x^2}\right)$.

14. $\dfrac{2}{(1 - x)^2 \cos^2\left(\dfrac{1 + x}{1 - x}\right)}$. **15.** $(2x + 3)\cos(x^2 + 3x + 2)$.

16. $3x^2/\sqrt{\{1 - (3 + x^3)^2\}}$. **17.** -1. **18.** 1.

19. $\dfrac{\sqrt{2}}{x}(x^{\sqrt{2}} + x^{-\sqrt{2}})$. **20.** $\sqrt[3]{5}\cos(x + 7)\{\sin(x + 7)\}^{\sqrt[3]{5}-1}$.

21. $-\dfrac{a\alpha \sin x}{\sqrt{\{1 - (a\cos x + b)^2\}}} \cdot \{\text{arc}\sin(a\cos x + b)\}^{a-1}$.

§ 5, p. 166.

1. (a) Max. for $x = -\sqrt{2}$, min. for $x = \sqrt{2}$, infl. for $x = 0$.

(b) Max. for $x = \frac{2}{5}$, min. for $x = 0$, infl. for $x = -\frac{1}{10}$.

(c) Max. for $x = 1$, min. for $x = -1$, infl. for $x = 0, \pm\sqrt{3}$.

(d) Max. for $x = \sqrt[4]{3}$, min. for $x = -\sqrt[4]{3}$,

infl. for $x = 0, \pm\sqrt[4]{6 \pm \sqrt{33}}$.

(e) Max. for $x = (n + \frac{1}{2})\pi$, min. for $x = n\pi$, infl. for $x = \dfrac{2n + 1}{4}\pi$.

2. Max. for $x = -\sqrt{-p}$; min. for $x = \sqrt{-p}$; infl. for $x = 0$. No maxima or minima when $p \geqq 0$. Roots are all real, or two complex and one real, according as $q^2 + 4p^3 \leqq 0$ or > 0.

3. The point $(0, 1)$.

4. Equation of line is $(y - y_0)/(x - x_0) = -\sqrt[3]{(y_0/x_0)}$.

5. $\sqrt{189}$ ft.

6. The point dividing the line ab in the ratio $\sqrt[3]{a} : \sqrt[3]{b}$.

7. The square.

8. The rectangle with corners $x = \pm a/\sqrt{2}$, $y = \pm b/\sqrt{2}$.

9. The right-angled triangle, i.e. $c^2 = a^2 + b^2$.

10. The side of rectangle opposite to g must be at the distance $\frac{1}{2}\{\sqrt{(8r^2 + h^2)} + h\}$ from the centre.

11. The cylinder whose height is equal to the diameter of its base.

13. If φ is the angle of the prism and n its index of refraction, the angle of incidence must be arc $\sin\left(n \sin \frac{\varphi}{2}\right)$.

14. $x = \left(\sum_i a_i\right)/n$.

15. Find the minimum of $x^p - px$.

16. Find the minima of $x - \sin x$ and $\sin x - \frac{2}{\pi} x$ in the interval $0 \leqq x \leqq \frac{\pi}{2}$. Or, show that $\frac{\sin x}{x}$ is monotonic in that interval.

18. $\left(\dfrac{a_1 + a_2 + \ldots + a_{n-1}}{n-1}\right)^{\frac{n-1}{n}} \dfrac{1}{\sqrt[n]{(a_1 a_2 \ldots a_{n-1})}}$.

§ 6, p. 177.

1. $0 \cdot 693$.   **2.** $\log x$.   **3.** $1/x \log x$.   **4.** $1/\sqrt{(1+x^2)}$.

5. $\dfrac{1 - 2x\sqrt{1 + \log x} \cos x}{2x\sqrt{1 + \log x}(\sqrt{1 + \log x} - \sin x)}$.

6. $x/(x^2 + 1) - 1/3(2 + x)$.

7. $\dfrac{\sqrt[3]{7x^2 + 1}}{\sqrt{x - 2}\sqrt{x^4 + 1}}\left(\dfrac{14x}{3(7x^2 + 1)} - \dfrac{1}{4(x - 2)} - \dfrac{2x^3}{(x^4 + 1)}\right)$.

11. $x = 1/\lambda$, provided $\lambda \neq 0$; if $\lambda = 0$, no maximum exists.

12. $(\log a)^2 \cdot a^{(a^x)} \cdot a^x$.

13. $a^{\sin x \, (\log x)^2} \cdot \log a \left\{ \cos x (\log x)^2 + \dfrac{2 \sin x \log x}{x} \right\}$.

§ 7, p. 183.

1. (a) Keep x fixed and differentiate with respect to y; then put y equal to zero.

(b) Evaluate $f(x)$ first for rational x, and then by continuity for irrational x.

2. Differentiate with respect to y, and then put y equal to 1.

3. 2315 years.

4. (a) $y = \beta + ce^{ax}$; (b) $y = -\dfrac{\beta}{\alpha} + ce^{ax}$, $\alpha \neq 0$; $y = \beta x + c$, $\alpha = 0$.

(c) $y = \beta x e^{ax} + ce^{ax}$; (d) $y = \dfrac{\beta}{\gamma - \alpha} e^{\gamma x} + ce^{ax}$, $\gamma \neq \alpha$.

§ 8, p. 189.

1. $\sinh a - \sinh b = 2 \cosh\left(\dfrac{a + b}{2}\right) \sinh\left(\dfrac{a - b}{2}\right)$.

$\cosh a + \cosh b = 2 \cosh\left(\dfrac{a + b}{2}\right) \cosh\left(\dfrac{a - b}{2}\right)$.

$\cosh a - \cosh b = 2 \sinh\left(\dfrac{a + b}{2}\right) \sinh\left(\dfrac{a - b}{2}\right)$.

2. $\tanh(a \pm b) = \dfrac{\tanh a \pm \tanh b}{1 \pm \tanh a \tanh b}.$

$\coth(a \pm b) = \dfrac{1 \pm \coth a \coth b}{\coth a \pm \coth b}.$

$\sinh \tfrac{1}{2}a = \sqrt{\left(\dfrac{\cosh a - 1}{2}\right)}; \quad \cosh \tfrac{1}{2}a = \sqrt{\left(\dfrac{\cosh a + 1}{2}\right)}.$

3. (a) $\sinh x + \cosh x$;   (b) $-4 \dfrac{e^{\tanh x + \coth x}}{\cosh 4x - 1}$;

(c) $(1 + \sinh 2x) \coth(x + \cosh^2 x)$;   (d) $1/\sqrt{(x^2 - 1)} + 1/\sqrt{(x^2 + 1)}$;

(e) $\alpha \sinh x / \sqrt{(\alpha^2 \cosh^2 x + 1)}$;   (f) $2/(1 - x^2)$.

4. $\sinh b - \sinh a.$

§ 9, p. 195.

1. (a) Higher than x^N;   (b) lower than x^ϵ;   (c) same as 1;   (d) higher than x^N;   (e), (f) higher than $x^{\frac{1}{2}-\epsilon}$, lower than $x^{\frac{1}{2}+\epsilon}$;   (g) same as x; (h) higher than x^N;   (j) lower than x^ϵ.

2. Higher than e^{ax}, $(\log x)^\alpha$, same as e^{x^β};   (b) lower than e^{ax}, e^{x^α}; (c) bounded;   (d) same as e^x, lower than e^{x^α}, higher than $(\log x)^\alpha$;   (e), (f), (g) lower than e^{ax}, e^{x^α}, higher than $(\log x)^\alpha$;   (h) higher than $e^{x^{1-\epsilon}}$, lower than $e^{x^{1+\epsilon}}$, higher than e^{ax}, $(\log x)^\alpha$;   (j) same as $\log x$, lower than e^{ax}, e^{x^α}.

3. (a) Same as x^β;   (b) lower than $\left(\dfrac{1}{x}\right)^\epsilon$;   (c) same as x;   (d) same as x; (e) same as $x^{5/2}$;   (f) same as $x^{3/2}$;   (g) higher than x^N;   (h) higher than $x^{1-\epsilon}$, lower than x;   (j) lower than $\left(\dfrac{1}{x}\right)^\epsilon$.

4. Yes; 0.   **5.** 0, 1.

6. $\lim\limits_{x \to 0} \dfrac{f(x)}{x} = f'(0) = 0.$ **8, 9.** Use result of Ex. **7.**

Appendix, p. 203.

1. $f''[g\{h(x)\}]g'^2\{h(x)\}h'^2(x) + f'[g\{h(x)\}]g''\{h(x)\}h'^2(x)$
$\qquad + f'[g\{h(x)\}]g'\{h(x)\}h''(x).$

2. (a) $x^{\sin x}\left(\dfrac{\sin x}{x} + \log x \cdot \cos x\right).$

(b) $(\cos x)^{\tan x}\left(-\tan^2 x + \dfrac{\log \cos x}{\cos^2 x}\right).$

(c) $\dfrac{u'(x)}{u(x) \log v(x)} - \dfrac{v'(x) \log u(x)}{v(x)\{\log v(x)\}^2}.$

4. (a) $e^{ax}[\alpha^n x^3 + 3n\alpha^{n-1}x^2 + 3n(n-1)\alpha^{n-2}x + n(n-1)(n-2)\alpha^{n-3}];$

(b) $\dfrac{2(-1)^n(n-1)!}{x^n}\left(\sum\limits_{\nu=1}^{n-1}\dfrac{1}{\nu} - \log x\right);$

(c) $\dfrac{(-1)^m}{2}\{\cos x - 3^{2m}\cos 3x\}, \quad \text{for } n = 2m;$

$\quad \dfrac{(-1)^m}{2}\{3^{2m+1}\sin 3x - \sin x\}, \quad \text{for } n = 2m + 1.$

(d) $\dfrac{(-1)^l}{2}[(m+k)^{2l}\sin(m+k)x - (m-k)^{2l}\sin(m-k)x], \quad \text{for } n = 2l;$

$\quad \dfrac{(-1)^l}{2}[(m+k)^{2l+1}\cos(m+k)x - (m-k)^{2l+1}\cos(m-k)x], \quad \text{for}$

$\quad n = 2l + 1.$

(e) $e^x\left[\left(\sum\limits_{l=0}^{2l\leqq n}(-1)^l\binom{n}{2l}2^{2l}\right)\cos 2x\right.$

$\quad\quad \left.+ \left(\sum\limits_{l=0}^{2l+1\leqq n}(-1)^{l+1}\binom{n}{2l+1}2^{2l+1}\right)\sin 2x\right] = 5^{n/2}e^x\cos(2x + n\alpha),$

where $\tan\alpha = 2$ (expanding $(1 + 2i)^n$ by the binomial theorem, and grouping real and imaginary terms).

(f) $e^x . \sum\limits_{\nu=0}^{6}\binom{6}{\nu}\binom{n}{\nu}(1 + x)^{6-\nu}.$

5. Let $y = \arcsin x$. Then

$$\frac{d^n y}{dx^n} = \frac{d^{n-1}}{dx^{n-1}}\left(\frac{1}{\sqrt{(1-x^2)}}\right) = \frac{d^{n-2}}{dx^{n-2}}\left(\frac{x}{(1-x^2)^{3/2}}\right).$$

Apply Leibnitz's rule to this last expression:

$$\frac{d^n y}{dx^n}\bigg|_{x=0} = (n-2)\frac{d^{n-3}}{dx^{n-3}}\left(\frac{1}{(1-x^2)^{3/2}}\right)_{x=0}$$

$$= 3.(n-2)\frac{d^{n-4}}{dx^{n-4}}\left(\frac{x}{(1-x^2)^{5/2}}\right),$$

and continue the process:

$$\frac{d^n y}{dx^n}\bigg|_{x=0} = 1.3.5\ldots(2\nu-1).(n-2)(n-4)\ldots(n-2\nu+2)\frac{d^{n-2\nu}}{dx^{n-2\nu}}\left(\frac{x}{(1-x^2)^{(2\nu+1)/2}}\right).$$

If $n = 2l$, $\dfrac{d^n y}{dx^n}\bigg|_{x=0} = 0$; if $n = 2l+1$, $\dfrac{d^n y}{dx^n}\bigg|_{x=0} = 1^2.3^2.5^2\ldots(2l-1)^2.$

$$\frac{d^{2l}}{dx^{2l}}(\arcsin x)^2\bigg|_{x=0} = \sum\limits_{k=0}^{l-1}\binom{2l}{2k+1}1^2.3^2\ldots(2k-1)^2.1^2.3^2\ldots(2l-2k-3)^2.$$

$$\frac{d^{2l+1}}{dx^{2l+1}}(\arcsin x)^2\bigg|_{x=0} = 0.$$

6. Differentiate $(1 + x)^n$ twice and put $x = 1$.

CHAPTER IV

§§ 2, 3,* p. 217.

1. $\frac{1}{2}e^{x^2}$.   **2.** $-\frac{1}{4}e^{-x^4}$.   **3.** $\frac{2}{9}(1+x^3)^{3/2}$.   **4.** $\frac{1}{2}(\log x)^2$.

5. $-\dfrac{1}{n-1}\left(\dfrac{1}{\log x}\right)^{n-1}$.

6. Hint: write denominator in the form $(3x-1)^2+1$: arc tan $(3x-1)$.

7. $\log\left\{\dfrac{x-1}{2}+\sqrt{1+\left(\dfrac{x-1}{2}\right)^2}\right\}$.

8. Hint: $6x/(2+3x)=2-4/(2+3x)$: $2x-\frac{4}{3}\log|2+3x|$.

9. arc $\sin x - \sqrt{(1-x^2)}$.   **10.** $\log\left\{\dfrac{x+1}{2}+\sqrt{1+\left(\dfrac{x+1}{2}\right)^2}\right\}$.

11. arc $\sin\dfrac{x+1}{2}$.   **12.** $\frac{1}{2}\log(x^2-x+1)+\dfrac{1}{\sqrt{3}}$ arc tan $\dfrac{2x-1}{\sqrt{3}}$.

13. 2 ar $\cosh\left(\dfrac{x-2}{\sqrt{3}}\right)+\sqrt{(x^2-4x+1)}$.

14. $-\frac{1}{3}\sqrt{(2+2x-3x^2)}+\dfrac{4}{3\sqrt{3}}$ arc $\sin\dfrac{3x-1}{\sqrt{7}}$.

15. $\dfrac{2}{\sqrt{3}}$ arc tan $\dfrac{2x+1}{\sqrt{3}}$.   **16.** $\dfrac{2}{\sqrt{3}}$ arc tan $\dfrac{2x-1}{\sqrt{3}}$.

17. $\dfrac{1}{\sqrt{(b-a^2)}}$ arc tan $\dfrac{x+a}{\sqrt{(b-a^2)}}$, if $b-a^2>0$; $-\dfrac{1}{x+a}$, if $b-a^2=0$;

$-\dfrac{1}{\sqrt{(a^2-b)}}$ ar tanh $\dfrac{x+a}{\sqrt{(a^2-b)}}$, if $b-a^2<0$.

18. $-x^4/4-x^3/3-x^2/2-x-\log|x-1|$.

19. Hint: $\sin^3 x\cos^4 x=\sin x\cos^4 x(1-\cos^2 x)=\sin x\cos^4 x-\sin x\cos^6 x$:

$-\dfrac{\cos^5 x}{5}+\dfrac{\cos^7 x}{7}$.

20. $\dfrac{\sin^3 x}{3}-2\dfrac{\sin^5 x}{5}+\dfrac{\sin^7 x}{7}$.   **21.** $\frac{1}{9}(1-x^2)^{9/2}-\frac{1}{7}(1-x^2)^{7/2}$.

22. $\frac{1}{2}$ arc $\sin x-\frac{1}{2}x\sqrt{(1-x^2)}$.   **23.** $\pi^2/32$.

24. $\dfrac{1+(-1)^n}{n+1}$.   **25.** 2.   **26.** $\dfrac{1}{2(1+a^2)}-\dfrac{1}{2(1+b^2)}$.

27. $\frac{1}{3}(a^3-b^3)+\frac{1}{2}(a^2-b^2)+(a-b)+\log\dfrac{a-1}{b-1}$.

28. $\frac{1}{4}\left(1-\cos\dfrac{\pi^2}{2}\right)$.   **29.** Cf. Ex. 8, p. 88: $1/(n+1)$.

* Here and elsewhere the constants of integration are omitted.

§ 4, p. 225.

1. Take $f = x$, $g' = \cos x/\sin^2 x$: $-x/\sin x + \log \tan x/2$.

2. Take $f = x^4/4$, $g' = 4x^3/(1 - x^4)^2$: $x^4/4(1 - x^4) + \frac{1}{4} \log |1 - x^4|$.

3. $(x^2 - 2) \sin x + 2x \cos x$.

4. $-\frac{1}{2}(x^2 + 1)e^{-x^2}$. **5.** $4\pi(-1)^n/n^2$. **6.** 0.

7. $\frac{1}{2}(x^2 \sin x^2 + \cos x^2)$. **8.** $\frac{1}{32} \sin 4x - \frac{1}{4} \sin 2x + \frac{3}{8}x$.

9. $\frac{1}{192} \sin 6x + \frac{3}{64} \sin 4x + \frac{15}{64} \sin 2x + \frac{5}{16}x$.

10. Put $x = \cos \theta$: $x\sqrt{1 - x^2}(-\frac{1}{16} - \frac{1}{24}x^2 + \frac{1}{6}x^4) + \frac{1}{16} \arc \sin x$.

11. $e^x(x^2 - 2x + 2)$. **12.** $-\dfrac{1}{(n - 1)x^{n-1}} \log x - \dfrac{1}{(n - 1)^2 x^{n-1}}$.

13. $\dfrac{x^{m+1}}{m + 1} \log x - \dfrac{x^{m+1}}{(m + 1)^2}$. **14.** $\frac{1}{3}x^3\{(\log x)^2 - \frac{2}{3} \log x + \frac{2}{9}\}$.

16. Put $x^2 = t$, then use Ex. 15.

17. Integrate by parts repeatedly.

19. Use mathematical induction: assume that the n-th iterated integral $f_n(x)$ is given by $\dfrac{1}{(n - 1)!} \displaystyle\int_0^x f(u)(x - u)^{n-1} du$, and expand the integrand by the binomial theorem. Then $f_{n+1}(x) = \displaystyle\int_0^x f_n(t) dt$; integrate each term by parts.

§ 5, p. 234.

1. $\log \sqrt{\left|\dfrac{x}{2 - 3x}\right|}$. **2.** $\log \left|1 - \dfrac{1}{x}\right|$.

3. $\log \left|\dfrac{x}{x + 1}\right|^3 + \dfrac{3}{x + 1} + \dfrac{3}{2(x + 1)^2}$.

4. $\dfrac{x}{3} - \frac{1}{8} \log |x + 1| + \frac{49}{72} \log |3x - 5|$.

5. $-\dfrac{1}{2(x - 1)} + \log \sqrt[4]{\dfrac{1 + x^2}{(1 - x)^2}}$. **6.** $\dfrac{-1}{2(x - 1)} - \log \sqrt[4]{\dfrac{1 + x^2}{(1 - x)^2}}$.

7. $\log \dfrac{1}{\sqrt[3]{|x - 1|}} + \frac{1}{6} \log(x^2 + x + 1) + \dfrac{1}{\sqrt{3}} \arc \tan \dfrac{2x + 1}{\sqrt{3}}$.

8. $\log \sqrt[3]{|x + 1|} - \frac{1}{6} \log |x^2 - x + 1| + \dfrac{1}{\sqrt{3}} \arc \tan \dfrac{2x - 1}{\sqrt{3}}$.

9. $\log \dfrac{1}{\sqrt[5]{(x - 2)^2}} + \log \sqrt[5]{1 + x^2} + \frac{9}{5} \arc \tan x$.

10. $\frac{2}{3} \log |x + 2| + \frac{5}{6} \log |x - 1| - \frac{3}{2} \log |x + 1|$.

11. $-\dfrac{x^3}{3} + \log \sqrt[4]{\left|\dfrac{x + 1}{x - 1}\right|} - \frac{1}{2} \arc \tan x$.

12. $\frac{1}{3}$ arc tan $x + \frac{\sqrt{3}}{12}$ log $\frac{x^2 + \sqrt{3}x + 1}{x^2 - \sqrt{3}x + 1} + \frac{1}{6}$ arc tan $(2x + \sqrt{3})$

$\qquad + \frac{1}{6}$ arc tan $(2x - \sqrt{3})$.

13. $\frac{1}{6}$ log $\frac{x-1}{x+1} + \frac{\sqrt{2}}{3}$ arc tan $\frac{x}{\sqrt{2}}$.   14. $-\frac{3x^2 + 2}{2x(x^2 + 1)} - \frac{3}{2}$ arc tan x.

§ 6, p. 241.

1. $-\dfrac{2}{1 + \tan \dfrac{x}{2}}$.   2. $\tan \dfrac{x}{2}$.   3. $\dfrac{2}{\sqrt{3}}$ arc tan $\left(\dfrac{2 \tan \dfrac{x}{2} + 1}{\sqrt{3}}\right)$.

4. $\dfrac{1}{8}\left(\tan^2 \dfrac{x}{2} - \cot^2 \dfrac{x}{2}\right) + \dfrac{1}{2}$ log $\left|\tan \dfrac{x}{2}\right|$.   5. log $\left|\dfrac{\tan \dfrac{x}{2} + 1}{\tan \dfrac{x}{2} - 1}\right|$.

6. $\dfrac{1}{\sqrt{2}}$ arc tan $\dfrac{1}{2}\sqrt{2}$.   7. $\dfrac{1}{\sqrt{2}}$ arc tan $\dfrac{\tan x}{\sqrt{2}}$.

8. $\dfrac{1}{2\sqrt{3}}$ arc tan $\dfrac{2 \tan x}{\sqrt{3}}$.   9. $\dfrac{1}{2 \cos^2 x} +$ log cos x.

10. $\dfrac{1}{\sqrt{2}}$ log $\left|\dfrac{\tan \dfrac{x}{2} - 1 + \sqrt{2}}{\tan \dfrac{x}{2} - 1 - \sqrt{2}}\right|$.

11. $\dfrac{1}{4}$ log $\dfrac{\cos^2 x - \cos x + 1}{(\cos^2 x + \cos x + 1)^3} + \dfrac{1}{2\sqrt{3}}$ arc tan $\dfrac{2 \cos x - 1}{\sqrt{3}}$

$\qquad - \dfrac{1}{2\sqrt{3}}$ arc tan $\dfrac{2 \cos x + 1}{\sqrt{3}}$.

12. $\dfrac{1}{2} x\sqrt{x^2 - 4} - 2$ ar cosh $\dfrac{x}{2}$.

13. $\dfrac{1}{2} x\sqrt{4 + 9x^2} + \dfrac{2}{3}$ ar sinh $\dfrac{3}{2} x$.

14. 2 arc tan $\sqrt{\dfrac{x - 3}{x - 1}}$.

15. $\frac{1}{3}\sqrt{(x^2 + 4x)^3} - (x + 2)\sqrt{(x^2 + 4x)} + 4$ ar cosh $\dfrac{x + 2}{2}$.

16. $\sqrt{x} - \sqrt{(1 - x)} + \dfrac{1}{2\sqrt{2}}$ log $\left|\dfrac{(\sqrt{x} - \sqrt{\frac{1}{2}})(\sqrt{1 - x} + \sqrt{\frac{1}{2}})}{(\sqrt{x} + \sqrt{\frac{1}{2}})(\sqrt{1 - x} - \sqrt{\frac{1}{2}})}\right|$.

17. log $\left|x \cdot \dfrac{\sqrt{1 + x} - \sqrt{1 - x}}{\sqrt{1 + x} + \sqrt{1 - x}}\right| + \sqrt{(1 - x^2)}$.

18. $\frac{1}{2}$ ar cosh $(2x - 2a + 1) + \sqrt{(x-a)^2 + (x-a)} - 2\sqrt{x-a}$.

19. $\frac{2}{3(b-a)}$ $(\sqrt{(x-a)^3} - \sqrt{(x-b)^3})$.

§ 8, p. 254.

1. Div.   **2.** Conv.   **3.** Conv.   **4.** Conv.   **5.** Div.

6. Conv.   **7.** Conv.   **8.** Div.   **9.** Conv.   **10.** Conv.

11. Conv.   **14.** (a) For $0 < s < 1$.   (b) For $0 < s < 2$.

15. Yes.

Miscellaneous Examples IV, p. 255.

1. Put arc $\sin x = t$: $\frac{1}{2}e^{\text{arc} \sin x}(x + \sqrt{1 - x^2})$.

2. $\frac{1}{9}\cos^9 x - \frac{1}{7}\cos^7 x$.

3. $x\{(\log x)^2 - 2\log x + 2\}$.   **4.** $\frac{1}{4}\log\dfrac{2 - \cos x}{2 + \cos x}$.

5. Put $\sqrt{(1 - e^{-2x})} = t$: $x - \sqrt{(1 - e^{-2x})} + \log\{1 + \sqrt{(1 - e^{-2x})}\}$.

6. 0.   **7.** 0.   **10.** 0.

12. Consider the function $1/x$ for the interval $1 \leq x \leq 2$. Subdivide the interval into n equal parts and form the lower sum as in Chap. II, § 1 (p. 76 *et seq.*). This turns out to be α_n. Now let $n \to \infty$. The result is $\log 2$.

13. Compare with $1/\sqrt{(1 - x^2)}$ at $x = 0, 1/n, 2/n, \ldots, (n-1)/n$: $\pi/2$.

14. Evaluate

$$\lim_{n \to \infty} \log \sqrt[n]{\frac{n!}{n^n}} = \lim_{n \to \infty} \frac{1}{n}\left[\log 1 + \log\left(1 - \frac{1}{n}\right) \ldots + \ldots + \log\left(1 - \frac{n-1}{n}\right)\right],$$

using the definition of the definite integral.

15. $1/(1 + \alpha)$.

CHAPTER V

§ 1, p. 267.

1. $(x^2 + y^2)^3 = a^2(x^2 - y^2)^2$.

2. Take c as rotating with constant velocity, and measure the time so that at time $t = 0$ the point P is in contact with the circle C: $x = (R + r)\cos\theta - r\cos\{(R + r)\theta/r\}$, $y = (R + r)\sin\theta - r\sin\{(R + r)\theta/r\}$.

3. $x = 2R\cos\theta(1 - \cos\theta) + R$, $y = 2R\sin\theta(1 - \cos\theta)$.

4. $x = (R - r)\cos\theta + r\cos\{(R - r)\theta/r\}$, $y = (R - r)\sin\theta$ $- r\sin\{(R - r)\theta/r\}$.

6. Take rectangular co-ordinates so that the origin is at the centre of C and the point P lies on the x-axis at time $t = 0$: $x^{2/3} + y^{2/3} = R^{2/3}$.

7. $x = 3at/(1 + t^2)$, $y = 3at^2/(1 + t^2)$.

10. $\alpha = \arctan\left(\dfrac{r(f' - g')}{r^2 + f'g'}\right)$.

11. $x = \dfrac{f'(y_0 g' + x_0 f') - g'(gf' - fg')}{f'^2 + g'^2}$, $y = \dfrac{g'(y_0 g' + x_0 f') + f'(gf' - fg')}{f'^2 + g'^2}$.

12. (a) C itself; (b) the cardioid of the circle with diameter PM, having its vertex at P.

§§ 2, 3, p. 290.

1. $\frac{2}{5}(b^{5/2} - a^{5/2})$.   **2.** $3a^2/4$.   **3.** $\frac{1}{6}a^2(\theta_2{}^3 - \theta_1{}^3)$.   **4.** $6\pi R^2$.

5. $6\pi r^2$.   **6.** $\pi(1 + \frac{1}{2}x_0{}^2)$.   **7.** $\frac{1}{2}\pi(a^2 + b^2 + x_0{}^2)$.

8. $x = R + s(1 - s/2R + s^2/32R^2)(1 - s/16R)$,
$\qquad y = R(s/R - s^2/16R^2)^{3/2}(1 - s/8R)$,   for $0 \leqq s \leqq 16R$

9. $x = 2a \arccos(1 - s/4a) - (1 - s/4a)\sqrt{\{s(1 - s/8a)/2a\}}$,
$\qquad y = s - s^2/8a$,   for $0 \leqq s \leqq 8a$.

10. $s = \sqrt{(4/9 + x)^3} - 8/27$.   **11.** $6R$.

12. (a) $\frac{1}{2}a\{\operatorname{ar\,sinh}\theta + \theta\sqrt{(1 + \theta^2)}\}$.

    (b) $\dfrac{\sqrt{(1 + m^2)}}{m}(e^{m\theta} - e^{m\theta_0})$.

    (c) $8R(1 - \cos\frac{1}{2}\theta)$.   (d) $a\{\frac{1}{3}(\theta^3 - \theta_0{}^3) + \theta - \theta_0\}$.

13. (a) $\frac{1}{2}(1 + 4x^2)^{3/2}$: min. $\frac{1}{2}$ at $x = 0$.

    (b) $(a^2\sin^2\varphi + b^2\cos^2\varphi)/ab$:   if $a > b$,   min. b/a at $\varphi = 0$, π, max. a/b at $\varphi = \pi/2$, $3\pi/2$.

14. $\rho = 1/\sqrt{t}$.

17. Vol. $\pi r^2(h_2 - h_1) - \frac{1}{3}\pi(h_2{}^3 - h_1{}^3)$.   Surface $2\pi(h_2 - h_1)r$.

18. If ρ is the radius of circle and r the distance of its centre from the line, the volume is $2\pi^2 r\rho^2$, the surface area $4\pi^2 r\rho$.

19. $\pi(x_1 - x_0) + \dfrac{\pi}{2}(\sinh 2x_1 - \sinh 2x_0)$.

20. $k = \pi s$.

21. $y = -\operatorname{ar\,cosh}\dfrac{1}{x} + \sqrt{(1 - x^2)} + \text{const.}$; $s = \log\left(\dfrac{x}{x_0}\right)$;

$\qquad x = e^s$, $y = -\operatorname{ar\,cosh}e^{-s} + \sqrt{(1 - e^{2s})} + \text{const.}$

22. Let ds, ds' be the lengths of arc, l, l' the total lengths, A, A' the areas, and k, k' the curvatures of the curve and the parallel curve respectively. Then
$$ds' = (1 + pk)ds; \qquad k' = k/(1 + pk);$$
$$A' = A + lp + \pi p^2; \qquad l' = l + 2\pi p.$$

23. (a) $\xi = r(\sin\varphi_2 - \sin\varphi_1)/(\varphi_2 - \varphi_1)$,
$\qquad \eta = -r(\cos\varphi_2 - \cos\varphi_1)/(\varphi_2 - \varphi_1)$,

where φ_1, φ_2 are the θ-co-ordinates of the extremities of the arc.

(b) $\xi = (x_2 \sinh x_2 - x_2 \sinh x_1 - \cosh x_2 + \cosh x_1)/(\sinh x_2 - \sinh x_1)$

$\eta = \{2(x_2 - x_1) + \sinh 2x_2 - \sinh 2x_1\}/4(\sinh x_2 - \sinh x_1)$,

where (x_1, y_1), (x_2, y_2) are the extremities of the arc.

24. $(\alpha^2 + \beta^2)(b - a) + \frac{2}{3}(\beta^3 - \alpha^3)$.

25. (a) $\sinh x_2 - \sinh x_1 + \frac{1}{3}(\sinh^3 x_2 - \sinh^3 x_1)$,

(b) $(x_2{}^2 + 2) \sinh x_2 - (x_1{}^2 + 2) \sinh x_1 - 2x_2 \cosh x_2 + 2x_1 \cosh x_1$,

if $0 \leqq x_1 \leqq x_2$.

§ 4, p. 298.

1. $\dfrac{dx}{dt} = -\dfrac{r \sin(2t/r)}{2\sqrt{(l^2 - r^2 \sin^2(t/r))}} - \sin\dfrac{t}{r}$;

$\dfrac{d^2x}{dt^2} = -\dfrac{l^2 \cos(2t/r) + r^2 \sin^4(t/r)}{\sqrt{(l^2 - r^2 \sin^2(t/r))^3}} - \dfrac{1}{r}\cos\dfrac{t}{r}$.

2. Horizontal.

3. $u = v_0/(1 + ksv_0)$, $t = s/v_0 + \frac{1}{2}ks^2$.

4. (a) $x = 4 \arctan e^t - \pi$; $x = \pi$.

5. (a) $t = \dfrac{1}{\sqrt{(2\mu M)}} \cdot (y_0\sqrt{(y_0 - y)} - y_0{}^{3/2} \arctan \sqrt{\{y/(y_0 - y)\}} + \frac{1}{2}\pi y_0)$.

(b) $\sqrt{\{2\mu M(1/R - 1/y_0)\}}$; (c) $\sqrt{\dfrac{2\mu M}{R}}$.

6. $\theta = at$, $r = \dfrac{k}{1 - e \cos at}$, where $a = \dfrac{(1 - e)^2}{k^2}\sqrt{ck}$;

period $= \dfrac{2\pi}{a} = \dfrac{2\pi}{(1 - e)^2 c^{1/2}} \cdot k^{3/2}$.

CHAPTER VI

§ 1, p. 319.

1. $0 \cdot 28$.   **2.** $0 \cdot 182$.   **3.** Impossible; series not valid.

§ 2, p. 325.

2. $\dfrac{1}{1 - x}$: $\theta = \dfrac{1 - (1 - x)^{1/(n+2)}}{x}$.

$\dfrac{1}{1 + x}$: $\theta = \dfrac{(1 + x)^{1/(n+2)} - 1}{x}$.

§ 3, p. 330.

1. $1 + \frac{1}{2}x - \dfrac{1}{4(1 + \theta x)^{3/2}}$; $-\frac{1}{4} < R < -\dfrac{\sqrt{2}}{16}$.

2. $1 \cdot 5$; error just over 6%.   **3.** $1 + \frac{1}{3}x$; $|x| < 0 \cdot 3$,

4. $1 + \frac{1}{3}x - \frac{1}{9}x^2$; $\frac{5}{81} \times 10^{-3}$.

5. (a) $1 + \frac{x}{n}$; $\frac{1}{2n}\left(\frac{1}{n} - 1\right) \times 10^{-2}$.

    (b) $1 + \frac{x}{n} + \frac{1}{2n}\left(\frac{1}{n} - 1\right)x^2$; $\frac{1}{6n}\left(\frac{1}{n} - 1\right)\left(\frac{1}{n} - 2\right) \times 10^{-3}$.

6. $0\cdot0100$.   **7.** (a) $0\cdot9999$;   (b) $5\cdot0133$;   (c) $9\cdot8489$.   **8.** $0\cdot515$.

9. $x^2 - \frac{x^4}{3} + \frac{2x^6}{45} + \frac{x^8}{8!}\,(-128\cos(2\theta x))$.

10. $1 - \frac{3x^2}{2} + \frac{7x^4}{8} + \frac{3}{4}\frac{x^6}{6!}\,(243\cos(3\theta x) + \cos(\theta x))$.

11. $-\frac{1}{2}x^2 - \frac{1}{12}x^4 - \frac{1}{45}x^6$

       $- 16\frac{x^8}{8!}\,(17 + 248\tan^2(0x) + 756\tan^4(\theta x) + 840\tan^6(\theta x)$
                   $+ 315\tan^8(\theta x))$.

12. $x + \frac{1}{3}x^3 + \frac{2}{15}x^5$

       $+ 16\frac{x^7}{7!}\,(17 + 248\tan^2(\theta x) + 756\tan^4(\theta x) + 840\tan^6(\theta x)$
                   $+ 315\tan^8(\theta x))$.

13. $\frac{1}{2}x^3 + \frac{1}{12}x^4 + \frac{1}{45}x^6$

       $+ 16\frac{x^8}{8!}\,(17 + 248\tan^2(\theta x) + 756\tan^4(\theta x) + 840\tan^6(\theta x)$
                   $+ 315\tan^8(\theta x))$.

14. $1 - x^2 + \frac{1}{2}x^4 - \frac{x^6}{3!}e^{-\theta^2 x^2}$.

15. $1 + \frac{1}{2}x^2 + \frac{5}{24}x^4$

       $+ \frac{x^6}{6!}\,(720\sec^7(\theta x) - 840\sec^5(\theta x) + 182\sec^3(\theta x) - \sec(\theta x))$.

16. $-\frac{1}{3}x - \frac{1}{45}x^3 - \frac{2}{945}x^5 - \cdots$.

17. $\frac{1}{6}x + \frac{7}{360}x^3 + \frac{31}{15120}x^5 + \cdots$.

18. $x - \frac{3}{2}x^2 + \frac{11}{6}x^3 + \frac{x^4}{4!}\frac{1}{(1 + \theta x)^5}\,(-50 + 24\overline{\log 1 + \theta x})$.

19. $1 + x + \frac{1}{2}x^2 - \frac{3}{24}x^4 + \frac{x^5}{5!}e^{\sin\theta x}\,(\cos^5(\theta x) - 10\cos^3(\theta x) + \cos(\theta x)$

       $-10\sin(\theta x)\cos^3(\theta x) + 15\sin(\theta x)\cos(\theta x) + 6\sin^2(\theta x)\cos(\theta x))$.

20. $x + \frac{1}{3}x^3$; $0 < x < \pi/4$.

21. (a) $y = x^2 + x^4 + 2x^6 + \cdots$;   (b) $y = 1 - x^2 - x^4 - 2x^6 - \cdots$;

    (c) $y = x^3 + x^9 + \cdots$.

§ 4, p. 335.

1. 2.   **2.** 4.   **3.** $a = 8/3$, $b = 16/3$, $c = -5/3$, $d = -5/3$.

4. Third order and also zero order at $(0, 0)$; zero order at $(\frac{1}{2}, \frac{1}{2})$.

5. Third order at $(0, 0)$.

7. Take P as origin and the tangent to the curve at P as x-axis. Let the co-ordinates of Q be (x, y). Then the centre of the circle in question lies on the y-axis at the point $\eta = \dfrac{y}{2} + \dfrac{x^2}{2y}$; use Ex. 6.

8. Take axes as in Ex. 7; let the slope of the curve at Q be y'. Then the two normals intersect on the y-axis at the point $\eta = y + \dfrac{x}{y'}$. Now write $y = \dfrac{y''(0)}{2!} x^2 + \dots$, and let $x \to 0$.

9. At a point P where $\rho = \dfrac{(1 + y'^2)^{3/2}}{y''}$ is a maximum or minimum, we necessarily have $y''' = \dfrac{3y'y''}{(1 + y'^2)}$. Take axes as in Ex. 7; then $y'''(0) = 0$, so that the equation of the curve in the neighbourhood of $x = 0$ is $y = \dfrac{1}{2\rho} x^2 + ax^4 + \dots$. The equation of the osculating circle is $y = \dfrac{1}{2\rho} x^2 + bx^4 + \dots$, and the contact is at least of order **3.**

10. Minimum at $x = 0$.

Appendix, p. 341.

    1. na^{n-1}.   **2.** 1/6.   **3.** 1/30.   **4.** 2.   **5.** 1.

    6. Write expression as $\cot x/\cot 5x$: 1/5.

    7. 1/2.   **8.** 1/3.   **9.** Take logarithms: **1.**

    10. e.   **11.** 2.   **12.** -2.

CHAPTER VII

§ 1, p. 348.

    1. (a) 3·14; (b) 3·1416.   **2.** ·89.

    3. 0·93.

§ 2, p. 355.

    1. Error $< -\cdot03$ metre, $< 0\cdot007\%$.   **2.** 0·693.   **3.** 1·609438.

    4. 3·14159.

§ 3, p. 360.

    1. 1·0755.   **2.** 4·4934.   **3.** 1·475.

    4. 0, 1·90, $-1\cdot90$.   **5.** 1·045.

    6. Write equation in form $x = 1 + 0\cdot3x^3 - 0\cdot1x^4$; 1·519.

    7. $-1\cdot2361$, 3·2361, 5·0000.

CHAPTER VIII

§ 1, p. 376.

1. Use the fact that $\dfrac{1}{\nu(\nu+1)} = \dfrac{1}{\nu} - \dfrac{1}{\nu+1}$.

2. Split up $1/x(x+1)(x+2)$ into partial fractions: in the result substitute $x = 1$, $x = 2, \ldots, x = \nu$ in turn and add.

4. Convergent for $\alpha > 0$.

5. Put $\Sigma a_\nu = A$. For every positive ε, $|s_n - A| < \varepsilon$ if n is greater than a certain m. Write

$$\frac{s_1 + \ldots + s_N}{N} = \frac{s_1 + \ldots + s_m}{N} + \frac{N-m}{N}\,\frac{s_{m+1} + \ldots + s_N}{N-m}$$

and let $N \to \infty$.

6. Yes.   **7.** No.

§ 2, p. 382.

1. Convergent.

2. Prove first that $n!/n^n \leqq 2/n^2$ when $n > 2$: convergent.

3. Divergent.   **4.** Cf. Chap. III, § 9, p. 189: divergent.

5. Note that $(\log n)^{\log n} = n^{\log\,(\log n)}$ and $\log\,(\log n) > 2$ when n is large: convergent.

6. Convergent.   **7.** $1/(n+1)^2$.

8. Error $= \dfrac{1}{(n+1)!}\left(1 + \dfrac{1}{n+2} + \dfrac{1}{(n+2)(n+3)} + \cdots\right)$

$$< \frac{1}{(n+1)!}\left(1 + \frac{1}{n+1} + \frac{1}{(n+1)^2} + \cdots\right)$$

$$< \frac{1}{(n+1)!}\,\frac{1}{1 - \dfrac{1}{n+1}} < \frac{1}{n \cdot n!}.$$

9. Error $= \dfrac{1}{(n+1)^{n+1}} + \dfrac{1}{(n+2)^{n+2}} + \cdots$

$$< \frac{1}{(n+1)^{n+1}} + \frac{1}{(n+1)^{n+2}} + \frac{1}{(n+1)^{n+3}} + \cdots < \frac{1}{n(n+1)^n}.$$

10. Error $= \dfrac{n+1}{2^{n+1}} + \dfrac{n+2}{2^{n+2}} + \cdots$. Now for $n > 1$,

$$n + 2 < \tfrac{3}{2}(n+1), \quad n+3 < \tfrac{3}{2}(n+2) < (\tfrac{3}{2})^2(n+1), \ldots;$$

hence

$$\text{Error} < \frac{n+1}{2^{n+1}}\,(1 + \tfrac{3}{4} + (\tfrac{3}{4})^2 + \cdots) < \frac{n+1}{2^{n-1}}.$$

12. Convergent.   **13.** Compare with $\int \dfrac{dx}{x\,(\log x)^a}$.

14. Compare with $\int \dfrac{dx}{x\,\log x\,(\log\log x)^a}$.

16. Use Schwarz's inequality.

17. $1 + \tfrac{1}{2} - \tfrac{2}{3} + \ldots - \dfrac{2}{3n+3} = \overset{3n+3}{\underset{\nu=1}{\Sigma}} \dfrac{1}{\nu} - 3\overset{n+1}{\underset{\nu=1}{\Sigma}} \dfrac{1}{3\nu} = \overset{3n}{\underset{\nu=n+2}{\Sigma}} \dfrac{1}{\nu}$;

then use the formula on p. 381,

$$1 + \tfrac{1}{2} + \tfrac{1}{3} + \ldots + \frac{1}{n} = \log n + C + \varepsilon_n,$$

where $\lim\limits_{n \to \infty} \varepsilon_n = 0$.

18. Take the sum from $\nu = 1$ to $\nu = mn$:

$$\overset{mn}{\underset{\nu=1}{\Sigma}} \frac{\alpha_\nu}{\nu} = \underset{\nu \neq kn}{\Sigma} \frac{1}{\nu} - \underset{\nu=kn}{\Sigma} \frac{n-1}{\nu} = \overset{mn}{\underset{\nu=1}{\Sigma}} \frac{1}{\nu} - \overset{m}{\underset{k=1}{\Sigma}} \frac{n}{kn} = \overset{mn}{\underset{\nu=m+1}{\Sigma}} \frac{1}{\nu}.$$

§§ 3, 4, p. 397.

3. (a) $\lim\limits_{n \to \infty} f_n(x) = \begin{cases} 0 & \text{if } x = 0 \\ 1 & \text{if } x \neq 0. \end{cases}$

(b) $\lim\limits_{n \to \infty} f_n(x) = \begin{cases} 0 & \text{if } x = 0 \\ 1 & \text{if } x \neq 0 \end{cases} \quad (\alpha > 0).$

Convergence is non-uniform, and $\lim\limits_{n \to \infty} \int_{-1}^{1} f_n(x)\,dx = \int_{-1}^{1} \lim\limits_{n \to \infty} f_n(x)\,dx.$

4. $\lim\limits_{n \to \infty} f_n(x) = \begin{cases} 0 & \text{if } |x| < 1 \\ \tfrac{1}{2} & \text{if } |x| = 1 \\ 1 & \text{if } |x| > 1. \end{cases}$

9. Consider $\lim\limits_{n \to \infty} \sqrt[2n]{(1 - x^{2n})}$ for $-1 < x < +1$ and $\lim\limits_{n \to \infty} \sqrt[2n]{(1 - y^{2n})}$ for $-1 < y < +1$.

10. Let $\varepsilon > 0$. Divide up the interval by points $x_0 = a$, $x_1, \ldots$, $x_m = b$ into sub-intervals of length less than $\varepsilon/3M$. At each point x_i we can choose n_i so large that $|f_n(x_i) - f_m(x_i)| < \varepsilon/3$ when n and $m > n_i$. Let N be the greatest of $n_0, n_1, \ldots, n_m$. Then prove by the mean value theorem that in each sub-interval the inequality $|f_n(x) - f_m(x)| < \varepsilon$ holds when n and $m > N$.

§§ 5, 6, p. 409.

Note on Ex. 1–20: in most of these problems the ratio test is effective, but for Ex. 12–15 the root test is preferable.

1. $|x| < 1.$ **2.** $|x| < 1.$ **3.** $|x| < 1.$

4. $|x| < 1.$ **5.** $|x| < 1.$ **6.** $-\infty < x < +\infty.$

7. $|x| < 1.$ **8.** $|x| < 1.$ **9.** $|x| < 1.$

10. $|x| < 1.$ **11.** $|x| < 1.$ **12.** $|x| < 1/a.$

13. $|x| < 1.$ **14.** $|x| < 1.$ **15.** $-\infty < x < +\infty.$

16. $|x| < 4.$ **17.** $|x| < 1.$ **18.** $|x| < 1$ or a, whichever is the greater.

19. $|x| < 1.$

20. Note that $1/n^{1+1/n}$ lies between n^{-1} and n^{-2}: $|x| < 1.$

21. $\displaystyle\sum_{\nu=0}^{\infty} \frac{(\log a)^\nu}{\nu!} x^\nu.$

22. $\displaystyle -\frac{1}{2} - \frac{x}{3} - \frac{x^2}{4} - \ldots - \frac{x^n}{n+2} - \ldots = -\frac{1}{x^2}\sum_{\nu=2}^{\infty}\frac{x^\nu}{\nu}.$

23. Write $\sin^2 x = \frac{1}{2} - \frac{1}{2}\cos 2x$: $\displaystyle\sum_{\nu=1}^{\infty}\frac{(-1)^{\nu-1}2^{2\nu-1}}{(2\nu)!}x^{2\nu}.$

24. $\displaystyle 1 + \sum_{\nu=1}^{\infty}\frac{(-1)^\nu 2^{2\nu-1}}{(2\nu)!}x^{2\nu}.$

25. $\displaystyle\sum_{\nu=3}^{\infty}\frac{(-1)^{\nu-1}(2x)^{2\nu}}{32(2\nu)!}(15 + 3^{2\nu} - 6 \cdot 2^{2\nu}).$

26. $\displaystyle x^3 + \frac{1}{2}\frac{x^9}{3} + \frac{1\cdot3}{2\cdot4}\frac{x^{15}}{5} + \ldots = x^3 + \sum_{\nu=2}^{\infty}\frac{(x^3)^{2\nu-1}}{2\nu-1}\cdot\frac{1\cdot3\ldots(2\nu-3)}{2\cdot4\ldots(2\nu-2)}.$

27. $1 \cdot 4142.$

28. (a) $\displaystyle 1 - \frac{1}{3\cdot3!} + \frac{1}{5\cdot5!} - \frac{1}{7\cdot7!} + - \ldots.$

    (b) $\displaystyle\frac{1}{2} + \frac{1}{320} + \frac{1}{3\cdot2^{12}} + \ldots.$

    (c) $\displaystyle 1 - \frac{1}{2^2} + \frac{1}{3^2} - \frac{1}{4^2} + \frac{1}{5^2} - + \ldots.$

    (d) Put $x = 1/t$: $\displaystyle\frac{1}{10} - \frac{2^5-1}{10^6} + \frac{2^9-1}{24\cdot10^9} - + \ldots.$

29. (a) $\displaystyle x + x^2 + \frac{x^3}{3}.$ (b) $\displaystyle x^2 - x^3 + \frac{11x^4}{12}.$

    (c) $\displaystyle x + \frac{x^2}{2} + \frac{13x^3}{24} + \frac{19x^4}{48}.$ (d) $\displaystyle x^2 - \frac{x^4}{3}.$

31. $|x| < \rho.$ **32.** $f(x) = 4e^x - x - 1.$

Appendix, p. 423.

1. Break off the series at the n-th term; then

$$\frac{1}{2}x + \frac{1}{2\cdot4}x^2 + \frac{1\cdot3}{2\cdot4\cdot6}x^3 + \ldots + \frac{1\cdot3\ldots(2n-3)}{2\cdot4\ldots2n}x^n < 1 - \sqrt{(1-x)} \leqq 1.$$

Put $x = 1$: the partial sums all $\leqq 1$.

2. Use Ex. 1. Show that the greatest error occurs when $x = 1$ and that it can be made less than ε.

3. Write $|t| = \sqrt{t^2} = \sqrt{\{1 - (1 - t^2)\}}$: then put $x = 1 - t^2$ in Ex. 2.

4. The substitution $x = a + (b - a)t$ transforms the function $f(x)$ into a function $\varphi(t)$, $0 \leqq t \leqq 1$. Approximate to $\varphi(t)$ by a polygonal function $\psi(t)$ to within $\varepsilon/2$ (cf. Ex. 2, p. 70). Represent $\psi(t)$ as a sum of the form $a + bt + \Sigma c_i \, |t - t_0|$. Approximate to this by a polynomial (cf. Ex. 3) and replace t by its expression in terms of x.

7. If there were only a finite number of primes, the identity would be valid for any positive s, in particular for $s = 1$. (Multiplication of absolutely convergent series.)

8. First prove by induction that

$$(1 - x) \prod_{\nu=0}^{n-1} (1 + x^{2^\nu}) = 1 - x^{2^n}.$$

CHAPTER IX

§§ 1, 2, p. 437.

3. $\sum\limits_{\nu=1}^{n} \sin \nu\alpha =$ imaginary part of $\sum\limits_{\nu=0}^{n} e^{i\nu\alpha}$: $\sin\left(\dfrac{n+1}{2}\right)\alpha \, \sin\dfrac{n}{2}\,\alpha / \sin\dfrac{1}{2}\,\alpha$.

4. Use the formula $\sigma_n(\alpha) = \frac{1}{2}(1 - e^{i\alpha})^{-1}(e^{-in\alpha} - e^{(n+1)i\alpha})$ on p. 436.

5. Evaluate $\dfrac{1}{\pi}\displaystyle\int_{-\pi}^{\pi} \sigma_k(\alpha)\,d\alpha$, and then use expression for $s_m(\alpha)$ in terms of $\sigma_k(\alpha)$.

§§ 3, 4, p. 446.

1. (a) $\dfrac{e^{a\pi} - e^{-a\pi}}{\pi}\left\{\dfrac{1}{2a} + \sum\limits_{\nu=1}^{\infty} \dfrac{(-1)^\nu}{a^2 + \nu^2}\,(a \cos \nu x - \nu \sin \nu x)\right\}$.

(b) $\dfrac{8}{15}\,\pi^4 - 48 \sum\limits_{\nu=1}^{\infty} \dfrac{(-1)^\nu}{\nu^4} \cos \nu x$.

(c) $\dfrac{\sin a\pi}{a} \sum\limits_{\nu=1}^{\infty} (-1)^\nu \nu \left[\dfrac{1}{\nu^2 - (a+1)^2} + \dfrac{1}{\nu^2 - (a-1)^2} - \dfrac{2}{\nu^2 - a^2}\right] \sin \nu x$,

if a is not an integer; $\frac{1}{2}\sin(a - 1)x + \sin ax + \frac{1}{2}\sin(a + 1)x$, if a is an integer.

(d) $\dfrac{b - a}{2\pi} + \dfrac{1}{\pi} \sum\limits_{\nu=1}^{\infty} \left(\dfrac{\sin \nu b - \sin \nu a}{\nu} \cos \nu x - \dfrac{\cos \nu b - \cos \nu a}{\nu} \sin \nu x\right)$.

2. Apply the transformation $x = -\pi + 2\pi t$ to § 4, No. 2 (p. 440).

3. $B_2(t) = t^2 - t + \frac{1}{6}$; $B_3(t) = t^3 - \frac{3}{2}t^2 + \frac{1}{2}t$; $B_4(t) = t^4 - 2t^3 + t^2 - \frac{1}{30}$.

4. $B_1(t)$ has already been given in Ex. 2. By (b) of the definition in Ex. 3, the other expansions are obtained by successive integration. The constants of integration must be proved to be zero.

5. In the results for $B_2(t)$ and $B_4(t)$ in Ex. 3 and 4 put $t = 0$.

6. In the results for $B_3(t)$ in Ex. 3 and 4 put $t = \frac{1}{4}$.

8. $\cos \pi x = \prod\limits_{\nu=1}^{\infty} \left(1 - \dfrac{x^2}{(\nu + \frac{1}{2})^2}\right).$

CHAPTER X

§ 2, p. 465.

3. (a) Discontinuous on the line $x = 0$; (b) discontinuous for $x = y = 0$; (c) discontinuous on the line $x = -y$; (d) discontinuous for $y = -x^2$.

§ 3, p. 472.

1. (a) $\dfrac{\partial f}{\partial x} = \dfrac{2x}{3\sqrt{(x^2 + y^2)^2}}, \quad \dfrac{\partial f}{\partial y} = \dfrac{2y}{3\sqrt[3]{(x^2 + y^2)^2}}.$

(b) $\dfrac{\partial f}{\partial x} = 2x \cos(x^2 - y), \quad \dfrac{\partial f}{\partial y} = -\cos(x^2 - y).$

(c) $\dfrac{\partial f}{\partial x} = e^{x-\nu}, \quad \dfrac{\partial f}{\partial y} = -e^{x-\nu}.$

(d) $\dfrac{\partial f}{\partial x} = -\dfrac{1}{2\sqrt{(1 + x + y^2 + z^2)^3}}, \quad \dfrac{\partial f}{\partial y} = -\dfrac{y}{\sqrt{(1 + x + y^2 + z^2)^3}},$

$\dfrac{\partial f}{\partial z} = \dfrac{-z}{\sqrt{(1 + x + y^2 + z^2)^3}}.$

(e) $\dfrac{\partial f}{\partial x} = yz \cos(xz), \quad \dfrac{\partial f}{\partial y} = \sin(xz), \quad \dfrac{\partial f}{\partial z} = xy \cos(xz).$

(f) $\dfrac{\partial f}{\partial x} = \dfrac{x}{1 + x^2 + y^2}, \quad \dfrac{\partial f}{\partial y} = \dfrac{y}{1 + x^2 + y^2}.$

2. (a) $\dfrac{\partial f}{\partial x} = y, \quad \dfrac{\partial f}{\partial y} = x, \quad \dfrac{\partial^2 f}{\partial x^2} = \dfrac{\partial^2 f}{\partial y^2} = 0, \quad \dfrac{\partial^2 f}{\partial x \partial y} = 1.$

(b) $\dfrac{\partial f}{\partial x} = \dfrac{1}{x}, \quad \dfrac{\partial f}{\partial y} = \dfrac{1}{y}, \quad \dfrac{\partial^2 f}{\partial x^2} = -\dfrac{1}{x^2}, \quad \dfrac{\partial^2 f}{\partial x \partial y} = 0, \quad \dfrac{\partial^2 f}{\partial y^2} = -\dfrac{1}{y^2}.$

(c) $\dfrac{\partial f}{\partial x} = \dfrac{1 + y^2}{(1 - xy)^2}, \quad \dfrac{\partial f}{\partial y} = \dfrac{1 + x^2}{(1 - xy)^2}, \quad \dfrac{\partial^2 f}{\partial x^2} = \dfrac{2(1 + y^2)y}{(1 - xy)^3},$

$\dfrac{\partial^2 f}{\partial x \partial y} = \dfrac{2(x + y)}{(1 - xy)^3}, \quad \dfrac{\partial^2 f}{\partial y^2} = \dfrac{2(1 + x^2)x}{(1 - xy)^3}.$

(d) $\dfrac{\partial f}{\partial x} = yx^{y-1}, \quad \dfrac{\partial f}{\partial y} = x^\nu \log x, \quad \dfrac{\partial^2 f}{\partial x^2} = y(y - 1)x^{y-2},$

$\dfrac{\partial^2 f}{\partial x \partial y} = x^{y-1}(1 + y \log x), \quad \dfrac{\partial^2 f}{\partial y^2} = x^\nu (\log x)^2.$

(e) $\dfrac{\partial f}{\partial x} = yx^{y-1}e^{(x^y)}$, $\dfrac{\partial f}{\partial y} = x^y \log[xe^{(x^y)}]$,

$$\frac{\partial^2 f}{\partial x^2} = yx^{y-2}e^{(x^y)}(yx^y + y - 1),$$

$$\frac{\partial^2 f}{\partial x \partial y} = x^{y-1}e^{(x^y)}(1 + y \log x + yx^y \log x),$$

$$\frac{\partial^2 f}{\partial y^2} = x^y (\log x)^2 e^{(x^y)}(1 + x^y).$$

3. Differentiate $\varphi(x^2 + y^2) = \psi(x)\psi(y)$ partially with respect to x and with respect to y. Eliminate $\varphi'(x^2 + y^2)$, put $y = 1$, and solve the resulting differential equation: $f(x, y) = ae^{b(x^2+y^2)}$.

§ 4, p. 479.

1. (a) $\dfrac{\partial f}{\partial x} = -\dfrac{x + y \cos z}{\sqrt{(x^2+y^2+2xy \cos z)^3}}$, $\dfrac{\partial f}{\partial y} = -\dfrac{y + x \cos z}{\sqrt{(x^2+y^2+2xy \cos z)^3}}$,

$\dfrac{\partial f}{\partial z} = \dfrac{xy \sin z}{\sqrt{(x^2 + y^2 + 2xy \cos z)^3}}$;

(b) $\dfrac{\partial f}{\partial x} = \dfrac{1}{\sqrt{(z^2 + 2zy^2 + y^4 - x^2)}}$, $\dfrac{\partial f}{\partial y} = -\dfrac{2xy}{(z+y^2)\sqrt{(z^2+2zy^2+y^4-x^2)}}$,

$\dfrac{\partial f}{\partial z} = -\dfrac{x}{(z + y^2)\sqrt{(z^2 + 2zy^2 + y^4 - x^2)}}$;

(c) $\dfrac{\partial f}{\partial x} = 2x\left(1 + \dfrac{y}{1 + x^2 + y^2 + z^2}\right)$,

$\dfrac{\partial f}{\partial y} = \log(1 + x^2 + y^2 + z^2) + \dfrac{2y^2}{1 + x^2 + y^2 + z^2}$,

$\dfrac{\partial f}{\partial z} = \dfrac{2yz}{1 + x^2 + y^2 + z^2}$;

(d) $\dfrac{\partial f}{\partial x} = \dfrac{1}{2(1 + x + yz)\sqrt{(x + yz)}}$, $\dfrac{\partial f}{\partial y} = \dfrac{z}{2(1 + x + yz)\sqrt{(x + yz)}}$,

$\dfrac{\partial f}{\partial z} = \dfrac{y}{2(1 + x + yz)\sqrt{(x + yz)}}$.

2. (a) $\dfrac{\partial f}{\partial x} = x^{(x^x)}x^x\left(\log x + (\log x)^2 + \dfrac{1}{x}\right)$; (b) $\dfrac{\partial f}{\partial x} = \dfrac{1}{x^{3+1/x^2}}(2 \log x - 1)$.

5. $z_x = 3$, $z_y = 1$; $z_r = z_x \cos\theta + z_y \sin\theta$, $z_\theta = -z_x r \sin\theta + z_y r \cos\theta$.

7. (a) $ad - bc$; (b) $1/r$; (c) $4xy$.

§ 5, p. 485.

2. (a) $-\dfrac{5}{4}$; (b) $-\dfrac{\pi}{2}$; (c) -1; (d) -1.

3. $(a) - \dfrac{21}{32}$; (b) π; (c) 2; $(d) - \dfrac{19}{3}$.

4. Max. value 6, min. value -6.   **5.** $\partial z/\partial x = -1$, $\partial z/\partial y = -1$.

§ 6, p. 499.

1. (a) $a^2b^2(a^2 - b^2)/8$; (b) -4; (c) $\log 2$; (d) $e^{ab}/b - 1/b - a$;
  (e) $\pi/16$; (f) $4/3$.

2. 2π.

3. Use the fact that the figure is symmetrical; $\frac{1}{16}$ of the volume lies above the triangle with vertices $(0, 0)$, $(1, 0)$, $(1, 1)$ and below the surface $x^2 + z^2 = 1$; $16/3$.

4. $\frac{1}{3}\pi(r - h)^2(2r + h)$.

		Area	Centre of Gravity	Moment about x-axis	Moment about y-axis	Moment of Inertia about x-axis	Moment of Inertia about y-axis
5.	(a)	$\frac{1}{2}r^2$	$(0, 4r/3\pi)$	$\frac{2}{3}r^3$	0	$\pi r^4/8$	$\pi r^4/8$
	(b)	ab	$(\frac{1}{2}a, \frac{1}{2}b)$	$\frac{1}{2}ab^2$	$\frac{1}{2}a^2b$	$\frac{1}{3}ab^3$	$\frac{1}{3}a^3b$
	(c)	$4ab$	$(0, 0)$	0	0	$4ab^3/3$	$4a^3b/3$
	(d)	πab	$(0, 0)$	0	0	$\pi ab^3/4$	$\pi a^3b/4$
	(e)	$\frac{1}{2}ab$	$(\frac{1}{3}a, \frac{1}{3}b)$	$\frac{1}{6}ab^2$	$\frac{1}{6}a^2b$	$ab^3/12$	$a^3b/12$

		Volume	Centre of Gravity	Moment of Inertia about x-axis	Moment of Inertia about y-axis	Moment of Inertia about z-axis
6.	(a)	abc	$(\frac{1}{2}a, \frac{1}{2}b, \frac{1}{2}c)$	$\frac{1}{3}abc(b^2 + c^2)$	$\frac{1}{3}abc(c^2 + a^2)$	$\frac{1}{3}abc(a^2 + b^2)$
	(b)	$\frac{2}{3}\pi a^3$	$(0, 0, 3a/8)$	$4\pi a^5/15$	$4\pi a^5/15$	$4\pi a^5/15$
	(c)	$\frac{1}{6}abc$	$(\frac{1}{4}a, \frac{1}{4}b, \frac{1}{4}c)$	$abc(b^2 + c^2)/60$	$abc(c^2 + a^2)/60$	$abc(a^2 + b^2)/60$

CHAPTER XI

§ 2, p. 509.

1. $c_1e^t + c_2e^{2t}$; $e^{2t} - e^t$.

2. $c_1e^{-t} + c_2e^{-2t}$; $e^{-t} - e^{-2t}$.

3. $c_1e^{\frac{1}{2}t} + c_2e^{-t}$; $\frac{2}{3}(e^{\frac{1}{2}t} - e^{-t})$.

4. $c_1e^{-2t} + c_2te^{-2t}$; te^{-2t}.

5. $c_1e^{-\frac{1}{2}t} + c_2te^{-\frac{1}{2}t}$; $te^{-\frac{1}{2}t}$.

6. $e^{-\frac{1}{2}t}\left(c_1 \cos \dfrac{\sqrt{3}}{2} t + c_2 \sin \dfrac{\sqrt{3}}{2} t\right) = ae^{-\frac{1}{2}t} \cos \dfrac{\sqrt{3}}{2} (t - \delta)$;

$\dfrac{2}{\sqrt{3}} e^{-\frac{1}{2}t} \sin \dfrac{\sqrt{3}}{2} t$; $\nu = \sqrt{3}/2$, $T = 4\pi/\sqrt{3}$, $a = 2/\sqrt{3}$, $\delta = \pi/\sqrt{3}$.

7. $\sqrt{2}e^{-\frac{1}{2}t} \cos\frac{1}{2}(t + \frac{1}{2}\pi)$; $a = \sqrt{2}$, $\delta = -\pi/2$, $\nu = \frac{1}{2}$.

§ 3, p. 519.

1. $-\dfrac{e^{-t}}{1+\omega^2}+\dfrac{2e^{-2t}}{4+\omega^2}+\dfrac{(2-\omega^2)\cos\omega t+3\omega\sin\omega t}{(1+\omega^2)(4+\omega^2)}$;

$$\alpha=\frac{1}{\sqrt{(1+\omega^2)(4+\omega^2)}},\quad \tan\omega\delta=\frac{3\omega}{2-\omega^2},\quad \omega=0.$$

2. $e^{-\frac{1}{2}t}\left((\omega^2-1)\cos\dfrac{\sqrt3}{2}t-\dfrac{1}{\sqrt3}(\omega^2+1)\sin\dfrac{\sqrt3}{2}t\right)$
$$\overline{\hspace{3cm}1-\omega^2+\omega^4\hspace{3cm}}$$
$$+\frac{(1-\omega^2)\cos\omega t+\omega\sin\omega t}{1-\omega^2+\omega^4};$$

$$\alpha=\frac{1}{\sqrt{(1-\omega^2+\omega^4)}},\quad \tan\omega\delta=\frac{\omega}{1-\omega^2},\quad \omega=\frac{1}{\sqrt2}.$$

3. $e^{-\frac{1}{2}t}\left(\omega\cos\dfrac{\sqrt3}{2}t+\dfrac{1}{\sqrt3}\omega(2\omega^2-1)\sin\dfrac{\sqrt3}{2}t\right)$
$$\overline{\hspace{3cm}1-\omega^2+\omega^4\hspace{3cm}}$$
$$+\frac{(1-\omega^2)\sin\omega t-\omega\cos\omega t}{1-\omega^2+\omega^4};$$

$\alpha,\quad \tan\omega\delta,\quad \omega$ as in Ex. 2.

4. $\dfrac{-e^{-t}((1-2\omega^2)\cos\frac{1}{2}t+(1+2\omega^2)\sin\frac{1}{2}t)}{1+4\omega^4}$
$$+\frac{(1-2\omega^2)\cos\omega t+2\omega\sin\omega t}{1+4\omega^4};$$

$$\alpha=\frac{1}{\sqrt{(1+4\omega^4)}},\quad \tan\omega\delta=\frac{2\omega}{1-2\omega^2},\quad \omega=0.$$

5. $e^{-2t}\left(\dfrac{\omega^2-4}{(\omega^2+4)^2}-\dfrac{2t}{\omega^2+4}\right)+\dfrac{(4-\omega^2)\cos\omega t+4\omega\sin\omega t}{(\omega^2+4)^2}.$

§ 4, p. 527.

1. $\log(1+y^2)(y+\sqrt{(y^2+1)})+2(1+x)^{-1/2}=c.$

2. $(y^3-2x^3)/y=c.$

3. $\log y-\displaystyle\int^{x/y}\frac{v\,dv}{\log v-v^2}=c.$

4. $1/y=\log x+1+cx.$

5. $x=\operatorname{arc\,tan}y-1+ce^{-\operatorname{arc\,tan}y}.$

6. $y^2=\sin x-\cos x+ce^{-x}.$

7. $cy^2=e^{xy-1/xy}.$

8. $y^3=x-2+ce^{-x}.$

9. $\cos x\cdot\cos y=c.$

10. $ye^{x/y} + x = c.$

11. $x = c_1 e^t + c_2 t e^t + c_3 t^2 e^t.$

12. $x = c_1 e^{3t} + c_2 t e^{3t} + c_3.$

13. $y = c_1 \cos x + c_2 \sin x + c_3 x \cos x + c_4 x \sin x.$

14. $y = c_1 + c_2 e^{-x} \cos \dfrac{\sqrt{3}}{2} x + c_3 e^{-x} \sin \dfrac{\sqrt{3}}{2} x.$

15. $y = c_1 e^x + c_2 x e^x + c_3 e^{-x} + c_4 x e^{-x}$
$\qquad\qquad + c_5 \cos x + c_6 \sin x + c_7 x \sin x + c_8 x \cos x.$

16. $y = c_1 + c_2 e^{x/a}.$

17. $e^{-y} = c_1 \sec (x + c_2).$

18. $y = c_1 + c_2 x + c_3 e^x + c_4 e^{-x}.$

19. $y = c_1 \arctan x + c_2.$

20. $x = \displaystyle\int^y \dfrac{dy}{2 \log |y - 1| + c_1} + c_2.$

21. $x = -1/(c_1 t + c_2).$

22. $s = (\arcsin t)^2 + c_1 \arcsin t + c_2.$

23. $t + c_2 = \dfrac{1}{c_1} \sqrt{(c_1 s^2 - 2ks)} - \dfrac{2k}{c_1} \arcsin \sqrt{\left(\dfrac{c_1 s}{2k} \right)}.$

MISCELLANEOUS EXAMPLES

CHAPTER I

1. Use § 5, No. 7 (p. 34).

2. $39 = 1 \cdot 3^3 + 1 \cdot 3^2 + 1 \cdot 3 + 0$, hence the required answer is 1110.

3. (a) 10011100; (b) 2130.

4. (a) 758; (b) 5954; (c) 10,000; (d) ·2; (e) ·023; (f) ·2497.

5. (a) $1 \cdot 41 < \sqrt{2} < 1 \cdot 42$; (b) $2 \cdot 65$.

6. (a) $x \leqq \dfrac{-3 - \sqrt{5}}{2}, \; x \geqq \dfrac{-3 + \sqrt{5}}{2}.$

    (b) All values of x.

    (c) $x \leqq -3 - 2\sqrt{2}; \; -3 + 2\sqrt{2} \leqq x \leqq 3 - 2\sqrt{2}; \; x \geqq 3 + 2\sqrt{2}.$

    (d) $x \geqq -2.$

7. Square both sides. Equality only if $a = b$.

8. Use Ex. 7. Equality only if $a = b$.

9. (a) Add the three inequalities $a^2 + b^2 \geqq 2ab, \; b^2 + c^2 \geqq 2bc,$ $c^2 + a^2 \geqq 2ca$.

    (b) Multiply together the three inequalities

$$\frac{a + b}{2} \geqq \sqrt{ab}, \quad \frac{b + c}{2} \geqq \sqrt{bc}, \quad \frac{c + a}{2} \geqq \sqrt{ca}.$$

    (c) Add together the inequalities of the type $a^2b^2 + b^2c^2 \geqq 2b^2ac$.

10. Apply Schwarz's inequality to the numbers x_1, x_2, x_3 and 1, 1, 1.

11. From the relationship $(a_i - a_j)(b_i - b_j) \geqq 0$ we obtain

$$a_i b_i + a_j b_j \geqq a_i b_j + a_j b_i;$$

sum for all integral values of i and j from 1 to n.

12. (a) Expand $(1 - 1)^n$ by the binomial theorem.

    (e) In the identity $(1 + x)^n (1 + x)^n = (1 + x)^{2n}$, expand and collect terms in x^n.

14. $n^2(n + 1)^2/4$.

15. (a) Write $\dfrac{1}{\nu(\nu + 1)(\nu + 2)} = \dfrac{1}{2}\left(\dfrac{1}{\nu(\nu + 1)} - \dfrac{1}{(\nu + 1)(\nu + 2)}\right)$ and sum from $\nu = 1$ to n; $\dfrac{1}{4} - \dfrac{1}{2n(n + 1)};$

    (b) $\dfrac{n(3n + 5)}{2(n + 1)(n + 2)};$ (c) $\dfrac{7n^2 + 21n + 8}{36(n + 1)(n + 2)}.$

16. (a) $\frac{1}{2}(n^2 - n + 2)$;   (b) $\frac{1}{6}(5n^3 - 18n^2 + n - 30)$.

17. (a) $n(n^2 + 5)/6$;   (b) $n(n - 5)(5n^2 + 11n + 26)/24$.

18. Assume the theorem to be true for $n = m$, and then multiply by $(a + b)$, obtaining the theorem for $n = m + 1$. Verify the theorem for $n = 1, 2$.

19. (a) 1;   (b) $\frac{1}{4}$;   (c) ∞.

25. (c) If $m > n$, $\left| a_m - a_n \right| = \dfrac{1}{(n+1)!} + \dfrac{1}{(n+2)!} + \cdots + \dfrac{1}{m!}$

$$= \frac{1}{(n+1)!} \left(1 + \frac{1}{n+2} + \frac{1}{(n+2)(n+3)} + \cdots + \frac{1}{(n+2)\ldots m} \right)$$

$$< \frac{1}{(n+1)!} \left(1 + \frac{1}{n+1} + \frac{1}{(n+1)^2} + \cdots + \frac{1}{(n+1)^{m-n-1}} \right)$$

$$< \frac{\cdot 1}{(n+1)!} \ \frac{1}{1 - \dfrac{1}{n+1}} < \frac{1}{n \cdot n!}$$

(d) Similar to (c).

26. Let $c_n = \sum\limits_{\nu=0}^{n} \dfrac{1}{\nu!}$, $d_n = \sum\limits_{\tau=0}^{n} \dfrac{(-1)^\tau}{\tau!}$.

$$c_n d_n = \sum_{\nu, \tau = 0}^{n} \frac{(-1)^\tau}{\tau! \, \nu!}, \text{ and setting } \tau + \nu = \mu, \text{ we have}$$

$$c_n d_n = \sum_{\mu=n+1}^{2n} \sum_{\tau=0}^{n} \frac{(-1)^\tau}{\tau!(\mu - \tau)!} + \sum_{\mu=0}^{n} \sum_{\tau=0}^{\mu} \frac{(-1)^\tau}{\tau!(\mu - \tau)!}.$$

Now $\sum\limits_{\tau=0}^{\mu} \dfrac{(-1)^\tau}{\tau!(\mu - \tau)!} = 0$ if $\mu > 0$, so that

$$\left| c_n d_n - 1 \right| = \left| \sum_{\mu=n+1}^{2n} \sum_{\tau=0}^{n} \frac{(-1)^\tau}{\tau!(\mu - \tau)!} \right| < \sum_{\mu=n+1}^{2n} \frac{2^\mu}{\mu!}.$$

$$< \frac{2^{n+1}}{(n+1)!} \left(1 + \frac{2}{n+1} + \frac{2^2}{(n+1)^2} + \cdots \right)$$

$$< \frac{2^{n+1}}{(n+1)!} \ \frac{1}{1 - \dfrac{2}{n+1}} < \frac{2^{n+1}}{(n-1) \cdot n!}.$$

Since $\dfrac{2^n}{n!} \to 0$ as $n \to \infty$, $c_n d_n \to 1$ and $\lim\limits_{n \to \infty} d_n = \dfrac{1}{e}$.

27. (a) The sequence is monotonically increasing and is bounded above by 2, since if $a_n < 2$, $a_{n+1} = \sqrt{2 + a_n} < \sqrt{4} < 2$.

(b) Let $\lim\limits_{n \to \infty} a_n = a$. Then use the relation $a_{n+1} = \sqrt{2 + a_n}$ to obtain $a = \sqrt{2 + a}$ or $a = 2$.

33. (a) 1;   (b) 1;   (c) $1/e$.

35. (a) $\dfrac{1}{11}$;   (b) $\dfrac{1}{1001}$;   (c) $\dfrac{\varepsilon}{1 + \varepsilon}$.

36. (a) $4\varepsilon/(1 + 2\varepsilon)$; (b) $\varepsilon/7$; (c) $\arccos(1 - \varepsilon)$.

39. Use the fact that if x is rational, $n!x$ is an even integer for all sufficiently large values of n.

40. (a) Continuous; (b) Discontinuous at $x = 0$; (c) Discontinuous at $x = 0, \pm 1, \pm 2, \ldots$; (d) Discontinuous for all values of x.

42. Yes; consider signs at $x = 0$ and at $x = \pi/5$.

44. Let ε be arbitrary; then $|f(x') - f(x'')| < \varepsilon$ provided only that $|x' - x''| < \delta$. In particular, $|f(x') - f(x'')| < \varepsilon$ if $|x' - a| < \delta$, $|x'' - a| < \delta$, which is Cauchy's criterion for convergence.

45. (a) $(x^2 + y^2 - bx)^2 = a^2(x^2 + y^2)$.

(b) $3x^2 - 4x - 4 + 4y^2 = 0$.

(c) $x^3 = y^2(2a - x)$.

(d) $x^3 + y^3 = 3axy$.

47. (a) Circle with centre at $-\frac{5}{3}i$ and radius $\frac{4}{3}$.

(b) If $k > 1$, circle with centre at $-\dfrac{1}{k^2 - 1}\alpha + \dfrac{k^2}{k^2 - 1}\beta$ and radius $\dfrac{k}{k^2 - 1}|\beta - \alpha|$; if $k < 1$, interchange α and β; if $k = 1$, the perpendicular bisector of the line joining α, β.

(c) Consider the three possibilities $k < 1$, $= 1$, > 1.

48. The "triangle inequality": the sum of two sides of a triangle is greater than the third side.

49. The sum of the squares on the diagonals of a parallelogram is equal to the sum of the squares on all the sides of the parallelogram.

CHAPTER II

52. $\sin\dfrac{1}{x} - \dfrac{1}{x}\cos\dfrac{1}{x}$.

53. $f'(x) = (1 + 2x)\sin\dfrac{1}{x} - \left(1 + \dfrac{1}{x}\right)\cos\dfrac{1}{x}$, $x \neq 0$; $f'(0)$ does not exist, but the difference quotient $\dfrac{f(x) - f(0)}{x}$ as $x \to 0$ has the upper and lower limits 1 and -1 respectively.

54. $f''(x) = \left(\dfrac{2}{x^3} - \dfrac{1}{x}\right)\sin x - \dfrac{2}{x^2}\cos x$, $x \neq 0$; $f''(0) = -\dfrac{1}{3}$.

55. Use the mean value theorem.

56. Use the mean value theorem.

57. Consider $\varphi(x) = \{f(x + h) - f(x)\}/h$. Prove that for small (fixed) values of h this assumes values above and below μ; hence for some value of x, $\varphi(x) = \mu$. Then use the mean value theorem.

58. Find the equation $y = g(x)$ of the tangent; apply the mean value theorem to $f'(x) - g'(x)$, and use the result of Ex. 55.

59. Find the equation $y = g(x)$ of the chord joining $x = x_1$, $x = x_2$ of the curve; consider $h(x) = f(x) - g(x)$, $h''(x) = f''(x) \geqq 0$. If $h(x) > 0$ somewhere in the interval $x_1 \leqq x \leqq x_2$, there would be a point ξ with $h'(\xi) = 0$, $h(\xi) > 0$; then use Ex. 58.

60. Use Ex. 59.

61. 0·006.

62. (a) $\frac{1}{2}x^{-1/2}$; (b) $\sec^2 x$.

63. Use Ex. 62: (a) 2; (b) 1.

66. Let $\mu = \dfrac{1}{a}\displaystyle\int_0^a u(t)\,dt$. Find the equation $y = g(x)$ of the tangent to the curve $y = f(x)$ at the point $x = \mu$. Then $f(x) \geqq g(x)$ for all values of x (cf. Ex. 58). Put $x = u(t)$ and integrate.

67. Suppose that the acceleration is everywhere less than 4. Then $v < 4t$, and similarly $v < 4 - 4t$. Then the distance traversed, $s = \displaystyle\int_0^1 v\,dt$, is less than 1.

CHAPTER III

68. (a) $\left(\dfrac{2\tan x}{\cos^2 x} + \cot x\right)e^{\tan^2 x + \log \sin x}$.

(b) $4(x+2)^3\,(x^2+1)^{5/7}\sqrt[3]{(1-x^2)} - \dfrac{2x}{3\sqrt[3]{(1-x^2)^2}}(x+2)^4(x^3+1)^{5/7}$

$\qquad + \tfrac{10}{7}x(x^2+1)^{-2/7}(x+2)^4\sqrt[3]{(1-x^2)}$.

(c) $-x\sin x + \cos x + 3x^2\sin x + x^3\cos x - \dfrac{3x^2}{\sin x} + \dfrac{x^3\cos x}{\sin^2 x}$.

69. The denominator must not vanish for any real value of x; consider its discriminant. Also, the numerator of the derivative must not vanish. The conditions are $ac - b^2 > 0$, $a > 0$, $\alpha b - a\beta = 0$, $\alpha \neq 0$, or $a = b = 0$, $\alpha \neq 0$, $c \neq 0$.

70. Max. for $x = -1/e$, min. for $x = 1/e$, infl. at the point $(0, 1)$ and at the point given by $(2 + \log x^2)^2 + 2/x = 0$.

74. Let T be the triangle of given area and least perimeter, and let b be any side of it. Then, keeping b fixed, T must be the triangle of given base b and given area having the least perimeter. Hence T must be isosceles, and the two sides of T other than b are equal to one another. But b is any side, and T is therefore equilateral.

Analytically, we need consider isosceles triangles only. Let the co-ordinates of the vertices be $(-x, 0)$, $(x, 0)$ and $\left(0, \dfrac{A}{x}\right)$; then the perimeter is $2x + \dfrac{2}{x}\sqrt{(x^4 + A^2)}$. Equate the first derivative to zero, and find the second derivative.

75. In virtue of Ex. 71, consider isosceles triangles only.

76. In virtue of Ex. 72, consider isosceles triangles only.

77. (a) The derivative of $(1 + x)e^x$ is always positive for $x \geqq 0$; the minimum for $x \geqq 0$ is when $x = 0$, namely 1; (b) integrate (a) from 0 to x; (c) integrate (b) from 0 to x.

78. Let $f(\theta) = \dfrac{[\theta a^p + (1 - \theta)b^p]^{1/p}}{[\theta a^q + (1 - \theta)b^q]^{1/q}}$; then $f(0) = f(1) = 1$. Find $f'(\theta)$ and show that either $f'(\theta) \equiv 0$ or $f'(\theta) = 0$ for exactly one value of θ in the interval from 0 to 1. In the latter case, show that $f(\theta)$ is never equal to 1 for $0 < \theta < 1$. Then evaluate $f'(0)$; it is equal, except for a positive factor, to

$$b^q \frac{a^p - b^p}{p} - b^p \frac{a^q - b^q}{q} = \int_b^a b^p x^{p-1}[b^{q-p} - x^{q-p}]\,dx,$$

which is negative unless $a = b$. Therefore $f(\theta) \leqq 1$.

79. Equality sign holds only if $f'(0) = 0$, or $a = b$.

82. Make $a^{-\theta}b^{\theta-1}(\theta a + (1 - \theta)b)$ a minimum.

85. (a) Higher; (b) the same; (c) lower; (d) higher.

86. Integrate the left-hand side, sum, then differentiate again.

89. $\dfrac{d^{n+1}}{dx^{n+1}}\,(e^{x^2/2}) = \dfrac{d^n}{dx^n}\,(xe^{x^2/2}) = x\,\dfrac{d^n}{dx^n}\,(e^{x^2/2}) + n\,\dfrac{d^{n-1}}{dx^{n-1}}\,(e^{x^2/2})$ by Leibnitz's rule.

90. Eliminate u_{n+1} from both equations; $nu_{n-1} = u_n{}'$; differentiate one of the equations and use this relation.

91. $u_n(x) = x^n + \dfrac{n(n-1)}{2}\,x^{n-2} + \dfrac{n(n-1)(n-2)(n-3)}{2.4}\,x^{n-4} + \cdots$.

92. Apply Leibnitz's rule to

(a) $\dfrac{d^{n+2}}{dx^{n+2}}\,(x^2 - 1)^{n+1} = \dfrac{d^{n+2}}{dx^{n+2}}\,[(x^2 - 1).(x^2 - 1)^n]$;

(b) $\dfrac{d^{n+2}}{dx^{n+2}}\,(x^2 - 1)^{n+1} = \dfrac{d^{n+1}}{dx^{n+1}}\,[2(n + 1)x.(x^2 - 1)^n]$.

(c) Equate the two expressions for P'_{n+1} in (a) and (b).

93. $P_n(x) = \dfrac{(2n)!}{2^n(n!)^2}\left(x^n - \dfrac{n(n-1)}{2(2n-1)}\,x^{n-2} + \dfrac{n(n-1)(n-2)(n-3)}{2.4.(2n-1)(2n-3)}\,x^{n-4}\cdots\right)$.

94. Same as in Ex. 93.

95. By the binomial theorem, $\sum\limits_{n=0}^{p} \lambda_{n,\,p}(x) = (x+1-x)^p = 1.$

Also, differentiating

$$(a+x)^p = \sum_{n=0}^{p} \binom{p}{n} a^{p-n} x^n$$

k times, we have

$$\binom{p}{k}(a+x)^{p-k} = \sum_{n=k}^{p} \binom{p}{n} \binom{n}{k} a^{p-n} x^{n-k}.$$

Multiplying by x^k and putting $a = 1 - x$, we have

$$\binom{p}{k} x^k = \sum_{n=k}^{p} \binom{n}{k} \binom{p}{n} (1-x)^{p-n} x^n = \sum_{n=k}^{p} \binom{n}{k} \lambda_{n,\,p}(x).$$

CHAPTER IV

96. $\frac{12}{13} x^{13/12} - \frac{6}{5} x^{5/6} + \frac{4}{3} x^{3/4} + \frac{12}{7} x^{7/12} - 2x^{1/2} - 3x^{1/3} + 4x^{1/4} + 12 x^{1/12}$
$-2\log(1+x^{1/4}) - 4\log(1+x^{1/12}) - 4\sqrt{3}\ \text{arc tan}\ \dfrac{2}{\sqrt{3}}\ (x^{1/12} - \tfrac{1}{2}).$

97. $\frac{4}{7}(1+e^x)^{7/4} - \frac{4}{3}(1+e^x)^{3/4}.$

98. $-6\sqrt[3]{(1+x)^2}(\tfrac{1}{4} + \tfrac{1}{5}\sqrt[6]{(1+x)} + \tfrac{1}{6}\sqrt[3]{(1+x)} + \tfrac{1}{7}\sqrt{(1+x)}$
$\qquad\qquad\qquad + \tfrac{1}{8}\sqrt[3]{(1+x)^2} + \tfrac{1}{9}\sqrt[6]{(1+x)^5}).$

99. Put $x + \dfrac{1}{x} = t$: $\dfrac{1}{2} \log \dfrac{x^3 - x + 1}{x^2 + x + 1}.$

100. $\dfrac{1}{n}\ \text{arc cos}\ \dfrac{1}{x^n}.$

101. $\dfrac{1}{n!}\left[\log x - \binom{n}{1} \log(x+1) + \binom{n}{2} \log(x+2) - + \cdots \right.$
$\qquad\qquad\qquad\qquad\qquad \left. \pm \binom{n}{n} \log(x+n) \right].$

102. $\dfrac{(n-1)(n-3)\ldots 1}{n(n-2)\ldots 2} \cdot \dfrac{\pi}{2}$ if n is even; $\dfrac{(n-1)(n-3)\ldots 2}{n(n-2)\ldots 3}$ if n is odd.

103. $2^{12}/(1 . 3 . 5 . 7 . 9 . 11 . 13).$

104. $\dfrac{(2n)!}{2^{2n}(n!)^2} \cdot \dfrac{\pi}{2}.$

105. $\dfrac{2^{2n}(n!)^2}{(2n+1)!}.$

106. $\pi/16.$

107. $\pi/32.$

108. $\displaystyle\int x^a (\log x)^m\, dx = \dfrac{x^{a+1} (\log x)^m}{a+1} - \dfrac{m}{a+1} \int x^a (\log x)^{m-1}\, dx.$

109. $\displaystyle\int x^n e^{ax} \sin bx\, dx = \dfrac{x^n e^{ax}}{a^2 + b^2} (a \sin bx - b \cos bx)$
$\qquad\qquad - \dfrac{an}{a^2 + b^2} \int x^{n-1} e^{ax} \sin bx\, dx + \dfrac{bn}{a^2 + b^2} \int x^{n-1} e^{ax} \cos bx\, dx.$

110. $\int x^n e^{ax} \cos bx \, dx = \dfrac{x^n e^{ax}}{a^2 + b^2}(a \cos bx + b \sin bx)$

$$-\frac{an}{a^2 + b^2}\int x^{n-1}e^{ax}\cos bx\,dx - \frac{bn}{a^2 + b^2}\int x^{n-1}e^{ax}\sin bx\,dx.$$

111. $\int e^{ax}\sinh bx\,dx = \dfrac{e^{ax}}{b^2 - a^2}(b \cosh bx - a \sinh bx).$

112. $\int e^{ax}\cosh bx\,dx = \dfrac{e^{ax}}{b^2 - a^2}(b \sinh bx - a \cosh bx).$

114, 115, 116. Integrate by parts. **117.** $2^{n+1}(n!)^2/(2n+1)!$.

118. Convergent. **119.** Convergent.

120. Convergent if $n > -1$; divergent if $n \leqq -1$.

121. Convergent if $n > -1$, $m > -1$; otherwise divergent.

122. Convergent if $n > 0$, $m > -1$; otherwise divergent.

123. Convergent. **124.** Divergent.

125. Convergent. **126.** Convergent.

127. Convergent if $n > 0$; divergent if $n \leqq 0$.

128. Convergent if $m > n - 1$; divergent if $m \leqq n - 1$.

129. Convergent. Consider

$$\int_{\nu\pi}^{(\nu+1)\pi}\frac{dx}{1 + x^4 \sin^2 x} = \left(\int_{\nu\pi}^{(\nu+\epsilon)\pi} + \int_{(\nu+\epsilon)\pi}^{(\nu+1-\epsilon)\pi} + \int_{(\nu+1-\epsilon)\pi}^{(\nu+1)\pi}\right)\frac{dx}{1 + x^4 \sin^2 x}.$$

In the first and last integrals the integrand < 1, and in the second integral the integrand $< \dfrac{1}{\pi^4 \nu^4 \sin^2 \epsilon \pi}$, so that

$$\int_{\nu\pi}^{(\nu+1)\pi}\frac{dx}{1 + x^4 \sin^2 x} < 2\epsilon\pi + \frac{\pi}{\pi^4 \nu^4 \sin^2 \epsilon\pi}.$$

Choose $\epsilon = \dfrac{1}{\nu^{4/3}}$; then $\sin\epsilon\pi > \tfrac{1}{2}\epsilon\pi$, and

$$\int_{\nu\pi}^{(\nu+1)\pi}\frac{dx}{1 + x^4 \sin^2 x} < \frac{k}{\nu^{4/3}} < k\int_{\nu-1}^{\nu}\frac{dx}{x^{4/3}},$$

where k is a constant. Finally,

$$\int_A^B \frac{dx}{1 + x^4 \sin^2 x} < \int_{n\pi}^{m\pi}\frac{dx}{1 + x^4 \sin^2 x} < k\int_{(n-1)\pi}^{(m-1)\pi}\frac{dx}{x^{4/3}}$$

$$= \frac{3k}{\pi^{1/3}}\left[\frac{1}{\sqrt[3]{n-1}} - \frac{1}{\sqrt[3]{m-1}}\right] < \frac{3k}{\pi^{1/3}\sqrt[3]{n-1}} \to 0 \text{ as } n \to \infty.$$

Or, $\displaystyle\int_{\nu\pi}^{(\nu+1)\pi}\frac{dx}{1 + x^4 \sin^2 x} < \int_{\nu\pi}^{(\nu+1)\pi}\frac{dx}{1 + (\nu\pi)^4 \sin^2 x} < \frac{\pi}{\sqrt{1 + (\nu\pi)^4}} < \frac{k}{\nu^2}.$

130. $\displaystyle\int_0^A \frac{x\,dx}{1+x^2\sin^2 x} > \int_0^A \frac{x\,dx}{1+x^2} > \tfrac{1}{2}\log(1+A^2)$; divergent.

131. Convergent if $\beta < -2$, $\beta + 1 < \alpha < -1$ or $\beta > 0$, $-1 < \alpha < \beta/2 - 1$; otherwise divergent.

Suppose that $\beta \leqq 0$. Then $\displaystyle\int^\infty \frac{x^\alpha\,dx}{1+x^\beta\sin^2 x}$ converges only if $\alpha < -1$;

$\displaystyle\int_0 \frac{x^\alpha\,dx}{1+x^\beta\sin^2 x}$ behaves like $\displaystyle\int_0 \frac{x^\alpha\,dx}{1+x^{\beta+2}}$, i.e. if $\beta + 2 \geqq 0$, then $\alpha > -1$, contrary to the preceding; if $\beta + 2 < 0$, $\alpha - \beta - 2 > -1$.

Suppose that $\beta > 0$. Then $\displaystyle\int_0 \frac{x^\alpha dx}{1+x^\beta\sin^2 x}$ converges only if $\alpha > -1$. Furthermore,

$$\frac{\nu^\alpha \pi^{\alpha+1}}{\sqrt{1+(\nu+1)^\beta\pi^\beta}} = \int_{\nu\pi}^{(\nu+1)\pi} \frac{(\nu\pi)^\alpha dx}{1+(\nu+1)^\beta\pi^\beta\sin^2 x}$$

$$< \int_{\nu\pi}^{(\nu+1)\pi} \frac{x^\alpha dx}{1+x^\beta\sin^2 x} < \int_{\nu\pi}^{(\nu+1)\pi} \frac{(\nu+1)^\alpha\pi^\alpha dx}{1+(\nu\pi)^\beta\sin^2 x} < \frac{(\nu+1)^\alpha\pi^{\alpha+1}}{\sqrt{1+(\nu\pi)^\beta}}$$

or $\qquad k_1\nu^{\alpha-\beta/2} < \displaystyle\int_{\nu\pi}^{(\nu+1)\pi} \frac{x^\alpha dx}{1+x^\beta\sin^2 x} < k_2\nu^{\alpha-\beta/2}$.

Hence $\displaystyle\int_\pi^\infty \frac{x^\alpha dx}{1+x^\beta\sin^2 x}$ converges if, and only if, $\alpha - \beta/2 < -1$.
The integral may also be estimated by the method of Ex. 129.

132. $\displaystyle\int_a^\infty \frac{f(\alpha x) - f(\beta x)}{x}\,dx = \int_{a\alpha}^\infty \frac{f(x)}{x}\,dx - \int_{a\beta}^\infty \frac{f(x)}{x}\,dx$

$$= \int_{a\alpha}^{a\beta} \frac{f(x)}{x}\,dx = L\log\frac{\beta}{\alpha} + \int_{a\alpha}^{a\beta} \frac{f(x)-L}{x}\,dx.$$

Show that this last integral tends to zero as $a \to 0$.

134. Consider $\displaystyle\int_a^b \frac{f(\alpha x) - f(\beta x)}{x}\,dx$, and proceed as in Ex. 132.

135. In the formula $\Gamma(n) = \displaystyle\int_0^\infty e^{-t}t^{n-1}dt$ substitute $t = x^2$ and $t = \log\dfrac{1}{x}$ respectively.

CHAPTER V

136. (a) $x^5 + y^5 = 5ax^2y^2$; (b) $x = a \arccos\dfrac{a-y}{b} + \sqrt{b^2 - (a-y)^2}$.

138. (a) $x^2 + y^2 = \sqrt{(a^2x^2 + b^2y^2)}$.   (c) $x(x^2 + y^2) + py^2 = 0$.
     (b) $x^2 + y^2 = \sqrt{(a^2x^2 - b^2y^2)}$.   (d) $x = 0$.

141. $x = t - \dfrac{l\sqrt{p}}{\sqrt{t+p}}$, $y^2 = 4pt\left(1 + \dfrac{l}{2\sqrt{p}\sqrt{t+p}}\right)^2$.

142. $5a^2/2$.   **143.** $\pi b(2a + b)(a - b)^2/(2a^2)$.

144. $\dfrac{4b(a+b)}{a}\left(1-\cos\dfrac{a}{2b}t\right)$.

146. Choose the axes so that the curve touches the x-axis at the origin, and express the ordinate y as a function of the angle which the tangent at the point (x, y) makes with the x-axis.

147. (a) $l^3/12$; (b) $l^3/3$; (c), (d) $l(l^2/12 + d^2)$.

148. $r = ce^{\cot a \cdot \theta}$.   **149.** $(x - c)^2 + y^2 = k^2$.

151. $(x - c_1)^2 + y^2 = c_2^2$.   **152.** $y = a \cosh \dfrac{x - b}{a}$.

153. The length of a straight line joining the points (r_ν, φ_ν), $(r_{\nu+1}, \varphi_{\nu+1})$ of the curve is

$$\sqrt{\{(r_{\nu+1} - r_\nu)^2 + 2r_\nu r_{\nu+1}(1 - \cos(\varphi_{\nu+1} - \varphi_\nu))\}},$$

and the length of a polygonal line inscribed in the curve is

$$\sum_{\nu=0}^{n-1} \sqrt{\{(\Delta r_\nu)^2 + r_\nu r_{\nu+1}(\Delta\varphi_\nu)^2 + r_\nu r_{\nu+1}(\Delta\varphi_\nu)^4 \cdot R_\nu\}},$$

where the $|R_\nu|$'s are all bounded. Letting the maximum of $\Delta\varphi_\nu$ tend to zero, we obtain

$$\int \sqrt{\left\{\left(\frac{dr}{d\varphi}\right)^2 + r^2\right\}}\, d\varphi.$$

CHAPTER VI

157. $x^2 - \dfrac{1}{3}x^4 + \dfrac{2}{45}x^6 + \ldots$; $(\sin x)^2 = \left(x - \dfrac{x^3}{3!} + \dfrac{x^5}{5!} - x^7 R\right)^2$

$$= x^2 - \frac{1}{3}x^4 + \frac{2}{45}x^6 + x^8 R',$$

where R and R' remain bounded as $x \to 0$.

158. $x + \dfrac{x^3}{3} + \dfrac{2}{15}x^5 + \ldots$; $\dfrac{\sin x}{\cos x} = \dfrac{x - \dfrac{x^3}{3!} + \dfrac{x^5}{5!} - x^7 R}{1 - \dfrac{x^2}{2!} + \dfrac{x^4}{4!} - x^6 S}$

$$= x + \frac{1}{3}x^3 + \frac{2}{15}x^5 + x^7 T,$$

where R, S, T are bounded as $x \to 0$.

159. $1 - \dfrac{x^2}{4} - \dfrac{x^4}{96} - \ldots$; $\sqrt{\cos x} = \left(1 - \dfrac{x^2}{2!} + \dfrac{x^4}{4!} - x^6 R\right)^{1/2}$

$$= 1 + \frac{1}{2}\left(-\frac{x^2}{2!} + \frac{x^4}{4!} - x^6 R\right) - \frac{1}{8}\left(-\frac{x^2}{2!} + \frac{x^4}{4!} - x^6 R\right)^2$$

$$+ \left(-\frac{x^2}{2!} + \frac{x^4}{4!} - x^6 R\right)^3 S = 1 - \frac{x^2}{4} - \frac{x^4}{96} + x^6 T,$$

where R, S, T are bounded as $x \to 0$.

160. (a) $1 - \dfrac{x^2}{3} - \dfrac{x^4}{45} - \dfrac{2x^6}{945} \ldots$ (b) $1 - \dfrac{x^2}{12} + \dfrac{x^4}{1440} - \dfrac{x^6}{23712} \ldots$

(c) $1 + \dfrac{x^2}{2} + \dfrac{5}{24} x^4 + \dfrac{61}{720} x^6 + \ldots$ (d) $1 + x + \dfrac{x^2}{2} - \dfrac{x^3}{8} \ldots$

(e) $e + ex + ex^2 + \dfrac{5}{6} ex^3 + \ldots$ (f) $-\dfrac{x^2}{6} - \dfrac{x^4}{180} - \dfrac{x^6}{2835} \ldots$

161. $x + \dfrac{1}{2}\dfrac{x^3}{3} + \dfrac{1.3}{2^2.2!}\dfrac{x^5}{5} + \dfrac{1.3.5}{2^3.3!}\dfrac{x^7}{7} + \ldots$

162. $\displaystyle\sum_{\tau=0}^{\infty} \left(\sum_{\nu=0}^{\tau} \binom{2n}{n}\binom{2\tau - 2n}{\tau - n} \frac{1}{(2n+1)(2\tau - 2n + 1)} \right) \frac{x^{2\tau + 2}}{2^{2\tau}}.$

163. (a) $\displaystyle\sum_{\nu=0}^{\infty} (-1)^\nu \frac{1.3.5.\ldots.(2\nu - 1)}{2.4.6.\ldots.2\nu} \frac{x^{2\nu+1}}{2\nu + 1};$

(b) $\displaystyle\sum_{\nu=0}^{\infty} \frac{(-1)^\nu}{\nu!} \frac{x^{2\nu+1}}{2\nu + 1};$ (c) $\displaystyle\sum_{\nu=0}^{\infty} \frac{(-1)^\nu}{(2\nu + 1)!} \frac{x^{2\nu+1}}{2\nu + 1}.$

164. (a) $\dfrac{(2n)!\, x^{2n+1}}{2^{2n}(n!)^2(2n+1)};$ (b) $\dfrac{x^{2n+1}}{n!(2n+1)};$ (c) $\dfrac{x^{2n+1}}{(2n+1)!(2n+1)}$

166. $e - \dfrac{e}{2}\left(\dfrac{1}{x}\right) + \dfrac{11e}{24}\left(\dfrac{1}{x}\right)^2 - \ldots$

167. (a) $-\dfrac{e}{2};$ (b) $\dfrac{11e}{24};$ (c) $0;$ (d) $e^{-1/6};$ (e) $1.$

169. (a) Minimum at $x = 0$;   (b) maxima and minima at points where $\tan\dfrac{1}{x} = \dfrac{1}{x}$, which occur once in each interval $\dfrac{1}{(n+\frac{1}{2})\pi} < x < \dfrac{1}{(n-\frac{1}{2})\pi}$, $n = \pm 1, \pm 2, \ldots$; maxima and minima alternately.

CHAPTER VII

170. $5.881a.$ **171.** $11.$

172. $0.82247.$ **173.** $.175, .302, 3.490.$

174. Since $\log(\alpha + x)$ is convex downward, and $\alpha > 0$,

$$\log(\alpha + 1) + \ldots + \log(\alpha + n) > \int_{\frac{1}{2}}^{n+\frac{1}{2}} \log(\alpha + x)\, dx$$

$$= (n + \tfrac{1}{2} + \alpha) \log(n + \tfrac{1}{2} + \alpha) - (\alpha + \tfrac{1}{2}) \log(\alpha + \tfrac{1}{2}) - n,$$

or $\alpha(\alpha + 1) \ldots (\alpha + n) > \alpha \dfrac{(n + \frac{1}{2} + \alpha)^{n+\frac{1}{2}+\alpha}}{(\alpha + \frac{1}{2})^{\alpha+\frac{1}{2}}} e^{-n} > k(\alpha)\, n!\, n^\alpha,$

where $k(\alpha)$ is a positive number depending on α. Furthermore,

$$\frac{a_n}{a_{n-1}} = \left(1 + \frac{\alpha}{n}\right)\left(1 - \frac{1}{n}\right)^c = 1 - \frac{\alpha(\alpha + 1)}{2}\frac{1}{n^2} + \frac{R}{n^3},$$

where R remains bounded as $n \to \infty$. Therefore, for sufficiently large values of n, $a_n < a_{n-1}$ and the sequence is monotonically decreasing.

175. $c + (n + \frac{1}{2})\log n - \sum\limits_{\nu=1}^{l}(n_\nu + \frac{1}{2})\log n_\nu$.

CHAPTER VIII

178. If $\lim a_n \leqq 1$, the terms do not tend to zero. If $\lim a_n > k > 1$, compare the series with $\Sigma \frac{1}{k^n}$.

179. For any ε, $\sum\limits_{\nu=n}^{m} a_\nu < \varepsilon$ for every n, m sufficiently large. But $\sum\limits_{\nu=n}^{m} a_\nu > (m - n)a_m$, or $ma_m < \varepsilon + na_m$. Keeping n fixed, choose m so large that $na_m < \varepsilon$; for every such m, $ma_m < 2\varepsilon$.

180. Apply Ex. 179.

181. Let s_n denote the partial sums of $\sum\limits_{\nu=1}^{\infty} a_\nu$, s the sum, and let $\sigma_n = s_n - s$. Then

$$\sum\limits_{\nu=n}^{m} a_\nu b_\nu = \sum\limits_{\nu=n}^{m}(\sigma_\nu - \sigma_{\nu-1})b_\nu = \sum\limits_{\nu=n}^{m}\sigma_\nu(b_\nu - b_{\nu+1}) - \sigma_{n-1}b_n + \sigma_m b_{m+1}.$$

For every sufficiently large ν, $|\sigma_\nu| < \varepsilon$, and

$$\left|\sum\limits_{\nu=n}^{m} a_\nu b_\nu\right| < \varepsilon\sum\limits_{\nu=n}^{m}|b_\nu - b_{\nu+1}| + \varepsilon|b_n| + \varepsilon|b_{m+1}|$$
$$< \varepsilon|b_n - b_{m+1}| + \varepsilon|b_n| + \varepsilon|b_{m+1}|.$$

This is in turn less than $4B\varepsilon$, where B is a bound for $|b_\nu|$, and the series $\sum\limits_{\nu=1}^{\infty} a_\nu b_\nu$ converges.

182. Proceed as in Ex. 181:

$$\sum\limits_{\nu=n}^{m} a_\nu b_\nu = \sum\limits_{\nu=n}^{m}(s_\nu - s_{\nu-1})b_\nu = \sum\limits_{\nu=n}^{m} s_\nu(b_\nu - b_{\nu+1}) - s_{n-1}b_n + s_m b_{m+1}$$

and use the monotonic character of b_n, the fact that $b_n \to 0$, and that $|s_\nu| < s$ for every ν.

183. (a), (b), (d), (f) Convergent; (c) convergent if $\theta \neq 2n\pi$; (e) convergent if $\theta \neq (2n + 1)\pi$.

184. (a) $\frac{1}{2}\log 2$; (b) $\log 2$.

185. (a) $\alpha = 1$; (b) $\alpha \geqq 1$.

186. (a) Diverges; (b) converges.

188. If $|a_n| < \dfrac{1}{n^{1+\epsilon}}$ for every sufficiently large n, then

$$\log \frac{1}{|a_n|} > (1 + \epsilon)\log n \quad \text{or} \quad \frac{\log 1/|a_n|}{\log n} > 1 + \epsilon.$$

Reverse the argument: $\dfrac{\log 1/|a_n|}{\log n} > 1 + \epsilon$ implies $|a_n| < \dfrac{1}{n^{1+\epsilon}}$. Similarly for divergence.

189. Apply Ex. 188.

190. Proceed as in Ex. 188.

191. The n-th root test may be written as follows: if $\dfrac{\log 1/|a_n|}{n} > \epsilon$, the series converges; if $< -\epsilon$, the series diverges. Write

$$\frac{\log 1/|a_n|}{\log n} = \frac{n}{\log n}\,\frac{\log 1/|a_n|}{n}.$$

192. If $\left|\dfrac{a_{n+1}}{a_n}\right| < \dfrac{b_{n+1}}{b_n}$ for every $n \geqq N$, then

$$|a_{n+1}| < \frac{b_{n+1}}{b_n}|a_n| < \frac{b_{n+1}}{b_n}\frac{b_n}{b_{n-1}}|a_{n-1}| < \ldots < \frac{|a_N|}{b_N}b_{n+1};$$

therefore $\Sigma\,|a_\nu|$ converges if Σb_ν does. Similarly for divergence.

194. Use Ex. 192, comparing with $\overset{\infty}{\underset{\nu=1}{\Sigma}} \dfrac{1}{\nu^\alpha}$. The series $\Sigma\,|a_\nu|$ converges if

$$\frac{|a_n|}{|a_{n+1}|} > \left(1 + \frac{1}{n}\right)^\alpha > 1 + \frac{\alpha}{n} + \frac{R}{n^2},$$

where $\alpha > 1$. Then

$$n\left(\frac{|a_n|}{|a_{n+1}|} - 1\right) > \alpha + \frac{R}{n} > 1 + \epsilon.$$

Reverse the argument:

$$n\left(\frac{|a_n|}{|a_{n+1}|} - 1\right) > 1 + \epsilon$$

implies the convergence of $\Sigma\,|a_\nu|$. Similarly for divergence.

195. $\Sigma\,|a_\nu|$ converges if

$$\frac{|a_n|}{|a_{n+1}|} > \left(1 + \frac{1}{n}\right)\left(1 + \frac{\log\left(1 + \frac{1}{n}\right)}{\log n}\right)^\alpha > 1 + \frac{1}{n} + \frac{\alpha}{n\log n} + \frac{R}{n^2\log n},$$

where $\alpha > 1$. Then

$$n\log n\left(\frac{|a_n|}{|a_{n+1}|} - 1 - \frac{1}{n}\right) > \alpha + \frac{R}{n} > 1 + \epsilon.$$

Reversal of this argument gives the convergence test; similarly for divergence.

197. (a) Converges if $\beta - \alpha > 1$, diverges if $\beta - \alpha \leqq 1$.

(b) Converges if $\gamma > \alpha + \beta$, diverges if $\gamma \leqq \alpha + \beta$.

198. (a) If $x \geqq 1 + \varepsilon$, $\overset{\infty}{\underset{\nu=1}{\Sigma}} \dfrac{1}{\nu^x} \leqq \overset{\infty}{\underset{\nu=1}{\Sigma}} \dfrac{1}{\nu^{1+\epsilon}}$. Similarly for (b).

199. The partial sums of $\Sigma \cos \nu x$ are uniformly bounded for every x in $\varepsilon \leqq x \leqq 2\pi - \varepsilon$. (Write $\cos \nu x = \dfrac{e^{i\nu x} + e^{-i\nu x}}{2}$ and $\overset{n}{\underset{\nu=0}{\Sigma}} \cos \nu x = \tfrac{1}{2} \overset{n}{\underset{\nu=-n}{\Sigma}} e^{i\nu x}$).
Then prove the theorem analogous to Ex. 182 for uniform convergence.

200. If x lies in the interval $\varepsilon \leqq x \leqq N$, then $y = \dfrac{x-1}{x+1}$ lies in the interval $-1 + \dfrac{2\varepsilon}{1+\varepsilon} \leqq y \leqq 1 - \dfrac{2}{N+1}$.

201. (a) $-1 < x < 1$; (b) $-4 < x < 4$; (c) $x > 1$; (d) $x > 0$; (e) any x; (f) no x; (g) $x > 1$; (h) $-1 < x < 1$.

202. If $\overset{\infty}{\underset{\nu=1}{\Sigma}} \dfrac{a_\nu}{\nu^{x_0}}$ converges, write $\overset{\infty}{\underset{\nu=1}{\Sigma}} \dfrac{a_\nu}{\nu^x} = \overset{\infty}{\underset{\nu=1}{\Sigma}} \dfrac{a_\nu}{\nu^{x_0}} \cdot \dfrac{1}{\nu^{x-x_0}}$, and use Ex. 181 or 182. If $\overset{\infty}{\underset{\nu=1}{\Sigma}} \dfrac{a_\nu}{\nu^{x_0}}$ diverges, $\overset{\infty}{\underset{\nu=1}{\Sigma}} \dfrac{a_\nu}{\nu^x}$ cannot converge for $x < x_0$, by what has just been proved.

203. Write $\Sigma \dfrac{a_\nu \log \nu}{\nu^x} = \Sigma \dfrac{a_\nu}{\nu^{x_0}} \cdot \dfrac{\log \nu}{\nu^{x-x_0}}$.

204. Clearly $\overset{\infty}{\underset{\nu=0}{\Sigma}} a_\nu x^\nu < \overset{\infty}{\underset{\nu=0}{\Sigma}} a_\nu$ for $x < 1$. On the other hand,

$$\lim_{x \to 1} \overset{\infty}{\underset{\nu=0}{\Sigma}} a_\nu x^\nu > \lim_{x \to 1} \overset{N}{\underset{\nu=0}{\Sigma}} a_\nu x^\nu = \overset{N}{\underset{\nu=0}{\Sigma}} a_\nu; \quad \text{or} \quad \lim_{x \to 1} \overset{\infty}{\underset{\nu=0}{\Sigma}} a_\nu x^\nu \geqq \overset{\infty}{\underset{\nu=0}{\Sigma}} a_\nu.$$

205. As in Ex. 204, $\lim\limits_{x \to 1} \overset{\infty}{\underset{\nu=0}{\Sigma}} a_\nu x^\nu \geqq \overset{\infty}{\underset{\nu=0}{\Sigma}} a_\nu$ and hence is ∞.

206. Write $\overset{\infty}{\underset{\nu=0}{\Sigma}} a_\nu x^\nu = \overset{\infty}{\underset{\nu=0}{\Sigma}} a_\nu X^\nu \left(\dfrac{x}{X}\right)^\nu$. Then prove the theorem analogous to Ex. 181 for uniform convergence: if $\overset{\infty}{\underset{\nu=0}{\Sigma}} a_\nu$ converges, and if the sequence $b_0(x), b_1(x), \dots, b_n(x), \dots$ is monotonic for every x and uniformly bounded for every x in a certain interval, then $\overset{\infty}{\underset{\nu=0}{\Sigma}} a_\nu b_\nu(x)$ converges uniformly in that interval.

207. This follows from the uniform convergence of the series $\overset{\infty}{\underset{\nu=0}{\Sigma}} a_\nu x^\nu$ in the interval $0 \leqq x \leqq X$. For then $\overset{\infty}{\underset{\nu=0}{\Sigma}} a_\nu x^\nu$ is continuous in that interval.

208. (a) $x(1+x)/(1+x^2)$; (b) $(1-x^2)/(1-x+x^2)^2$.

209. (a) The series is equal to $\dfrac{d}{dx}\left(\dfrac{e^x - 1}{x}\right)\Big|_{x=1}$;

(b) The series is equal to $\dfrac{\sqrt{(1+x)} - \sqrt{(1-x)}}{2}\Big|_{x=1}$.

CHAPTER IX

211. $\pi x \cot \pi x = 1 - 2x^2 \sum\limits_{\nu=1}^{\infty} \dfrac{1}{\nu^2 - x^2} = 1 - 2x^2 \sum\limits_{\nu=1}^{\infty} \dfrac{1}{\nu^2} \left(\sum\limits_{m=0}^{\infty} \dfrac{x^{2m}}{\nu^{2m}} \right)$

$$= 1 - 2 \sum\limits_{m=1}^{\infty} \left(\sum\limits_{\nu=1}^{\infty} \dfrac{1}{\nu^{2m}} \right) x^{2m}.$$

214. (a) $\displaystyle\int_0^1 \dfrac{\log x}{1 - x}\, dx = - \sum\limits_{\nu=1}^{\infty} \dfrac{1}{\nu^2}$; (b) $\displaystyle\int_0^1 \dfrac{\log x}{1 + x}\, dx = \sum\limits_{\nu=1}^{\infty} \dfrac{(-1)^\nu}{\nu^2}$.

216. (a) $\sqrt{2}$; (b) $\sqrt{3}$.

217. $\coth \pi x - \dfrac{1}{\pi x} = \dfrac{2x}{\pi} \left(\dfrac{1}{1^2 + x^2} + \dfrac{1}{2^2 + x^2} + \dfrac{1}{3^2 + x^2} + \cdots \right).$

CHAPTER XI

218. $x + c = \sqrt{(a^2 - y^2)} - a \log \dfrac{a + \sqrt{(a^2 - y^2)}}{y}.$

219. $\tfrac{1}{2} k y^2 + x = c.$

221. $l = \dfrac{E_0}{\sqrt{(\rho^2 + \omega^2 \mu^2)}} \sin(\omega l - \varphi) + \dfrac{A}{\sqrt{(\rho^2 + \omega^2 \mu^2)}}\, e^{-\rho t/\mu}$, where $\tan \varphi = \dfrac{\omega \mu}{\rho}.$

222. $x^2 = a^2 - \dfrac{k t^2}{a^2}$; time of descent is $a^2/\sqrt{k}.$

223. Differentiate with respect to x and solve the resulting differential equation for p in terms of x:

$$y = - \dfrac{1}{4x^2} \quad \text{and} \quad y = \dfrac{c}{x} + c^2.$$

224. $x = -1 + \sqrt{(2y + c)} + \log(-1 + \sqrt{(2y + c)}).$

INDEX

Wiley Classics Library